The Wave–Particle Dualism

## Fundamental Theories of Physics

*A New International Book Series on The Fundamental Theories of Physics: Their Clarification, Development and Application*

# The Wave–Particle Dualism

*A Tribute to Louis de Broglie on his 90th Birthday*

*edited by*


S. Diner, D. Fargue,
G. Lochak and F. Selleri

*Fondation Louis de Broglie, Paris, France*


FONDATION
Louis de Broglie


D. Reidel Publishing Company

A MEMBER OF THE KLUWER ACADEMIC PUBLISHERS GROUP

Dordrecht / Boston / Lancaster

**Library of Congress Cataloging in Publication Data**
Main entry under title:

CIP

The wave–particle dualism.

(Fundamental theories of physics)
Proceedings of an international symposium held in Perugia Apr. 22–30, 1982.
Includes index.
1. Wave-particle duality–Congresses. 2. Broglie, Louis de, 1892– –Congresses. I. Broglie, Louis de, 1892– II. Diner, Simon. III. Series.
QC476.W38W38 1983 539.7'21 83-15989
ISBN 90-277-1664-1

---

Published by D. Reidel Publishing Company
P.O. Box 17, 3300 AA Dordrecht, Holland.

Sold and distributed in the U.S.A. and Canada
by Kluwer Academic Publishers,
190 Old Derby Street, Hingham, MA 02043, U.S.A.

In all other countries, sold and distributed
by Kluwer Academic Publishers Group,
P.O. Box 322, 3300 AH Dordrecht, Holland.





Printed in The Netherlands

CONTENTS

# PREFACE

The Louis de Broglie Foundation (which was created in 1973, for the fiftieth anniversary of the discovery of wave mechanics) and the University of Perugia, have offered an international symposium to Louis de Broglie on his 90th birthday. This publication represents the Proceedings of this conference which was held in Perugia on April 22-30, 1982.
It was an opportunity for the developing of physical conceptions of all origins, which may serve to throw light on the mysterious power of the quantum theory. Quantum Mechanics has reached maturity in its formalism and although no experiment yet has come to challenge its predictions, one may question the limits of its validity. In fact the true meaning of this vision of the microphysical world remains the subject of endless debating, at the heart of which lies "the foundational myth" of wave-particle dualism.
Albert Einstein and Louis de Broglie are the two discoverers of this fundamental duality, which they always considered as a deep physical reality rather than a phenomenological artifice.
During the conference a survey has been given of the essential recent experimental results in corpuscular and quantum optics and the most up-to-date theoretical aspects of the specificity of microphysical phenomena : various interpretations of quantum mechanics, "alternative theories" and hidden parameters theories, probabilistic and axiomatic questions and tentative crucial experiments.
The conference took place in the magnificent atmosphere of the villa Colombella lent to us by the Università per Stranieri di Perugia.
Without the organizational activity of Professor V. Aquilanti and his colleagues, especially Professors G. Liuti and G.G. Volpi, this conference would not have met such a success.
Our thanks are for all the organizations which have in a way or another helped us in conducting this entreprise : Università per Stranieri di Perugia, Consiglio Nazionale delle Ricerche, Regione dell'Umbria, Provincia di Perugia, Università di Perugia, Azienda Autonoma del Soggiorno di Perugia, Comune di Perugia.

# DE BROGLIE'S INITIAL CONCEPTION OF DE BROGLIE WAVES

Georges Lochak

Fondation Louis de Broglie, 1 rue Montgolfier
F-75003 PARIS

## 1. INTRODUCTION

In this paper my intention is to present de Broglie's personal view on "matter waves", with some consequences on the theory of measurement. This conception of "matter waves" was at the very origin of wave Mechanics and was, later on, developed by de Broglie and his collaborators during the last 25 years. There are two papers on this subject in this book. The present one combines my own two lectures given during the Perugia Symposium and the second paper (by Daniel Fargue) is especially devoted to the double solution theory and the fundamental idea of using non-linear equations. In connection with this last problem, let me note that I am very glad to see how the approach of physicists has changed during the last few years : 20 or 25 years ago, when we were the only ones to speak about non linear equations and the possibility of generalizing Quantum Mechanics on the basis of this idea, all the official scientific commissions considered that it was an utterly foolish idea devoid of any interest, which only demonstrated that de Broglie had definitely became an old man. Today the possibility of non linear effects in the diffraction or interferences of matter-waves is largely recognized and this is certainly a very important evolution, which may help to find an intelligible feature of the dualistic properties of matter. Nevertheless, it must be confessed that, just like a quarter of century ago, the main difficulty remains to find <u>one</u> non linear equation, on the basis of a general principle and not to stay with an infinity of possible equations, among which the right one may exist. So, it would be a criminal illusion to ignore the fact that, in spite of some practical progress, the path to a true physical solution to the problem of wave-particle dualism is, most probably, as difficult as it was 25



*S. Diner et al. (eds.), The Wave-Particle Dualism, 1–25.*

years ago. The correct non linear equation will probably only be found as a result of many years of work ... unless (as sometimes happens in the history of science) some young student who was, of course, not invited to Perugia, is in the process of finding a very simple and absolutely new solution to the problem !

## 2. DE BROGLIE'S MAIN IDEA

Let us come, now, to the very subject of this lecture. De Broglie's main idea is not dualism, but coexistence between waves and particles. Here lies the difference between Bohr and de Broglie : Bohr believed in a kind of double faced physical being which appears to us, in certain circumstances as a particle and in others as a wave ; on the contrary, de Broglie considered that there is only one thing which is always (at the same time) a particle and a wave and which is such that the properties of the particle which we observe are guided -commanded- by the wave structure of the system. And it was his idea from the beginning of wave mechanics. We shall try to show this in a short survey of de Broglie's famous first three papers which appeared in the Comptes Rendus de l'Académie des Sciences in 1923.

The idea was the following. De Broglie first considered a particle ; to this particle, he supposed that (following his own words : "as a consequence of a great law of nature"), one can associate a frequency defined by the equality between "Einstein's energy" and "Planck's energy" written in the proper system :

$$m_0 c^2 = h \nu_0 \tag{1}$$

At this stage it is supposed that the "proper frequency belongs to the particle and not to a wave. But de Broglie remarked that if an observer looks at the particle moving before him with a velocity $\underline{v}$, the mass is transformed according to the well known formula :

$$m = \frac{m_0}{\sqrt{1 - \frac{v^2}{c^2}}} \tag{2}$$

whereas the frequency $\nu_0$, which must be considered as being internal to the particle, obeys the relativistic law of the slowing down of clocks and becomes :

$$\nu_1 = \nu_0 \sqrt{1 - \frac{v^2}{c^2}} \ . \tag{3}$$

So that the equality (1) no longer works ! It is not relativistically invariant. De Broglie was deeply puzzled by this failure and he only reached the following hypothesis after a long period of reflection : he supposed that in the proper system there is not only an "internal clock", with a frequency $\nu_0$, associated to the particle, but also a stationary wave, with the same frequency and the same phase as the clock and written :

$$e^{i\nu_0 t_0} \tag{4}$$

But now, in another galilean system, with a relative velocity v, the Lorentz transformation of time will give :

$$t_0 = \frac{1}{\sqrt{1 - \frac{v^2}{c^2}}}(t - \frac{vx}{c^2})$$

and the wave will become :

$$e^{i \frac{\nu_0}{\sqrt{1-v^2/c^2}}(t - \frac{vx}{c^2})} \tag{5}$$

Now, we can see that the frequency $\nu$ of this wave is not the same as the frequency $\nu_1$ defined in (3) but :

$$\nu = \frac{\nu_0}{\sqrt{1-v^2/c^2}} \quad . \tag{6}$$

The variance of the wave's frequency $\nu$ is, thus, the same as the variance (2) of the mass and, using (1), (2) and (6), we may write, invariantly :

$$mc^2 = h\nu \quad , \tag{7}$$

which would be impossible with the frequency $\nu_1$ of the "internal clock" of the particle. Besides, we get through (5) the phase velocity V and de Broglie's well know formula :

$$Vv = c^2 \; . \tag{8}$$

As we can see, V is greater than c, but we know that the associated group velocity may be shown to be equal to v, i.e. equal to the velocity of the particle[1]

After the discovery of the important properties (6) and (7) of the wave frequency $\nu$, de Broglie could have abandoned his initial idea of an internal clock frequency $\nu_1$. But this was not to be the case because he made a second important remark :

In the frame of a galilean observer who sees the particle moving before him with the velocity v, de Broglie calculated the phase of the internal clock of the particle when the latter is in a point x at the instant t. In virtue of (1) and (3), one gets :

$$\phi_{(clock)} = \nu_1 t = \frac{m_0 c^2}{h}\sqrt{1 - \frac{v^2}{c^2}}\,\frac{x}{v} \tag{9}$$

Now, de Broglie calculated at the same moment t and in the same point x at which the particle lies, the value of the phase of the wave. Using (5) and (6) we will find this value to be (remember that x = vt !) :

$$\phi_{wave} = \nu(t - \frac{vx}{c^2}) = \frac{m_0c^2}{h}\frac{1}{\sqrt{1-\frac{v^2}{c^2}}}(\frac{x}{v} - \frac{vx}{c^2}) = \frac{m_0c^2}{h}\sqrt{1-\frac{v^2}{c^2}}\,\frac{x}{v} \qquad (10)$$

It is the same phase ! In other words, we know that the frequency of the internal clock of the particle is smaller than the frequency of the wave, but at the same time, the velocity of the particle is smaller than the phase velocity of the wave, and the result is a kind of compensation of these two discrepancies, so that we can assert : "For any galilean observer, the phase of the "internal clock" of the particle is, at each instant equal to the value of the phase of the wave calculated at the same point at which the particle lies".

This is de Broglie's law of accordance (or harmony) of phases. It is interesting to underline here that de Broglie considered this law and not the discovery of the famous wave-length to be his fundamental achievement. Ten years ago, during the celebration of his eightieth birthday at the Academy of Sciences, he said to us, recalling Bergson's words : "A man never has more than one great idea in his life" and he added : "If I ever had such an idea, it is certainly the law of phase harmony which was expressed in the first chapter of my Thesis in 1924".

The wave-length became famous after the discovery of the diffraction of material corpuscles and thanks to the proeminent role played by Schrödinger's equation. But in fact, the forgotten law of phase harmony, which is no longer quoted in text books, constitutes the basis of the whole problem of wave-particle dualism and contains, in the opinion of de Broglie, the deep mystery which has to be solved in the first place if one is to understand quantum mechanics. De Broglie never considered that, in stating this law, he had given any explanation of wave particle dualism : he only found an important formula which follows from the laws of relativity. But the question is : what property is hidden behind this formula ? What is this mysterious balance between corpuscle and wave (similar to the balance between a surf-rider and a sea wave), which is expressed in the formula ?

In order to explain de Broglie's mode of reasoning it might be interesting to recall how, thanks to his wave, he reached the quantized levels of Bohr's atom. Everybody knows how easy it is to find this result in an elementary way, by stating a resonance condition on de Broglie's wave-length along a circular Kepler's orbit. But this was not the way chosen by de Broglie, who did not make use of the wave-length ! His reasoning was more complicated because it was based on the phase-harmony but in this way he made use of wave-particle dualism.

He said the following : Consider a circular Bohr orbit. The electron starts from a point O and describes the orbit with a velocity v, while the wave starts at the same moment, but with the much greater phase velocity $V = v/c^2$, so that, at a certain time $\tau$ the wave will overtake the particle in a point O'. At the instant $\tau$, the particle will have described an arch OO' and for the wave, we must have the same arch plus an entire orbit, from which we get :

$$V\tau = \frac{c^2}{v}\tau = v(\tau+T) \qquad \tau = \frac{v^2}{c^2-v^2}T, \tag{11}$$

where T is the period of the electron on the orbit. From that, we find the internal phase of the particle (the phase of the internal clock !), at the instant $\tau$ of the overtaking of the particle by the wave (we make use of (1), (3) and (11)) :

$$2\pi\nu_1\tau = 2\pi\frac{m_0c^2}{h}\sqrt{1-\frac{v^2}{c^2}} \times \frac{v^2}{c^2-v^2}T \tag{12}$$

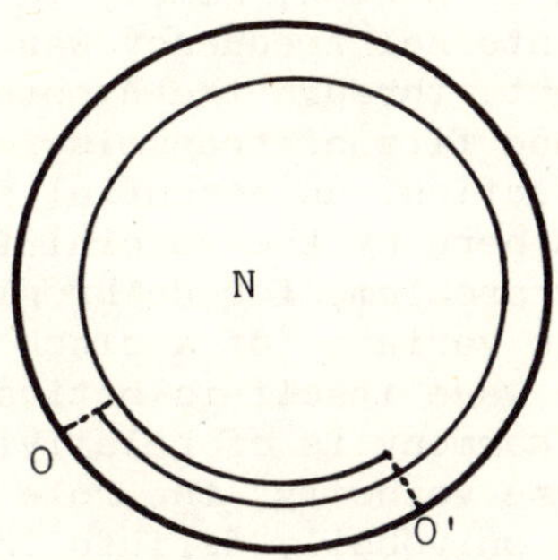


Fig.1 : During the time the electron needs to describe the arch of trajectory OO', it is overtaken by its own wave, the velocity of which is much greater than the velocity of the corpuscle.

Then de Broglie imposed the principle of phase harmony between the internal phase and the phase of the overtaking wave, which is ensured by the condition :

$$2\pi\nu_1\tau = 2n\pi \quad ; \qquad n \in \mathbb{N}. \tag{13}$$

But, from (12) and (13) it follows that :

$$m_0v^2T/\sqrt{1-v^2/c^2} = nh, \tag{14}$$

which is exactly Bohr's condition in a relativistic form.

Introducing the length L = vT of the orbit and the wave-length $\lambda=h/p$, we obtain immediately the formula (which could be set at once as a condition) :

$$L = n\lambda . \qquad (15)$$

Of course, de Broglie verified this formula later but it is interesting to note that, even in his thesis, he gave the formula $\lambda = h/p$ one hundred pages further than the Bohr's atom problem (incidentally and at the non-relativistic approximation). For him, the fundamental quantities were not the wave-length but the frequencies of the wave and the clock, and the phase and group velocities (the latter equal to the velocity of the particle) -in other words, the ingredients that enter in the law of phase harmony.

It is important to realize that, in a certain sense, the principal element of de Broglie's reasoning is not the wave, but the clock. And this is why de Broglie always considered that the first fundamental idea of wave mechanics was not in the work of Huygens but in the work of Newton, namely in the famous "theory of fits", in which an internal frequency was already introduced in the particles of light, through the hypothesis of alternacies of fits of reflection and fits of transmission.

To conclude this section, an essential point has to be emphasized : the role played here by the special theory of relativity. We noted that the first problem, for de Broglie, was a discrepancy between the relativistic variance of a clock frequency and a mass. The introduction of the wave itself is motivated by this problem. The principle of phase harmony is of relativistic essence. Let us not forget that the phase velocity, the role of which is absolutely fundamental, is only univocally defined in a relativistic theory for instance, in the Schrödinger equation, this velocity remains undefined.

The importance of the theory of relativity in de Broglie's theory appears in the same manner in his analogy between mechanics and optics.

In de Broglie's thesis there was another way of obtaining a wave formulation of mechanics : it was the analogy between the principle of least action and the principle of Fermat. Although this problem is far better known than the previous one, we must recall it for the very purpose of showing the role of relativity.

Let us write the quadridimensional momentum :

$$J_\alpha = m_0 c\, u_\alpha + e\, A_\alpha \qquad (\alpha=1,2,3,4) \qquad (16)$$

where $u_\alpha$ is the quadri-velocity and $A_\alpha$ the electromagnetic potential. The principle of Hamilton takes the form :

$$\delta \int_P^Q J_\alpha \, dx^\alpha = 0 \qquad (\alpha=1,2,3,4), \qquad (17)$$

where P and Q are the ends of the universe line of the motion. After that, de Broglie took the case of a conservative system and wrote the principle in the old form of Maupertuis :

$$\delta \int_A^B J_i \, dx^i = 0 \qquad (i=1,2,3) \tag{18}$$

where A and B are, now, the ends of the space trajectory.

Now, de Broglie compared these formulae with the principle of the shortest path in optics. He first wrote the principle in a relativistic way :

$$\delta \int_P^Q d\phi = \delta \int 2\pi O_\alpha dx^\alpha \; (\alpha=1,2,3,4), \tag{19}$$

where P and Q are again the end points of a line of universe (the quadri-opticray) and $\phi$ the phase of the wave ; $O_\alpha$ is the universe wave-vector. If we consider the case of an index of refraction independant of time, this will be the equivalent of the conservative case in mechanics and we shall get, instead of the "Hamilton form" (19), the "Maupertuis form" :

$$\delta \int_A^B O_i dx^i \qquad (i=1,2,3), \tag{20}$$

which is nothing but the integral of Fermat (of course, A and B are points in $\mathbb{R}^3$).

Now, de Broglie considered the two universe vectors : the quadri-momentum $J_\alpha$ and the quadri-wave vector $O_\alpha$ and he noted that, if we postulate that Planck's formula has a universal meaning, we can find a relation between these vectors. As a matter of fact if they are written more explicitely, we have (dropping here the potential A for the sake of simplicity, but this is not necessary):

$$J_i = m_0 c u_i = \frac{m_0 v_i}{\sqrt{1-v^2/c^2}} = p_i \quad (i=1,2,3) \; ; \; J_4 = \frac{w}{c} \tag{21}$$

($v_i$ = velocity of the particle ; w = energy)

$$O_i = \frac{\nu}{V} n_i \quad (i=1,2,3) \quad ; \quad O_4 = \frac{\nu}{c} \tag{22}$$

($\nu$ = frequency ; V = phase velocity ; $n_i$ = unitary wave vector). We see immediately (in virtue of Planck' law) that :

$$w = h\nu \Rightarrow J_4 = h \, O_4 . \tag{23}$$

Thus, postulating the relativistic invariance of this relation, de Broglie claimed that we must have :

$$J_\alpha = h \, O_\alpha \qquad (\alpha=1,2,3,4). \tag{24}$$

And so, by comparing (21) and (22), we find de Broglie's formula :

$$p = \frac{h\nu}{V} = \frac{h}{\lambda} \tag{25}$$

and the equivalence of the principles of Fermat and Maupertuis.

This result is generally considered as the true great achievement of de Broglie and the corner stone of wave mechanics. But this is not the opinion of de Broglie himself. He sees a radical difference between his two resonings (the principle of phase harmony and the equivalence of the two principles of minimum) and he prefers the first. He considers, indeed, that the latter is strictly confined to the geometrical optics limit and classical mechanics, whereas the law of phase harmony has a general meaning and covers not only the classical approximation, but the whole of wave mechanics, including all the features of wave propagation.

Is this opinion correct or not ? We are to this day unable to answer this question, which remains open for the future. But we may consider that the scientific testament of Louis de Broglie consists essentially in the hope that, one day, somebody will explain the profound nature of this strange link between waves and particles, which he discovered 60 years ago.

## 3. DE BROGLIE'S THEORY OF MEASUREMENT

In this part, we shall briefly explain some consequences on the theory of measurement of de Broglie's conceptions on wave-particle dualism. These consequences were developed by de Broglie and his co-workers during the last 30 years.

We shall essentially speak about three questions :

The first will be the special role played in the process of measurement by the localisation of particles, i.e. the preeminence of the observation of the position among all other physical quantities which may be measured in microphysics.

The second problem, strongly linked to the first, will be the definition of three kinds of probabilities in quantum mechanics : the so called present, predicted and hidden probabilities.

Lastly, I shall say a few words about the controversial (and even explosive) problem of Bell's inequality.

First let us rapidly recall some well known points of quantum mechanics concerning the measurement of physical quantities. We know the principles suggested by Max Born at the very beginning of quantum mechanics : a) when a normalized  function is given, we obtain directly the probability $|\psi|^2 dv$ of finding the particle in any elementary volume of space.

b) Secondly, if a physical quantity A is represented by an operator A in a Hilbert space, the proper values $a_i$ of A will be the possible values obtained by a measurement of the quantity A. Now if we expand $\psi$ on the normalized proper functions $\phi_i$ of A :

$$\psi = \sum_i c_i \phi_i , \qquad (26)$$

the square modulus $c_i$ gives the probability of finding precisely the corresponding value $a_i$ when A will be measured.

The most important point about the difference between the current interpretation of quantum mechanics and de Broglie's one concerns the theory of transformations.

Classically, it is admitted in quantum mechanics that a certain symmetry, an equality of meaning exists between all the possible representations of a physical system. For example, we can write the q and p representations, with :

$\delta(q-q_0)$ proper function of $Q_{op} = q$

$e^{-\frac{i}{\hbar}pq}$ " " " $P_{op} = i\hbar \frac{\partial}{\partial q}$

and we can express a $\psi$ function, equivalently, in the two representations as :

$$\psi(q) = \int \psi(q_0)\,\delta(q-q_0)dq_0 = \frac{1}{\sqrt{\hbar}} \int c(p)\, e^{-\frac{i}{\hbar}pq} dp \qquad (27)$$

In most text books, starting with the famous one by Dirac, the equivalence of all representations is considered both as an evidence and an elegance of the quantum theory, which is, of course, from a purely mathematical point of view, absolutely correct. But from a physical point of view, this is a fundamental aspect on which de Broglie disagrees with the current interpretation. Thirty years ago, he made this remark : among all the quantities which can actually be measured, there is in fact a strong dissymetry between the position, on one side, and the whole set of all possible variables on the other side. In general, the position is the sole physical quantity you are able to observe without any preparation of the physical system. You just have to register what happens when a wave falls, say on a wall, and using a photographic film, or any other registering device, you will know where, i.e. in what point, of the wave, the particle was localized.

Generally, one is unable to answer directly any other question about any quantity characterizing the system than the question of localization. We may even say more. If you intend to measure another physical quantity and if you thoroughly examine all the devices we are presently able to construct in order to do so, you will see that they are always founded on the same procedure : this consists in the observation of the position of a particle in conditions such that the sole registration of the particle in a certain region of space leads you to a univocal conclusion on the value of the physical quantity you wanted to measure.

The most general measurement process obeys the scheme given on Fig.2.

We see that the process consists of a spatial separation of the wave packets corresponding to the different proper states (and thus, proper values) of A, so that the registration of the particle in one of those packets (say $a_k$)². The necessity of a spectral analyser (of such or such a kind, according to the type of experiment), which must be adapted to the physical quantity we want to measure and which cannot be adapted at once to all the quantities, is the true origin of the Heisenberg uncertainties. More generally it also lies at the basis of the mysterious strangeness of "quantum probabilities" which has constituted such a delectable subject for theoreticians during the last fifty years. This brings us to the problem of the distinction between two kinds of probabilities: present and predicted. What does it mean ?

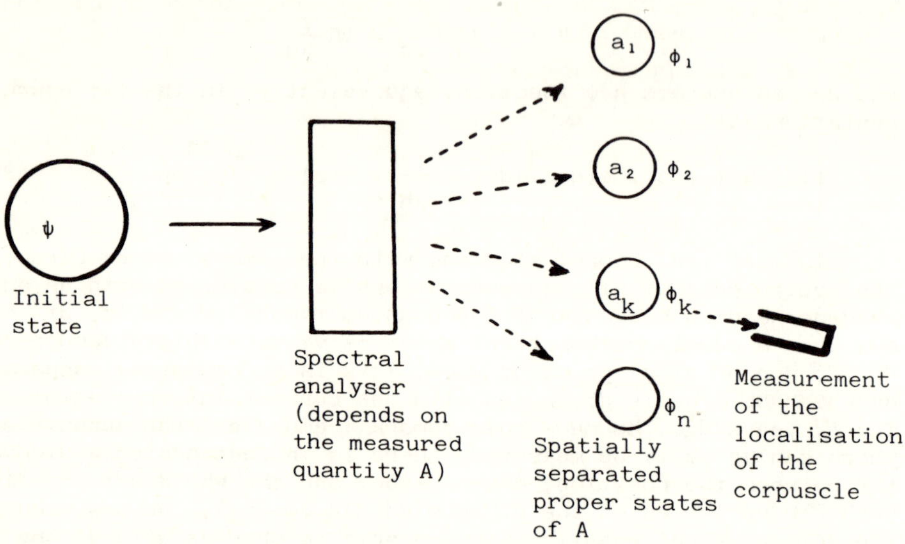


Fig.2 : General scheme of a measurement process.

In de Broglie's terminology, a present probability concerns an event which may be observed without any preparation. This will be the case for all "classical" probabilities, such as those which appear in the problem of games (dice game, heads or tails etc.). And it will also be the case for the probabilities concerning the measurement of the position in microphysics and, more generally, of any other physical quantity which may be measured by a simple inspection of realized events, i.e. for all the quantities for the measurement of which the preparation is already performed. In other words, the probabilities will be present for the values of a physical quantity after the system has passed through the spectral analyser.

But, on the contrary, it is not the case, in general, for most physical quantities, precisely because you need to prepare the system before the measurement, owing to an analyser which modifies the state of the system in such a way (as we have seen) that the measurement will be reduced to a localisation of the particle. This brings about a new type of probabilities : they are no longer present, but only predicted. Let us explain this important point more accurately. Let us suppose that we have two series of events, respectively numbered i and k. We shall write : $P_i$ and $P_k$ for the probabilities of i and k, $P_i^{(k)}$ and $P_k^{(i)}$ for the conditional probabilities of actualization of an event i (resp. k) when k (resp. i) is known, and $P_{ik}$ for the joint-probability of realization of i and k. We know that the classical scheme of probability calculus is based on the formulae :

$$P_{ik} = P_i P_k^{(i)} = P_k P_i^{(k)} \tag{$28_1$}$$

$$P_i = \sum_k P_{ik} \quad ; \quad P_k = \sum_i P_{ik} \tag{$28_2$}$$

$$P_i = \sum_k P_k P_i^{(k)} \quad ; \quad P_k = \sum_i P_i P_k^{(i)} \tag{$28_3$}$$

(The third one is, of course, implied by the first two and let us note, also, that we shall confine the major part of our analysis to the case of discrete systems, which does not make our remarks less general).

The essential point is the existence of the joint probability $P_{ik}$ which supposes the possibility of a simultaneous actualization of events i and k, without any interaction between the two of them ; the hypothesis stands thus : the fact that i (resp. k) is observed does not constitute an obstacle for the observation of k (resp. i), in the same conditions and with the same result as would have been obtained if i (resp. k) had not been observed.

Now, let us consider the quantum case of the observation, on a system in a state $\psi$, of two quantities represented by operators A and B, with respective eigen states $\phi_i$ and $\chi_k$, and eigen values $\alpha_i$ and $\beta_k$.

Each vector $\phi$ or $\chi$ may be transformed into the other by a unitary transformation d :

$$\phi_i = \sum_k d_{ik} \chi_k \quad ; \quad \chi_k = \sum_i d^*_{ki} \phi_i , \tag{29}$$

and we may expand $\psi$ in two ways :

$$\psi = \sum_i c_i \phi_i = \sum_{ik} c_i d_{ik} \chi_k . \tag{30}$$

We know that the probability of finding the system in a state $\phi_i$, with the eigen value $\alpha_i$ for the quantity A, is :

$$P_i = |c_i|^2 \tag{31}$$

In the same way, the probability of finding the system in the state $\chi_k$ and the eigen value $\beta_k$ for the quantity B, may be written, using (30)

$$P_k = |\sum_i c_i\, d_{ik}|^2 \tag{32}$$

Now we can infer from (29) the following conditional probabilities (the meaning is obvious) :

$$P_k^{(i)} = P_i^{(k)} = |d_{ik}|^2 \tag{33}$$

Thanks to (31), (32), (33), we can define two expressions :

$$P_{i,k} = P_i\, P_k^{(i)} = |c_i|^2\, |d_{ik}|^2 \qquad (34_1)$$

$$P_{k,i} = P_k\, P_i^{(k)} = |\sum_j c_j\, d_{jk}|^2\, |d_{ik}|^2 \qquad (34_2)$$

But, except in the case of commuting operators, we have :

$$P_{i,k} \neq P_{k,i} \tag{35}$$

Thus, <u>we cannot define a joint probability $P_{ik}$</u> ; $(28_1)$ will be violated and so will $(28_2)$ and <u>a fortiori</u> $(28_3)$ because :

$$\sum_i P_i\, P_k^{(i)} = \sum_i |c_i|^2\, |d_{ik}|^2 \neq P_k \qquad (36_1)$$

$$\sum_k P_k\, P_i^{(k)} = \sum_{jk} |c_j\, d_{jk}|^2\, |d_{ik}|^2 \neq P_i \qquad (36_2)$$

We are confronted here with the central problem of quantum probabilities. As we know, the impossibility of defining a joint probability is equivalent to the existence of Heisenberg's uncertainties ; both are related to the presence of interference terms in (32) and these terms are nothing but the expression of the fact that on the measurement scheme given on Fig.2, it is <u>impossible to find a single spectral analyser</u> -and, therefore a single splitting of wave packets- which would allow us to measure simultaneously by <u>all</u> the physical quantities, when A,B etc. do not commute. Thus when A and B do not commute, one is not able to answer the two simultaneous questions : what is the value of A <u>and</u> what is the value of B, when the physical system is in a given system $\psi$. In such a case, we will equivalently get Heisenberg's uncertainties, different spectral analysers for the measurement of A and B on Fig.2 and the inequality (35) which destroys the classical probabilistic scheme.

Now, pay attention to the fact that we are speaking here about real waves propagating in <u>physical space</u> and not only about functions defined in an abstract space. The spectral analyser we considered above is, of course, a real physical device and not a Hamiltonian (although we obviously need such a Hamiltonian for a mathematical description of the device). But we are not speaking only about real waves but also about real particles and we have implicitly admitted the fundamental hypothesis that a particle is permanently localized in its wave. In other words, before being observed in a certain wave packet (on the right hand-side of Fig.2), it is supposed that the particle was already localized at every moment in some very small region of space : this means that we admit the existence of a hidden trajectory which went from a certain point of the initial wave packet $\psi$, through the analyser, up to the packet $\phi_k$, where the particle was finally observed. This means that we suppose that the measuring apparatus (i.e. the counter, scintillating screen, photographic film etc.) set on the right side of Fig.2, only allows us to observe a pre-existing circumstance, namely the presence of the particle in such or such wave packet. Thus, it is supposed that the presence of the particle is only revealed by the observation, but in no way "created" or "brought on" by it.

Now we may -we must- ask this question : is this hypothesis compatible or not with the orthodox interpretation of quantum mechanics ? In other words, is the assertion of a preeminence of the position among all observable physical quantities compatible with the orthodox quantum theory of measurement ? However, we shall put this question aside : the problem has been treated in this same book by Francis Fer.

Up to now, we have only met with two of the three kinds of probabilities described by de Broglie : the present probabilities, i.e. those which correspond to a classical probabilistic set of observable events, and the predicted probabilities, i.e. those which concern events which are not observable by simple inspection, because they need a special preparation (spectral splitting) to be observed and, what is more, a preparation which cannot be the same for all the events. Now we shall meet with the third kind of probabilities : the hidden probabilities.

It is well known that in his famous book on the "Mathematical Foundations of Quantum Mechanics", von Neumann proved a theorem which claims that there are no pure states without statistical dispersions. This result is, in itself, intuitively obvious because the absence of dispersions in a pure state would mean that it would be possible to measure simultaneously the physical quantities attached to a system described by this pure state. But in fact we know that this is impossible for non commuting quantities. In this sense, the theorem is nothing but a consequence of Heisenberg's uncertainties. But from this theorem one may conclude that it is not possible to describe a pure state as a mixture of states without dispersions : this is also obvious, because such states do not exist ! This implies that one cannot consider a pure state of quantum mechanics as something like a probabilisable ensemble submitted to the classical scheme (28) of the probability theory : this is exactly what we were speaking about above. Now if you cannot do this, you are allowed to come to the conclusion that it is impossible to find a set of hidden parameters, obeying the laws of quantum mechanics and forming a statistical mixture with a classical probabilistic scheme (28), which would be able to give an account of the quantum probabilistic laws. The contradiction is made obvious by the preceding results ! But von Neumann goes further and claims that, in general, it is impossible to conceive any hidden parameters theory behind quantum mechanics.

De Broglie's answer consists essentially in asserting that if any hidden parameters do exist, they cannot obey quantum mechanics, because if you try to imagine hidden parameters, it is, of course, in order to restore the classical scheme of probabilities. Thus, since you cannot abandon this requirement, you are obliged, for the hidden motions of these parameters, to abandon the possibility of using quantum laws. Now, if you need a classical scheme of probabilities for the objective (but hidden) values of the physical quantities which are introduced in quality of hidden parameters, these probabilities cannot be the probabilities observed on the result of a measurement : simply because the observed probabilities do obey the quantum scheme and not the classical one !

So, de Broglie was led to the introduction of a third kind of probabilities : the so called hidden probabilities, that is, probabilities of a classical kind, which satisfy the scheme (28) but which concern the statistical distribution of hidden variables and which are hidden themselves because when you try to carry out a measurement, you must prepare the system, i.e. modify its state in a suitable manner so that the hidden classical statistics are broken and make way for quantum statistics.

To conclude, we must realize that if we intend to introduce hidden parameters into the quantum theory, the statistics which governs the distribution of these parameters must be hidden too and in no case can such a statistical distribution coïncide with the statistical distribution of the measurement results.

In our opinion, this is the basic point for every attempt to construct a hidden parameters theory. But unfortunately, this idea remains ignored and is never considered by the authors who work on hidden variables problems and, especially on related statistical problems.

For example, the correct conclusion to von Neumann's theorem should only be that all the probabilities of wave mechanics cannot be simultaneously present, which is, as we have said, a consequence of the existence of non commuting quantities. For this reason it is correct to assert that because these probabilities violate the classical scheme, they cannot be introduced into a hidden parameters theory with the aim of restoring the classical determinism : in this sense von Neumann's assertion was correct. But when he claimed that a hidden parameter theory is impossible, he was wrong because he neglected the fact that in such a theory the probabilities must be classical, but also need to be hidden (instead of being present) : his mistake consisted in attempting to assign to hidden parameters the predicted quantum probabilities instead of hidden classical ones.

Since the time when a refutation of von Neumann's theorem was published by de Broglie (for instance in his book on the Theory of measurement in wave mechanics), several other refutations have been published. All of them are based on arguments of logical incoherence of the theorem and some of them exhibit mathematical errors. But although the indication of a mathematical error is obviously sufficient to invalidate a theorem, de Broglie's refutation is farther reaching, because his reasoning is of a general physical character and forewarns us against other attempts of the same type. For instance, the refutation of Bell's inequality which I gave a few years ago (1976) was based on de Broglie's reasoning. It is really not difficult to show (see the following section) that Bell's inequality implies a classical probabilistic scheme for the statistical distribution of the measurement results and thus, contains a radical contradiction with a feature of the quantum theory which is notoriously confirmed by observed physical facts : namely quantum statistical previsions. It is not surprising that many experiments performed up to now (including that carried out by Aspect) have invalidated Bell's inequality. And I must add that this problem has absolutely nothing to do with the problem of locality, which is basically the aim of experiments undertaken in order to test the inequality.

## 4. A FEW WORDS ABOUT BELL'S INEQUALITY

When I showed that Bell's inequality necessarily implies a classical scheme for measurement statistics, in contradiction with quantum mechanics, I was surprised to see that some people strongly opposed this assertion and many others ignored it. I am glad to see that many authors have now rediscovered this pretty obvious fact and I wish to demonstrate it here, once again in two slightly different ways : first I would like to reintroduce the proof I presented in 1976 (in a more favorable atmosphere !) and then I shall give a refutation of a more recent proof of Bell's inequality suggested by d'Espagnat.

a) My 1976 reasoning. In the original proof (somewhat modified by Clauser, Horne, Shimony and Holt) of Bell's inequality, an integral is introduced :

$$P(\vec{a},\vec{b}) = \int d\lambda \ \rho(\lambda) \ A(\vec{a},\lambda)B(\vec{a},\lambda), \qquad (37)$$

which represents the correlation between the measurements of the spins, along two directions $\vec{a}$ and $\vec{b}$, of two particles which had previously interacted. $\underline{A}$ and $\underline{B}$ are the results of the two measurements ; $\lambda$ is a set of hidden parameters (given in the initial state of the system) of which it is supposed that the measurements depend deterministically ; $\rho(\lambda)$ is the statistical distribution of these parameters, with the help of which are calculated all the mean values of the measurement results. $\rho(\lambda)$ is taken (like $\lambda$) in the initial state of the system, so that the statistics introduced in the measurement results do not depend on the measurement process itself. This is the very hypothesis that contradicts quantum mechanics, as we can easily see even in the case of a single physical quantity, for instance the spin of one of the two particles. With the help of this distribution $\rho(\lambda)$, we can easily define the probabiliy $P_{\vec{a}}(\alpha)$ for finding the value $A(\vec{a},\lambda) = \alpha(\alpha = \pm 1)$ as a result of the measurement of the spin of the particle a in the direction $\vec{a}$ ; and we can find a similar probability related to the measurement of the spin of the same particle in another direction $\vec{a}'$.

Indeed, if we define in the configuration space $E\{\lambda\}$ of the hidden parameters the following two subspaces :

$$
\begin{aligned}
E_{\alpha} &= E_{A(\vec{a},\lambda)=\alpha} = E\{\lambda \mid A(\vec{a},\lambda) = \alpha\}, \qquad \alpha = \pm 1 \\
E_{\alpha'} &= E_{A(\vec{a}',\lambda)=\alpha'} = E\{\lambda \mid A(\vec{a}',\lambda) = \alpha'\}, \qquad \alpha' = \pm 1
\end{aligned}
\tag{38}
$$

then we obtain the probabilities

$$
\begin{aligned}
P_{\vec{a}}(\alpha) &= \Pr\{A(\vec{a},\lambda) = \alpha\} = \int_{E_{\alpha}} \rho(\lambda)d\lambda \\
P_{\vec{a}'}(\alpha') &= \Pr\{A(\vec{a}',\lambda) = \alpha'\} = \int_{E_{\alpha'}} \rho(\lambda)d\lambda
\end{aligned}
\tag{39}
$$

Now, we shall also be able to define the quantity

$$
P_{\vec{a},\vec{a}'}(\alpha,\alpha') = \Pr\{A(\vec{a},\lambda) = \alpha, A(\vec{a}',\lambda) = \alpha'\} = \int_{E_{\alpha}\cap E_{\alpha'}} \rho(\lambda)d\lambda \tag{40}
$$

and this is nothing but the probability of finding value $\alpha$ for the spin measure of particle a in the direction $\vec{a}$ and value $\alpha'$ for the spin measure of the same particle in another direction $\vec{a}'$ : it is the joint probability !

Since a measurement of a spin can only give the value $\pm 1$ (reduced here to unity), we shall have

$$
E_{\alpha=-1} \cup E_{\alpha=1} = E\{\lambda\} \ ; \ E_{\alpha'=-1} \cup E_{\alpha'=1} = E\{\lambda\}. \tag{41}
$$

Therefore,

$$
\int_{E_{\alpha}\cap E_{\alpha'=1}} \rho(\lambda)d\lambda + \int_{E_{\alpha}\cap E_{\alpha'=-1}} \rho(\lambda)d\lambda = \int_{E_{\alpha}} \rho(\lambda)d\lambda \tag{42}
$$

whence, according to (39) and (40), we deduce

$$
P_{\vec{a}}(\alpha) = \sum_{\alpha'=\pm 1} P_{\vec{a},\vec{a}'}(\alpha,\alpha') \ ; \ P_{\vec{a}}(\alpha') = \sum_{\alpha=\pm 1} P_{\vec{a},\vec{a}'}(\alpha,\alpha'). \tag{43}
$$

This is the formula ($28_2$) and now we may also write ($28_1$), by introducing conditional probabilities :

$$P_{\vec{a}}^{(\vec{a}')}(\alpha,\alpha') = \frac{P_{\vec{a},\vec{a}'}(\alpha,\alpha')}{P_{\vec{a}'}(\alpha')} \; ; \; P_{\vec{a}'}^{(\vec{a})}(\alpha,\alpha') = \frac{P_{\vec{a},\vec{a}'}(\alpha,\alpha')}{P_{\vec{a}}(\alpha)} \qquad (44)$$

And lastly, using (43), we get :

$$P_{\vec{a}}(\alpha) = \sum_{\alpha'=\pm 1} P_{\vec{a}}^{(\vec{a}')}(\alpha,\alpha')P_{\vec{a}'}(\alpha') \; ; \; P_{\vec{a}'}(\alpha') = \sum_{\alpha=\pm 1} P_{\vec{a}'}^{(\vec{a})}(\alpha,\alpha')P_{\vec{a}}(\alpha), \qquad (45)$$

i.e. ($28_3$). Thus, we see that the hypothesis which is at the origin of Bell's inequalities implies a classical statistical scheme for measurement results, that is, implies (in contradiction with Heisenberg's uncertainties and with experimentation) the possibility of a simultaneous measure of two spin components of the same particle.

In other words Bell's mistake is opposite to von Neumann's one : the latter ascribed the quantum statistics of measurement results to hidden parameters whereas the former ascribed the classical statistics of hidden parameters to measurement results. They both neglect the difference pointed out by de Broglie between hidden classical probabilities and predicted quantum probabilities.

b) D'Espagnat's problem of the twin students (1979). In order to demonstrate the theorem in a simple general way, the author imagined an ingenuous example : he replaced spins by twins !

Let us consider with him a university where each student has a twin brother (a real one, of course) and where the only subjects taught and tested are three languages : Latin, Greek and Chinese.

First of all, it is easy to prove that in any case (without taking into account the "twin" hypothesis) we may divide the students into three "statistical samples" and assert that after the exams :

"The number of graduates in Latin and Greek in sample 1 will not be larger than the number of graduates in Latin and Chinese in sample 2 increased by the number of graduates in Greek who failed in Chinese in sample 3".

This is exactly Bell's inequality ! But one may object that these languages are difficult, that different exams undergone by a single student may perturb each other (these are perturbations caused by measurements !) and this is why d'Espagnat introduced the twins, so that one exam may be undergone by one brother and the second one by the second brother (without perturbing the first one), but both will be credited with the same mark since they are supposed to be perfect twins.

It seems that we will thus find Bell's inequality without supposing that the same student may undergo two examinations, that is without forgetting the measurement perturbations. This is wrong ! Let us try to give a real proof of the inequality.

In order to do so, let us begin by separating the twins so

that we get two "twin sets" each of which will be divided into three samples (resp. : 1,2,3 and 1',2',3'). In each sample, we shall denote l, g, c the number of "direct" graduates (i.e. having undergone the exam themselves) in Latin, Greek, and Chinese, $\tilde{l}$, $\tilde{g}$, $\tilde{c}$ the number of those who "directly" failed and by $\underline{l}$, $\underline{g}$, $\underline{c}$ (resp. $\underline{\tilde{l}}$, $\underline{\tilde{g}}$, $\underline{\tilde{c}}$) the number of "indirect" graduates who passed (resp. failed) through an examination undergone by their twin brother. Of course, we have by hypothesis $l=\underline{l}$, $g=\underline{g}$, $c=\underline{c}$ etc.

Now consider the following figure :

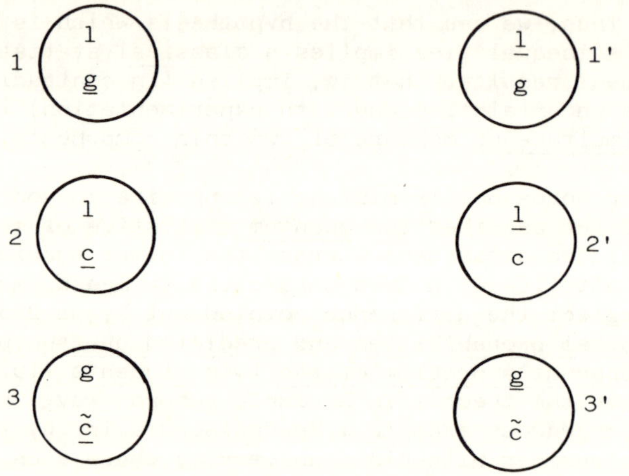


Fig.3

If we denote, for instance, $(l,\underline{g})_1$ the number of students in sample 1 who are "direct" graduates in Latin and "indirect" graduates in Greek and if in the same way we introduce analogous brackets, the previous expression of Bell's inequality will be written :

$$(l,\underline{g})_1 \leqslant (l,\underline{c})_2 + (g,\underline{\tilde{c}})_3. \qquad (46)$$

Now, in order to prove this inequality, we can write, on sample 1, the trivial decomposition :

$$(l,\underline{g})_1 = (l,\underline{g},\underline{c})_1 + (l,\underline{g},\underline{\tilde{c}})_1 \qquad (47)$$

But the three samples are statistically equivalent, whence :

$$(l,\underline{g})_1 = (l,\underline{g},\underline{c})_2 + (l,\underline{g},\underline{\tilde{c}})_3. \qquad (48)$$

Now, in each sample, we have trivially :

$$(1,\underline{g},\underline{c})_2 \leqslant (1,\underline{c})_2 \;;\; (1,\underline{g},\underline{\tilde{c}})_3 \leqslant (\underline{g},\underline{\tilde{c}})_3 \tag{49}$$

and introducing (49) in (48) we get (46).

Unfortunately, the decomposition we have introduced in (47) (and what else could we do in order to find an inequality ?) implies that it is physically meaningful to talk about students whose knowledge has been directly tested in two languages (for instance in Greek and Chinese). This means that finally, in order to prove the inequality, we have been obliged so to speak "to perform on the same object measurements of two quantities that we supposed were not simultaneously measurable". In other words we ended up supposing that the measurement statistics were classical.

## 5. IN SEARCH OF A NEW SOLUTION

One may pose the following question : is it really possible to build a model with hidden parameters obeying hidden classical probabilities which would nevertheless be in agreement with the quantum probabilistic scheme for measurement results ? In order to answer the question, we shall give a spin measurement model based on the ideas of de Broglie. Of course, it will just be sketched, not as a true theory but only as an example.

The main idea will be that the corpuscle is permanently localized in the wave (although we do not know exactly where) and that the spin is well defined at each point of the wave but cannot be measured without a special preparation because a measurement implies the following :

a) Detection of the particle.

b) Realization of a state such that a relatively rough determination of the position will lead to a univocal determination of the spin.

For the sake of simplicity, we shall consider the nonrelativistic case, so that the wave associated with the particle is represented by a spinor with two components obeying the Pauli equation. The hidden parameters which Bell denoted by $\lambda$ are now the position coordinates $\vec{r}$ of the particle in the wave ; therefore the density $\rho(\lambda)$ will be written

$$\rho(\vec{r}) = \psi^*(\vec{r})\psi(\vec{r}) = |\psi_1(\vec{r})|^2 + |\psi_2(\vec{r})|^2 \tag{50}$$

with

$$\psi = \binom{\psi_1}{\psi_2}, \qquad \int \rho(\vec{r})dv = 1 \tag{51}$$

In any state $\psi$ of the particle, we shall define a spin

$$\vec{s}(\vec{r}) = \frac{\psi^*\vec{\sigma}\psi}{\psi^*\psi} \tag{52}$$

where $\vec{\sigma}$ denotes the set of the three Pauli matrices

$$\sigma_1 = \begin{pmatrix} 0 & 1 \\ 1 & 0 \end{pmatrix}, \quad \sigma_2 = \begin{pmatrix} 0 & -i \\ i & 0 \end{pmatrix}, \quad \sigma_3 = \begin{pmatrix} 1 & 0 \\ 0 & -1 \end{pmatrix} \tag{53}$$

Formula (52) means that if the particle is found at point $\vec{r}$ of the wave $\psi(\vec{r})$, it then has the spin $\vec{s}(\vec{r})$. But, for the time being, the position $\vec{r}$ of the particle remains unknown to us, and we know only the probability density $\rho(\vec{r})$ given by (50) ; the spin is thus a <u>hidden variable</u>.

Let us consider two different directions $\vec{a}$, $\vec{a}'$ in space. <u>Before</u> any measurement is made, if we suppose the particle to be <u>localized</u>, at a given instant, at a given point $\vec{r}$, in the wave, we will be led by the preceding formulae to attribute the following values to the spin :

$$\vec{s}(\vec{a},\vec{r}) = \frac{\psi^*\vec{\sigma}.\vec{a}\psi}{\psi^*\psi} \quad ; \quad \vec{s}(\vec{a}',\vec{r}) = \frac{\psi^*\vec{\sigma}.\vec{a}'\psi}{\psi^*\psi}, \tag{54}$$

in the directions $\vec{a}$ and $\vec{a}'$. But pay attention ! The two quantities $\vec{s}(\vec{a},\vec{r})$ and $\vec{s}(\vec{a}',\vec{r})$ are <u>simultaneously defined</u> but <u>not simultaneously measurable</u> : they are not to be mistaken, as in Bell's formula, with what he denotes by $A(\vec{a},\lambda)$ and $A(\vec{a}',\lambda)$, i.e. with the results of a measurement. <u>At this stage</u>, these quantities remain hidden quantities (we shall examine the question of measurement below) and they obey a classical statistics which is easy to construct, but which will be hidden too, and which is not to be confused with the measurement statistics.

In order to build the hidden statistics let us write more briefly r, s, s' instead of $\vec{r}$, $\vec{s}(\vec{a},\vec{r})$, $\vec{s}(\vec{a}',\vec{r})$, and let us denote as R, S, S' random variables whose possible values are respectively r, s, s'. First we will have the densities of conditional probabilities ($\delta$ = Dirac function) :

$$\rho_S^{(R)}(\alpha,r) = \delta(\alpha - s(r)), \quad \rho_{S'}^{(R)}(\alpha',r) = \delta(\alpha' - s(r')) \tag{55}$$

which simply means that <u>if</u> R takes the value r then, by hypothesis, S certain by takes the value s(r) and S' the value s'(r). The probability densities for S and S' will be :

$$\rho_S(\alpha) = \int \rho(r)\delta(\alpha - s(r))dr \; ; \; \rho_{S'} = \int \rho(r)\delta(\alpha - s'(r))dr \tag{56}$$

Here we see that the <u>position</u> r of the particle in the wave plays the role played by <u>$\lambda$ for Bell</u> ; $\rho(r)$ is given by (50), and don't forget that this probability is not hidden, but <u>present</u>, as we said previously. Of course, we may also define the joint probability that S will take a value in $[\alpha, \alpha+d\alpha]$ <u>and</u> S' in $[\alpha',\alpha'+d\alpha]$ :

$$\rho(\alpha,\alpha') = \int \rho(r)\delta(\alpha - s(r))\delta(\alpha' - s'(r))dr \tag{57}$$

and we can immediately verify that :

$$\rho_S(\alpha) = \int \rho(\alpha,\alpha')d\alpha' \qquad \rho_{S'}(\alpha') = \int \rho(\alpha,\alpha')d\alpha \tag{58}$$

Thus, we have the classical statitical scheme (28), from which Bell's inequalities may be deduced : but this scheme is hidden and so are Bell's inequalities, because if we try to measure a spin component, we cannot do so without disturbing the form of the wave and without changing the density $\rho$ and thereby all the other probability distributions.

For us to be sure, let us first consider the measurement of a spin component parallel to a certain direction $\vec{a}$. This can be done by letting the particle enter an inhomogeneous magnetic field oriented along direction $\vec{a}$. This field will split the wave into two distinct wave trains which shall be such that, if the particle is found to be in one of them, we may then be sure that $\vec{s}(\vec{a},\vec{r}) = 1$ and, if it is found in the other, then we know that $\vec{s}(\vec{a},\vec{r}) = -1$.

Let us then take the direction $\vec{a}$ as coinciding with the z axis. We shall have according to (50), (52) :

$$\vec{s}(\vec{a},\vec{r}) = s_3(\vec{r}) = \frac{\psi^*\sigma_3\psi}{\psi^*\psi} = \frac{\psi_1{}^*\psi_1 - \psi_2{}^*\psi_2}{\psi_1{}^*\psi_1 + \psi_2{}^*\psi_2} \tag{59}$$

The wave train for which $s_3 = 1$ will then be such that

$$\psi^+ = \begin{pmatrix}\psi_1\\ 0\end{pmatrix} \tag{60}$$

and the one for which $s_3 = -1$ will be

$$\psi^- = \begin{pmatrix}0\\ \psi_2\end{pmatrix} \tag{61}$$

The total wave on leaving the magnetic field will be $\psi = \psi^+ + \psi^-$, but, since the two components are separated in space, filling two distinct regions $R^+$ and $R^-$, the probability of getting the value +1 of the component $S_3$ of the spin will be

$$P_3(+) = \int_{R^+} \psi^*\psi dv = \int |\psi_1|^2 dv \tag{62}$$

while the probability of getting the value -1 will be

$$P_3(-) = \int_{R^-} \psi^*\psi dv = \int |\psi_2|^2 dv \tag{63}$$

The normalization of the wave ensures that

$$P_3(+) + P_3(-) = 1 \tag{64}$$

It is understood that only the $s_3$ component of the spin has its values distributed according to the elementary probability law given by $P_3(+)$ and $P_3(-)$, and one also sees that such a law is different from the law which yielded the hidden distributions before the measurement.

Let us now assume that we wanted to measure the spin component along another direction distinct from $\vec{a}$ : Let us denote it by $\vec{a}'$.

We should then set up another apparatus in all ways similar to the one above, except that its magnetic field would then be orientated along direction $\vec{a}'$, which would again split the initial wave into two wave trains (obviously different from those we had obtained above). Each of these wave trains would correspond to one of the values +1 or -1 of the component $S(\vec{a}',\vec{r})$. We could then get for these two values certain probabilities $P_3'(+)$ and $P_3'(-)$ with expressions similar to those we considered above ; but in order to do this, it is necessary to change the reference frame and take for the z axis no longer the direction $\vec{a}$ but the direction $\vec{a}'$ ; in Eqs.(62) and (63) we obviously have now new functions $\psi_1'$ and $\psi_2'$.

It is thus seen that if one assumes that the hidden parameters are the coordinates of the particle, then the measurement of two different components of a spin requires two partitions of the initial wave train that are different and incompatible with each other. It clearly follows that the probability density $\rho(r)$, which pertains to the initial wave, will give rise, when resulting from the magnetic field, to two different densities $\rho(\vec{a},\vec{r})$ and $\rho(\vec{a}',\vec{r}')$, corresponding to the measurements of components $S(\vec{a})$ and $S(\vec{a}')$ of the spin.

We are clearly unable to write here any integral like (31) with A or B representing measurement results : THE VERY IDEA OF BELL'S INEQUALITY IS ABSOLUTELY EXCLUDED FROM SUCH A HIDDEN VARIABLES THEORY.

Some conclusions

Among de Broglie's physical ideas the main conception which must above all be underlined is his profound belief in the existence of a physical material wave, in the permanent localization of the corpuscle within this wave, and in the preeminent role he assigns to the relation between the frequency of the wave and the clock-frequency of the particle. This is, briefly summarized, de Broglie's scientific legacy.

Now, his theory of the double solution must only be considered as a possible model, imagined by him in order to construct a theory on the basis of these ideas. In fact, in spite of the existence of some attractive features (such as the possibility of a visualization of the wave particle dualism) and some remarkable results in the problems of propagating waves (such as the description of diffraction phenomena), the theory remains unfinished.

Nevertheless, even in its present state, the theory has given rise to two important conceptions, namely : the introduction (thirty years ago) of solitons in quantum mechanics, and a physical (and not only operational) quantum measurement theory, even if the latter is still confined to measurements of the first kind (i.e. without intervention of correlated particles) and only to special cases of measurements of the second kind (i.e. with correlations).

This measurement theory is based on a general idea concerning hidden parameters which is at once so important and so obstinately ignored by most physicists that I am not afraid of emphasizing it too much : I am speaking, of course, about the absolute necessity of a distinction between hidden, predicted and present probabilities, if we are to speak about hidden parameters. Given that I am sure that most of the contributions that will be devoted to the same problem in this book will diverge from my opinion I want to insist and to claim once more that I consider Bell's inequalities as being in radical contradiction with statistical statements of quantum mechanics which are certainly in agreement with physical facts in all the experiments carried out in order to test these inequalities. For this reason, the experimental violation of Bell's inequalities has nothing to do, in my opinion, with the so called "non locality" or "non separability". This violation simply signifies that quantum probabilities are not classical probabilities ! This is a fact which was discovered more than half a century ago, which it is nice to see confirmed again, but which perhaps does not need so many sophisticated and beautiful experiments in order to be accepted. The true interest of such experiments is, in my opinion, to confirm the exactness of the quantum statement that two particles (having previously interacted) may be considered as constituting a pure state : this statement was originally used for very small systems (such as molecules) but it was a priori not obvious at all that it should remain true for particles which are separated by more than ten meters. The fact that this is true constitutes a genuine progress. But as puzzling as this fact may be, in no way does it prevent the possible existence of hidden parameters, for the elementary reason that such an assertion is based on an inequality which supposes a statistical hypothesis which is false even for one particle, so that the inequality looses any physical meaning.

I cannot conclude these few remarks without drawing attention to the fact that the best one can do at a given stage of the history of science is to provide more or less good arguments in favour of such or such a theoretical hypothesis. One must be very cautious about the alleged final character of a scientific conclusion, because any scientific conclusion is always subject to revision. The claim of having definitively closed a door, in science, even if no objection can momentarily he opposed (which is not presently the case !) is a very dangerouse one : such an attitude does not belong to science but to ideology and it is rather apt to arouse the laughter of future generations. What I consider to be true for other people is also true, of course, for me : I never forget that it is quite possible that a new microphysics may be elaborated without any reference to hidden parameters, but there is a great difference between considering a possibility and accepting the extravagant pretention of a final scientific prohibition.

1 From (6) and (8) we get immediately de Broglie's wave-length, for :

$$\lambda = V/\nu = (c^2/v)/(\frac{m_0c^2}{h}/\sqrt{1 - \frac{v^2}{c^2}}) = h/(m_0v/\sqrt{1 - \frac{v^2}{c^2}}) = h/p$$

2 A classical objection is that this is true for devices like a prism, a mass-spectrograph, an apparatus of Stern-Gerlach etc, but not for measurements obtained, for instance, by a resonance in a molecular beam. This objection is not valid because the resonance in the molecular beam is in fact tested by the intensity of the emerging beam, i.e. the presence or the absence of a particle in a certain region of space.

BIBLIOGRAPHY

(1) L. de Broglie :

a) Some important papers :

C.R.A.S., 177, p. 517, p. 548, p. 630, 1923 ; Philosophical Mag., XLVII, p. 446, 1924 ; J. Phys., série VI, t. VII, n° 11, p. 321, 1926 ; J. Phys., série VI, t. VIII, n° 5, p. 225, 1927 ; J. Phys., 20, p. 963, 1959 ; C.R.A.S., 277, série B, p. 71, 1973 ; Found. of Phys. (with G. Lochak, J.A. Beswick and J. Vassalo Pereira), 6, p. 3, 1976.

b) Books (among 44 titles) :

- Recherches sur la théorie des quanta (Thesis, 1924) : first published in : Annales de Physique, 10e série, t. III, p. 22-128, 1925 ; german translation : Akademische Verlaggesellschaft, Leipzig, 1927 ; reprinted by Masson, Paris, 1963.
- Selected papers on wave mechanics (with Léon Brillouin), Blackie & Son, Glasgow, 1928.
- Une tentative d'interprétation causale et non linéaire de la mécanique ondulatoire : la théorie de la double solution, Gauthier-Villars, Paris, 1956 ; english translation : Elsevier, Amsterdam, 1960.
- La théorie de la mesure en mécanique ondulatoire (interprétation usuelle et interprétation causale), Gauthier-Villars, Paris, 1960.
- Etude critique des bases de l'interprétation actuelle de la mécanique ondulatoire, Gauthier-Villars, Paris, 1963 ; english translation, Elsevier, Amsterdam, 1965.
- La réinterprétation de la mécanique ondulatoire, Gauthier-Villars, Paris, 1971.
- Contribution to : Foundations of quantum mechanics, Proc. Int. Schools Enrico Fermi, Varenna, 1971.
- Les incertitudes d'Heisenberg et l'interprétation probabiliste de la mécanique ondulatoire, Gauthier-Villars, Paris, 1982 (with a long preface of G. Lochak on The evolution of the ideas of Louis de Broglie on the interpretation of wave-mechanics : translated in English in the Special issue of Found. of Physics devoted to

Louis de Broglie, 1982.
(2) J. Andrade e Silva : Found. of Phys., 2, 245, 1972 ; contribution to Foundations of quantum mechanics, Proc. Int. Schools Enrico Fermi, Varenna, 1971.
(3) G. Lochak, Found. of Phys., 6, p. 3, 1976 (with de Broglie et al. : see above) ; 6, p. 173, 1976 ; several contributions to the Written Symposium on Hidden Variables and Quantum Uncertainty, since 1975 (Association Ferdinand Gonseth, Case postale 1081, CH-2501 BIENNE) ; Revue de Métaphysique et de Morale, 88ème année, n° 1, p. 85, 1983.
(4) J. Bell, Rev. Mod. Phys. 38, p. 447, 1966 ; contributions to the School of Physics, Varenna, 1971 (see above) and to the Written Symposium (cited above).
(5) J.F. Clauser and A. Shimony, Reports on Progress in Physics, 41, p. 1881-1927, 1978.
(6) B. d'Espagnat, A la recherche du réel, Gauthier-Villars, Paris, 1979.

# WAVE-PARTICLE DUALITY OF LIGHT: A CURRENT PERSPECTIVE

P. W. Milonni

Department of Physics
University of Arkansas
Fayetteville, Arkansas 72701

ABSTRACT.

Ideas of wave-particle duality of light are traced through various stages, from Einstein to quantum optics and recent developments in the conceptual foundations of quantum mechanics.

## I. INTRODUCTION

One of the remarkable things about quantum mechanics is the extent to which people disagree on what the theory says about physical reality. Even the founding fathers of wave-particle duality, Albert Einstein and Louis de Broglie, ultimately rejected the prevalent (Copenhagen) interpretation of quantum mechanics, as did Schrödinger. Perhaps "if one percent of the effort spent on physics were devoted to clarification, we could soon teach the basic concepts of quantum mechanics to the general run of nine-year-olds!"(1) At the present time, however, it seems safe to say, without too much exaggeration, that "...nobody understands quantum mechanics." (2)

The most positivistic advocates of the Copenhagen interpretation find the notion of an objective physical reality to be irrelevant and obsolete. Those of us of an Einsteinian mind-set are of course encouraged by the fact that such great scientists as de Broglie reject the Copenhagen philosophy. However, we must admit that recent evidence points against a "naive realism." In any case these questions will be debated for a long time to come, and new ideas may be encouraged by recording the thoughts of many people. In this vein I hope this survey will not be entirely superfluous.



*S. Diner et al. (eds.), The Wave-Particle Dualism, 27–67.*

## II. EARLY IDEAS

Great ideas have their precursors. However, I will begin with Einstein, for his work provided the first clear indication that light can simultaneously display properties of waves and particles.

### A. Einstein (1905)

In the introduction to his paper "Concerning an Heuristic Point of View Toward the Emission and Transformation of Light," (3) Einstein wrote:

> The wave theory of light, which operates with continuous spatial functions, has worked well in the representation of purely optical phenomena and will probably never be replaced by another theory. It should be kept in mind, however, that the optical observations refer to time averages rather than instantaneous values. In spite of the complete experimental confirmation of the theory as applied to diffraction, reflection, refraction, dispersion, etc., it is still conceivable that the theory of light which operates with continuous spatial functions may lead to contradictions with experience when it is applied to the phenomena of emission and transformation of light.
>
> It seems to me that the observations associated with blackbody radiation, fluorescence, the production of cathode rays by ultraviolet light, and other phenomena connected with the emission or transformation of light are more readily understood if one assumes that the energy of light is discontinuously distributed in space. In accordance with the assumption to be considered here, the energy of a light ray spreading out from a point source is not continuously distributed over an increasing space but consists of a finite number of energy quanta which are localized at points in space, which move without dividing, and which can only be produced and absorbed as complete units.

That paper is best known for Einstein's prediction concerning the photoelectric effect ("the production of cathode rays by ultraviolet light"). However, only a short discussion near the end of the paper was devoted to the photoelectric effect; most of the paper was concerned with blackbody radiation, particularly the Wien limit ($h\gamma >> kT$) of the Planck distribution.

Let $\rho(\gamma)\,d\gamma$ be the average energy density of blackbody radiation in the frequency interval $[\gamma, \gamma + d\gamma]$. Under some general assumptions the entropy of the radiation may be written

$$S = V\int_0^\infty d\nu\, \phi(\rho, \nu) \tag{2.1}$$

where V is the volume of the cavity. In writing (2.1) Einstein assumed that the different frequency components are independent and that the radiation is fully characterized by $\rho(\gamma)$. Now the conditions of maximum entropy and fixed average energy lead easily to the equation

$$\frac{\partial\phi}{\partial\rho} = \frac{1}{T} \tag{2.2}$$

for the function $\phi$. It follows from this and the Wien Law

$$\rho(\nu) = \alpha\nu^3 e^{-\beta\nu/T} \tag{2.3}$$

that

$$S = -\frac{V}{\beta}\int_0^\infty d\nu\, \frac{\rho(\nu)}{\nu}\left[\log\left(\frac{\rho(\nu)}{\alpha\nu^3}\right) - 1\right] \tag{2.4}$$

For radiation in the frequency interval $[\gamma, \gamma + d\gamma]$ with energy $E_\gamma = V\rho(\gamma)d\gamma$, therefore, we have from (2.4) the entropy

$$S_\nu = -\frac{E_\nu}{\beta\nu}\left[\log\left(\frac{E_\nu}{V\alpha\nu^3 d\nu}\right) - 1\right] \tag{2.5}$$

If the energy is held fixed while changing reversibly the volume of the cavity from $V_1$ to $V_2$, the entropy change of this radiation is found from (2.5) to be

$$S_\nu(2) - S_\nu(1) = \frac{E_\nu}{\beta\nu}\log\left(\frac{V_2}{V_1}\right) \tag{2.6}$$

Einstein justified his "heuristic point of view" by comparing Eq. (2.6) to the entropy change of an ideal gas whose volume is isothermally changed from $V_1$ to $V_2$:

$$S_g(2) - S_g(1) = Nk\log\left(\frac{V_2}{V_1}\right) \tag{2.7}$$

where N is the number of particles. Equations (2.6) and (2.7) are of precisely the same form. The two expressions are in fact identical if

$$E_\nu = \beta\nu Nk = Nh\nu \tag{2.8}$$

Thus the radiation satisfying Wien's law (2.3) "behaves thermodynamically as though it consisted of a number of independent energy quanta of magnitude [$h\nu$]." (3) Einstein regarded this heuristic picture as a new idea, not a direct consequence of Planck's work. (4)

Using the heuristic viewpoint motivated by (2.8), Einstein showed that Stokes' rule followed trivially by conservation of energy. He then predicted the linear relation between radiation frequency and stopping potential in the photoelectric effect. He wrote:

> As far as I can see, there is no contradiction between these conceptions and the properties of the photoelectric effect observed by Herr Lenard. If each energy quantum of the incident light, independently of everything else, delivers its energy to electrons, then the velocity distribution of the ejected electrons will be independent of the intensity of the incident light; on the other hand the number of electrons leaving the body will, if other conditions are kept constant, be proportional to the intensity of the incident light. (3,5)

Einstein's prediction was fully corroborated by Millikan's experiments in 1916. (6) Millikan himself, however, considered Einstein's heuristic picture "reckless," and even implied that it had been "pretty generally abandoned." (7)

It is generally believed that the photoelectric effect provides incontrovertible evidence for photons. (8) Actually the situation is considerably more subtle than is generally indicated. This point will be discussed in Section 4.

Einstein was careful to point out that "non-Wien radiation" might not conform to his heuristic picture. In fact he would later find that blackbody radiation in the Rayleigh-Jeans limit ($h\nu/kT \ll 1$) conformed better to an interpretation based on classical waves rather than particles.

B. Taylor (1909)

In a very short paper in the Proceedings of the Cambridge Philosophical Society (9), G. I. Taylor mentioned a theory of light in which "There are regions of maximum energy widely separated by large undisturbed areas. When the intensity of light is reduced these regions become more widely separated, but the amount of energy in any one of them does not change; that is, they are indivisible units." He described an experiment he had performed to see whether for very dim light these indivisible units would modify ordinary diffraction phenomena. In one case he used light so dim that at any given time we might say there was not more than a single photon arriving at the photographic plate; the exposure time was about three months! What Taylor observed was an interference pattern of the same form regardless of the intensity. From today's perspective we can also say that Taylor observed (indirectly) the build-up of a diffraction pattern by successive photons, each of which produces on the plate not the entire pattern but only a spot on the pattern.

With a photomultiplier it is possible to demonstrate Taylor's effect in a similar experiment more directly, registering individual pulses of photoelectrons while observing the entire diffraction pattern:

> It's a unique experience, sitting there, watching the pattern, one photon coming in at a time, knowing that you are, at that moment, probably as close to understanding as you will ever come -- so close and yet so far. (10)

C. Einstein (1909)

In 1909 Einstein provided the first clear indication of the wave-particle duality of light. (11) Now Einstein used Planck's formula for $\rho(\gamma)$, not Wien's. He applied a result he had derived in 1904 for the energy fluctuations of a system in thermal equilibrium (12):

$$\langle E^2\rangle - \langle E\rangle^2 \equiv \langle \Delta E^2\rangle = kT^2 \frac{\partial\langle E\rangle}{\partial T} \tag{2.9}$$

For thermal radiation in the frequency interval $[\gamma, \gamma + d\gamma]$ $(\langle E_\gamma\rangle = V\rho(\gamma)d\gamma)$ Einstein obtained, from Planck's formula and (2.9),

$$\langle \Delta E_\nu^2\rangle = \left[h\nu\rho(\nu) + \left(\frac{c^3}{8\pi\nu^2}\right)\rho^2(\nu)\right]V d\nu \tag{2.10}$$

a result now called the <u>Einstein fluctuation formula</u>.

In Eq. (2.10) $\rho(\gamma)$ is the Planck spectrum,

$$\rho(\nu) = \frac{8\pi h\nu^3/c^3}{e^{h\nu/kT} - 1} \tag{2.11}$$

If $\rho(\gamma)$ instead is given by the Rayleigh-Jeans formula,

$$\rho(\nu) = \frac{8\pi\nu^2}{c^3} kT \qquad (2.12)$$

then (2.9) leads to

$$\langle \Delta E_\nu^2 \rangle_{waves} = \frac{c^3}{8\pi\nu^2} \rho^2(\nu) V d\nu \qquad (2.13)$$

instead of (2.10). This result is obtained in a purely classical electromagnetic wave approach to thermal radiation, as was shown by Lorentz. (13) If, on the other hand, we use the Wien law (2.3), we obtain

$$\langle \Delta E_\nu^2 \rangle_{particles} = h\nu \rho(\nu) V d\nu \qquad (2.14)$$

a result characteristic of the classical Poisson statistics of (distinguishable) particles. Thus the Einstein fluctuation formula (2.10) may be written

$$\langle \Delta E_\nu^2 \rangle = \langle \Delta E_\nu^2 \rangle_{waves} + \langle \Delta E_\nu^2 \rangle_{particles} \qquad (2.15)$$

The first term was expected from ordinary classical electromagnetic theory. The second suggested "the point of view of the Newtonian emission theory." (11) Einstein's formula (2.15) indicated that both the wave and particle pictures are oversimplifications.

Using (2.11), we may write (2.10) as

$$\langle \Delta E_\nu^2 \rangle = h\nu\rho(\nu)\left[1 + \frac{1}{e^{h\nu/kT} - 1}\right] V d\nu \tag{2.16}$$

In the "Wien limit" $h\nu/kT >> 1$, we have

$$\langle \Delta E_\nu^2 \rangle_{Wien} \cong h\nu\rho(\nu) V d\nu = \langle \Delta E_\nu^2 \rangle_{particles} \tag{2.17}$$

In the "Rayleigh-Jeans limit" $h\nu/kT << 1$, on the other hand,

$$\langle \Delta E_\nu^2 \rangle_{\substack{Rayleigh-\\Jeans}} = \langle \Delta E_\nu^2 \rangle_{waves}$$

D. Einstein (1917)

Einstein's formula (2.15) is a consequence of Planck's spectrum. He made several attempts to derive the Planck spectrum in a consistent fashion, but failed. (14) In 1916 he returned to the problem, this time using the Bohr postulates (1913) (15). His derivation of the Planck spectrum using the A and B coefficients is now standard textbook material.

What the textbooks focus on is the first part of Einstein's paper, dealing with rate equations for the energy level populations of the atoms. What Einstein himself apparently considered more important was the second part of the paper, which leads to the conclusion that radiation quanta have linear momentum $h\gamma/c$, without which he could not obtain Planck's formula in a consistent manner:

> Most important, however, seems to me to be the result concerning the momentum transfer to the molecule due to the absorption and emission of radiation...

> If a [radiation quantum] has the effect that a molecule struck by it absorbs or emits a quantity of energy hγ in the form of radiation..., then a momentum hγ/c is always transferred to the molecule. For an absorption of energy, this takes place in the direction of propagation of the radiation bundle, for a [stimulated emission] in the opposite direction...
>
> If the molecule undergoes a loss in energy of magnitude hγ without external excitation, by emitting this energy in the form of [spontaneous emission], then this process, too, is directional...the molecule suffers a recoil of magnitude hγ/c in a direction which is only determined by "chance", according to the present state of the theory.
>
> These properties of the elementary processes...make the formulation of a proper quantum theory of radiation appear almost unavoidable. The weakness of the theory lies on the one hand in the fact that <u>it does not get us any closer to making the connection with wave theory</u>; on the other, that it leaves the duration and direction of the elementary processes to "chance". Nevertheless I am fully confident that the approach chosen here is a reliable one.

The recoil of an atom in emission or absorption brings atomic motion and kinetic theory into the picture. Einstein had included recoil and translational degrees of freedom in a classical model he explored with L. Hopf. (14) The "atoms" in the Einstein-Hopf model interact with radiation as electric dipole oscillators. The interaction with the electromagnetic field adds to the kinetic energy of the particles as a result of momentum transfer (recoil) from the field. When an atom is moving, however, it suffers a net drag force resulting from momentum transfers (14-16). In thermal equilibrium these two effects cancel, giving an expression for the spectral energy density of the thermal radiation field. Einstein and Hopf obtained in this way the Rayleigh-Jeans spectrum.

The form of the drag force in Einstein's 1917 paper is the same as in his paper with Hopf. However, the term associated with the gain of kinetic energy due to atomic recoil is different. The difference is essentially the "particle" term in the Einstein fluctuation formula (2.15), although Einstein did not describe it this way. <u>The presence of the "particle" term implies that an atom recoils when it undergoes spontaneous emission</u>. This effect does not arise in the classical Einstein-Hopf theory, because a classical oscillating point dipole does not recoil as it radiates (17). This absence of recoil in the classical theory is easily inferred from the inversion symmetry of the dipole field. Recoil that would be associated with radiation propagating in a direction $\vec{r}$ from the dipole is cancelled by the equal contribution from radiation in the direction $-\vec{r}$.

Einstein thus drew the rather bold conclusion that "Outgoing radiation in the form of spherical waves does not exist." (15) We will return to this point in Section 3.

E. Debye and Compton (1923)

Debye (18) and Compton (19) took Einstein's radiation quanta seriously and used the idea in a kinematical description of the scattering of radiation by electrons. They wrote equations expressing conservation of energy and momentum, assuming radiation quanta carry energy $h\gamma$ and momentum $h\gamma/c$. Compton's experiment (19) evidently showed "very convincingly that a radiation quantum carries with it directed momentum as well as energy," and was regarded as a direct experimental confirmation of Einstein's hypothesis of radiation quanta.

As a result of a paper by Bohr, Kramers, and Slater (20), which raised the possibility that energy and momentum are conserved only statistically, coincidence experiments were performed by Bothe and Geiger (21) to test whether the scattered photon and recoil electron appear simultaneously. Their results supported causality and put an end to the Bohr-Kramers-Slater program.

F. Bose and Einstein (1924, 1925)

Bose (22) gave a derivation of Planck's spectrum based entirely on particle notions. However, the particles (radiation quanta) were not treated according to classical statistics. Instead of counting particles to determine the thermodynamic probability of a macrostate, he counted cells in phase space. Einstein (23) derived an equivalent expression by counting the number of ways in which N indistinguishable particles can be distributed among Z cells of phase space. This counting procedure gives "Bose-Einstein statistics," in which the mean number of particles in state j is

$$\langle n_j \rangle = \frac{1}{e^{E_j/kT} - 1} \tag{2.19}$$

for (nonconserved) radiation quanta.

Bose wrote to Einstein that he "tried to deduce the coefficient $8\pi\gamma^2/c^3$ in Planck's law independent of classical electrodynamics, only assuming that the ultimate elementary region in the phase space has the content $h^3$." (23,24) The derivation is simple. Let $Z_p$ be the number of cells, each of size $h^3$, in the phase space of a particle in volume V with momentum in the interval [p, p + dp]:

$$Z_p = h^{-3}\int_V d^3x\, d^3p = 4\pi V h^{-3} p^2 dp \qquad (2.20)$$

For radiation quanta of momentum $h\gamma/c$, therefore,

$$Z_\nu = \frac{4\pi V}{c^3} \nu^2 d\nu \qquad (2.21)$$

Multiplication by two to account for two independent polarization states of the radiation quanta gives the number of states in the frequency interval [$\gamma$, $\gamma + d\gamma$]. And this is the same as the number of modes of the electromagnetic field in that frequency interval. Combining this with (2.19) gives the Planck formula for the spectral energy density of thermal radiation:

$$\rho(\nu) = \frac{8\pi\nu^2}{c^3} (h\nu) \langle n_\nu \rangle \qquad (2.22)$$

and we may write Einstein's formula (2.10) as the well-known result

$$\langle \Delta n_\nu^2 \rangle = \langle n_\nu \rangle + \langle n_\nu \rangle^2 \qquad (2.23)$$

for Bose-Einstein statistics.

For large $\langle n_\nu \rangle$ the second term on the right in (2.23) -- the "wave" term -- dominates. The fluctuations are not smoothed away as $n_\nu$ increases. This circumstance, which is much different from classical particle statistics, reflects the way in which states are counted in Bose-Einstein statistics. If, for instance, we have 10 balls and 5 boxes, there are 5 ways in which all the balls can be put into a single box, but only one way to distribute them uniformly over the boxes if we do not distinguish arrangements differing only by the exchange of balls.

We have

$$\langle E_\nu \rangle = Z_\nu \langle n_\nu \rangle h\nu \equiv \langle N_\nu \rangle h\nu \qquad (2.24)$$

where $N_\nu = Z_\nu \langle n_\nu \rangle$ is the average number of "particles" in the portion of phase space associated with V and $d\nu$. The Wien limit is characterized by $\langle n_\nu \rangle \ll 1$, or $N_\nu \ll Z_\nu$, i.e., the average number of particles is small compared with the number of cells of size $h^3$ in phase space. When $N_\nu \gg Z_\nu$, however, the difference between Bose-Einstein and Maxwell-Boltzmann counting is crucial, and this is the limit where the wave-like fluctuations dominate. Thus, in addition to uncovering a failure of classical statistics, the work of Bose and Einstein indicated a resemblance between indistinguishability and classical wave notions.

G. de Broglie and Einstein (1923-1925)

Einstein (23) combined (2.19) and (2.20) to obtain an expression like (2.23) for a quantum gas of material particles. He noted that, as in the case of radiation quanta, there are both wave-like and particle-like contributions. To explain the wave term he appealed to de Broglie's idea of associating waves with particles, in a sense providing both the first support and the first application of that profound thesis.

Of course de Broglie's theory was very accurately confirmed a few years later in the experiment of Davisson and Germer (1927). By then, however, de Broglie's thesis had also stimulated Schrödinger and the development of wave mechanics. The next major chapter in the wave-particle duality of light was the one Einstein (Section 2, D) predicted would be "unavoidable" from his inference of atomic recoil in emission and absorption: "The formulation of a proper quantum theory of radiation..." showing "the connection with wave theory."

## III. THE QUANTUM THEORY OF RADIATION

### A. Born, Heisenberg, and Jordan (1926)

The electromagnetic radiation field is canonically equivalent to a system of uncoupled harmonic oscillators. Born, Heisenberg, and Jordan (25) used this equivalence to treat the radiation field according to Heisenberg's matrix form of quantum mechanics. (26) They showed that the quantum theory of radiation leads naturally to the Einstein fluctuation formula containing wave-like and particle-like contributions.

It is easy to see how the Einstein fluctuation formula is explained by the quantum theory of radiation. Consider for simplicity a single mode (of frequency $\gamma$) of the field. The mean-square fluctuation of the photon number is

$$\langle \Delta n_\nu^2 \rangle = \langle a^\dagger a\, a^\dagger a \rangle - \langle a^\dagger a \rangle^2 \qquad (3.1)$$

where a is the lowering operator for the harmonic oscillator, i.e., the photon annihilation operator, and $n = a^\dagger a$ is the photon number operator. For a thermal field the density operator is

$$\rho = (1 - e^{-h\nu/kT}) \sum_{m=0}^{\infty} e^{-mh\nu/kT} |m\rangle\langle m|$$

$$= \sum_{m=0}^{\infty} p(m) |m\rangle\langle m| \qquad (3.2)$$

where

$$p(m) = \left(\frac{1}{\langle n_\nu \rangle + 1}\right)\left(\frac{\langle n_\nu \rangle}{\langle n_\nu \rangle + 1}\right)^m \qquad (3.2)$$

and $\langle n_\gamma \rangle$ is given by (2.19). For a thermal field, then, we obtain from (3.1) - (3.3) the fluctuation formula (2.23).

It is interesting to use the commutation relation $[a, a^\dagger] = 1$ to write (3.1) as

$$\begin{aligned} \langle \Delta n_\nu^2 \rangle &= \langle a^\dagger (a^\dagger a + 1) a \rangle - \langle a^\dagger a \rangle^2 \\ &= \langle a^\dagger a \rangle + (\langle a^\dagger a^\dagger a a \rangle - \langle a^\dagger a \rangle^2) \\ &= \langle n_\nu \rangle + \langle \Delta n_\nu^2 \rangle_{\text{waves}} \end{aligned} \tag{3.4}$$

where, as the notation suggests, the second term has the character of the intensity fluctuations of a classical wave. The first term -- the particle-like term -- thus arises from the noncommutation of a and $a^\dagger$ (or q and p). Since the zero-point energy of the harmonic oscillator may also be attributed to this noncommutation, i.e.,

$$\begin{aligned} H &= \tfrac{1}{2}(p^2 + 4\pi^2\nu^2 q^2) = \tfrac{1}{2} h\nu (a^\dagger a + a a^\dagger) \\ &= h\nu (a^\dagger a + \tfrac{1}{2}) = h\nu n_\nu + \tfrac{1}{2} h\nu \end{aligned} \tag{3.5}$$

it follows that the particle term is connected with zero-point energy. Born, Heisenberg and Jordan noted this: "The reasons leading to the appearance [of the particle term] are obviously closely connected with the reasons for occurrence of a zero-point energy." (27)

Born, Heisenberg and Jordan also showed that the second term in (3.4) could be associated with wave behavior. (28) The essence of their argument was later developed independently by several authors in connection with the Brown-Twiss effect (Section 4). For a heuristic model of thermal radiation consider the electric field produced by N independently radiating sources of equal amplitude:

$$E = i \sum_{j=1}^{N} (\alpha e^{-i\theta_j} - \alpha^* e^{i\theta_j}) \tag{3.6}$$

The phases $\theta_j$ are regarded as uniformly distributed random variables over the interval $[0, 2\pi]$. The measured intensity is proportional to the average of $E^2$ over these random phases:

$$\begin{aligned} \langle I \rangle &\propto \langle E^2 \rangle \\ &= \langle - \sum_{i=1}^{N} \sum_{j=1}^{N} [\alpha^2 e^{-i(\theta_i + \theta_j)} - |\alpha|^2 e^{-i(\theta_i - \theta_j)} \\ &\quad - |\alpha|^2 e^{i(\theta_i - \theta_j)} + \alpha^* \alpha^* e^{i(\theta_i + \theta_j)}] \rangle \\ &= 2N |\alpha|^2 \end{aligned} \tag{3.7}$$

Similarly $I^2 \propto E^4 = 8N^2 |\alpha|^4$, and thus we have

$$\langle I^2 \rangle - \langle I \rangle^2 = \langle \Delta I^2 \rangle = \langle I \rangle^2 \tag{3.8}$$

If we write $I \propto \langle n_\gamma \rangle$, we have a classical wave interpretation of the "wave" term in the fluctuation formula (2.23).

In the quantum theory of radiation, photon annihilation and creation operators appear in place of classical c-number amplitudes $\alpha$ and $\alpha^*$. We obtain both the wave and particle terms in the fluctuation formula, the particle term stemming from the noncommutation of annihilation and creation operators.

B. Dirac (1927)

Dirac (29) applied the quantum theory of radiation to the

emission and absorption of radiation by an atom, and derived Einstein's A and B coefficients. It was a major accomplishment to describe spontaneous emission, for it showed that the creation of a particle (photon) could be described by the quantum theory. In another remarkable paper, Landau (30) also treated "The Damping Problem in Wave Mechanics," and obtained some results for spontaneous emission that are usually attributed to Weisskopf and Wigner (31). Landau also displayed a formal correspondence of quantum-mechanical field operators for spontaneous emission to the classical field of radiation reaction. Dirac, incidentally, had written that "The present theory, since it gives a proper account of spontaneous emission, must presumably give the effect of radiation reaction on the emitting system, and enable one to calculate the natural breadths of spectral lines...." (32)

Dirac's paper established the foundations of quantum electrodynamics. Fermi (33) reviewed the theory in 1932 and considered several examples of its application. He showed, for instance, that the Doppler effect is a consequence of conservation of energy and momentum, requiring the recoil of the radiating atom; then he showed how the Dirac formalism accounts for this. He also described, *inter alia*, the Lippman interference effect in terms of the quantum theory of radiation.

The explanation of optical interference effects in the quantum theory was particularly gratifying, for it showed that the theory incorporates naturally the wave-particle duality of light. The electric field operator in the Coulumb gauge, for example, is

$$\vec{E}(\vec{x},t) = i \sum_k (h\nu_k)^{1/2} \left[ a_k(t) \vec{F}_k(\vec{x}) - a_k^\dagger(t) F_k^*(\vec{x}) \right] \tag{3.9}$$

where the (transverse) vector function $\vec{F}_k(\vec{x})$ satisfies the Helmholtz equation and is chosen subject to the *classical* electromagnetic boundary conditions. (34) The annihilation and creation operators, $a_k$ and $a_k^\dagger$, act on field mode $k$ with frequency $\gamma_k$. Spatial interference effects determined by the mode functions $\vec{F}_k(\vec{x})$ are just the same as in the classical theory. A classical intensity pattern, however, refers in the quantum theory to the probability distribution for *single-photon* detection, or else there would be a violation of (microscopic) energy conservation. In the famous words of Dirac, "Each photon ... interferes only with itself. Interference between two different photons never occurs." (35)

Dirac's statement has sometimes been taken out of context, leading to unnecessary confusion. The Brown-Twiss experiment, for example, was thought by some physicists to stand in violation of Dirac's statement; some even thought a hoax was being perpetrated! (36) Part of the confusion lies in the tendency to think of photons propagating, prior to detection, like the proverbial billiard balls. The quantum theory of radiation says no such thing. In fact, there is no spatial coordinate operator that can be associated with a photon or any other particle without rest mass. (37)

## IV. QUANTUM OPTICS

One might expect that quantum optics, in which photon-counting experiments play an important role, must deal exclusively with quantized electromagnetic fields. This is not true. People calling themselves quantum opticians regularly think (or at least calculate) in classical electromagnetic terms, treating only material particles quantum mechanically. This approximation, called semiclassical radiation theory, enjoys great success. In fact it is so successful that some people have questioned the need, or even the validity, of the quantum theory of radiation. The issue was hotly contested from the early sixties to the mid-seventies (38).

In my view the most significant outcome of the debate was a deeper appreciation of both the semiclassical and quantum theories of radiation. Phenomena like the photoelectric effect, long considered to be "obviously" inexplicable without photons, were found to demand much closer scrutiny before semiclassical explanations could be discarded with good conscience. And the subtler predictions of quantum electrodynamics for these phenomena were found to be verifiably correct. I will now discuss some examples that are particularly relevant to the wave-particle duality of light.

### A. Failures of the Classical Wave Picture

An important element of our discussion in Section 1 is the recoil of an atomic emitter or absorber. In particular, Einstein's analysis of the blackbody problem implied that atomic recoil is associated with spontaneous emission, and I mentioned that this recoil is not expected in a classical wave picture of spontaneous emission. Recoil associated with spontaneous emission was apparently observed by Frisch in 1933 (39). Frisch writes that

> ...[Bohr] took me by the waistcoat button and said: "I hope you will come and work with us sometime; we like people who can carry out 'thought experiments.'!" You see, the recoil of an atom had been discussed a good deal in the early days, but few people had thought that it could be measured. (40)

More recent observations of recoil in spontaneous emission have been reported by Picque and Vialle (41) and Scheider, et al. (42) These observations, and others involving deflection of atoms by laser beams, provide direct confirmation of the truth in Einstein's hypothesis of 1917, including the assertion that "Outgoing radiation in the form of spherical waves does not exist."

Most textbooks point to the photoelectric effect as clear evidence for photons (43), and this fits nicely with the historic role played by Einstein's 1905 paper (3). However, the experimental evidence cited has in fact been explained in semiclassical radiation theory with Fermi's Golden Rule (44).

Almost everything about the photoelectric effect -- at least everything in the textbooks -- may be explained semiclassically. I say "almost" because a delicate experiment done by Clauser "resurrects the photoelectric effect as a phenomenon requiring quantization of the electromagnetic field." (45) The essence of the experiment is to test whether a single excited atom can excite more than one other similar atom in a single spontaneous decay, a possibility allowed by semiclassical radiation theory but precluded by quantum electrodynamics.

In the quantum theory of radiation it is easy to show that the two-photon coincidence rate, for the field radiated by a single unperturbed atom, vanishes. This rate is found, in a Heisenberg-picture calculation, to be proportional to

$$\langle \sigma^{\dagger} \sigma^{\dagger} \sigma \sigma \rangle = 0 \tag{4.1}$$

where $\sigma$ is the lowering operator for the two-state system associated with the transition. In fact all n-photon coincidence rates, $n \geq 2$, vanish, although the radiation intensity, proportional to $\langle \sigma^{\dagger} \sigma \rangle$, does not vanish (for an initially excited atom), and indeed has the same spatial pattern as computed in the semiclassical theory. The "wave" in the quantum-electrodynamical case, however, is the probability amplitude for the arrival of a single photon.

If the radiation field is treated in a consistent classical fashion, however, the $\sigma$ operators in (4.1) are in effect replaced by their expectation values (46), and the n-photon coincidence rate does not vanish. This implies that energy conservation is violated in individual emission-absorption events, but not necessarily in a statistical, macroscopic sense (45).

Clauser's results show no trace of the "extra" coincidences predicted by semiclassical radiation theory, which deals only with ordinary electromagnetic waves without any wave-particle dualism.

It may seem surprising that spontaneous emission, which is the fundamental process responsible for most of the light around us, cannot be explained within the context of the classical Maxwell equations (i.e., using semiclassical radiation theory). The "neoclassical" theory of spontaneous emission (47) put forth by Jaynes and his collaborators showed that several aspects of spontaneous emission could be accounted for by a semiclassical theory. Perhaps the most damning evidence against any standard semiclassical theory of spontaneous emission was discussed by Clauser (48) in 1972. The experiment in question (49) measures the photon polarization correlations in an atomic cascade. As in the preceding example, (45) the experiment measures a two-photon coincidence rate. The semiclassical description fails, in essence, because there is no counterpart in the theory of the "all-or-nothing" nature of photon polarization predicted by the quantum theory of radiation (50). Instead, it allows for any fraction of an incident "photon's" energy to be passed, depending on the orientation of the polarizer (i.e., the classical law of Malus).

The polarization correlations may be calculated from several points of view, e.g., with the (second-order) field correlation function used in quantum optics, or the more general idea of interference of probability amplitudes (46).

A subject which has received considerable attention in the past decade is the spontaneous emission from an atom driven by a resonant coherent field (51). Experiments (52) have shown excellent agreement with theoretical predictions (53) for the spectrum of the resonance fluorescence. In addition, the resonance fluorescence was found, as predicted (54), to display photon antibunching (55).

This effect has a simple qualitative explanation (54,55): The probability of photon emission from a two-state atom is proportional to the upper-state occupation probability. When a photon is emitted the atom makes a "quantum jump" to the lower state, and a second photon cannot be emitted until the driving field has pumped the atom back into the upper state. The field radiated by the driven atom thus displays an intensity anticorrelation effect, i.e., $\langle I(t)I(t+\tau)\rangle$ is zero for delay time $\tau = 0$. For large delay times there are no intensity correlations. This is the opposite of the Brown-Twiss or "photon-bunching" effect, where there is a positive intensity correlation for short delay times.

Photon anti-bunching in resonance fluorescence may thus be simply understood according to a particle (photon) picture of the emitted radiation, but it is impossible to understand using a classical wave theory. It is interesting to note that three of the four effects we have mentioned -- the observation that a photon emitted by an atom cannot "split" and excite two detectors, the polarization correlations in a cascade decay, and now photon anti-bunching -- involve a second-order correlation function of the type $\langle a^\dagger a^\dagger aa\rangle$. In each case the two-photon coincidence rate (at the same place or time), which is proportional to this second-order correlation function, vanishes according to the quantum theory of radiation, but not according to the (semi-)classical theory.

Obviously some rather subtle effects are necessary to demonstrate the failure of the classical wave picture on experimental grounds. What about other "proof" of the particle nature of light, such as the Compton effect? Such effects usually deserve more careful consideration than is generally admitted. Thus, for instance, the Compton formula relating $\Delta\lambda$ and $\theta$ was obtained semiclassically by Schrödinger in 1927 (56). In Schrödinger's approach the wave functions for the electron before and after scattering produce a total distribution $|\Psi|^2$ (Schrödinger's interpretation of the wave function) which acts as a grating for the incident light, so the process is considered to be like a Bragg reflection. Alternatively, the incident and scattered light waves may be supposed to act as a grating for the $\Psi$ wave (56).

B. The Brown-Twiss Effect

One of the earliest and best-known experiments in quantum optics was reported by Brown and Twiss in 1956 (57). The Brown-Twiss effect, or "wave interaction noise," (58) is frequently called photon bunching. It may be regarded in part as an experimental proof for the "wave term" of the Einstein fluctuation formula for thermal radiation. Many people were stunned by the existence of such an effect (36), some suggesting that a revision of quantum mechanics would be necessary if the effect were real. However, Brown and Twiss (57, 58) explained the effect as a consequence of classical interference in the field of a noisy source (59). Their arguments were identical in spirit to the earlier explanations by Lorentz (13) and Born, et. al. (25) of the Einstein wave term. (Section 3, A)

Figure 1 shows the basic idea of the Brown-Twiss experiment. We are again interested in a second-order field correlation function, the average over t of the product of the intensities $I_1(t)$ and $I_2(t + \tau)$. (60, 61) Recorded data as a function of $\tau$ might take the form of the curve of Figure 2, showing "photon bunching" (i.e., positive correlation) for small $\tau$. The photodetectors do not respond independently near $\tau = 0$; instead "the photons have a distinct tendency to arrive in pairs." (62)

(What Brown and Twiss actually recorded was the correlation of fluctuations of the intensities at the two detectors. This was necessary to observe the photon bunching because the response time in their experiment was much larger than the coherence time of the light. Only in the limit of very short resolving time is the Bose-Einstein formula (2.23) "directly" applicable to an interpretation of the photon bunching.) (60)

The Brown-Twiss effect is sometimes regarded as a simple consequence of the fact that photons are bosons and as such tend to bunch together. But photons do not always tend to bunch together, for they have been observed in one case to anti - bunch! The fluctuation formula (2.23), for instance, is not applicable to fields from a non-thermal source (e.g., a laser, or a driven two-state atom). On the other hand, the tendency for thermal fields to exhibit photon bunching does reflect, from the particle viewpoint, the boson nature of photons.

Some physicists thought the Brown-Twiss photon bunching was incompatible with Dirac's statement that "Interference between two different photons never occurs." (35) It seemed that different photons were interfering, for they were, after all, tending to arrive in pairs. As mentioned earlier, this difficulty resulted from a misunderstanding of Dirac's statement. The objection that different photons do not interfere might just as well have been raised against the Einstein fluctuation formula, for the wave term indicates that photons in the same cell of phase space are not uncorrelated.

What are really "interfering" in any case are not photons but probability amplitudes for indistinguishable processes. The two-slit Young experiment is one example: A photon can go through one slit or the other, the probability amplitudes for the two possibilities are added, and the probability distribution at the observation plane therefore has the form of an intensity interference pattern. In a one-photon field the intensity pattern gives the probability distribution for photon detection on the screen. There is no danger in thinking of the photon as "interfering with itself." If there are two photons, however, we cannot assume that they interfere with each other. Otherwise we would have probabilities for detecting four photons at some points on the screen, and this would violate energy conservation. The error lies in thinking of photons interfering rather than probability amplitudes for different paths. (63)

If we have a two-photon field in the two-slit experiment, the amplitude for one photon to arrive at a point $\vec{x}_1$ on the screen and the other to arrive at $\vec{x}_2$ must be added to the amplitude for the process with the two photons (assumed indistinguishable) interchanged. This leads to the same shape for the "intensity" pattern, except that the probability of photon detection at any point is twice that for the single-photon case. Photons are not interfering, only probability amplitudes.

To describe the Brown-Twiss correlations in this way we must consider probability amplitudes for processes leading to a two-photon coincidence. There are interference terms associated with the indistinguishable (if the photons are in the same cell of phase space) processes in which the roles of the two photons are interchanged. This is the interference leading to the "photon bunching." (64)

Consider, for example, the Brown-Twiss interferometer (65) (Figure 3). According to these ideas, we expect an excess coincidence rate (photon bunching) if the detector separation D is small enough that

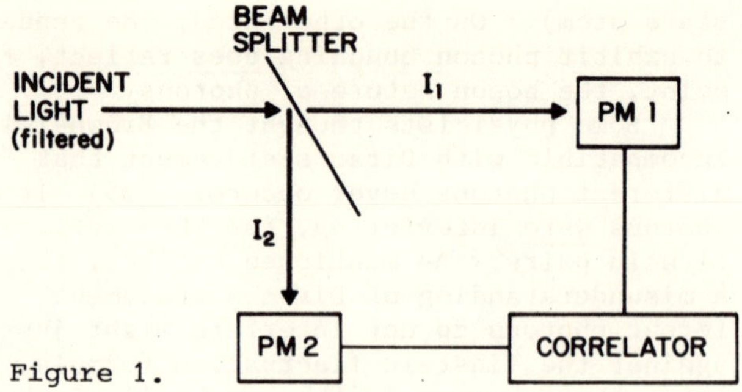


Figure 1.

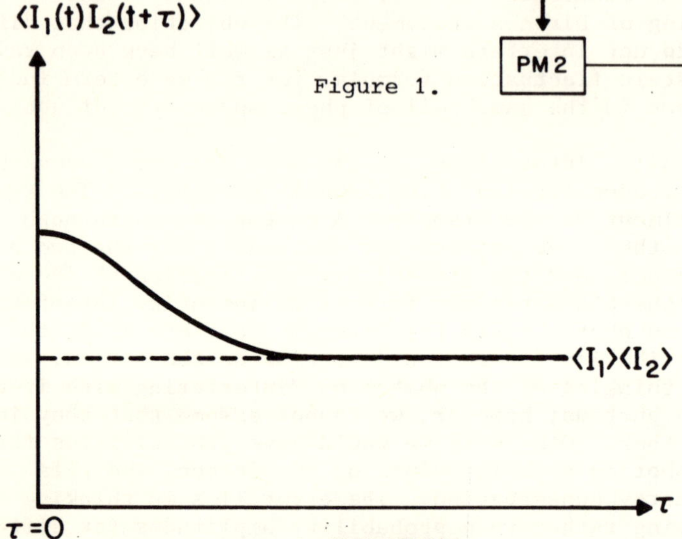


Figure 2.

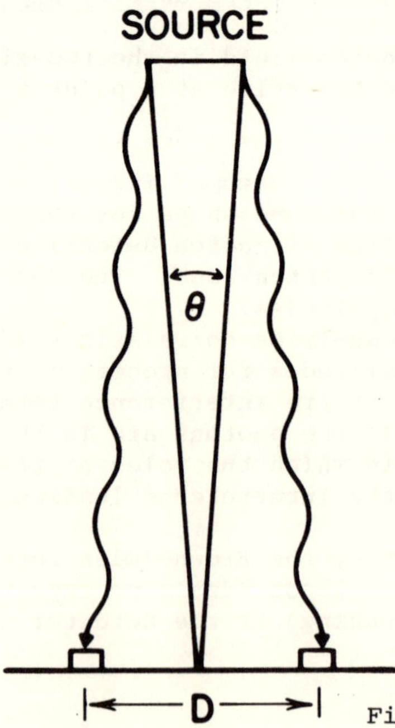


Figure 3.

$$h \gtrsim \Delta x \Delta p = D \Delta p \sim D\left(\frac{h\nu}{c}\right)\theta$$

or

$$D \lesssim \frac{\lambda}{\theta} \qquad (4.2)$$

This has been used to measure stellar angular diameters. (The Michelson stellar interferometer may be discussed from the same point of view.) The two-photon coincidences vanish for large separation D because then the alternative photon paths are in principle distinguishable. In the Brown-Twiss experiment (Figure 1) the two-photon correlations similarly decrease for delay times large compared with the coherence time of the light (inverse frequency bandwidth), because then the photons can be distinguished on average by their different frequencies.

Slightly more involved arguments lead to the concepts of coherence length, coherence area, and coherence volume (66). The latter may be shown to be the projection onto real space of a unit cell of phase space. Coherence, in the sense of "ability to display interference," can thus be connected with indistinguishability.

It should be noted that the "degeneracy parameter" (66) $\langle n_\gamma \rangle$, being the average photon number in a unit cell of phase space, is usually very small for thermal sources and optical frequencies (67). In the experiments of Brown and Twiss the degeneracy parameter was roughly $10^{-3}$. It is a tribute to their ingenuity, of course, that they could still discriminate against particle-like fluctuations and detect wave-like fluctuations.

The photon-counting experiment of Pfleegor and Mandel (68) (Figure 4a) is particularly relevant to our discussion. The demonstration by Magyar and Mandel (69), that the fields from two independent lasers can produce interference fringes, once again sparked discussions over Dirac's remark that different photons do not interfere. As in the case of the Brown-Twiss effect, there is no difficulty in understanding the results according to classical wave theory; in this case the explanation is obvious.

Quantum mechanically, the interference effect is an interference between probability amplitudes for indistinguishable alternatives (64, 68). The two paths indicated in Figure (4b) are indistinguishable if P and Q lie within a coherence area and the obser-

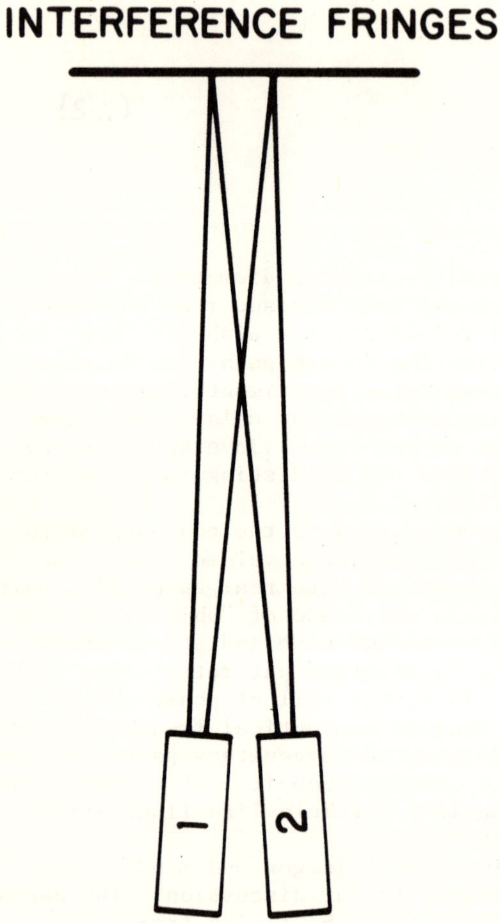


Figure 4[a]

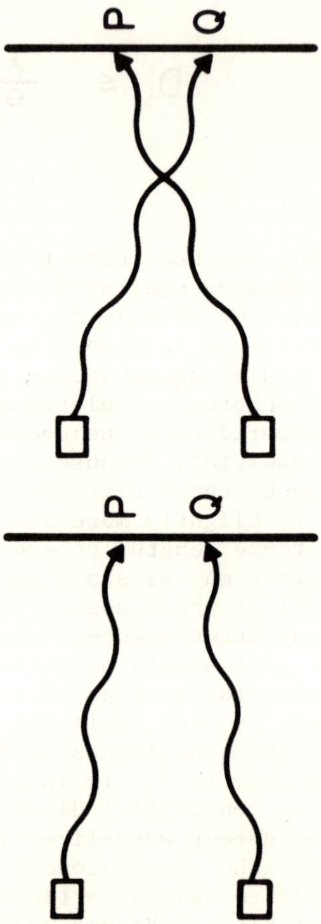


Figure 4[b]

vation time is less than the coherence time. Then we are dealing with a coherence volume, associated with a unit cell of phase space. Transient fringes are observed, the number of photons corresponding to the (average) number of photons in a cell of phase space (i.e., the degeneracy parameter).

The experiment was performed by Pfleegor and Mandel with highly attenuated beams. In their experiment the time interval between photon absorptions (∿150 nsec) was much greater than the transit time from the lasers to the detectors. In other words, "with high probability, one photon is absorbed before the next one is emitted by one or the other source." (68) The important point is that interference fringes were nevertheless observed. (They were inferred from intensity correlations between pairs of points in trials registering about 10 photons each.)

Photon-counting experiments are discussed in the authoritative review of Mandel and Wolf on optical coherence and photon statistics (66,70). We will now briefly discuss the latter subject as it relates to the wave-particle duality of light.

## C. Photon Statistics

As emphasized by Glauber, "The variety of fields encountered in the quantum theory is simply much larger than that allowed by classical theory." (62) Consider, for example, a single-mode field described by the n-photon state $|n_\gamma\rangle$, sometimes called a Fock state. For such a state we obviously have $\langle \Delta n_\gamma^2 \rangle = 0$. From equation (3.4) then, it follows that

$$\langle \Delta n_\gamma^2 \rangle_{waves} < 0 \quad . \qquad (4.3)$$

Such states obviously do not show the bunching effect, which therefore is not "simply" a consequence of the fact that photons are bosons.

Our single-mode field described by $|n_\gamma\rangle$, being monochromatic, would exhibit perfect first-order coherence. For example, we would find interference fringes of "visibility" one in a Michelson interferometer. This is just a consequence of monochromaticity. We would also observe such sharp fringe contrast wtth a single mode of a thermal field. However, the second-order coherence properties of these fields, associated with second-order field correlation functions, are much different; for instance, the Fock state will not show positive intensity correlations. $\langle n_\gamma \rangle$ might be the same for the two cases, but we could distinguish them by looking at two-photon coincidence rates. In other words, we must distinguish various orders of coherence. Monochromaticity guarantees only first-order coherence.

This leads to a consideration of states of the field that are coherent to all orders. They have the same coherence properties as the idealized "classical stable wave." As such, they do not show any "photon bunching" associated with a noisy source. In fact there are no intensity correlations, because the field is perfectly stable in the classical sense. Thus, for instance,

$$\langle \Delta n_\gamma^2 \rangle_{\text{waves}} = 0 \tag{4.4}$$

for these coherent states (62, 66, 70, 71). They may be defined formally as eigenstates of the photon annihilation operator a.

The field from an ideal laser is described by a coherent state. As such, the laser produces radiation with statistical properties dramatically different from those of ordinary light sources. No thermal source, no matter how narrowly filtered in frequency, can show the same coherence properties as laser radiation. Before the advent of lasers, "all of the delicate and ingenious techniques of optics [were] exercises in the constructive use of noise." (62)

Thermal sources typically have degeneracy parameters $\sim 10^{-3}$, whereas values $>10^{14}$ can be achieved with lasers (66, 67, 72). In a laser, in other words, we can have an extremely large photon number in a unit cell of phase space, providing spectacular proof of the boson character of photons. In ordinary thermal sources the inapplicability of the Pauli exclusion principle to photons is not so strikingly obvious.

It may be shown that the photon number distribution for an ideal laser follows a Poisson distribution, in contrast to the Bose-Einstein distribution (3.3) (62, 66). Both distributions have been very nicely verified (73).

The whole field of photon statistics owes its existence to the development of fast detectors that can respond to single photons. The probability of counting n photoelectrons in a time interval T is a good measure of the photon number distribution if T is small enough (compared with the coherence time of the light) (60,66).

## V. QUANTUM REALITY

I mentioned earlier that Bose and Einstein's statistics of radiation quanta connected indistinguishability with wave-like behavior. Einstein recognized in the statistics "a mutual influence ...which for the time being is of a quite mysterious nature." (23)

It was a vague premonition of things to come: In the formalism of quantum mechanics this "mutual influence" appears as the postulate of symmetrization of the wave function.

This sort of correlation leads to a deeper and more far-reaching "wave" behavior involving the interference of probability amplitudes for indistinguishable processes. It is an oversimplification to say that particles have a wave-particle duality about them. If the wave character of a particle is expressed by $\Psi(\vec{x}_1,t)$, then for a two-particle system it is presumably expressed by $\Psi(\vec{x}_1,\vec{x}_2,t)$, not an ordinary "wave" in three-dimensional space. We have seen, furthermore, that the tendency to think of particles themselves interfering has led to considerable confusion.

When analyzed more carefully, however, the idea of interfering probability amplitudes leads to a surprising conclusion. Recall the idealized two-slit experiment with an electron. The fact that there is an interference pattern for the electron probability distribution on the screen means we cannot say the electron has gone through either one slit or the other (74). For if this were true, we should just add the probability distributions associated with the two separate slits. But then we do not obtain the correct interference pattern, for we have ignored the interference of the probability amplitudes.

This leads to the notion that it is operationally meaningless to ask what the electron "really" does in the two-slit experiment. The only way to answer that is to actually determine by a measurement the path of the electron. But then we are no longer looking at the same physical system, for our intervention has inevitably disturbed it, and we no longer have indistinguishable processes. Because of the uncertainty principle, we cannot determine which alternative occurs and still have an interference pattern.

Intuitively, we tend to think the electron "really" does go through one slit or the other. But a calculation based on that assumption leads to incorrect predictions. It is not so much that we need not or should not think of the electron passing through either one slit or the other. We simply cannot do so and maintain a correct theory. It might even be said that "The doctrine that the world is made up of objects whose existence is independent of human consciousness turns out to be in conflict with quantum mechanics and with the facts established by experiment."(75) Such is the importance of interfering probability amplitudes.

Of course these profound conclusions only follow if quantum mechanics is a complete and correct theory. The possibility that quantum mechanics might not provide a "complete" description of a physical system has led to the consideration of what are generally called hidden-variable theories. These theories assume there is an objective reality (e.g., the electron goes through either one slit or the other). They are generally constructed to give predictions as close as possible to those of quantum mechanics (which, after all, gives remarkably accurate predictions).

Bell's theorem (76) shows that so-called local hidden-variable theories lead to statistical predictions that in some cases are different from those of quantum mechanics. And experiments (77-80) have confirmed in such cases the predictions of quantum mechanics. We must apparently conclude that any local theory assuming an objective physical reality is incompatible with experimental results.

A. Correlation and Completeness?

The Einstein-Podolsky-Rosen (EPR) paper (81) focused on two-particle correlated states to discuss the question, "Can quantum-mechanical description of physical reality be considered complete?"

There are several misconceptions about the EPR argument. One of them is that EPR denied the possibility of correlation of spatially separated particles. Another is that EPR argued from the standpoint of a deterministic philosophy, when in fact determinism *per se* plays little role in their argument (82).

Let us briefly recall Bohm's EPR-type *Gedankenexperiment* (83). We imagine a system of two spin-½ particles in a state with spin part

$$|\psi\rangle = 2^{-1/2}\left[\,|+\hat{n}\rangle_1|-\hat{n}\rangle_2 - |-\hat{n}\rangle_1|+\hat{n}\rangle_2\,\right] \qquad (5.1)$$

i.e., the singlet state. Here $|\pm\hat{n}\rangle_1$ is the state for which particle 1 has spin up or down, respectively, in the direction $\hat{n}$, and $|\pm\hat{n}\rangle_2$ refers in the same way to particle 2. The direction $\hat{n}$ is arbitrary, i.e. $|\Psi\rangle$ is spherically symmetric.

The particles described by (5.1) may have interacted in some way in the past, but we imagine a situation in which they are now widely separated and no longer interacting in any way. Specifically, let us imagine that they are moving in opposite directions, particle 1 along the $+\hat{z}$ direction and particle 2 along the $-\hat{z}$ direction. If we measure the spin component of particle 1 along the $\hat{x}$ axis and find spin "up", then according to (5.1) we must find spin "down" for particle 2. Alternatively, if we find spin "down" for particle 1, we must find spin "up" for particle 2. The same conclusion applies if we choose to measure the spin component of 1 in, for instance, the $\hat{y}$ direction.

To examine this *Gedankenexperiment* in the context of the EPR argument, let us first recall the principal presuppositions of Einstein, Podolsky, and Rosen. First they consider that a necessary condition for a theory to be complete is that *every element of the physical reality must have a counterpart in the physical theory*. (81)

Then they are careful to say what they mean by "an element of physical reality": If, without in any way disturbing a system, we can predict with certainty ( i.e., with probability equal to unity) the value of a physical quantity, then there exists an element of physical reality corresponding to this physical quantity. (81)

Consider now the Bohm Gedankenexperiment. The spin of particle 2 in the $\hat{x}$ direction can be predicted with certainty by measuring only the spin of particle 1 in the $\hat{x}$ direction, i.e., "without in any way disturbing" particle 2. There is thus an element of physical reality corresponding to the x-component of the spin of particle 2. However, we may choose instead to measure the spin of 1 in the $\hat{y}$ direction, in which case we can conclude that there must be an element of physical reality corresponding to the y-component of the spin of particle 2. Therefore the x- and y-components of the spin of particle 2 must have simultaneous reality, because we can predict either of them with certainty, without ever touching the particle. Since quantum theory does not allow the x- and y-components of the spin of particle 2 to have simultaneously predetermined values, it does not provide a complete description of physical reality.

It is important to note that the two particles may be very far apart when a measurement on particle 1 is made. If we assume that a measurement on particle 1 might in fact disturb particle 2 and so render the EPR argument inapplicable, we must evidently have to assume an acausal action at a distance. (Einstein was unwilling to seriously entertain this possibility!)

The EPR problem has a vast literature. Much of it is devoted to explaining why there is no "paradox." In my opinion such discussions often miss the whole point of the EPR argument. EPR saw no paradox as such. For them quantum mechanics might well be correct. They merely asserted that quantum mechanics could not be a complete theory because it does not account for every element of physical reality. The correlation of two systems (e.g., as in Eq. (5.1)) is not disputed. It is only used to argue that quantum mechanics does not provide a complete description of physical reality.

Many physicists feel that the EPR argument was successfully rebutted by Bohr (84). However, it seems that few people have very clear ideas about what Bohr meant. (85) Clauser and Shimony, for example, "are not convinced that Bohr ever succeeded in giving a coherent statement of his philosophical position." (80, 86) In any case Einstein was never persuaded, and the papers at this symposium prove that the issue is still very much alive.

For Einstein the only satisfactory interpretation of the wave function is in terms of ensembles, not individual systems:

> ...the concept that the Ψ-function completely describes the physical behavior of the individual single system is untenable. But one can well make the following claim: if one regards the Ψ-function as the description of

> an ensemble, it furnishes statements which--as far as we can judge--correspond satisfactorily to those of classical mechanics and, at the same time, account for the quantum structure of reality....In this interpretation the [EPR paradox] also disappears. Furthermore, it has the advantage that the description thus interpreted is an objective description whose concepts clearly make sense independently of the observation and of the observer. (87)

Einstein's arguments for the ensemble interpretation of the wave function were basically very simple. The essence of one of them (88) may be simply phrased in the language of laser spectroscopy: If we calculate that an atomic transition is saturated (i.e., zero population difference) by a strong resonant field, does this mean that any given single atom in the field has an energy midway between the lower and upper levels of the transition? If that were true, there would (according to Einstein) be a contradiction with experiments of the Franck-Hertz type. Therefore the predictions of quantum mechanics refer to ensembles of systems. Quantum mechanics does not provide a complete description of a single isolated system.

The view that quantum-mechanical probabilities refer to individual systems is sometimes called the "Born interpretation." I am not exactly sure how "probability" is to be associated with a single system. (Perhaps in terms of Heisenberg's Platonic notion of "potentiality"?) Nor am I sure that Born himself drew such a distinction:

> Einstein admits that one can regard the "probabilistic" quantum theory as final if one assumes that the Ψ-function relates to the ensemble and not to an individual case. This has always been my assumption as well, and I consider the frequent repetition of an experiment as the realization of the ensemble. (89)

If we agree with Einstein that quantum mechanics describes only ensembles and not individual systems, then the EPR argument is inapplicable. Quantum mechanics would not offer a complete description of reality. In the Bohm Gedankenexperiment, for example, the theory would not preclude particle 2 from having simultaneously "real" x and y spin components. (90)

## B. Bell's Theorem

It is often conjectured that the inability of quantum mechanics to make definite predictions for the results of a measurement might stem from our ignorance of the values of certain (unspecified) hidden variables. The values of these hidden variables for any individual system might uniquely determine the result of a measurement on the system. The quantum-mechanical wave function might merely represent a statistical distribution of hidden variables.

Bell (76) noted that the hidden-variable model of Bohm (91) could yield the same predictions as quantum mechanics in an EPR-type Gedankenexperiment, but that it demanded a nonlocal interaction between the separated systems (92). He proved that all hidden-variable theories in agreement with the statistical predictions of quantum mechanics must have this feature. It follows that local hidden - variable theories cannot reproduce in every case the statistical predictions of quantum mechanics. Various proofs and generalizations of Bell s theorem have been given by Bell and others (93)

If we accept the EPR definition of reality, then all the different components of the spin of each particle in the Bohm Gedankenexperiment have a simultaneous reality not accounted for by quantum mechanics. They are really "there" all along, so to speak, but quantum mechanics does not tell us what they are. There might, for example, be a hidden variable $\lambda_1$ whose value uniquely determines the x-component of the spin of particle 1. We assume that the spin of particle 1 in any direction $\hat{a}$ is a function $A(\hat{a}, \lambda_1)$ of $\hat{a}$ and $\lambda_1$. In order to agree with the prediction of quantum mechanics we must restrict $A(\hat{a}, \lambda_1)$ to two possible values:

$$A(\hat{a}, \lambda_1) = \pm \frac{1}{2} \tag{5.2}$$

The particular value of $\lambda_1$ determines the sign on the right-hand side of this equation. Similarly there might be a hidden variable $\lambda_2$ such that the spin of particle 2 in any direction $\hat{b}$ is a function $B(\hat{b}, \lambda_2)$ such that

$$B(\hat{b}, \lambda_2) = \pm \frac{1}{2} \tag{5.3}$$

Note an important assumption implicit in (5.2) and (5.3): $A(\hat{a}, \lambda_1)$ is independent of $\hat{b}$ and $B(\hat{b}, \lambda_2)$ is independent of $\hat{a}$. This defines a local hidden-variable theory. This assumption is reasonable in that a measurement of the spin of particle 1 in the direction of $\hat{a}$ should be independent of the direction $\hat{b}$ in which the spin of 2 is measured. (Remember that the two particles may be very far apart when a measurement is made. The assumption of a local theory rules out consideration of instantaneous interaction at a distance.)

Conditions (5.2) and (5.3) need to be supplemented by one more if we construct a hidden-variable theory in such a way as to agree as fully as possible with quantum mechanics. Namely, when $\hat{a} = \hat{b}$, the state (5.1) requires the measured spins to be opposite. Thus we assume that

$$A(\hat{a},\lambda_1) = -B(\hat{a},\lambda_2) \tag{5.4}$$

for all possible values $\lambda_1$ and $\lambda_2$ of the hidden variables, and all directions $\hat{a}$.

Let $P(\lambda_1, \lambda_2,)$ be the joint probability distribution for $\lambda_1$ and $\lambda_2$. Then

$$E(\hat{a},\hat{b}) = \int d\lambda_1 \int d\lambda_2\, P(\lambda_1,\lambda_2)\, A(\hat{a},\lambda_1)\, B(\hat{b},\lambda_2) \tag{5.5}$$

is the expectation value for the product of the spin component of 1 in the $\hat{a}$ direction and the spin component of 2 in the $\hat{b}$ direction. This is a measurable quantity. Bell showed that (5.2) - (5.5) lead to statistical predictions different from those of quantum theory. (Of course Bell's proof is more general than we are indicating. $\lambda_1$ and $\lambda_2$ may, for instance, represent sets of hidden variables.)

Bell considers $E(\hat{a}, \hat{b}) - E(\hat{a}, \hat{c})$. From (5.2) - (5.5) this may be written as

$$\begin{aligned} E(\hat{a},\hat{b}) - E(\hat{a},\hat{c}) &= \int d\lambda_1 \int d\lambda_2\, P(\lambda_1,\lambda_2) \times \\ &\quad [A(\hat{a},\lambda_1)B(\hat{b},\lambda_2) - A(\hat{a},\lambda_1)B(\hat{c},\lambda_2)] \\ &= \int d\lambda_1 \int d\lambda_2\, P(\lambda_1,\lambda_2)\, A(\hat{a},\lambda_1)\, B(\hat{b},\lambda_2) \times \\ &\quad [1 - 4B(\hat{b},\lambda_2)B(\hat{c},\lambda_2)] \\ &= \int d\lambda_1 \int d\lambda_2\, P(\lambda_1,\lambda_2)\, A(\hat{a},\lambda_1)\, B(\hat{b},\lambda_2) \times \\ &\quad [1 + 4A(\hat{b},\lambda_1)B(\hat{c},\lambda_2)] \end{aligned} \tag{5.6}$$

Since the quantity in brackets in the integrand is positive-definite, and $|A(\hat{a}, \lambda_1)\, B(\hat{b}, \lambda_2)| = \frac{1}{4}$, we obtain the Bell inequality

$$|E(\hat{a},\hat{b}) - E(\hat{a},\hat{c})| \leq \frac{1}{4} + E(\hat{b},\hat{c}) \tag{5.7}$$

A realistic theory constructed according to the postulates (5.2) - (5.5) must predict this inequality for the Bohm Gedankenexperiment.

The inequality (5.7) is not predicted by quantum mechanics. In quantum theory

$$E(\hat{a},\hat{b}) = -\frac{1}{4}\,\hat{a}\cdot\hat{b} \quad (\textit{quantum mechanics}) \tag{5.8}$$

and it is easily seen that the Bell inequality is violated.

In connection with our remarks about EPR and the interpretation of the wave function, it should be emphasized that the difference between the statistical prediction (5.7) and the corresponding statistical prediction of quantum mechanics is independent of whether $\psi$ provides a complete description of a single individual system. However we interpret the wave function, we are led to predictions different from those of a (local) realistic theory.

Bell's conclusion is similar to those of van Neumann (94) and others regarding hidden-variable theories. Bell's work, however, was the first to involve experimentally decidable propositions. (95)

One might reasonably guess that it is the deterministic aspect of the hidden variables that leads to the conflict between (5.7) and quantum mechanics. However, Bell (96) and Clauser and Horne (97) have shown that a stochastic (but still local) hidden-variable theory leads to similar inequalities. The general Bell theorem shows that essentially any theory formulated within the framework of reality, unless it admits nonlocal aspects in conflict with special relativity, must disagree with the predictions of quantum mechanics.

C. Experiments Support Quantum Mechanics

If certain entirely reasonable auxiliary assumptions are made, (97, 98) a test of the whole class of local hidden-variable theories may be made in two-photon atomic cascade experiments of the type mentioned earlier. (48, 49) In these experiments the correlated photon polarizations play the role of the correlated spins in

Bohm's Gedankenexperiment.

The experimental situation has been reviewed by Clauser and Shimony. (80) It seems almost certain that the experimental evidence strongly supports quantum mechanics and rules out explanations based on any local, realistic theory. "The conclusions are philosophically startling: Either one must totally abandon the realistic philosophy of most working scientists, or dramatically revise our concept of space-time." (80)

One possible loophole is that the polarizer orientations in the experiments might actually determine in part the polarizations of the emitted photons. (99) A quantum-mechanical analysis of spontaneous emission by an atom inside a cavity, for example, shows that only emission into the allowed cavity modes is possible. (100) However, in order for the cavity to so affect the emission, the distances between the atom and the cavity walls must be small compared with the speed of light times the atomic lifetime (i.e., the coherence length of the spontaneous emission). In the very accurate experiment reported recently by Aspect, et. al., (101) the polarizers were as large as 6.5 m away from the atomic emitter, and this is large compared with the coherence length (∿1.5m) associated with the lower transition of the cascade. Therefore this objection to the interpretation of the experiments appears to be satisfactorily answered.

From a purists' point of view the experiments are only confirming again the very nonclassical aspects of quantum-mechanical superpositions and probability amplitudes. We are back to the peculiarities of the two-slit experiment, "which has in it the heart of quantum mechanics; in reality it contains the only mystery." (102) Bell's theorem and the experiments it inspired show that quantum mechanics is mysterious indeed, but correct as far as anyone knows. The point is no longer that quantum mechanics does not account for certain elements of reality (EPR), but rather that those elements of reality do not exist.

## ACKNOWLEDGEMENT

I have many people to thank for helpful conversations or remarks, especially Richard J. Cook, Joseph H. Eberly, Edwin T. Jaynes, Michael Lieber, David Nesbitt, Gregory J. Salamo, and David Stoler. I also wish to thank Mei-Li Shih for invaluable assistance with the references, and T.W. Marshall for a copy of an essay, "Einstein's Reality - A Defense."

This work was supported in part by dual funding from the State of Arkansas and NSF EPSCOR grant ISP 8011447.

## REFERENCES

1. H. Bondi, "Logical Foundations of Physics," Perspectives in Quantum Theory, ed. by W. Yourgrau and A. van der Merwe (Dover, N.Y., 1979), p. 237.

2. R. P. Feynman, "The Character of Physical Law," (M.I.T. Press, Cambridge, Mass., 1978), p. 129.

3. A. Einstein, Ann. Physik 17, p. 132 (1905). I have used the English translation of A. B. Arons and M. B. Peppard, Amer. J. Phys. 33, p. 367 (1965).

4. M. J. Klein, "Einstein and the Development of Quantum Physics," in "Einstein. A Centenary Volume," ed. by A. P. French (Harvard University Press, Cambridge, Mass., 1979).

5. P. Lenard (Ann. Physik 8, p. 149 (1902)) had observed that the photoelectron energy was independent of the intensity of the incident light.

6. R. A. Millikan, Phys. Rev. 7, p. 18 (1916).

7. R. A. Millikan, Phys. Rev. 7, p. 355 (1916).

8. The term "photon" was coined by the chemist G. N. Lewis in 1926 (Nature 118, p. 874). His concept of these photons, however, did not attract much serious attention.

9. G. I. Taylor, Proc. Camb. Phil. Soc. 15, p. 114 (1909).

10. S. Parker, Am. J. Phys. 40, p. 1003 (1972).

11. A. Einstein, Phys. Z. 10, p. 185 (1909).

12. A. Einstein, Ann. Phys. (Leipzig) 14, p. 354 (1904).

13. H. A. Lorentz, "Les Théories Statistiques en Thermodynamique" (Teubner, Leipzig, 1916), p. 59.

14. See, for example, A. Einstein and L. Hopf, Ann. Phys. 33, p. 1105 (1910), and A. Einstein and O. Stern, Ann. Phys. 40, p. 551 (1913).

15. A. Einstein, Phys. Z. 18, p. 121 (1917).

16. T. H. Boyer, Phys. Rev. 182, p. 1374 (1969).

17. P. W. Milonni, Am. J. Phys. 49, p. 177 (1981).

18. P. Debye, Phys. Z. 24, p. 161 (1923).

19. A. H. Compton, Phys. Rev. 21, p. 483 (1923).

20. N. Bohr, H. A. Kramers, and J. C. Slater, Phil. Mag. 47, p. 785 (1924).

21. W. Bothe and H. Geiger, Z. Phys. 26, p. 44 (1924). Also A. H. Compton and A. W. Simon, Phys. Rev. 26, p. 889 (1925).

22. S. N. Bose, Z. Phys. 26, p. 178 (1924).

23. See M. Jammer, "The Conceptual Development of Quantum Mechanics" (McGraw-Hill, N.Y., 1966), p. 248.

24. "...a cell of volume $h^3$ in phase space is equivalent to a discrete state." P. A. M. Dirac, "The Principles of Quantum Mechanics" (Oxford, 1967), fourth edition, p. 239.

25. M. Born, W. Heisenberg, and P. Jordan, Z. Phys. 35, p. 557 (1926).

26. W. Heisenberg, Z. Phys. 33, p. 879 (1925); M. Born and P. Jordan, Z. Phys. 34, p. 858 (1925).

27. This point was also made in Reference (17). I was not aware then that the Born-Heisenberg-Jordan paper contained a discussion of the fluctuation formula.

28. The argument applied to vibrations of a string. Lorentz (Reference 13) had already done the calculation for electric waves.

29. P. A. M. Dirac, Proc. Roy. Soc. (London) A114, p. 243 (1927).

30. L. Landau, Z. Phys. 45, p. 430 (1927).

31. V. F. Weisskopf and E. Wigner, Z. Phys. 63, p. 54 (1930).

32. A description of spontaneous emission in terms of radiation reaction was given by J. R. Ackerhalt, P. L. Knight, and J. H. Eberly, Phys. Rev. Lett. 30, p. 456 (1973). A complementary description in terms of the zero-point field is also possible: P. W. Milonni, J. R. Ackerhalt, and W. A. Smith, Phys. Rev. Lett. 31, p. 958 (1973). See also P. W. Milonni, Phys. Rep. 25, p. 1 (1976).

33. E. Fermi, Rev. Mod. Phys. 4, p. 87 (1932).

34. For further discussion of this point see P. W. Milonni, Phys.

Rev. A25, p. 1315 (1982).

35. P. A. M. Dirac, Reference (24), p. 9.

36. E. T. Jaynes, private communication.

37. For a recent discussion with references to earlier work see R. J. Cook, Phys. Rev. A25, p. 2164 (1982).

38. The subject has been reviewed by several authors: P. W. Milonni, Phys. Rep. 25, p. 1 (1976); L. Mandel, in "Progress in Optics," ed. by E. Wolf (North-Holland, Amsterdam, 1976), vol. XIII, p. 27; I. R. Senitzky, ibid., vol. XVI, p. 415.

39. O. R. Frisch, Z. Phys. 86, p. 42 (1933).

40. O. R. Frisch, in "Niels Bohr," ed. by S. Rozenthal (North-Holland, Amsterdam, 1967).

41. J. L. Picque and J. L. Vialle, Opt. Commun. 5, p. 402 (1972).

42. R. Scheider, H. Walther, and L. Wöste, Opt. Commun. 5, p. 337 (1972).

43. An exception is A. E. Siegman, "An Introduction to Lasers and Masers," (McGraw-Hill, N.Y., 1971), p. 26.

44. L. Mandel, E. C. G. Sudarshan, and E. Wolf, Proc. Phys. Soc. Lond. 84, p. 435 (1964); P. A. Franken, in "Atomic Physics," ed. by B. Bederson, et. al. (Plenum, N.Y., 1969), p. 377. See also G. Wentzel, Z. Phys. 41, p. 828 (1927).

45. J. F. Clauser, Phys. Rev. D9, p. 853 (1974).

46. P. W. Milonni, Reference (38).

47. E. T. Jaynes, in "Coherence and Quantum Optics," ed. by L. Mandel and E. Wolf (Plenum, N.Y., 1973); ibid., 1978.

48. J. F. Clauser, Phys. Rev. A6, p. 49 (1972).

49. C. A. Kocher and Ed. D. Commins, Phys. Rev. Lett. 18, p. 575 (1967).

50. P. A. M. Dirac, Ref. (24), pp. 4-7.

51. For a review of the subject see C. Cohen-Tannoudji, in "Frontiers in Laser Spectroscopy," ed. by R. Balian, et. al. (North-Holland, Amsterdam, 1977), vol. I, pp. 3-102; P. L. Knight and P. W. Milonni, Phys. Rep. 66, p. 21 (1980).

52. F. Schuda, C. R. Stroud, Jr., and M. Hercher, J. Phys. B: Atom. Molec. Phys. 7, p. L198 (1974); H. Walther, in "Laser Spectroscopy," ed. by S. Haroche, et. al. (Springer-Verlag, Berlin, 1975); F. Y. Wu, R. E. Grove, and S. Ezekiel, Phys. Rev. Lett. 35, p. 1426 (1975).

53. A. I. Burshtein, Sov. Phys. JETP 21, p. 567 (1965); M. Newstein, Phys. Rev. 167, p. 89 (1968); B. R. Mollow, Phys. Rev. 188, p. 1969 (1969).

54. H. J. Carmichael and D. F. Walls, J. Phys. B: Atom. Molec. Phys. 9, p. L43 (1976); H. J. Kimble and L. Mandel, Phys. Rev. A13, p. 2123 (1976); C. Cohen-Tannoudji in "Frontiers in Laser Spectroscopy," ed. by R. Balian, S. Haroche, and S. Liberman (North-Holland, Amsterdam, 1977), vol. I, pp. 3-102.

55. H. J. Kimble, M. Dagenais, and L. Mandel, Phys. Rev. Lett. 39, p. 691 (1977).

56. E. Schrödinger, Ann. Phys. 82, p. 257 (1927). See also W. Gordon, Z. Phys. 40, p. 117 (1927).

57. R. H. Brown and R. Q. Twiss, Nature 177, p. 27 (1956).

58. R. H. Brown and R. Q. Twiss, Proc. Roy. Soc. A242, p. 300 (1957).

59. See also E. M. Purcell, Nature 178, p. 1449 (1956).

60. Of course what is really being measured is not intensity but photoelectron counts. A frequently used formula for the probability of counting n photons in a given time interval was derived by L. Mandel, Proc. Phys. Soc. 72, p. 1037 (1958); 74, p. 233 (1959).

61. An important point omitted in these idealized discussions is the role of spatial coherence in experiments of the Brown-Twiss type. The optical path differences from points on the source to points on the detector must be kept small to observe intensity correlations. In fact Brown and Twiss were concerned with the <u>decrease</u> of the intensity correlations with detector separation as a means of determining the angular diameter of the source (e.g., a star). See Eq. (4.2).

62. R. J. Glauber, "Optical Coherence and Photon Statistics," in "Quantum Optics and Electronics," ed. by C. DeWitt, et. al. (Gordon and Breach, New York, 1964).

63. In Feynman's approach to quantum mechanics the interference of probability amplitudes captures the very essence of the

theory. The most fundamental innovation of quantum mechanics was "the discovery that in nature the laws of combining probabilities were not those of the classical probability theory of Laplace." (R. P. Feynman, in "Proceedings of the Second Berkeley Symposium on Mathematical Statistics and Probability" (University of California Press, Berkeley, 1951), p. 533). The same view has been expressed by Dirac, Fields and Quanta 3, p. 154 (1972). See also W. Heisenberg, "The Physical Principals of the Quantum Theory" (University of Chicago Press, 1930), p. 59.

64. See also R. P. Feynman, "Theory of Fundamental Processes" (Benjamin, New York, 1962), pp. 4-6.

65. R. H. Brown and R. Q. Twiss, Phil. Mag. 45, p. 663 (1954); Nature 178, p. 1046 (1956).

66. L. Mandel and E. Wolf, Rev. Mod. Phys. 37, p. 231 (1965). See also R. J. Glauber, Ref. (62).

67. L. Mandel, J. Opt. Soc. Am. 51, p. 797 (1961). See also L. Mandel, J. Opt. Soc. Am. 69, p. 1038 (1979).

68. R. L. Pfleegor and L. Mandel, Phys. Rev. 159, p. 1084 (1967).

69. G. Magyar and L. Mandel, Nature 198, p. 255 (1963).

70. A clear introduction to the Brown-Twiss effect and photon statistics is given by R. Loudon, "The Quantum Theory of Light" (Clarendon Press, Oxford, 1973). For a recent review of non-classical aspects of photon statistics see also R. Loudon, Rep. Prog. Phys. 43, p. 913 (1980).

71. R. J. Glauber, Phys. Rev. 131, p. 2766 (1963).

72. Mandel's analysis (J. Opt. Soc. Am. 69, p. 1038 (1979)) is consistent with the picture of photons in a thermal field being roughly a typical wavelength $\tilde{\lambda}$ (proportional to $T^{-1}$) apart. The coherence volume is roughly $\tilde{\lambda}^3$, and it follows that the average photon number per unit cell of phase space, averaged over all frequencies, is of order unity.

73. F. T. Arecchi, Phys. Rev. Lett. 15, p. 912 (1965). See also, for example, F. T. Arecchi, A. Berne, and P. Burlamacchi, Phys. Rev. Lett. 16, p. 32 (1966).

74. R. P. Feynman, Ref. (63). See also R. P. Feynman, R. B. Leighton, and M. Sands, "The Feynman Lectures on Physics" (Addison-Wesley, Reading, Mass., 1965), vol. 3, Chapter 1.

75. B. d'Espagnat Sci. Am. 241, p. 158 (1979).

76. J. S. Bell, Physics 1, p. 195 (1965); Rev. Mod. Phys. 38, p. 447 (1966).

77. S. J. Freedman and J. F. Clauser, Phys. Rev. Lett. 28, p. 938 (1972).

78. J. F. Clauser, Phys. Rev. Lett. 36, p. 1223 (1976).

79. E. S. Fry and R. C. Thompson, Phys. Rev. Lett. 37, p. 465 (1976).

80. J. F. Clauser and A. Shimony, Rep. Prog. Phys. 41, p. 1881 (1978).

81. A. Einstein, B. Podolsky, and N. Rosen, Phys. Rev. 47, p. 777 (1935).

82. This point was emphasized in a letter from W. Pauli to M. Born in 1954. See "The Born-Einstein Letters" (Walker, New York, 1971), p. 221.

83. D. Bohm, "Quantum Theory" (Prentice-Hall, Englewood Cliffs, New Jersey, 1951), pp. 614-22.

84. N. Bohr, Phys. Rev. 48, p. 696 (1935).

85. The earlier "Bohr-Einstein debates" should not be confused with those relating specifically to EPR.

86. Bohr's position is discussed by C. A. Hooker in "Paradigms and Paradoxes," ed. by R. G. Colodny (University of Pittsburgh, Pittsburgh, Pennsylvania, 1972).

87. A. Einstein, letter to M. Born, 12 January 1954. See "The Born-Einstein Letters," Ref. (82), p. 215.

88. A. Einstein, J. Franklin Institute 221, p. 349 (1936). See p. 375.

89. M. Born, in "The Born-Einstein Letters," Ref. (82), p. 210.

90. For a discussion of the statistical (ensemble) interpretation of $\psi$ see L. E. Ballentine, Rev. Mod. Phys. 42, p. 358 (1970). See also R. G. Newton, Am. J. Phys. 48, p. 1029 (1980).

91. D. Bohm, Phys. Rev. 85, p. 169 (1952).

92. This was also recognized by Bohm.

93. See M. Jammer, "The Philosophy of Quantum Mechanics" (Wiley, New York, 1974), Chapter 7.

94. J. von Neumann, "Mathematical Foundations of Quantum Mechanics" (Princeton University, Princeton, New Jersey, 1955), pp. 307-25.

95. It is interesting that, although the EPR paper has sometimes been interpretated in part as a suggestion for hidden variables, the paper made no reference to von Neumann's work (1932) in this direction. It is more likely that EPR had no such intention. In a letter to Born in 1952, Einstein writes: "Have you noticed that Bohm believes (as de Broglie did, by the way, 25 years ago) that he is able to interpret the quantum theory in deterministic terms? That way seems too cheap to me." ("The Born-Einstein Letters," p. 192).

96. J. S. Bell, in "Foundations of Quantum Mechanics," Proceedings of the International School of Physics "Enrico Fermi," ed. by B. d'Espagnat (Academic Press, New York, 1971).

97. J. F. Clauser and M. A. Horne, Phys. Rev. D10, p. 526 (1974).

98. J. F. Clauser, M. A. Horne, A. Shimony, and R. A. Holt, Phys. Rev. Lett. 23, p. 880 (1969).

99. T. W. Marshall, Phys. Lett. 75A, p. 265 (1980).

100. See, for example, P. W. Milonni and P. L Knight, Opt. Commun. 9, p. 119 (1973).

101. A. Aspect, P. Grangier, and G. Roger, Phys. Rev. Lett. 47, p. 460 (1981).

102. R. P. Feynman, et. al., Ref. (74), p. 1.

# WAVE-PARTICLE DUALISM IN MATTER WAVE INTERFEROMETRY

H. Rauch

Atominstitut der Österreichischen Universitäten
Wien, Austria

## ABSTRACT

Neutron interferometry is a unique tool for investigations in the field of particle-wave dualism because massive elementary particles behave like waves within the interferometer. The invention of perfect crystal neutron interferometers providing widely separated coherent beams stimulated a great variety of experiments with matter waves in the field of basic quantum mechanics. The phase of the spatial and spinor wave function become a measurable quantity and can be influenced individually. High degrees of coherence and high order interferences have been observed by this technique. The $4\pi$-symmetry of a spinor wave function and the mutual modulation of nuclear and magnetic phase shifts have been measured in the past. Recent experiments dealt with polarized neutron beams, which are handled to realize the spin-superposition of two oppositionally polarized subbeams resulting in final polarization perpendicular to both initial beam polarizations. The different action on the coherent beams of static (DC) and dynamic (HF) flippers have been visualized. Monolithic multicrystal arrangements in Laue position can also be used to achieve an extremely high energy ($10^{-9}$ eV) or angular resolution (0.001 sec of arc). This feature is based on the Pendellösung interference within the perfect crystal. A transverse coherence length up to 6.5 mm is deduced from single slit diffraction experiments.

## 1. INTRODUCTION

The invention of perfect crystal neutron interferometry in 1974 (1,2) allowed the realization of experiments with matter



*S. Diner et al. (eds.), The Wave-Particle Dualism, 69–83.*

waves having widely separated beam paths. Thermal neutrons having a kinetic energy of about 0.025 eV, a velocity of about 2200 m/s and according to the de Broglie relation (3) an associated wave length of about 1.8 Å turned out to be an ideal probe for crucial test experiments of quantum mechanics.

A monolithic perfect silicon crystal serves as optical device for a coherent splitting, reflection and superposition of the neutron wave (Fig.1). The theoretical framework for these processes is well known from dynamical diffraction theory first developed for X-ray diffraction (4,5) but easily extended to neutron waves by solving the Schrödinger equation for a strictly periodical potential describing the perfect crystal (6,7).

The particle-wave dualism displays itself very directly because the neutron behaves inside the interferometer like a wave but its particle properties have to be transferred from the entrance to the exit, too. These particle properties are known to a high precision.

* Neutron mass (m = 1.674954320(61) g (8)) and the action of gravity on its path (9).
* Low energy neutron proton scattering is well-described by classical collision laws. The energy after collision depends on the scattering angle in laboratory frame like $E' = E_0 \cos^2\Theta$.
* Magnetic moment ($\mu = -1.91304211(88)\ \mu_K$ (10), $\mu_K$ ... nuclear magneton) and the spin value ($\vec{s} = 1/2\,\vec{\sigma}\hbar$, $\vec{\sigma}$ ... Pauli spin matrices).
* Charge and magnetic moment distribution indicate a mass radius of about 1 fm (11).
* Neutron half life for β-decay ($T_{1/2}$ = 623.6 (6.2) s) (12).

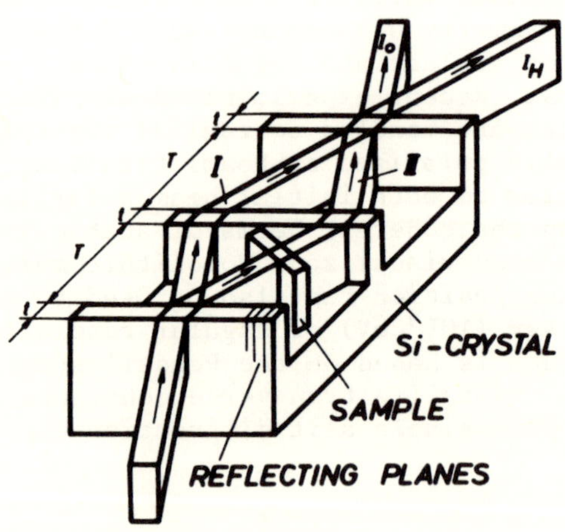


Figure 1. Principle of the perfect crystal interferometer

The wave properties are described by the Schrödinger equation in time-dependent

$$H\psi = i\hbar \frac{\partial\psi}{\partial t} \tag{1}$$

or time-independent form

$$H\psi = E\psi \tag{2}$$

The plane wave solution $\psi = \exp[i(\vec{k}\vec{r} - \omega t)]$ where $|\vec{k}| = 2\pi/\lambda$ does not define the particle in space and therefore a wave packet has to be constructed (13,14)

$$\psi = \frac{1}{(2\pi)^{3/2}} \int A(\vec{k}') \, e^{i(\vec{k}'\vec{r} - \omega t)} \, dk' \tag{3}$$

For a proper beam modelling $A(\vec{k}')$ can be assumed to have a Gaussian shape for all three dimensions centered around $\vec{k}_0$.

$$A(k') = \frac{1}{(2\pi\ \delta k^2)^{1/4}} \exp\left[-\frac{(k' - k_0)^2}{4\ \delta k^2}\right] \tag{4}$$

where $\delta k$ is related to the full half width $\Delta k$ (or $\Delta\lambda$) of the measurable momentum or wave length distribution $W(k)$ as $\delta k = \Delta k/2 \cdot \sqrt{2 \ln 2} = 0.4246\ \Delta k$. Using the Heisenberg uncertainty relation one gets $\delta x(t)$ the spreading of the wave packet in space at time t after defining the packet at $t = 0$ with $\delta x(0)$

$$[\delta x(t)]^2 = [\delta x(0)]^2 + \left[\frac{\hbar t}{2m(\delta x(0)}\right]^2 \tag{5}$$

It is worth-while to note that for thermal neutron beams in most cases the second term dominates the spatial dimension of the packet. $A(\vec{k})$ is usually determined by the resolution function $W(\vec{k}) = |A(\vec{k})|^2$ of the experimental setup. It is therefore defined for an ensemble of neutrons only. Nevertheless, it can be used to describe an individual neutron within a beam having this resolution function.

The wave and particle picture are linked together by the de Broglie relation (3)

$$\lambda = 2\pi/k = \frac{h}{mv} \tag{6}$$

where v is the group velocity.

## 2. FOUNDATION OF COHERENT CRYSTAL OPTICS

The solution of the stationary Schrödinger equation (equ.2) for a strictly periodical potential which, in the neutron case, can be written in the form of a Fermi pseudopotential

$$V(\vec{r}) = V(\vec{r}+\vec{R}_i) = \frac{2\pi\hbar^2}{m}\sum_i b_c \delta(\vec{r}-\vec{R}_i) \tag{7}$$

($\vec{R}_i$ is a lattice vector and $b_c$ the coherent scattering length of crystal nuclei) gives near a Bragg position two strong internal wave fields. The two related waves have slightly different group velocity because one wave field has its nodes at the atomic position of the lattice while the second one has its nodes between the atomic positions. Applying the boundary conditions for a flat crystal slab in Laue position one obtains the wave functions in forward and Bragg direction (6,7)

$$\psi_0 = v_0\psi_e = [\cos(A\sqrt{1+y^2}) + \frac{y}{\sqrt{1+y^2}}\sin(A\sqrt{1+y^2})]e^{iPt}\psi_e$$

$$\psi_H = v_H \exp(2\pi iyt/\Delta_0)\psi_e = [-i\frac{\sin(A\sqrt{1+y^2})}{\sqrt{1+y^2}}]e^{it(P+2\pi iy/\Delta_0)}\psi_e \tag{8}$$

For an even (e.g. (220)) Si-reflection in symmetrical Laue position the parameters are given as

$$P = -\frac{\pi y}{\Delta_0} - \frac{2\pi}{D_\lambda \cos\Theta_B}$$

$$\Delta_0 = \frac{\pi t}{A} = \frac{D_\lambda \cos\Theta_B}{2} = \frac{k\cos\Theta_B}{2b_c N} \tag{9}$$

$$y = \frac{(\Theta_B - \Theta)k^2 \sin 2\Theta_B}{4\pi b_c N}$$

where the small Debye-Waller correction ($e^{-W}$) has been omitted. In this equation $\Theta_B$ denotes the exact Bragg angle and N the particle density of the crystal.

For the interferometer crystal (Figure 1) relation (8) has to be used successively for the three crystal plates. The beams emerging from behind the interferometer are composed of two parts coming from beam path I and II, respectively (15,16)

$$\psi_0 = \psi_0^{I} + \psi_0^{II} = [v_0(y)\, v_H(y)\, v_{-H}(-y) + v_H(y)\, v_{-H}(-y)\, v_0(y)]\,.$$

$$\exp[-2\pi i y (T + t)/\Delta_0]\,.\psi_e \qquad (10)$$

and a similar relation holds for $\psi_H = \psi_H^{I} + \psi_H^{II}$. Combining equation 8 and 10 gives, for ideal geometry, the important relation

$$\psi_0^{I} = \psi_0^{II} \qquad (11)$$

and therefore

$$I_0 = |\psi_0^{I} + \psi_0^{II}|^2 = 4|\psi_0^{I}|^2 \qquad (12)$$

Particle conservation is guaranteed together with the beam in Bragg direction $I_H = |\psi_H^{I} + \psi_H^{II}|^2$. The inclusion of wave packets or the spherical wave treatment do not change the relations substantially.

For a real system the relation (11) becomes complex due to various imperfections of the crystal, its dimensions, low frequency vibrations etc.

$$\frac{\psi_0^{II}}{\psi_0^{I}} = a\, e^{i\varphi} \qquad (13)$$

and therefore

$$I_0 = |\psi_0^{I} + \psi_0^{II}| = (1 + a^2 + 2a\cos\varphi)|\psi_0^{I}|^2 \qquad (14)$$

and similar to optics a mutual coherence function can be defined

$$\gamma = \frac{2|\psi_0^{I}\, \psi_0^{II}|}{|\psi_0^{I}|^2 + |\psi_0^{II}|^2} \qquad (15)$$

## 3. COHERENCE MEASUREMENTS

For such measurements the phase relation between $\psi_0^{I}$ and $\psi_0^{II}$ has to be varied by the introduction of a sample having an index of refraction n and a thickness D. The index of refraction is given by the volume average of the interaction potential of equ. (7) ($\overline{V} = 2\pi\hbar^2 b_c N/m$)

$$n = \frac{k}{k_0} = \sqrt{1 - \frac{\overline{V}}{E}} \simeq 1 - \frac{\overline{V}}{2E} = 1 - \lambda^2 \frac{Nb_c}{2\pi} \tag{16}$$

which is independent of the sample structure and is, for thermal neutrons, in the range of $n \sim 1 - 10^{-5}$. The phase difference is governed by the optical path length difference

$$\psi_0^{II'} = \psi_0^I \, e^{-i(1-n)kD} = \psi_0^I \, e^{-i\lambda b_c ND} = \psi_0^I \, e^{i\chi} \tag{17}$$

and by varying D or N an intensity modulation

$$\begin{aligned} I_0 = |\psi_0^I + \psi_0^{II'}|^2 &= 2|\psi_0^I|^2 (1 + \cos N b_c \lambda D) \\ &= 2|\psi_0^I|^2 [1 + \cos(2\pi D/D_\lambda)] \end{aligned} \tag{18}$$

is effected. If a sample is rotated in both coherent beams D denotes the difference of the sample thickness in both beams. The inclusion of the experimental imperfections (equ. 13) causes a reduction of the full beam modulation (equ. 18), but for a very well balanced neutron interferometer a performance rather close to the ideal can be reached as shown in Figure 2 (17).

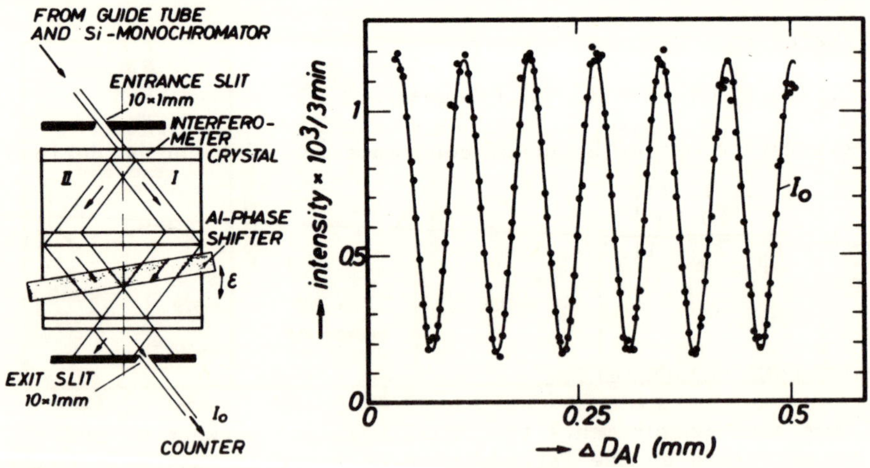


Figure 2. Environmental arrangement and maximal observed beam modulation.

The intensity modulation is about 90% and all characteristical interference phenomena can be observed, e.g. an increase of the intensity is observed at the minimum position if one coherent beam is absorbed. Similarly the interference pattern is disturbed if a detector is introduced to observe the beam path chosen by the neutron.

The index of refraction (equ. 16) and the intensity modulation (Equ. 18) depend on the neutron wave length and therefore an average about the wave length distribution has to be taken. For a Gaussian distribution function having a full half width $\Delta\lambda$ the calculated intensity modulation reads

$$\bar{I}_0 = \int I_0(\lambda) W(\lambda) d\lambda = 2|\psi_0^I|^2 \exp\left[-\left(\frac{1.886\, D\, \Delta\lambda}{D_\lambda\ \lambda}\right)^2\right] \cdot \left\{1 + \cos\frac{2\pi D}{D_\lambda}\right\} \qquad (19)$$

It is seen that the beam resolution $(\Delta\lambda/\lambda)$ reduces the modulation (coherence) especially at high orders $(n = D/D_\lambda)$ of the interference. Additionally, it should be mentioned that the same result is obtained if wave packets (equ. 3) are used for $\psi_0^I$ and $\psi_{02}^{II}$ in equ. (18) instead of $W(\lambda)$ in equ. (19) because $W(\lambda) = |A(\lambda)^2|$.

The usual beam attenuation $(\sigma_r = \sigma_a + \sigma_{incoh})$ due to absorption $(\sigma_a)$ and incoherent scattering $(\sigma_{incoh})$ processes has also to be accounted for. It is described by a complex interaction potential and therefore by a complex index of refraction (18)

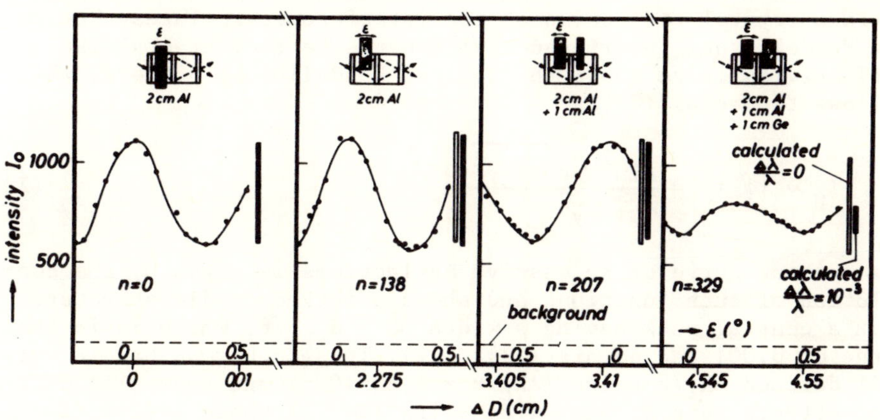


Figure 3. Observed high order interference compared to calculated values including attenuation effects and finite wave length spread of the neutron beam.

$$n \simeq 1 - \frac{\lambda^2 N}{2\pi} \sqrt{b_c^2 - \left(\frac{\sigma_r}{2\lambda}\right)^2} + i \frac{\sigma_r N\lambda}{2\pi} \tag{20}$$

Using equations (16), (17) and (19) together yields (19)

$$\overline{I}_0 = 2|\psi_0^I|^2 \exp[-\sigma_r ND/2] \exp\left[-\left(\frac{1.886\, D\, \Delta\lambda}{D_\lambda\, \lambda}\right)\right] \{\cosh(\sigma_r ND/2) + \cos(2\pi D/D_\lambda)\} \tag{21}$$

This gives a further limitation ($\sigma_r ND/2 \lesssim 1$) to the visibility of interference fringes. Related experiments have been performed using a highly monochromatic neutron beam ($\Delta\lambda/\lambda = 10^{-3}$) and up to 5 cm thick samples (Figure 3) (17), (20). The theoretical values of equ. (21) (full bar) agree rather well with the measured reduction of the intensity modulation. Interference fringes up to the 329th order could be observed and using equation (21) it can be seen from an extrapolation that the overall longitudinal coherence length of the wave packet is at least $500.\lambda = 875$ Å. This quantity is related to the monochromacy of the beam and the observable effects are equivalently describable by a distribution function or a wave packet representation of the wave function. The visibility of the interference pattern (equ. 21) depends on $\Delta\lambda$ (or $\Delta k$) and not on $\Delta x$ and therefore the spatial spreading (equ. 5) does not influence this effect.

Up till now we discussed the longitudinal coherence length. The transverse coherence length is determined by the multiple perfect crystal reflection at the interferometer plates. Due to the Pendellösung interference within the perfect crystal the reflectivity curve shows the typical Pendellösung pattern, which follows from equ. (8) as

$$I_H \propto |\psi_H|^2 = \frac{\sin^2 A\sqrt{1+y^2}}{1+y^2} \tag{22}$$

The rocking curve of successive reflections is given by the convolution of such functions and shows a marked needle structure with a central peak having a width $\Delta\alpha \sim d_{hkl}/t$, which is in the range of 0.001 sec of arc (21), (22). These neutrons have a very well defined $\Delta k_{\perp}/k$-value ($\Delta k_{\perp}/k = 4.3 \times 10^{-8}$) and show slit diffraction effects even at very wide slits ($a \gg \lambda$) whose diffraction pattern and broadening $\beta_{1/2}$ of the direct beam is given by the Frauenhofer formula (23)

$$I(\beta) = I_0 \frac{\sin^2[\pi a \sin \beta/\lambda]}{(\pi a \sin \beta/\lambda)^2} \tag{23}$$

$$\sin \beta_{1/2} = 0.888\ \lambda/a$$

To achieve the high angular resolution required to observe this narrow central peak a wedge shaped sample is rotated around the beam axis between the plates of a monolithic designed triple reflection crystal. The angular deflection $\delta$ depends on the index of refraction, the apex angle ($\eta$) of the wedge and the rotation angle ($\xi$) around the beam axis (22)

$$\delta = [2(1-n)\,\mathrm{tg}\frac{\eta}{2}]\,\sin\xi \tag{24}$$

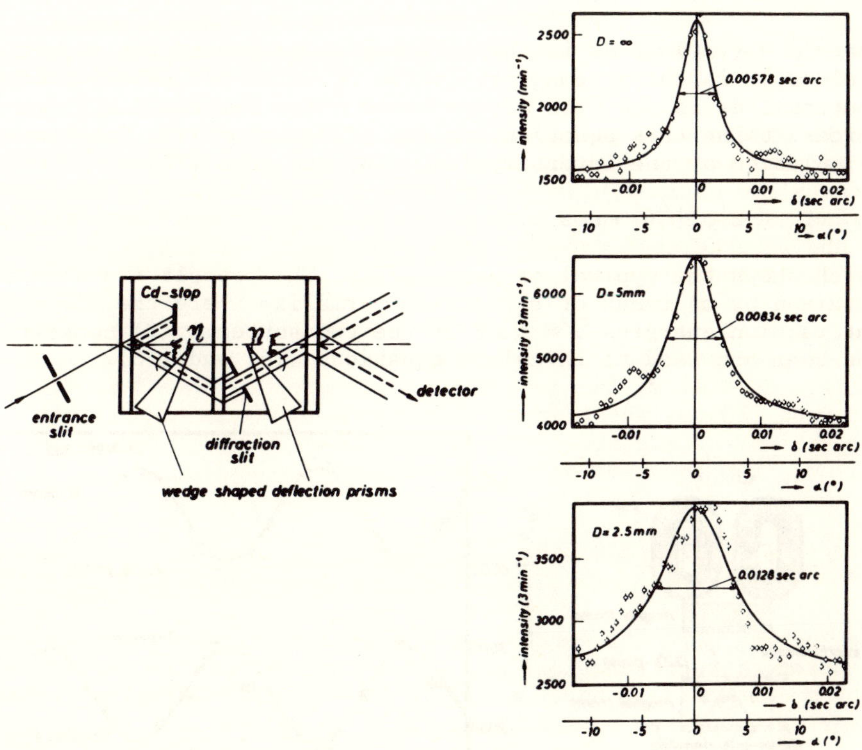


Figure 4. Experimental arrangement and experimental results of high resolution slit diffraction.

The experimental setup and the results are shown in Figure 4 (24). A beam broadening has been observed up to a slit width of 5 mm and a transversal coherence length of 6.5 mm can be deduced from these data. This value corresponds to $3.7 \times 10^7$ times the neutron wave length and is much larger than values reported before in the literature (23), (25).

## 4. SPINOR SYMMETRY MEASUREMENT

An important aim of neutron interferometry was the verification of the $4\pi$-symmetry of a spinor wave function, which was treated before as a kind of a "hidden" variable. The unitary rotation operator for a spinor wave function reads

$$\psi(\vec{\alpha}) = e^{-i\vec{\sigma}\vec{\alpha}/2}\,\psi(0) \tag{25}$$

where $\vec{\alpha} = (\gamma/v)\int \vec{B}\,ds$ is the rotation angle around the magnetic field $\vec{B}$, $\gamma$ is the gyromagnetic ratio of, v is the velocity of the neutron. $\int ds$ is the path integral and $\vec{\sigma}$ are the Pauli spin matrices. Using this equation instead of equation (17) for the calculation of the intensity (equ. 18) one gets (26)

$$I_0 = 2|\psi_0^I|^2(1+\cos\frac{\alpha}{2}) \tag{26}$$

which shows the typical $4\pi$ periodicity. This result can also be obtained by an index of refraction formalism where the different interaction energies $\overline{V} = \pm\mu B$ of the two subbeams ($\pm$) constituting the beam are used to calculate equations (16) and (18).

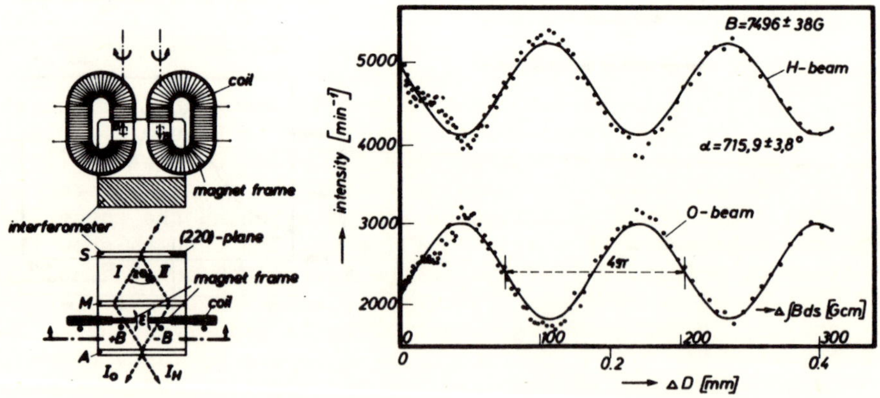


Figure 5. Experimental arrangement and results of the $4\pi$-spinor symmetry measurement.

Equation (26) holds for polarized and unpolarized beams as well, indicating the self-interference properties of the neutrons within the interferometer. The first verification of the $4\pi$-symmetry was achieved in 1975 using simple magnetic fields along one beam path (27), (28). The experimental setup and the results of a precision measurement using well defined magnetic fields within Mu-metal sheets is shown in Figure 5 (29). The periodicity factor extracted from these data using up-to-date values for the physical constants is (30)

$$\alpha_0 = 715.87 \pm 3.8 \text{ deg} \tag{27}$$

where the error bars indicate the simple $\sigma$-limits. In the meantime this $4\pi$-symmetry has been verified by other techniques too (31), (32), (33). Small correction terms to equ. (26) occur due to a slightly different transit time through the magnetic field of the two subbeams ($\pm$) (34). If nuclear (equ. 17) and magnetic phase shifts occur simultaneously additional effects concerning intensity and polarization modulation are predicted (26), (35) and experimentally verified (36).

An interesting feature exists for a polarized incident beam whose polarization within the coherent beam path I is rotated by an angle $\pi$. After coherent superposition a polarization perpendicular to both initial polarization states is predicted (37), (38), (26). In this case the unitary operator for nuclear and magnetic interaction (equations 17 and 25) reduces to

$$U = -i\sigma_x \exp(i\chi) \tag{28}$$

and the final polarization is calcualated as

$$\vec{P} = \frac{\psi_0^{I+}(1 + U^+)\vec{\sigma}(1 + U)\psi_0^I}{I_0} = \vec{e}_x \sin\chi - \vec{e}_y \cos\chi \tag{29}$$

This effect could be verified recently (39) and characteristical results and the experimental setup are shown in Figure 6. Polarized neutrons are first obtained from magnetic prism deflection (40) and then the polarization inversion within the interferometer and the three dimensional polarization analysis behind the interferometer are achieved with static DC-flippers. The results show that the final polarization lies within the (xy)-plane and the polarization vector can be rotated within this plane even by a scalar interaction $\chi$ only. Momentum conservation is preserved due to the beam polarization of the H-beam. If HF-flippers having a time-dependent interaction are applied for polarization reversal an additional phase factor related to the phase of the HF-field appears (41), (42) and a time-resolved measurement is required.

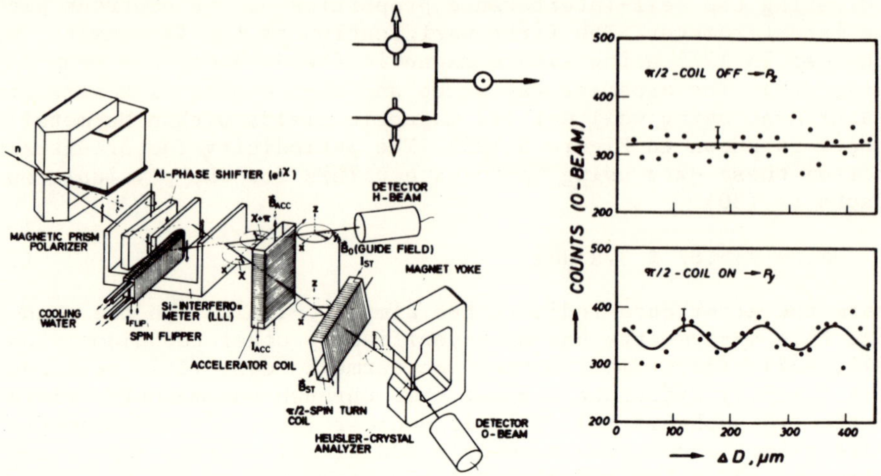


Figure 6. Experimental arrangement, spin handling and results of the spin superposition experiment.

5. SUMMARY

Neutron interferometry allows the realization of many textbook experiments of quantum mechanics whose interpretation has also been discussed in several review articles (43) - (47). Within the standard (Copenhagen) interpretation of quantum mechanics only probability amplitudes are relevant for the description of the coherent beams even the non-separable neutron mass and its magnetic moment is transported through the interferometer. It is easily shown that only self interference is important because even at a high flux reactor the mean time interval between two neutron counts is about 0.01 s while the mean time-of-flight through the interferometer is in the order of 30 μs. The wave packet (equations 3 and 4) describes a neutron but a neutron belonging to a certain ensemble and the parameters of the packet are determinable for the beam only. The introduction of so-called "empty waves" (48) does not conflict with this interpretation as long as no neutron detection along the coherent beam path is done, where the interpretation of the field collapse is different.

Figure 7 shows two Gedanken experiments where the first one is related to a Hanbury-Brown-Twiss experiment (49) and the second one to the Pisa experiment (48). In the Hanbury-Brown-Twiss case two detectors are operated in coincidence within the coherent beams and an antibunching effect is expected for zero delay time. The experimental difficulties arise for the condition that the detectors have to be sensitive to the coherence volume of the wave packet only, which is quite large in the longitudinal and

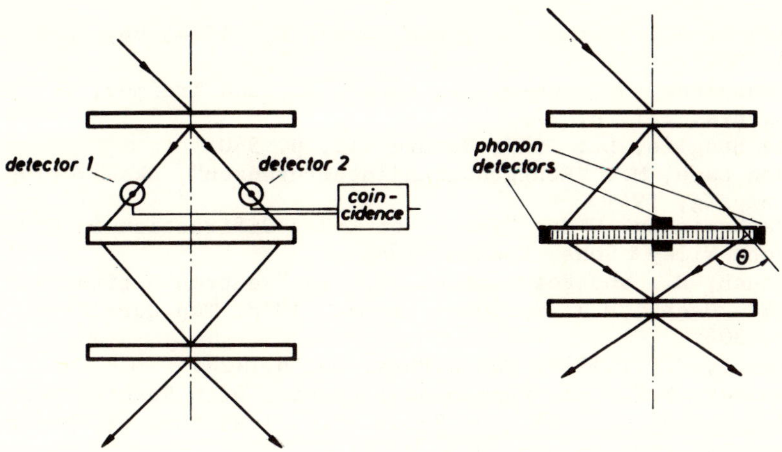


Figure 7. Gedanken experiments for neutrons.
Left: Hanbury-Brown-Twiss case; right: Pisa-experiment.

transversal direction (see chapter 3) but very small in the vertical direction (∿ 50 Å) where no beam collimation is applied. In the case of additional collimation the intensity decreases drastically. The second Gedanken experiment deals with the coherent creation or annihilation of phonons by coherent neutron interaction with the perfect crystal (50) and with a separate phonon detection (51). These effects are associated not only with a momentum change $\Delta Q$ as in the Bragg scattering case but also by an energy transfer $\Delta E$ which is described by ordinary coherent neutron-phonon spectroscopy (50). Therefore, phonon interferometry or phonon detection is possible besides the residual neutron interference and neutron detection. It is expected that neutron interference disappears as soon as the phonon detection allows the determination of the beam path chosen by the neutron.

Most of the experimental results have been obtained at the neutron interferometer setup at the High Flux Reactor at Grenoble which is operated by a Dortmund-Grenoble-Wien interferometer team.

REFERENCES

(1) Rauch, H., Treimer, W., and Bonse, U.: 1974, Phys.Lett. A47, p. 369
(2) Bauspiess, W., Bonse, U., Rauch, H., and Treimer, W.: 1974, Z.Phys. 271, p. 177
(3) de Broglie, L.: 1923, Nature 112, p. 540
(4) von Laue, M.: "Röntgenstrahlinterferenzen", Akad.Verlagsges. Leipzig, 1941
(5) Zachariasen, W.H.: "Theory of X-Ray Diffraction in Crystals", John Wiley & Sons, London, 1945
(6) Rauch, H., and Petrascheck, D.: in "Neutron Diffraction" (Ed. Dachs, H.), Springer Verlag, 1978, Top.Curr.Phys. 6, p. 303
(7) Sears, V.F.: 1978, Can.J.Phys. 56, p. 1261
(8) Ramsey, N.F.: in "Fundamental Physics with Reactor Neutrons and Neutrinos" (Ed. von Egidy, T.), Institute of Physics, Bristol and London, 1975, p. 61
(9) Koester, L.: 1976, Phys.Rev. D14, p. 907
(10) Greene, G.L., Ramsey, N.F., Mampe, W., Pendlebury, J.M., Smith, K., and Perrin, P.: 1977, Phys.Lett. 71B, p. 297
(11) Bilen'kaya, S.I., and Kazarinov, Yu.M.: 1980, Sov.J.Nucl.Phys. 32, p. 382
(12) Wilkinson, D.H.: 1982, Nucl.Phys. A377, p. 474
(13) Messiah, A.: "Quantum Mechanics", North Holland Publishing Comp., Amsterdam, 1965
(14) Hittmair, O.: "Lehrbuch der Quantenmechanik", Verlag K.Thiemig, München, 1972
(15) Petrascheck, D.: 1976, Acta Phys.Austr. 45, 217
(16) Bauspiess, W., Bonse, U., and Graeff, W.: 1976, J.Appl.Cryst. 9, p. 68
(17) Rauch, H.: in "Neutron Interferometry" (Ed. Bonse, U., and Rauch, H.), Clarendon Press, Oxford, 1979, p. 161
(18) Halpern, O.: 1952, Phys.Rev. 88, p. 1003
(19) Petrascheck, D., and Rauch, H.: "Theorie des Interferometers", AIAU-Report 76401, Atominstitut Wien, 1976
(20) Rauch, H., Badurek, G., Bauspiess, W., Bonse, U., and Zeilinger, A.: Proc.Int.Conf. Interaction Neutrons with Nuclei, Lowell/MA, July 1976, Vol. II, CONF-760716-P2, p. 1027
(21) Bonse, U., Graeff, W., Teworte, R., and Rauch, H.: 1977, phys.stat.sol.(a) 43, p. 487
(22) Bonse, U., Graeff, W., and Rauch, H.: 1979, Phys.Lett. 69A, p. 420
(23) Shull, C.G., 1969, Phys.Rev. 179, p. 752
(24) Rauch, H., Kischko, U., and Bonse, U.: to be published in Z.Phys. A
(25) Scheckenhofer, H., and Steyerl, A.: 1977, Phys.Rev.Lett. 39, p. 1310
(26) Eder, G., and Zeilinger, A.: 1976, Nuovo Cim. 34B, p. 76

(27) Rauch, H., Zeilinger, A., Badurek, G., Wilfing, A., Bauspiess, W., and Bonse, U.: 1975, Phys.Lett. 54A, p. 425
(28) Werner, S.A., Colella, R., Overhauser, A.W., and Eagen, C.F.: 1975, Phys.Rev.Lett. 35, p. 1053
(29) Rauch, H., Wilfing, A., Bauspiess, W., and Bonse, U.: 1978, Z.Physik B29, p. 281
(30) Rauch, H.: 1982, Hadronic J. 5, p. 729
(31) Klein, A.G., and Opat, G.I.: 1976, Phys.Rev.Lett. 37, p. 238
(32) Klempt, E.: 1976, Phys.Rev. D13, p. 3125
(33) Stoll, M.E., Wolff, E.K., and Mehring, M.: 1978, Phys.Rev. A17, p. 1561
(34) Bernstein, H.J.: in "Neutron Interferometry" (Ed. Bonse, U., and Rauch, H.), Clarendon Press, Oxford, 1979, p. 231
(35) Zeilinger, A.: 1976, Z.Physik B25, p. 97
(36) Badurek, G., Rauch, H., Zeilinger, A., Bauspiess, W., and Bonse, U.: 1976, Phys.Rev. D14, p. 1177
(37) Wigner, E.P.: 1963, Am.J.Phys. 31, p. 6
(38) Zeilinger, A.: in "Neutron Interferometry" (Ed. Bonse, U., and Rauch, H.) Clarendon Press, Oxford, 1979, p. 241
(39) Summhammer, J., Badurek, G., Rauch, H., and Kischko, U.: 1982, Phys.Lett. 90A, No. 3, p. 110
(40) Badurek, G., Rauch, H., Wilfing, A., Bonse, U., and Graeff, W.: 1979, J.Appl.Cryst. 12, p. 186
(41) Balcar, E.: in "Neutron Interferometry" (Ed. Bonse, U., and Rauch, H.), Clarendon Press, Oxford, 1979, p. 252
(42) Krüger, E.: 1980, Nukleonika 25, p. 889
(43) Werner, S.A.: Dec. 1980, Physics Today
(44) Greenberger, D.E., and Overhauser, A.W.: 1979, Rev.Mod.Phys. 51, p. 43
(45) Colella, R., and Overhauser, A.W.: 1980, Amer.Scientist 68, p. 70
(46) Bernstein, H.J., and Phillips, A.V.: Sep. 1981, Scientific American
(47) Silverman, M.P.: 1980, Eur.J.Phys. 1, p. 116
(48) Selleri, F.: "Can an Actual Existence be Granted to Quantum Waves", 1982, Ann.Foundation L. de Broglie 7, p. 45
(49) Hanbury-Brown, R., and Twiss, R.Q.: 1957, Proc.R.Soc. A242, p. 300 and 1957, Proc.R.Soc. A243, p. 291
(50) Marshall, W. and Lovesey, S.W.: "Theory of Thermal Neutron Scattering", 1971, Clarendon Press, Oxford
(51) Nieden, Th.z. and Weis, O.: 1975, Z.Physik B21, p. 11

# NEUTRON WAVE OPTICS STUDIED WITH ULTRACOLD NEUTRONS

A. Steyerl

Technische Universität München, D-8046 Garching, Germany

The paper discusses experiments demonstrating or utilizing the wave properties of neutrons with wavelengths of about 100 nm.

## 1. INTRODUCTION

The preparation of the present paper was difficult for me mainly because I tried at first to relate our experimental data on ultracold-neutron optics, which are simple, to that difficult subject: "non-linear wave mechanics", which has been created by de Broglie over 50 years ago (see /1/) and which seems to be resumed nowadays at a time of rapidly growing interest in non-linear phenomena (e.g. /2/). I must confess that for a simple-minded experimental physicist like me it turned out that this programme was too difficult. Therefore, I finally decided not to annoy you with, perhaps, unsound interpretations but present the plain data. I should add that this restraint was facilitated by the fact that, within experimental precision, we did not observe any deviation in the behaviour of ultracold-neutron waves from the predictions made within the framework of standard, linear, quantum mechanics, and therefore an urgent need of possible nonlinearities has not arisen.

## 2. WHAT ARE "ULTRACOLD NEUTRONS"?

The designation "ultracold" is commonly used for extremely slow neutrons with velocities below $\sim 10$ m/s. Their energies are of the order of $10^{-7}$ eV ($10^{-3}$ K) or below and their de Broglie wavelengths reach nearly macroscopic demensions, $\sim 10^{3}$ Å. Ultracold neutrons (UCNs) cannot penetrate matter as easily as faster



*S. Diner et al. (eds.), The Wave-Particle Dualism, 85–99.*

neutrons, and they suffer total reflection at the walls of suitable substances, at any angle of incidence. This unique property makes them suited for storage in closed cavities, the so-called "neutron bottles". The neutrons are contained in carefully prepared vessels for hundreds of seconds, travelling to and fro between the walls from which they rebound many thousands of times before loss processes due to reactions with the wall nuclei and the finite lifetime for beta-decay become significant /3 - 6/. Storage lifetimes reaching the limit of the $\beta$-decay lifetime have been obtained in a "magnetic storage ring" /7/, and an improvement on the lifetime value seems to be possible.

The main applications of UCNs envisaged until now are high-precision investigations of the properties of the neutron itself, like its lifetime for $\beta$-decay and searches for a possible electric charge and dipole moment. The low energy of UCNs seems to make them suitable for high-resolution spectroscopy, and a third branch of attractive physics with UCNs seems to be neutron optics which is the subject of the present paper.

## 3. PECULIARITIES OF UCN OPTICS

Dealing with neutrons of extremely low energy we must take into account a number of interactions which are usually negligible in thermal or fast neutron research because of their weakness, but play an important role at very low energies. These interactions include the gravitational potential ($mg \cong 10^{-7}$ eV per m of height), the interaction of the neutron magnetic dipole moment $\underline{\mu}$ with a magnetic field $\underline{B}$ ($\mu \cong 6 \times 10^{-8}$ eV/tesla), and the interaction with the "scattering potential" U which arises as an effect of multiple coherent scattering in the forward direction and gives rise to refraction and total reflection, as in light optics. For most substances, U is of the order of $10^{-7}$ eV. These various interactions may be incorporated in a spatially variable index of refraction, $n(\underline{r}) = [1 - (U(\underline{r}) \mp \mu B(\underline{r}) + mgz)/E_0]^{1/2}$. The sign of the magnetic term $\mp \mu B$ is negative for neutron spin antiparallel to B and positive for $\sigma \parallel B$. (The magnetic interaction can be included in this simple form whenever the "adiabatic condition" is satisfied, i.e., changes of magnetic field direction "sensed" by the neutron along its trajectory occur slowly compared with the frequency of Larmor precession in the local magnetic field.)

As a consequence of the gravitational interaction even empty space in the absence of magnetic fields may be considered as a refracting medium with a spatially variable index of refraction. Therefore, the optics of ultracold neutrons deals with curvilinear rays, and this introduces new features into the usual framework of optics.

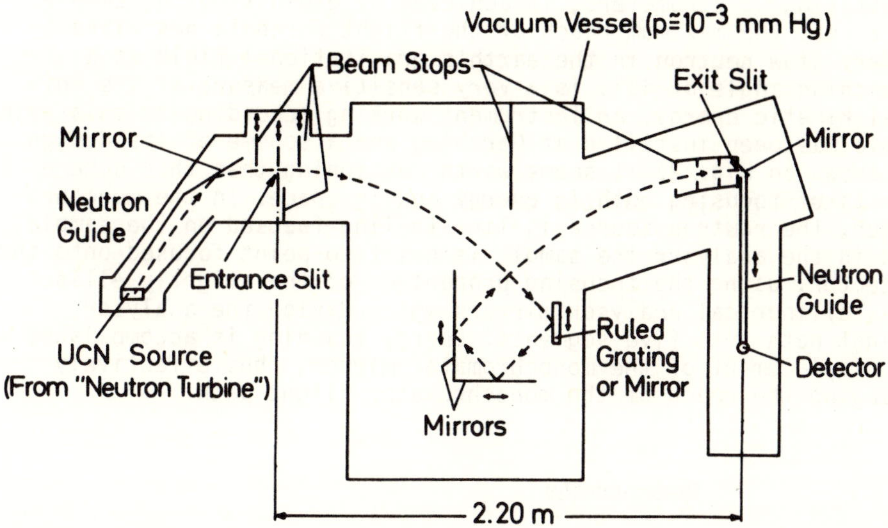


Fig. 1: Scheme of the "gravity diffractometer" for ultracold neutrons. The entrance slit is imaged onto the exit slit.

## 4. "GRAVITY FOCUSING" OF UCN BEAMS

The gravitational interaction has been utilized for UCN beam focusing in high-resolution experiments studying interference, diffraction and inelastic scattering phenomena.

Fig. 1 shows the scheme of the "UCN gravity diffractometer" /8,9/ installed at the Research Reactor at Garching near Munich. In this instrument, the vertical component of neutron velocity is precisely determined by the height of fall from the entrance slit to the horizontal mirror which may be replaced by a reflection sample or a combination of transmission sample and mirror. The beam spreading associated with the free fall from the entrance slit to the sample is undone in the phase of ascent from the sample to the exit slit. Thus the entrance slit is imaged onto the exit slit. The instrument admits of a resolution of 2 neV for the energy corresponding to the vertical neutron motion.

Before proceeding to the neutron-optical studies made possible with this high resolution, let me mention another scheme of

UCN focusing which we applied in a "gravity spectrometer" /10/. With this instrument quasi-elastic neutron scattering can be investigated with a resolution of a few neV in energy transfer. This resolution which is much higher than in conventional neutron scattering spectrometers, is achieved by exploiting the simple fact that the maximum reach of the flight parabola described by a very slow neutron in the earth's gravitational field at a launching angle of 45°, is a very sensitive measure of its initial kinetic energy. An instrument working according to this principle has been installed at Garching and a scheme of its design is shown on Fig. 2. It seems worth mentioning that this scheme admits of focusing both in energy and in space. In the monochromator, the neutron source is line-to-line focused on the sample, and in the analyser the sample is point-to-point focused onto the detector, using the focusing properties of the slightly elliptical, cylindrical analyser mirrors which divide the analyser flight path into five segments. Energy scanning is accomplished by displacement of the monochromator mirror, thus effectively changing the reach of the monochromator flight path.

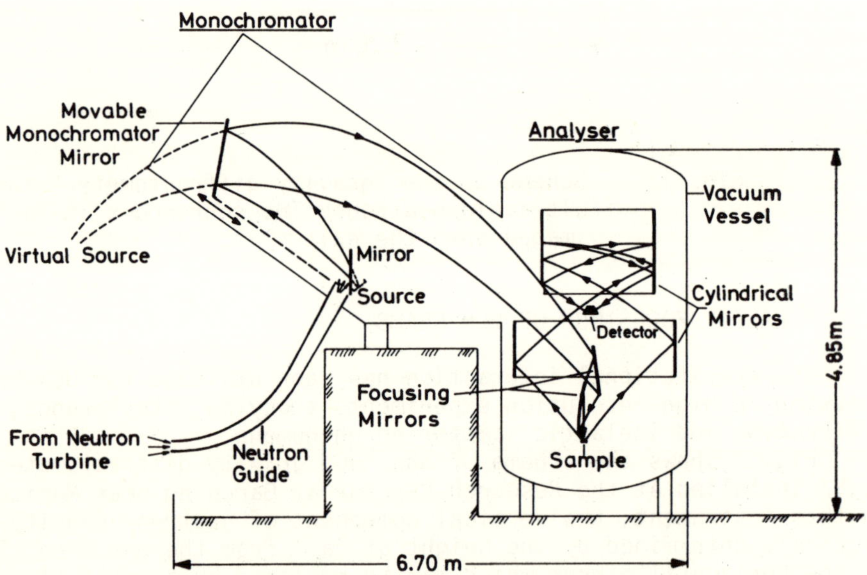


Fig. 2: Scheme of the "gravity spectrometer" NESSIE (for "NEutronen-Schwerkraft-SpektrometrIE"), in which the maximum reach of the flight parabola described by the ultracold neutrons is used for precise energy selection

The spectrometer is being used for measuring quasi-elastic line broadening (by only a few neV) in polymer solutions. Fig. 3 shows, as a first example, the quasi-elastic scattering from a polydimethylsiloxane (0.05 g/cm³) solution in deuterated benzene ($C_6D_6$) at 50° C (above the critical "θ"-point), and this data is compared to the resolution curve measured with a purely elastic scattering line (electro-graphite). The reason for showing this - preliminary - data in a talk on neutron optics is that it seems to demonstrate the precise functioning of the spectrometer in the way it was designed. The calculations for the fairly complicated neutron paths had all been based on classical mechanics and (curved-) ray optics, assuming an ideal geometry.

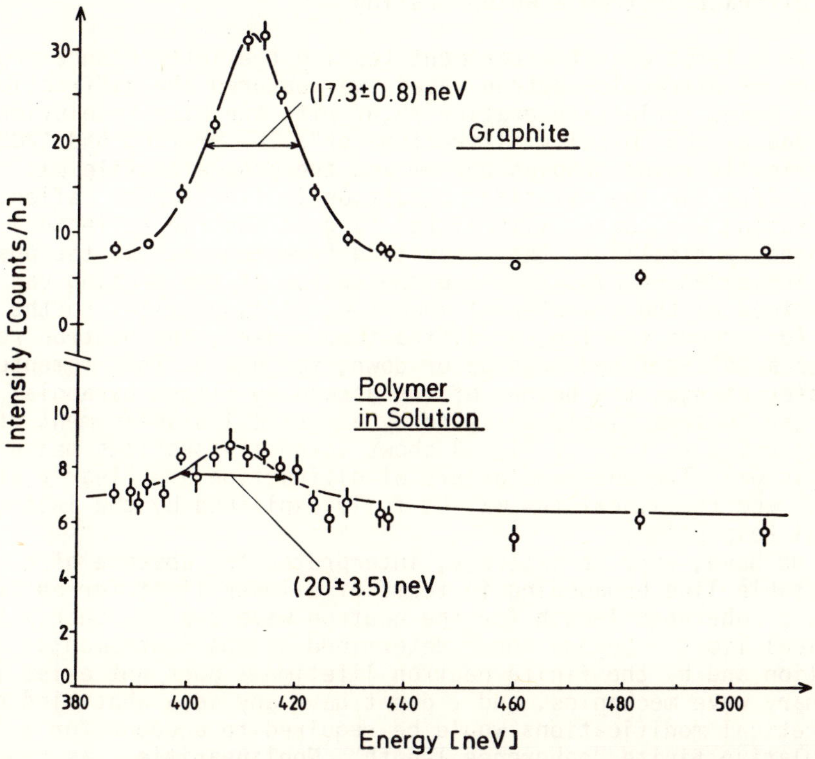


Fig. 3: The energy resolution curve of NESSIE measured with an elastically scattering graphite sample is compared to the slightly broadened scattering curve for a polydimethylsiloxane solution in deuterated benzene. The curves are plotted vs the incident neutron energy and are centred on the analyser energy of 412 neV.

## 5. DIFFRACTION AND INTERFERENCE EXPERIMENTS

The experiments on UCN diffraction and interference to be discussed now, had been initiated at a time when the early UCN storage experiments had yielded surprisingly short containment lifetimes and a number of suggestions and speculations had been put forward to account for the apparent anomaly. The most drastic idea was that the observed imperfection of neutron reflection from the bottle wall was perhaps an indication of a fundamental fault in the simple analysis in terms of standard wave mechanics (Ignatovich /11,12/).

### 5.1 Diffraction from a Ruled Grating

In a first direct experiment testing the interference properties of ultracold-neutron waves, we measured the diffraction from a ruled reflection grating /8,9/ with the high resolution provided by the "gravity diffractometer". The grating had 1200 mechanically ruled grooves per mm and the groove profile was appropriate for the "blazing" condition in first-order reflection. The grating was coated with nickel (a good neutron reflector) and arranged vertically in the "gravity diffractometer" at the position indicated on Fig. 1. Since the action of the grating can be understood as the transfer of momentum, hn/d, parallel to the surface (d: groove spacing, n: diffraction order), the neutron receives a well-defined push up or down, for $n \neq 0$. This momentum transfer changes the height of the ascending flight parabola, and this can be sensitively analysed by a vertical displacement of the scanning exit slit. Fig. 4 shows intensity profiles measured in this way. The observed orders of diffraction are clearly separated, and the linewidths may be fully explained by the instrumental resolution.

We have, very tentatively, interpreted the absence of a detectable line broadening in terms of a lower limit for an "intrinsic coherence length for the neutron wave train". Such a hypothetical limit - beyond those determined by the instrumental resolution and by the finite neutron lifetime - does not exist in ordinary wave mechanics, and I don't have any idea what kind of theoretical modifications would be required to account for a speculative finite "coherence length". Nonlinearities, as proposed by de Broglie /1/ or Bialynicki-Birula and Mycielski /2/ would certainly require drastic revisions in our present view of quantum phenomena. Thus, in the absence of a clear model even our interpretation of the absence of line broadening in terms of a lower limit for the "coherence length" seems doubtful, because even this interpretation may depend on the model. Therefore, the value of $\sim$0.1 mm derived from the minimum number of coherently illuminated grooves necessary for the observed linewidth, on the basis of the linear theory, should be taken only as a guide number.

(For the specific, logarithmic, non-linearity, $-b\psi \ln a|\psi|^2$, proposed by Bialynicki-Birula and Mycielski /2/ this would correspond to a limit for the constant b of $10^{-15}$ eV, similar to the limiting value recently reported by Gähler, Klein and Zeilinger /13/ and 100 times lower than the limit given by Shull et al. /14/. However, the proposed modification of the Schrödinger equation must be understood to be on the classical level, since this equation has, e.g., non-quantized oscillatory "breather" mode solutions, and it would, therefore, seem to require quantization. In the given form for the non-linear term, $\psi\Phi(|\psi|^2)$ (logarithmic or other), the diffraction process, or partial reflection or any scattering from fixed objects, would require energy for wave dispersion in space, which poses problems regarding the proposed probabilistic interpretation of the wave function. At any rate, a non-stationary solution of the non-linear equation for reflection from a potential wall, nothing to say about diffraction from a ruled grating, has apparently not yet been found. For all these reasons, I would rather hesitate to interpret our simple experiments in terms of something as complicated as non-linearities!)

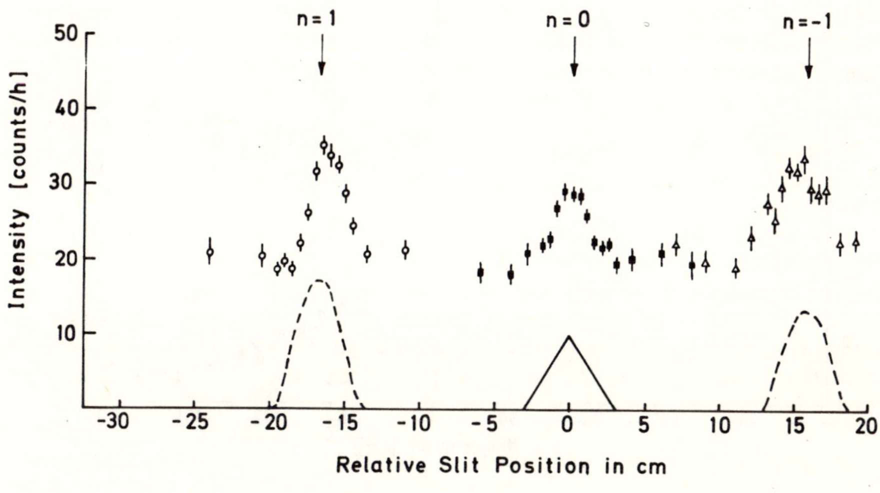


Fig. 4: Several orders of diffraction from a ruled grating (1200 grooves per mm), measured by vertical scanning of the exit slit in the "gravity diffractometer". Resolution curves are included for comparison.

## 5.2 Mirror Reflection

The main reason for mentioning the next, even simpler, experiment is that it seems to have provided one of the first direct clues as to the reason for the short UCN containment lifetimes in "bottles". We placed a float glass plate at the horizontal sample position in the "gravity diffractometer" (exchanging the grating by a vertical mirror). The reflection curve measured as a function of the neutrons' height of fall to the sample is shown in Fig. 5 /8/. It exhibits the typical steep edge at the critical height of fall which is determined by the limit for total reflection. However, the measured slope is considerably steeper than that calculated for reflection from a potential step function, i.e., a sharp transition from the vacuum to the bulk glass. The most plausible interpretation seems to be in terms of surface contamination.

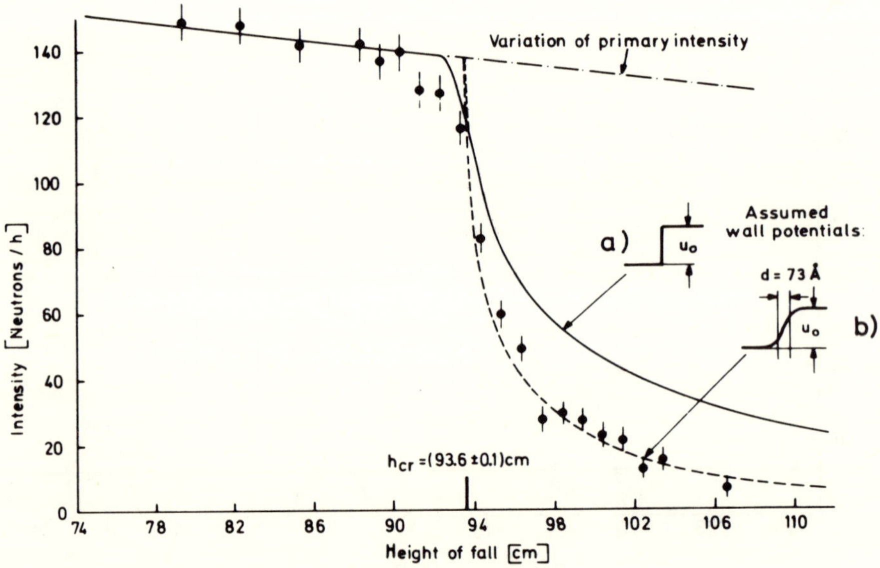


Fig. 5: Reflectivity vs neutron fall height in the "gravity diffractometer" for a float glass sample. The data points are compared to calculations for a) a step-function potential distribution, and, b) a potential distribution smoothed due to a hydrogenous surface contamination with a gradual change in composition.

Hydrogenous substances had for a long time been considered a prime candidate for UCN losses at the bottle walls, because of the extraordinarily high cross section for inelastic neutron scattering from protons, but some authors were reluctant to accept the large quantities of hydrogen necessary to explain the data (e.g., /15/). The reflection data in Fig. 5 can be explained by a smoothed wall potential due to an inhomogeneous contamination layer, and for a specific analytical form for the potential distribution near the surface /8/ an effective thickness of (73 ± 3) A for the transition is required to obtain a good fit to the data (dashed curve in Fig. 5). The corresponding amount of hydrogen is sufficient to account for the containment lifetimes reported in most of the experiments.

This conclusion was corroborated by independent experiments using different approaches /16, 17/. Recently, Mampe, Ageron and Gähler succeeded in directly demonstrating long containment lifetimes (several hundred seconds) in bottles after glow-discharge cleaning of the surfaces /18/.

## 5.3 Interferences at a Single Thin Film

As another approach to testing the interference properties of ultracold neutrons we examined the thin-film interference pattern for the reflection from a glass substrate with a 250 nm thick, homogeneous, evaporation film of gold /8/. The data taken at the "gravity diffractometer" is shown on Fig. 6. The interference pattern is clearly established and agrees with the simple plane-wave theory, allowing for resolution effects. The fitted numerical value of (268 ± 6) nm for the film thickness agrees with the quartz data taken during evaporation.

## 5.4 Neutron Resonances in More Complex Stratified Systems

An extension of the thin-film experiments to more complex stratified systems was undertaken with a view of possible applications as narrow-band interference filters. In addition, such systems offer the opportunity to study elementary quantum phenomena in synthetic, one-dimensional, potential distributions with almost macroscopic dimensions.

A simple example is a sequence of three thin films such that a layer of a material with a low scattering potential for neutrons (e.g. aluminium, for which U = 53 neV) is sandwiched between two identical layers of a medium with a high scattering potential (e.g. copper, for which U = 165 neV). Such a composite film constitutes a potential well in which neutrons with an appropriate energy of the motion perpendicular to the film plane may form quasi-bound states. These states can be studied by observing the resonance behaviour of reflection from, or transmission through, such samples, as in an optical Fabry-Perot interferometer. Similarly as in light optics, resonances are expected for neutron

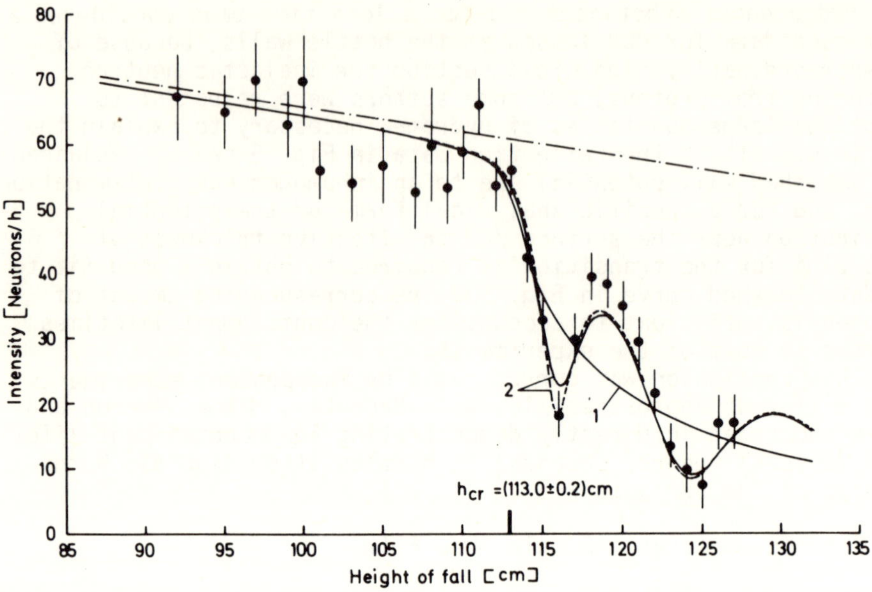


Fig. 6: High-resolution interference pattern for UCN reflection from a thin, homogeneous, gold film evaporated on glass. The solid curve 2 is a calculation with an adjusted film thickness.

wavelengths "matching" the width d of the potential well, and sharp resonances require, in addition, a high reflectivity of the "mirrors" (which are represented by the high-barrier copper films). The quasi-classical approximation yields the resonance condition

$$k_z \cong (\pi/d)\ (n + 1/2) \qquad [n = 0,1,\ldots]$$

for the perpendicular wavevector component $k_z$ within the well, and $k_z$ should lie within the region of total reflection for copper. Even in conditions of "total reflection" a slight transparency of the "mirrors" is ensured by the finite probability of tunneling through the thin copper films.

The samples were prepared by successive evaporation of the thin layers (~20 to 200 nm) on glass or silicon substrates, and experiments were performed both with cold neutrons ($\lambda \lesssim 10$ A) at grazing incidence and with ultracold neutrons ($\lambda \sim 10^3$ A) incident at large angles in the "gravity diffractometer" /19,20/.

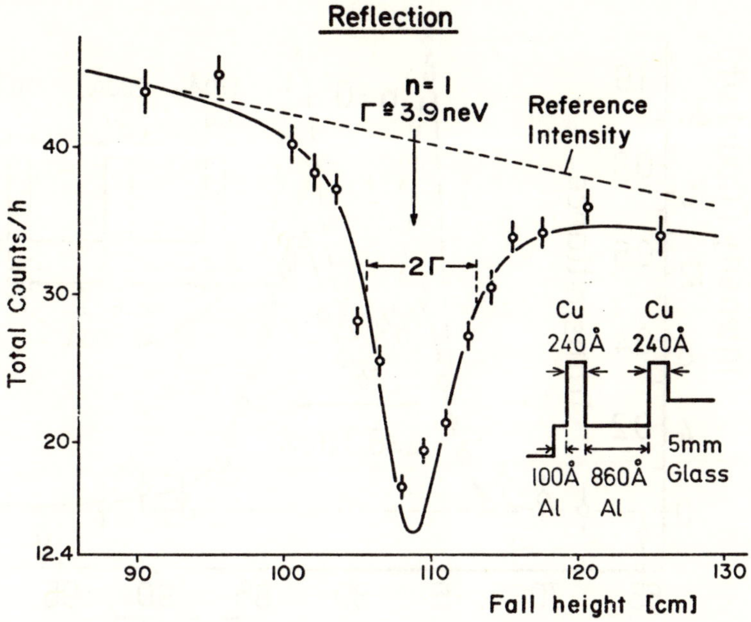


Fig. 7: Intensity reflected from a sample with the indicated layer sequence, which corresponds to a two-hump potential. The narrow minimum is due to the (n = 1) - state in the artificial potential well.

Fig. 7 shows the reflected intensity measured in the "gravity diffractometer" for a target with the following layers successively evaporated on glass: Cu (24 nm), Al (86 nm), Cu (24 nm), and, as a protection from oxidation, Al (10 nm). The potential function representing this sequence is shown in the inset. The observed pronounced reflectivity minimum at a fall height of 108.5 cm corresponds to n = 1 (the second stationary state). The data is well represented by the exact solution of the one-dimensional Schrödinger equation for the multistep potential, including the small instrumental resolution broadening (solid curve). The measured linewidth of $2\Gamma$ = 7.7 neV corresponds to a lifetime of $\tau = \hbar/2\Gamma \cong 10^{-7}$ s for the quasi-bound state. The most plausible explanation for the noticeable enhancement of minimum reflectivity at resonance seems to be a slight roughness of the film interfaces /21/. Similar deviations were observed with all the other samples.

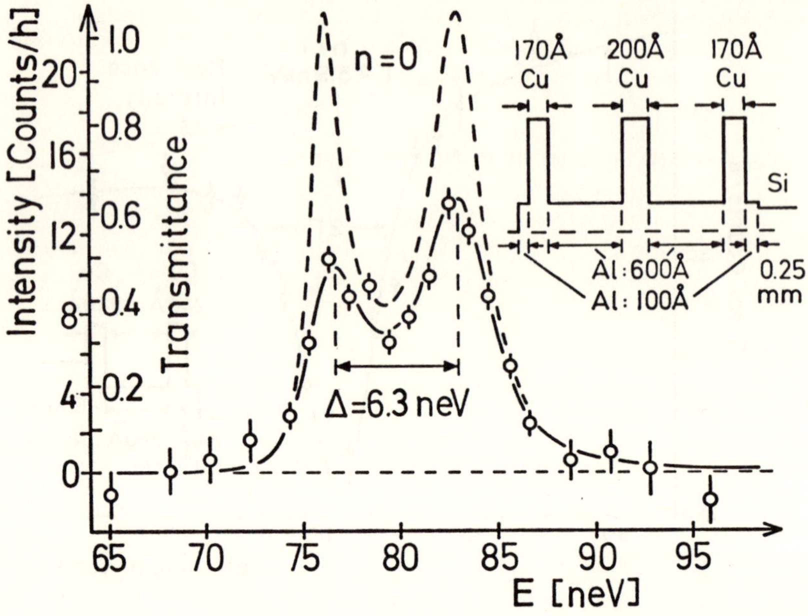


Fig. 8: Level splitting observed in the UCN transmission through a coupled-resonator potential created by the layer sequence indicated in the inset

As an extension of the above Fabry-Perot-type neutron-wave resonator we also studied a system of two coupled resonators, created by two identical potential wells arranged in series and weakly coupled by a common, sligthly transparent, potential barrier. Such a potential distribution may be generated by a layer sequence Cu-Al-Cu-Al-Cu, and we expect a "tunneling" splitting of the single-resonator states into two each. These states are characterized by a symmetric and an antisymmetric wavefunction, respectively. This is similar to the inversion splitting in the ammonia molecule, or, simpler, to the mechanical example of two coupled pendulums.

Fig. 8 shows the results for the UCN transmission through a target with the indicated structure, plotted vs the energy of vertical neutron motion in the "gravity diffractometer". The data clearly shows a splitting of the (n = 0) - state by only 6.3 neV. The measured points are compared to the calculation for monoenergetic neutrons on the basis of the multistep potential function, for the nominal film thicknesses and potential heights (dashed curve). An improved representation (the solid curve) is obtained by taking into account, mainly, the instrumental reso-resolution and a roughness scattering whose strength was adjusted to the data but appears to be plausible /20,21/. Again, we do not have to invoke a fundamental deviation from the elementary analysis in terms of plane waves.

## 6. IMAGE FORMATION WITH ULTRACOLD NEUTRONS

As a last example of UCN optics I should like to discuss, briefly, the possibility of image formation of an object and the prospects of a possible neutron microscope. The main incentive for work on the development of a neutron microscope arises from the interest in exploring the information to be obtained from the contrast in matter, as "seen" by neutrons - which is quite different from the contrast for light or electrons, and which is adjustable by proton-deuteron exchange. For thermal or cold neutrons, however, no satisfactory focusing systems are available, and it is only with ultracold neutrons, owing to their property of being reflected from mirrors even at large angles of incidence, that high magnifications at tolerable aberrations appear to be attainable. The wavelength limit to the resolution is of the order of 100 nm, i.e. about one order of magnitude below that for visible light, but with presently existing UCN sources a much more severe limitation is imposed by the small available intensity.

In designing systems for magnified image formation with UCNs we are faced with the problem of gravity and, hence, must deal with the optics of curvilinear rays. For example, a concave mirror, which is an achromatic system in light optics, would exhibit a chromatic aberration if used with UCNs because their ray curvature strongly depends on the wavelength.

A possibility of compensating this effect in a single-component magnifying system consists in "crossing" a concave mirror with a Fresnel-zone plate to obtain a reflecting, concave, "zone mirror" /22/, as sketched in Fig. 9. By choosing an appropriate geometry such a device can be made to achromatically diffract the UCN intensity incident from a point object into one image point. The zone mirror then exhibits much the same properties as a lens or a concave mirror in light optics. This has been demonstrated in an experiment where we obtained sharp images of an object slit with a magnification of up to six /23/.

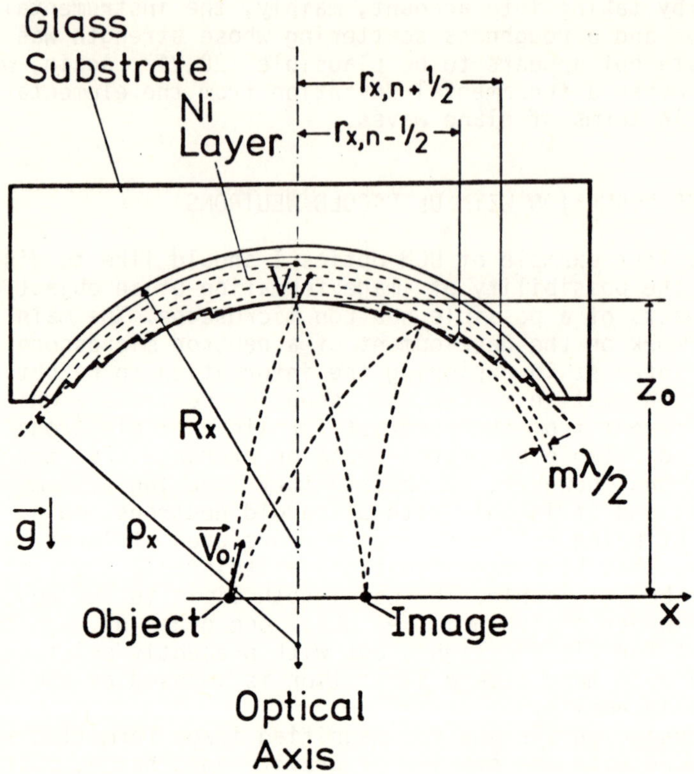


Fig. 9: Scheme of the "zone mirror", which provides achromatic image formation with ultracold neutrons.

At present we are testing, in an experiment at the Institut Laue-Langevin at Grenoble, a different scheme of achromatic imaging with UCNs, which should admit of much higher magnifications /24/. In this instrument the chromatic aberration of one, parabolic, mirror is compensated by that of a second, spherical, mirror. This scheme includes the possibility of using multilayer mirrors which reflect 2 to 3 times faster neutrons than ordinary mirrors, and this could help to increase the available beam intensity.

REFERENCES

1 de Broglie, L.: 1960, "Non-linear wave mechanics; a causal interpretation", Elsevier, Amsterdam, London, New York, Princeton
2 Bialynicki - Birula, I. and Mycielski, J.: 1970, Ann. of Physics 100, p.62
3 Groshev, L.V., Dvoretsky, V.N., Demidov, A.M., Panin, Yu.N., Lushchikov, V.I., Pokotilovky, Yu.N., Strelkov, A.V. and Shapiro, F.L.: 1971, Phys. Lett. B 34, p. 293
4 Steyerl, A. and Trüstedt, W.-D.: 1974, Z. Physik 267, p. 379
5 Steyerl, A.: 1977, "Very low energy neutrons", Springer Tracts in Mod. Phys., Vol. 80, p. 57
6 Golub, R. and Pendlebury, J.M.: 1979, Rep. Prog. Phys. 42, p. 439
7 Kügler, K.-J., Paul, W. and Trinks, U.: 1978, Phys. Lett. B 72, p. 422
8 Scheckenhofer, H. and Steyerl, A.: 1977, Phys. Rev. Lett. 39, p. 1310
9 Scheckenhofer, H. and Steyerl, A.: 1981, Nucl. Instr. and Methods 179, p. 393
10 Steyerl, A.: 1978, Z. Physik B 30, p. 231
11 Ignatovich, V.K.: 1974, Preprint Joint Inst. Nucl. Research, Dubna, E4-8039, unpublished
12 Ignatovich, V.K.: 1977, Preprint Joint Inst. Nucl. Research, Dubna, R4-10650, unpublished
13 Gähler, R., Klein, A.G. and Zeilinger, A.: 1981, Phys. Rev. A 23, p. 1611
14 Shull, C.G., Atwood, D.K., Arthur, J. and Horne, M.A.: 1980, Phys. Rev. Lett. 44, p. 765
15 Luschikov, V.I.: June 1977, Phys. Today 42, p. 42
16 Stoika, A.D., Strelkov, A.V. and Hetzelt, M.: 1978, Z. Physik B 29, p. 349
17 Lanford, W.A. and Golub, R.: 1977, Phys. Rev. Lett. 39, p. 1509; Bugeat, J.P. and Mampe, W.: 1979, Z. Physik B 35, p. 273
18 Mampe, W., Ageron, P. and Gähler, R.: 1981, Z. Physik B 45, p. 1
19 Steinhauser, K.-A., Steyerl, A., Scheckenhofer, H. and Malik, S.S.: 1980, Phys. Rev. Lett. 44, p. 1306
20 Steyerl, A., Ebisawa, T., Steinhauser, K.-A. and Utsuro, M.: 1981, Z. Physik B 41, p. 283
21 Steinhauser, K.-A.: 1981, Dissertation, Techn. Univ. München
22 Steyerl, A. and Schütz, G.: 1978, Appl. Phys. (Springer) 17, p. 45
23 Schütz, G., Steyerl, A. and Mampe, W.: 1980, Phys. Rev. Lett. 44, p. 1400
24 Herrmann, P.: 1982, Dipl. Thesis, Techn. Univ. München

"GESPENSTERFELDER"

F. Selleri

Istituto di Fisica - Università di Bari
INFN - Sezione di Bari

## 1. EINSTEIN'S "GESPENSTERFELDER"

Einstein's discovery of the equivalence between mass and energy[1] could have brought to a great unification of the scientific world-view. Since energy and momentum are always conserved while mass can be created and destroyed, the old idea that matter is an irreducible substance of the world had to be substituted with the idea that everything is really made of energy-momentum. This allowed one to understand in an unified way, at least in principle, matter itself, light, electromagnetic and gravitational fields and so on, briefly all of the objects of physical investigation. If this unification in practice did not take place it was because of good reasons, starting from the fact that Einstein himself in the same year 1905 was forced to admit the existence of something real (in the sense that it was assumed to propagate causally in space and time) but containing neither matter, nor energy-momentum. This was a necessary consequence of his dualistic picture of the electromagnetic radiation[2].

Einstein's objective dualism associated wave and corpuscles in such a way that the particle properties energy and momentum ($E$ and $\vec{p}$) and the wave properties frequency and wave number vector ($\nu$ and $\vec{k}$) were related by

$$E = h\,\nu \qquad ; \qquad \vec{p} = h\,\vec{k}$$

As it is well known de Broglie extended these relations to material particles (electrons, neutrons,...) so that they appear as a very fundamental property of nature.

A problem coming immediately to one's mind within Einstein's philosophy is the following : if the localized particle carries



*S. Diner et al. (eds.), The Wave-Particle Dualism, 101–128.*

all the energy and momentum in which sense can the wave be considered real ? This problem was felt so acutely by Einstein that he referred to these waves as Gespensterfelder (ghost fields) ; an object without energy and momentum is in fact unable to exert a pressure when hitting other bodies, which means that it does not have that quality which makes us call real a normal object. Still the equations of quantum mechanics describe this wave as propagating in space and time.

The difficulties associated to the conception of an empty wave (here and in the following these two words are taken as equivalent to "wave propagating in space and time but devoid of energy-momentum") have led many people to discard the idea as a scientific impossibility : if the wave is empty it cannot induce changes into physical objects ; hence it cannot be observed ; therefore it does not make any sense to postulate its existence.

This is probably one of the reasons of the rejection of Einstein's dualism from the large majority of the physicists, a rejection which displeased Einstein so much that he could write : "I must look like an ostrich hiding always his head in the relativistic sand for not having to face the ugly quanta[3]". The idea of the empty-wave was however not only Einstein's : it was entertained in different forms also by de Broglie, Bohr and Born and it can be said to belong to the line of thought of the Copenhagen school, even if Heisenberg preferred a "withdrawal into the mathematical scheme" to the ambiguous formulation of dualism contained in the complementarity principle.
As it is well known Louis de Broglie agreed largely with Einstein's dualism.

For our purposes we recall a 1977 paper[4] in which de Broglie discussed the Pfleegor-Mandel experiment[5] in which interferences between the beams emitted by two different lasers were observed in conditions in which the probability to have more than one quantum of energy $h\nu$ outside the lasers was negligibly small. He expressed his ideas in the following way : "In agreement with classical conceptions for me a particle is a very small object which is constantly localized in space and a wave is a physical process which propagates in space.." Later be added : "The particle is a very small region of high concentration of energy which is embodied in the wave in which it constitutes some kind of singularity generally in motion".

With these ideas the explanation of Pfleegor and Mandel's observations becomes elementary. Laser 1 and Laser 2 emit continuously two ondulatory beams which cross each other in the regior R where they interfere giving rise to interference maxima (shown as black stripes in FIG.1). If a particle is emitted from one of the two lasers (dot with arrow in FIG.1) it will be guided by the wave to reach with greater probability the regions in R where the wave amplitude is larger. Collecting many photons on a screen S put across the region R the interference figure produced by the wave will be revealed by the photon distribution. In this way the

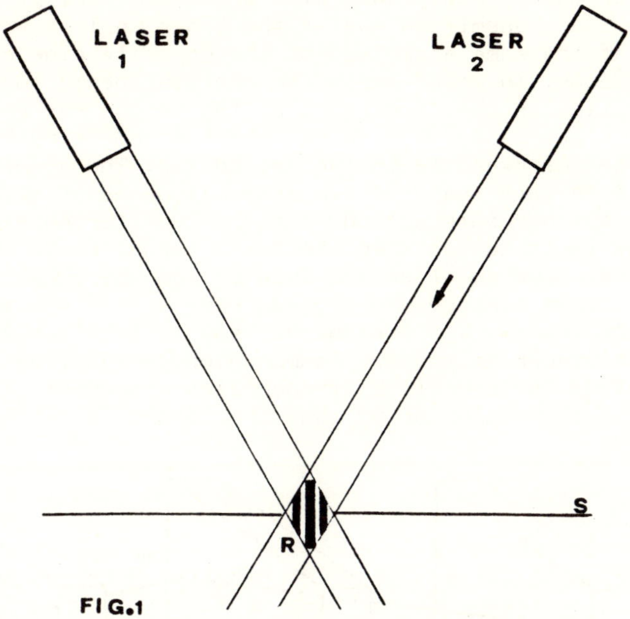


FIG.1 Set-up of the Pfleegor-Mandel experiment

role of the particles is only to allow the detection of the interference from the observation of their statistical distribution in the region where the interference exists.

## 2. EXPERIMENTS ON THE NATURE OF DUALISM

The review of some recent proposals which could conceivably give empirical meaning to the empty waves is left for the next section. Here we will rather check that the idea is not already excluded by existing experimental evidence by considering new and old experiments on self-interference (neutron interferometry[6] and the Janossy-Naray experiment[7] on optical photons).

The first neutron interferometer was operated in 1974 by Rauch, Treimer and Bonse at the Austrian Nuclear Institute in Vienna[8]. The interferometer is built starting from a single silicon crystal completely free of dislocations and other defects in the regular atomic structure. For example one can start from a cylindrical crystal about eight centimeters long and five centimeters in diameter and cut away part of it leaving three semicicular "ears" connected by the remainder of the cylinder.

When a beam of neutrons strikes the first ear in the interferometer at an angle $\theta$ from the normal to the surface (Bragg angle) of 20-30° it is scattered by planes of atoms perpendicular to the face of the crystal. This kind of scattering, gives rise to two beams : a transmitted one at the Bragg angle $\theta$ and a diffracted one at the same angle but on the opposite side of the scattering planes. In other words the emerging beams form a V whose vertex lies in the first ear. At the second ear each of these beams is again Laue-scattered, the four emerging beams forming a W whose vertexes lie in the second ear. The geometry of the apparatus is such that the two external beams of this W do not have any further interaction with the interferometer (FIG.2). The important facts happen when the two internal beams of the W converge in the same point on the face of the third ear : each of them is again Laue scattered and gives rise to a V coming out of the third ear. The two V's are however spacially superimposed in such a way that each of the two beams which emerge from the third ear results from the sum of the transmitted component of one beam and the diffracted component of the other beam.

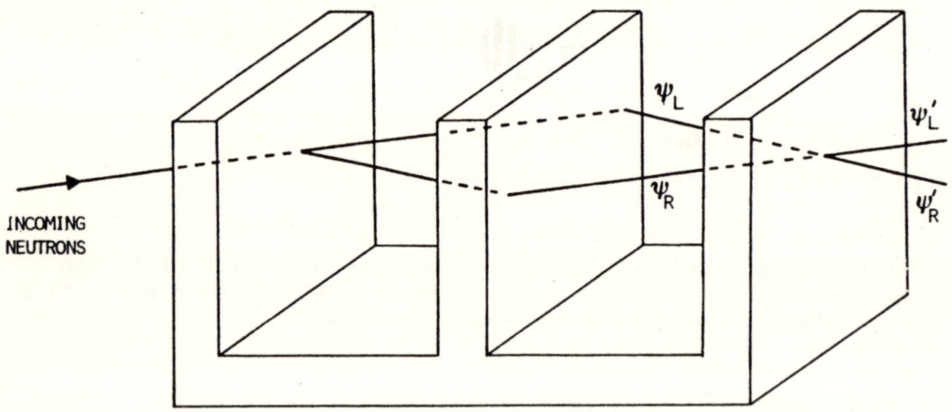


FIG.2 Neutron paths within a neutron interferometer. Not shown are the paths emerging from the second "ear" and leading outside the interferometer.

Let $\psi_R$ and $\psi_L$ be the two wave functions describing the beams which arrive on the <u>third</u> ear from right and from left respectively.

At the third ear each wave is split in a transmitted (T) and a diffracted (D) wave ; $\psi_R$ gives rise to $\psi_R^T$ and $\psi_R^D$, while $\psi_L$ gives rise to $\psi_L^T$ and $\psi_L^D$. The coherence of this phenomenon guarantees that $\psi_R$, $\psi_R^T$ and $\psi_R^D$ have the same phase and that the same holds for $\psi_L$, $\psi_L^T$ and $\psi_L^D$.

Now in the V which emerges from the third ear one has two waves which we call $\psi_R$ and $\psi_L$ and which result from the physical spacial superposition (and therefore the algebraic sum) of two different waves

$$\begin{cases} \psi'_R = \psi^T_L + \psi^D_R \\ \psi'_L = \psi^T_R + \psi^D_L \end{cases}$$

With some simplifying assumptions it is possible to show that

$$\begin{cases} |\psi'_R|^2 = \frac{1}{2}\,[|\psi_R|^2 + |\psi_L|^2 + 2|\psi_R|\cdot|\psi_L| \cos\alpha] \\ |\psi'_L|^2 = \frac{1}{2}\,[|\psi_R|^2 + |\psi_L|^2 - 2|\psi_R|\cdot|\psi_L|.\cos\alpha] \end{cases} \qquad (1)$$

where $\alpha$ is the phase difference between an R and an L wave. From the previous result one can see that probability is conserved and furthermore that the squared modulus of each beam emerging from the neutron interferometer depends on the relative phase of the right and left components. If this phase is known one can check if the famous probability law of Born is really valid : in fact $\psi'_R$ and $\psi'_L$ are spacially divided and the probability of a single neutron to choose the left or the right should be equal to $|\psi'_L|^2$ and $|\psi'_R|^2$ respectively. Repeating the experiment many times one should have emerging neutron fluxes proportional to $|\psi'_L|^2$ and $|\psi'_R|^2$. Practical experiments have been done with monochromatic neutron beams emerging from a nuclear reactor. Typically these have a flux of about 100 neutrons per second with an average time spacing of $10^{-2}$ sec. From a reactor emerge normally thermal neutrons with velocity of the order of $10^5$ cm/sec : therefore every neutron crosses the ten centimeter long interferometer in about $10^{-4}$ sec. The probability of having two neutrons in the apparatus is therefore of the order of $10^{-2}$. Obviously this allows one to conclude that self-interference is observed in these experiments. The experiments performed by using the most diverse physical means for varying the relative phase $\alpha$ (phase-shifting materials, magnetic fields, gravitational field of earth, earth rotation,...) have shown a very good consistency with eq.(1). Thus Born's statistical postulate (particle distribution given by $|\psi|^2$) is confirmed. Moreover every neutron interferes with itself (just as in the double slit experiment exposed in textbooks). Thus we conclude that in the case of every single neutron something propagates over both the possible paths within the interferometer.

This conclusion is apparently contradicted by a second experiment which one could do by putting neutron counters immediately behind the first ear : as it is well known, since such experiments have been performed many times with photons and electrons, every neutron would be revealed only by a single counter. Neutron counters are in last analysis sensitive to incoming energy-momentum : we can conclude that energy-momentum (and therefore mass) take always only one of the two paths which are possible when crossing an interferometer ear. On the other path one concludes so that a wave propagates which does not carry energy and

and momentum.

This wave devoid of energy-momentum which we extracted with a naïve phenomenological analysis of neutron interferometric experiments resembles obviously very much to the empty wave of Einstein.

The first observations of self-interference in experimentally certain conditions were made by L. Janossy at the Central Research Institute of Physics in Budapest in the fifties. The final experiment used an interferometer of the Michelson type and monochromatic photons coming from the 5461 Å (green) line of mercury[7]. The experiment had the following features (FIG.3) :

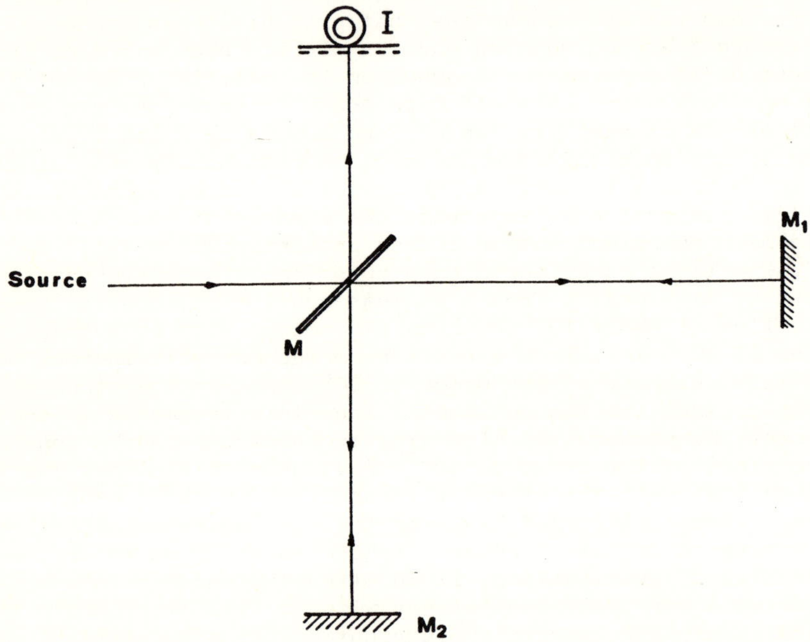


FIG.3 Set-up of the Janossy-Naray experiment. M is a semi-transparent mirror, $M_1$ and $M_2$ are normal mirrors and I is an interference detecting device.

1) Arm-length of about 14m. Since a photon can be considered to be a packet of length comparable with the coherence length of a spectral line, in the case of the green mercury line one would obtain longitudinal dimensions of about 1 m. If the dimensions of the interferometer were smaller than the "size" of the photon the interference phenomenon might be accounted for by thinking that the photon "floods" the whole of the arrangement simultaneously and thus comes under the influence of the two mirrors at the ends

of the arms. This picture cannot work if the arm-length is much larger than the photon coherence length, as 14 m is.

2) Photon intensity as low as $10^4$ photons/s. A 30 m long apparatus as the one used by Janossy and Naray is crossed by a photon in $10^{-7}$ sec. An intensity lower than $10^7$ photons/s is therefore required in order to insure that only one photon is present in the apparatus in any given time.

3) Underground tunnel cut in rock. In order to achieve the necessary stability the apparatus was built in an underground tunnel cut in rock 30 m under the surface of the earth. Even the presence and activity of a human being in the tunnel could disturb the regular functioning of the apparatus. The problem was solved by using remote control observing the interference pattern from outside by help of a periscope.

These experimental precautions allowed Janossy and Naray to reach important conclusions. Their paper ends with the following statement : "As the result of our measurements we have therefore to conclude that interference phenomena are perfectly normal even at such low intensities where at one time in average less than one photon is to be found in the arrangement ; this is true even if the dimensions of the arrangement greatly exceed the coherence length of the photons giving rise to the patterns".

An attempt at understanding this result in space-time terms would naturally assume that when the photon wave-packet arrives at the semitransparent mirror it is split in two parts which follow different trajectories, are reflected by different mirrors and recombine in the region where the interference pattern is observed. This possibility is however excluded by another experiment, just as in the case of neutrons.

If one sends a low intensity photon beam on a semitransparent mirror and puts on the two outgoing trajectories two photo-multipliers, coincidences should be observed if the energy-momentum of the incoming photon is split in two parts, for each classical wave packet of energy $h\nu$ is split into two similar wave packets of reduced amplitude by the beam splitter. If these attenuated wave packets are allowed to fall on two photodetectors placed in the path of each beam, they should sometimes give rise to coincident photoemissions at the two detectors, provided the energy $\frac{1}{2}h\nu$ is sufficient to cause photoemission. On the other hand, if $\frac{1}{2}h\nu$ is not sufficient energy for photoemission, no photoelectric counts should be registered at all.

An experiment to test these predictions is very easy to carry out. It is readily shown that no coincident pulses (other than accidental ones) are observed at the two detectors[9].

Also for photons, therefore, the existing experimental evidence does not contradict the idea that a quantum system is composed of an empty wave and a localized structure carrying energy and momentum and that a semitransparent mirror splits only the wave while energy and momentum make a well defined (albeit random) choice between the two trajectories.

The idea that the empty waves introduced in quantum theory by Einstein, de Broglie, Slater and Bohr could give rise to stimulated emissions in spite of the fact that they do not carry any energy-momentum has been expressed several times. Thus de Broglie wrote[10]
"If an arbitrary cause has generated the emission of a quantum of light... its wave... passing on nearby atoms will stimulate other emissions..."
Similarly Bohr wrote[11] about the action of his "virtual radiation field" :
"... an induced transition in an atom is not directly caused by a transition in a distant atom for which the energy difference between the initial and the final stationary state is the same. On the contrary, an atom which has contributed to *the induction of a certain transition in a distant atom through the virtual radiation field*... may nevertheless itself ultimately perform another of these transitions". (our italics). Another interesting statement is the one concerning the interference between the incident virtual plane wave and the outgoing spherical virtual waves radiated by "illuminated" atoms[12] :
"... these atoms will act as secondary sources of virtual wave radiation which interferes with the incident radiation. If the frequency of the incident waves coincides closely with the frequency of one of the virtual harmonic oscillators corresponding to the varius possible transitions, the amplitudes of the secondary waves will be especially large, and these waves will possess such phase relations with the incident waves that they will diminish or augment the intensity of the virtual radiation field, *and thereby weaken or strengthen its power of inducing transitions in other atoms*". (our italics).
The idea is thus very simple : it is *the wave* and not a localized particle -like structure which induces stimulated transitions in atoms. In this and in the next section we will discuss some indirect evidence in favour of this picture and present some possible experiments which could in principle allow to settle this question. Our reasoning will refer only to energy-momentum conserving processes since the BKS-idea of nonconservation is untenable after so many atomic and subatomic processes have always shown a complete consistence with the conservation laws. Therefore we will refer to stimulated *emissions*, where energy-momentum conservation is possible also if no external contribution is added to the energy of the emitting system.

Blake and Scarl[13] have investigated laser light amplification in a laser gain tube. A schematic representation of their apparatus is shown in FIG.4 : a beam of light generated by on He-Ne laser (L) at a wavelength of 6328 Å crosses an He-Ne laser gain-tube (LGT in the figure) capable of emitting at the same wavelength. The gain, defined as the ratio of the intensity of the laser beam with the discharge on, divided by the intensity with the discharge off (after the amplifier-tube spontaneous emission had been subtracted) was 1.30±0.02. At those low amplifications the gain is known to be linear for the intensity levels used in this experiment : physically, one can say that in first approximation it is possible to neglect the further amplification of the light generated in LGT by the laser beam. In these conditions one might expect the stimulating photon and the stimulated photon to be related in time, if the process of stimulated emission is generated by the energy-momentum of the e.m. field, since these physical quantities are always detected as localized in space and time. The amplified laser light, together with the unavoidable spontaneous emission from the LGT, crossed an angular-spatial filter consisting of two 0.425 mm pinholes and an interference filter F which allowed only light with wavelength (6328±5) Å to pass. A beamsplitter (BS) and two photomultipliers ($P_1$,$P_2$) were used as the two-photon detection system.

If the coincidence rate is defined as the number per unit time of photon detections by $P_1$ and $P_2$ having a tome difference less than 1 nanosecond, Blake and Scarl found that, after background subtraction, the coincidence rate was the same independently on the intensity of the laser light. This result was found in four experimental runs in which the ratios of amplified laser light to amplifier tube chaotic light in the same mode were 0, 0.5, 1 and 2. Therefore Blake and Scarl could conclude that "no correlations between incident and induced photons seem to be necessary in order to explain the results".

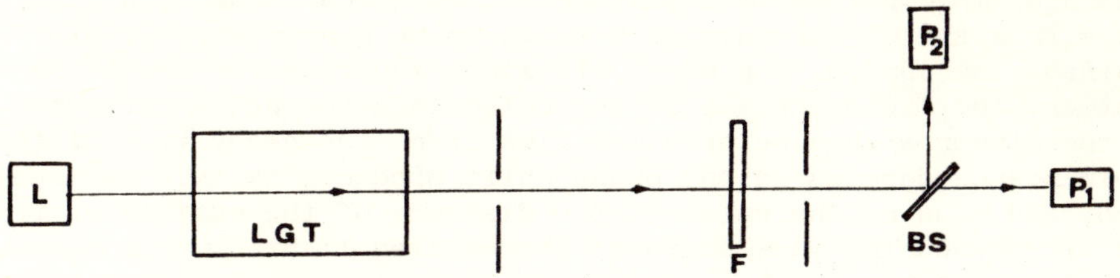


FIG.4 Schematic representation of the Blake-Scarl apparatus : L is a He-Ne laser and LGT an He-Ne laser gain tube. The beamsplitter M sends light to the phototubes $P_1$ and $P_2$.

This result was considered surprising by the authors who stated that "Stimulated emission illustrates one of the apparent difficulties of current photon descriptions of the electromagnetic field".

A natural explanation can be given in terms of the empty wave. Suppose a photon is a localized entity carrying energy-momentum and that it is accompanied by an extended empty wave. Suppose that the stimulated emission is due to the action of this wave. For laser light the wave has an infinitely long coherence length and has a constant probability of generating a stimulated photon. This is produced independently on the presence and position of photons in the stimulating wave. Thus no time correlation is expected between primary and secondary photons, as observed.

Another phenomenon whose existence is consistent with the empty-wave idea is the modification of fluorescence lifetimes in presence of mirrors. This observation shows that spontaneous emission of radiation has the interesting property of being affected by the source atom's local environment. This is hard to understand with different pictures of the wave-particle dualism, and Milonni and Knight[14] voiced the difficulty of many people by writing : "The experimental observation that fluorescence lifetimes depend upon the atom's surroundings was apparently met with some surprise : how could the spontaneous emission of a photon be affected by the local environment, since the atom can only "see" its surroundings by emitting a photon in the first place ?". In the following discussion Milonni and Knight conclude that the radiated field (or the photon "wave function") contains all the information needed for changing the atomic emission probability, provided only a time $t = 2\ell/c$ has elapsed since the atom entered in the region where the mirror is placed, at a distance $\ell$ from it.

The marked dependence on the distance from a mirror of the decay time of the luminescence of a molecule has been discussed by Kuhn[15] who has considered experimental data collected by Drexhage. The observed oscillations of the decay time with distance from the mirror have been shown to be quantitatively explainable by considering Hertz classical equation for dipole radiation. The field arrives at the molecule after being reflected by the mirror with a retardation time $2\ell/c$. The decay time of the luminescence depends in the phase shift produced by this retardation, and thus on the ratio of the distance of the oscillator from the mirror and the wavelength of the emitted light.

In quantum mechanical terms the phenomenon can be attributed to stimulation or inhibition of the emission of the light quantum, the stimulating field being the radiation field of the atom capable of emitting the quantum itself.

These results are extremely interesting since it appears virtually impossible to give any rational explanation of their existence in terms of a purely ondulatory (or of a purely corpuscular) description of radiation.

Such unilateral pictures are excluded also for several other reasons. Consider, for instance, a popular approach to electromagnetism which was the so-called "neoclassical theory" proposed by Jaines[16] and other authors. In this approach the e.m. field is treated in purely classical terms but quantum mechanics is used for the description of atoms and molecules. In this way several phenomena were thought to be explainable without the photon hypothesis. Such was the case of the photoelectric effect, of the

Hanbury Brown and Twiss effect[17], of so-called photon counting experiments, of photon bunching and of the random registration of photoelectric pulses generated by laser light. A careful conceptual and experimental analysis of these problem has however led Mandel[18] to the conclusion : "Although semiclassical theories have had considerable success in accounting for many observed effects,... they fail completely in other cases, and no evidence exists that should cause us to think of giving up Q.E.D. in favor of a semiclassical theory".

One of the fundamental weaknesses of the semiclassical approach is its inability to account for the short time delays in photoemission if energy conservation has to be true without exception.

Other approaches which tried to avoid dualism can similarly be excluded. Such is the case of the proposal put forward by Alfred Landé.

He developed a particular form of quantum theory which, in his opinion, demonstrates[19] "that wavelike looking atomic phenomena, such as diffraction of particles through crystals, are due to pure mechanical action of the diffractor under the rules of quantum mechanics".

According to Landé there does not exist any undulatory aspect of matter : "As thousands of experiments have shown, matter consists of discret particles, real, concrete, countable, with definite rest masses and charges, condensed in <u>small spaces</u>"[19]. Correspondingly, Landé's electromagnetic field is a pure wave and does not have any particle-like aspect. One can then conclude that Landé accepts a classical <u>picture</u> both of material particles and of the electromagnetic field. This is however supplemented with <u>quantum rules</u> which determine, for instance, the behaviour of a particle interacting with a slit, with a double slit, with the regular distribution of atoms in the crystal's lattice, and so on. Landé associates quantum rules to energy E, angular momentum $P_\phi$ and momentum P. The first rule controls the mutual exchanges of the energy in two bodies which are in frequency resonance $\nu$ or time period $\tau = 1/\nu$. It readt

$$\Delta E = h/\tau = h\nu \qquad \text{(Planck).}$$

A similar rule holds for the angular momentum

$$\Delta P_\phi = h/\phi \qquad \text{(Sommerfeld)}$$

where $\phi$ is the angular period. Every object has angular period $2\pi$ so that one always has the rule $\Delta P_\phi = h/2\pi$.

The third selection rule states that a body with a period $\ell$ in space changes its momentum parallel to $\ell$ only in amounts $h/\ell$ :

$$\Delta P = h/\ell \qquad \text{(Duane)}$$

This is the real novelty of Landé's approach, the first two rules being well known in standard quantum theory. The chief application of Duane's selection rule is to a crystal with a series of lattice planes at distance L and hence of period $\ell = L, L/2, L/3, \ldots$ Hence the crystal can change its momentum perpendicular to the planes only in quanta $\Delta p = nh/L$ and the particles interacting with it shall emerge along well defined quantized directions.

Even if one forgets Landé's theory difficulties for explaining Laue scattering in neutron interferometry, where for some reason unexplainable in the present approach (but very well explained by standard wave-theory) the neutrons emerge only along two directions, the whole scheme fails badly because of its inability to explain the oscillating intensities of the two final beams if a phase-shifting material of variable thikness is put on one of the two trajectories.

The proof of the latter statement is straightforward : with reference to FIG.1 let $I_L$ and $I_R$ be the intensities of the left and right internal beams, respectively. Let then $I'_L$ and $I'_R$ be the similar intensities of the two external beams. If one had to do with single particles propagating in one or the other of the two internal trajectories one should always have

$$\begin{cases} I'_L = I_L^D + I_R^T \\ I'_R = I_L^T + I_R^D \end{cases} \qquad (2)$$

where T and D denote the transmitted and diffracted parts of the intensity to which they refer. But $I_L^D$ and $I_L^T$ can be measured by putting an obstacle on the R-beam. Also $I_R^D$ and $I_R^T$ can similarly be measured. All the quantities entering in (2) can then be measured independently and one finds that (2) is often badly violated by existing experimental data of neutron interferometry.

It should therefore be evident that Landé's formulation of dualism is unable to explain some important empirical facts and that it should therefore be rejected as scientifically untenable.

## 3. STIMULATED EMISSION

When Einstein wrote in 1949 his Reply to Criticisms for AEPS he stated that the problem of wave/particle dualism was "probably the most interesting subject" to discuss[20]:

"I now come to what is probably the most interesting subject which absolutely must be discussed in connection with the detailed arguments of my highly esteemed collegues Born, Pauli, Heitler, Bohr and Margenau. They are all firmly convinced that the riddle of the double nature of all corpuscles (corpuscular and undulatory character) has in essence found its final solution in the statistical quantum theory ... In what follows I wish to adduce reasons which keep me from falling in line ..."

In the same book AEPS Bohr[21] refers to Einstein's picture of the electromagnetic field by recalling some discussions they had had on such problems :
"The discussions, to which I have often reverted in my thoughts, added to all my admiration for Einstein a deep impression of his detached attitude. Certainly, his favoured use of such picturesque phrases as "ghost waves (Gespensterfelder) guiding the photons" implied no tendency to mysticism, but illuminated rather a profound humour behind his piercing remarks".

The idea of the empty-wave was however not only Einstein's. It was entertained in diffrent forms also by de Broglie and Bohr and it can even be said to belong to the line of thought of the Copenhagen school. In the previous section it has been seen that some important experiments on the nature of dualism can naturally be interpreted in terms of an empty wave accompanying the propagation of energetic and localized photons. This is true in particular for the Janossy-Naray experiment and for the neutron interferometric experiments. In neutron interferometry one can interpret the existing evidence by assuming that the neutron wave is split in two different parts following macroscopically separated trajectories and that a neutron "particle" carrying energy, momentum and mass remains associated only to one of these parts.

How can one ever hope to reveal the presence of a wave which does not carry energy and momentum ? This problem can have an answer if one notices that we do not only measure energy changing processes but probabilities as well : therefore the wave could reveal its presence by modifying decay probabilities of an unstable system.

Such an effect could probably be seen, if it exists, with existing technology using a modified version of the Pisa experiment which we describe below[22].

An atomic source emits a well-defined line of wave-length $\lambda$, selected with an optical filter. The source is so built that photons are randomly emitted, one by one, in such a way that there is never more than one photon at a time in the whole apparatus. Each photon is made to hit on a semitransparent mirror M which "splits it" into two parts a transmitted and a reflected one. As it is well known no coincidences above the casual background are observed if two phototubes are put on the transmitted and reflected paths. This means that the property of discharging a photomultiplier is not split by M but is made to choose between the two paths. More accurate analyses reveal that the whole energy $E = hc/\lambda$ is always found on one of the two paths. Whatever is the carrier of energy we see then that it is not split by the mirror. Something is however split as shown by the low intensity permanence of the interference figure. We wish to study the possibility that what is split are just Einstein's Gespensterfelder (or Bohr's virtual waves). Therefore we assume that the mirror sends the whole energy in one direction while it splits the wave in a transmitted and a reflected part. Suppose we put a phototube $P_1$

(FIG.5) on the reflected part and concentrate our attention on the cases where this counter does tick : according to the previous picture we say therefore that only a virtual wave is present in the transmitted beam.

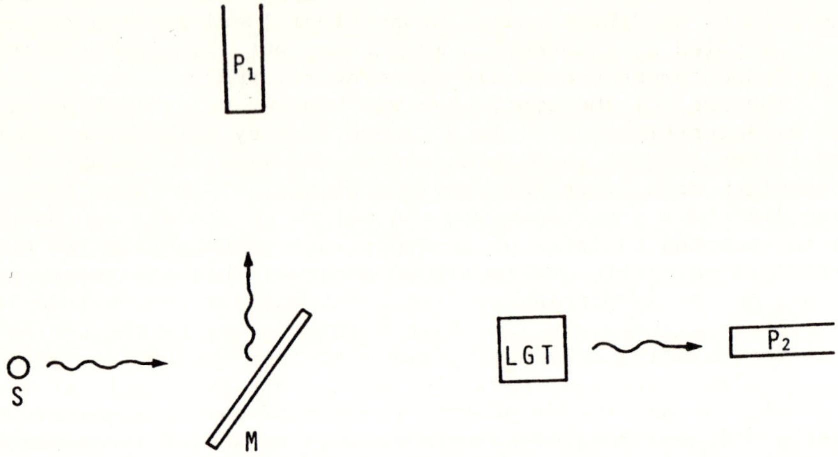


FIG.5 Experiment proposed by the author for the detection of empty waves : M is a beam splitter LGT a laser gain tube, $P_1$ and $P_2$ are two photomultipliers.

The transmitted beam is made to cross a laser gain tube (LGT) where the eventual wave-packet devoid of energy and momentum has a chance to reveal its existence by generating a zero energy-transfer stimulated emission. In fact, the molecules contained in LGT are kept in an excited band which includes the wave-length $\lambda$ of the incoming wave. The emitted photon could then be revealed by the phototube $P_2$ put after the LGT. In this way $P_1P_2$-coincidences would reveal the propagation of a zero-energy ondulatory phenomenon transmitted by M. The space-time propagation of this entity could be studied by checking that $P_1P_2$-coincidences disappear whenever an obstacle is put before LGT in the transmitted beam.

A positive answer from this experiment would definitely indicate that something propagates in space and time which does not carry energy-momentum but which can induce transitions. This would practically amount to the discovery of a new level of reality and would tend to put on a firmer basis the old controversy on the nature of the wave-particle dualism.

We will in the following review some further proposals for the detection of empty waves.

A proposal for the detection of empty waves was advanced in ref.22 and considered the use of superfluorescence (SF), that is of the coherent decay of many atoms predicted by Dicke[23] and studied experimentally by different groups[24]. Since SF concerns the coherent decay of something like $10^8$ atoms triggered by a single spontaneous emission due to quantum fluctuations of the electromagnetic field in vacuum, the natural idea would be to see whether an empty wave could give rise to SF-pulses. The basic apparatus would be different from the one shown in FIG.  only because the LGT would be substituted with a superfluorescent atomic cesium pencil.

Garuccio, Rapisarda and Vigier[25] presented a new experimental set-up, shown in FIG.6, which uses a very low intensity incoherent source of monochromatic photons. These photons are made to cross a semitransparent mirror $M_1$ : on the transmitted path a laser gain tube LGT is inserted for sake of amplification. Behind the LGT a second semitransparent mirror $M_2$ splits the beam again in two parts and one considers only those cases in which a phototube $PM_C$ put on the $M_2$-reflected part counts one photon.

This proposal depends on an idea opposite to that discussed in ref.22 namely on the assumption that the experiment proposed there has been performed and has failed (in the sense, at least, that a reasonably low upper limit has been set on the photon production rate from incoming empty waves).

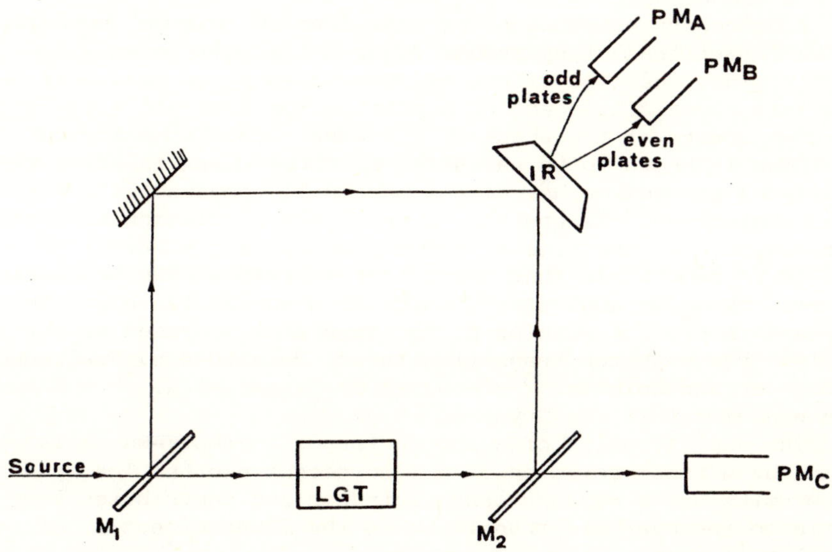


FIG.6 Set-up of the Garuccio-Rapisarda-Vigier experiment. $M_1$ and $M_2$ are semitransparent mirrors, $PM_A$, $PM_B$, $PM_C$, are photomultipliers and IR is an interference detecting apparatus.

If this were the case one could conclude that when a photon

has been detected by $PM_C$ no energetic photon is propagating on the $M_1$-reflected path. The photon detected by $PM_C$ is therefore necessarily connected to a situation where the single photon emitted by the source has been transmitted by $M_1$ and has crossed LGT. If it has generated an amplification it will happen rather often (33% of the cases, if only two photons come out of LGT) that the photon revealed by $PM_c$ is accompanied by a twin photon which is transmitted by $M_2$ and goes, following a certain optical trajectory, to an interference detecting system (IR) where it is superimposed to the $M_1$-reflected beam. Interferences should show up in IR if an empty wave has propagated in the $M_1$-reflected beam and could be detected by using in IR an instrument devised by Pfleegor and Mandel[26] and built with a stack of thin glass plates each of which has a thickness corresponding to a half fringe width. The plates are cut and arranged so that any photon falling on the odd plates are fed to one photomultiplier, $PM_A$, while photons falling on the even plates are fed to the other, $PM_B$. Detection of differences in the coincidence rates

$$PM_A \wedge PM_C \text{ versus } PM_B \wedge PM_C$$

would constitute evidence, in the stated conditions, that something not carrying energy-momentum is propagating in the $M_1$-reflected trajectory.

A necessary consequence of this idea is that an obstacle put on the latter trajectory should destroy the interference.

In order that the effect be observable it is obviously necessary that every single act of amplification does not give rise to a random change of the phase of the wave packet crossing LGT. This condition is known to be satisfied in those cases in which the amplifying gas volume has dimensions small compared with a radiation wavelength[27]. In general, a detailed calculation is needed in order to know the size of the phase-shifting effects.

It is remarkable that the latter experiment can be carried out with the same apparatus[28] which is used for the experiment proposed in [22]. The working of the apparatus is based on the Doppler free n-photon transition theory developed by Vasilenko, Chebotayev and Shishaev[29] discussed by Cagnac et al.[30] and verified experimentally by Grynberg et al.[31].

The amplification of a photon (or of an empty-wave) is revealed by a final emission of a fluorescent photon. The spontaneous emission is more or less isotropic and contributes very little to the over-all transition in the fixed direction of the stimulating beam.

A modification of the apparatus proposed by Garuccio, Rapisarda and Vigier can be used for implementing the experimental proposal put forward by Joâo and Maria Andrade e Silva[32].

The idea is again that the experiment proposed in ref.22 has been performed and has failed (no coincidences observed). One therefore concludes that when a photon has been detected by $PM_C$

no energetic photon is propagating on the $M_1$-reflected path. On the latter, however, two mirrors fend the wave to the interference device IR on which converge also the two beams generated by the further splitting of the $M_2$-reflected beam by a third semitransparent mirror $M_3$. With the numbering of the beams shown in FIG.7 we can therefore write for the over-all intensity I observed in the interference region

$$I = I_1 + I_2 + I_3 + 2\sqrt{I_1 I_2}\, \cos\delta_{12} \tag{3}$$

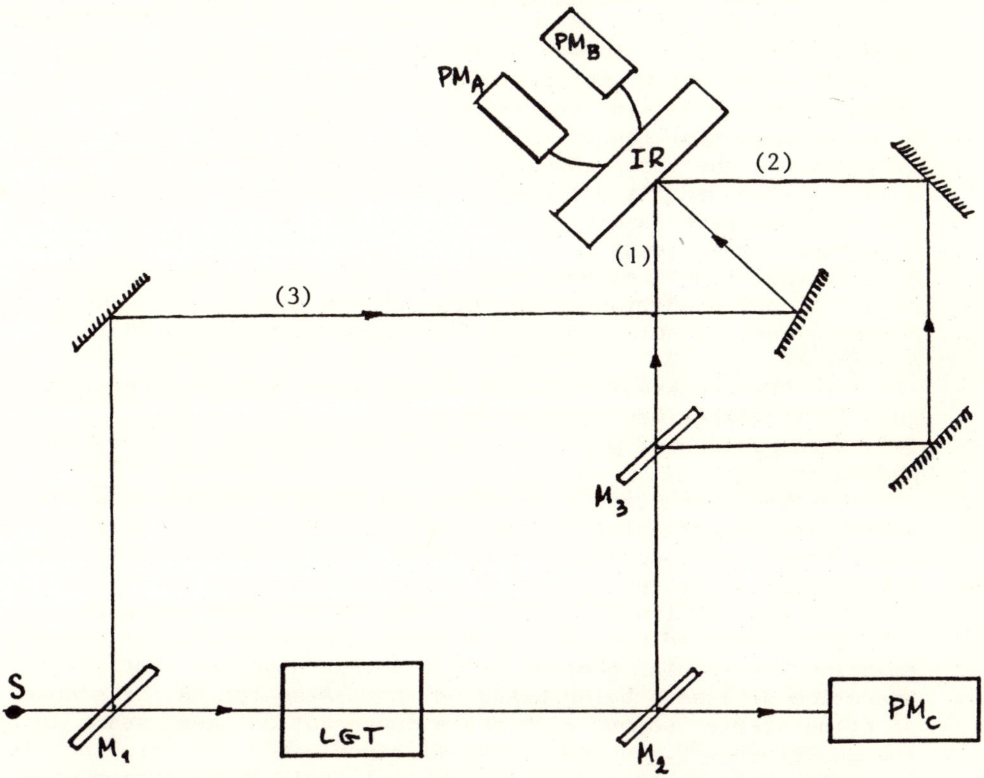


FIG.7 Set-up of the Andrade e Silva experiment. $M_1$, $M_2$ and $M_3$ are semitransparent mirrors, $PM_A$, $PM_B$ and $PM_C$ are photomultipliers and IR is an interference detecting apparatus.

where $I_i$ is the intensity of the i-th beam and $\delta_{ij}$ the relative phase-shift of the i-th and the j-th beams. In the previous equation the terms proportional to $\cos\delta_{13}$ and $\cos\delta_{23}$ have been put to zero, according to the idea that these phase-shifts assume random values in different events. Nevertheless the term $I_3$ must be present and has an observable effect on the so-called fringe visibility parameter V defined by

$$V = \frac{2\sqrt{I_1 I_2}}{I_1 + I_2 + I_3} \qquad (4)$$

which can be measured by dividing the coefficient of $\cos\delta_{12}$ by the non oscillating term in the interference figure.

Obviously V must change if also the $I_3$ term is made to disappear by putting an obstacle on the third beam.

Notice that in (3) and (4) $I_3$ is the average intensity of the third beam. The actual intensity for the events detected by the apparatus of FIG. is instead zero. Nevertheless, it appears physically plausible to assume that $I_3$ measures the "efficiency" of the corresponding empty wave to guide photons (coming from other trajectories) in the interference region.

The essential ingredient of the previous experiments on empty waves was the laser gain tube (LGT). Its main task was that of giving rise to additional photons generated by emissions stimulated by empty waves and/or by other photons. We could a-priori divide these experiments in three classes :

I- Only the stimulated emission of the photons is required with no additional condition on conservation of wave-length, phase and polarization. Of this class are the experiments proposed in [22] and [32].

II- Also the conservation of phase and wave-length is required : such is the experiment discussed in [25].

III- Conservation of polarization is required for the created photons.

Textbooks on stimulated emissions are not as sharp as one would wish in answering the question whether these three types of experiments can concretely be carried out. By and large it seems likely that all three types are actually possible. As we saw in the previous section stimulated emissions by empty waves are strongly indicated by existing experimental evidence. Stimulated emissions by regular photons are of course standard stuff, the operation of lasers being based on them. Also for the existence of stimulated emissions with phase conservation there seems to be enough evidence[33], so that class II experiments are in principle possible. Less discussed in the literature is the question whether polarization-conservation is valid in single photon amplification, even though any semiclassical model would lead to a straightforward positive answer.

It is the merit of Nick Herber[34] to have called attention to the startling consequences of the possibility of class III experiments.

Suppose that a very low intensity photon beam ("one photon at a time") is made to cross a LGT (FIG.8) where every incoming photon generates, say, hundred identical photons. This coherent flash of light is made to arrive on a neutral beam splitter, which does not alter the photon polarization, but simply operates a mechanical splitting of the beam into 50 transmitted and 50 reflected photons (on average). The transmitted photons are analyzed by $A_1$ in terms of two linear and orthogonal polarization states (x- and y- polarization) which are respectively detected by the counters $C_x$ and $C_y$. The reflected photons are instead analyzed by $A_2$ in terms of right-handed circular (R) and left-handed circular (L) polarization states detected by the counters $C_R$ and $C_L$, respectively, in the following way : the reflected photons cross a unitably oriented λ/2-plate which transforms circular polarization into linear, and viceversa. In this way incoming R(L) groups of 50 photons cross the calcite prism and are all transmitted (reflected) and then all detected by $C_R$ ($C_L$). Instead incoming x and y groups of 50 photons are split by the calcite prism in equally populated flashes of transmitted and reflected light (containing each 25 photons, on average). The situation is summarized in the following table :

| Single incoming photon (polarization state) | 100 photons outgoing from LGT (polarization state) | Response of counters | | | |
|---|---|---|---|---|---|
| | | $C_x$ | $C_y$ | $C_R$ | $C_L$ |
| x | x | 50 | 0 | 25 | 25 |
| y | y | 0 | 50 | 25 | 25 |
| R | R | 25 | 25 | 50 | 0 |
| L | L | 25 | 25 | 0 | 50 |

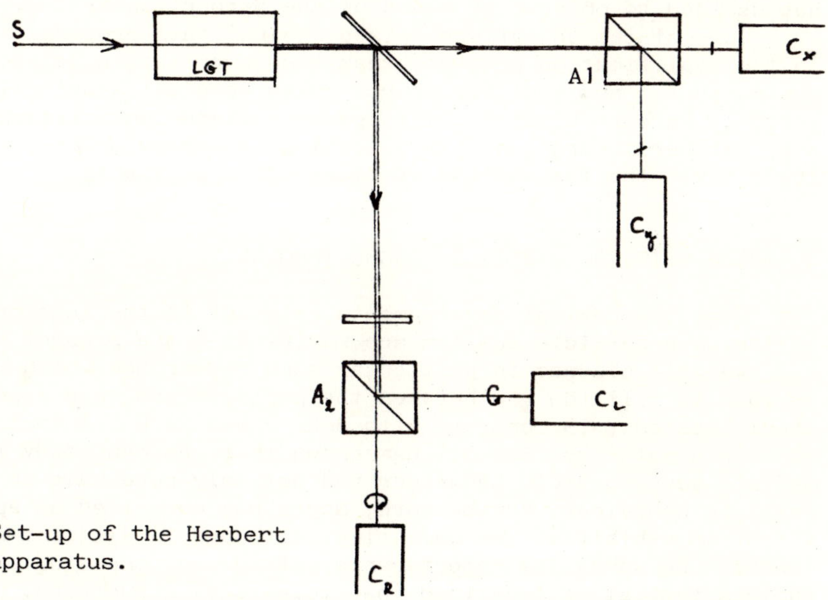


FIG.8 Set-up of the Herbert apparatus.

It thus appears evident that if class III experiments are

indeed possible one can distinguish experimentally between two unpolarized beams of light, one of them composed of 50 % x and 50 % y photons, and the other one composed of 50 % R and 50 % L photons even though there two beams are described quantum-mechanically by the same density-matrix.

In fact an R-L beam gives rise to symmetrical responses of $C_x$ and $C_y$ but to asymmetries of $C_R$ and $C_L$. Conversely an x-y beam generates symmetry (asymmetry) in $C_R$ and $C_L$ (in $C_x$ and $C_y$), as it is evident from the preceding table.

The insufficiency of the quantum mechanical treatment of unpolarized beams becomes obvious if one considers that class III experiments, if actually feasable, allow one to measure on a single photon (or at least on identical duplicates of a single photon) several non commuting observables. In the original Herbert proposal shown in fig.8 one can measure linear and circular polarizations, which are described quantum-mechanically by well-known noncommuting operators. More generally, one could conceive to measure several noncommuting observables "on the same initial photon" by inserting several mechanical beamsplitters on the trajectory of the amplified flath.

One of the amazing consequences of this hypothesis in the possibility of transmit superluminal signals if the quantum mechanical correlations for EPR-type experiments are correct : it is in fact enough that one of the two observers which make experiments on two correlated photons chooses to measure, for instance, a circular polarization in order to have the other observer capable to detect the type of polarization which the first one has decided to observe by means of the Herbert apparatus.

Some of the objections raised against this proposal (noise in the amplification process, weak time correlation between incoming photon and amplified burst) have been satisfactorily answered by Herbert himself : it appears nevertheless certain that some further theoretical clarification is needed before one can fully trust the feasability of class III experiments.

## 4. WHICH SLIT THE PARTICLE PASSES THROUGH ?

Some theoretical developments relevant to the interpretation of the wave-particle dualism are reviewed in the present section.

We will discuss in particular some researches which have done much to fill the gap between the particle and wave aspects of atomic systems, in contrast with Bohr's and Fock's formulations of complementarity. Bartell has shown that one can study intermediate wave-particle behaviour and not only pure wave or pure-particle behaviour. Furthermore, Noyes has conceived an apparatus which can exhibit at the same time, although in disjoint sets of events, the wave-like interference pattern and the single slit choices typical of localized (particle-like) structures.

The possibility to build an apparatus which can reveal with minor modifications (shifting of a screen) either the wave aspect minor modifications (shifting of a screen) either the wave aspect or the particle aspect of light has been discussed by Wheeler[35] and generalized by Bartell[36]. In Wheeler's proposal the experimenter is free to choose either to observe which slit has been traversed or to accumulate interference fringes. Bartell was able to modify the proposed apparatus in such a way as to detect intermediate particle-wave behaviour. A similar proposal has been put forward also by Wooters and Zurek[37]. Bartell's apparatus is shown in FIG.9.

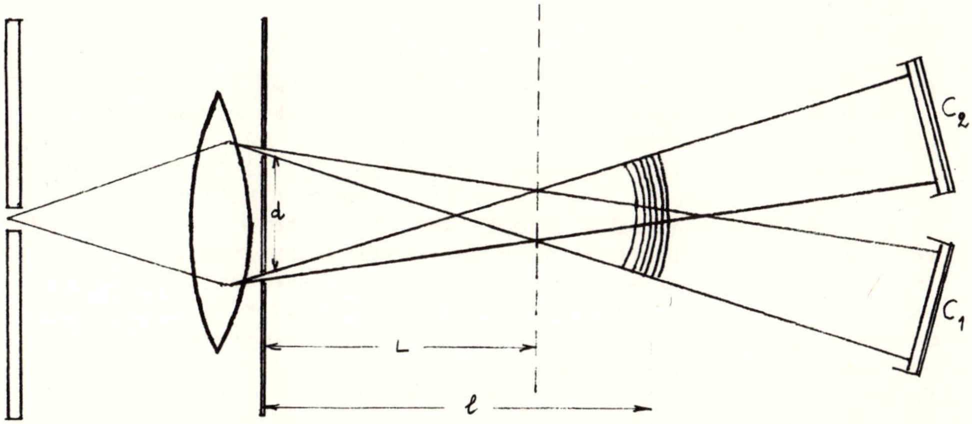


FIG.9 Bartell's apparatus. $C_1$ and $C_2$ are phototubes.

The interference screen is placed at the focal plane of the lens at a distance L from the slits 1 and 2. On it the Fraunhofer interference pattern can be observed. If, instead, it is desired to observe the trajectories of the photons, the screen is taken away and the photons are observed by the counters $C_2$ and $C_1$ which, because of the particular optical geometry of the apparatus, can only detect light from the lower and upper slit, respectively.

The apparatus incorporates Gaussian transmission filters which eliminate the secondary diffraction maxima characteristic of slits with abrupt edges.

If the movable screen is put at a distance $\ell > L$ from the slits one observes in general intermediate particle-wave behaviour. The intensity I(x) at a position x on the screen (the origin is the symmetry point) is predicted by elementary diffraction theory to be

$$I(x) = K[F(x)]^2 \cosh(ax)[1 + V(x) \cos bx] \qquad (5)$$

where K, a and b are constants dependent only on the geometry of the apparatus and, in the case of K, on the incoming light's intensity,

$$F(x) = \exp(-x^2/2\sigma^2)$$

is the slit form factor, $\sigma$ being a constant, and V(x) is the important "fringe-visibility" function which is related to the slits transmission probabilities $P_1(x)$ and $P_2(x)$ for radiation reaching coordinate x via

$$V(x) = 2[P_1(x)P_2(x)]^{1/2} \quad (6)$$

The expression (5) for I(x) is valid only in the case $\ell \gg d \gg \lambda$ where d and $\ell$ are defined in fig.12 and $\lambda$ is the radiation's wavelength.

If the slit discrimination ratio $P_1(x)/P_2(x)$ is selected to be 4 : 1, for example, the resultant fringe visibility is only 20 % lower than that of an ideal two-slit pattern. In other words, the interference figure at $\ell$ is very well visible in points where the probabilities of an incoming photon to have crossed slit 1 and slit 2, respectively, are 80 % and 20 %. Even if these probabilities are taken 95 % and 5 % the visibility parameter, as one can check from (6), reduces to about .45, a little more than a factor 2 below its optimal value 1. We see thus that in an experiment of this type ondulatory behaviour as reflected in the interference fringes goes together with an "almost" particle behaviour as reflected in the near unit value of $P_1(x)$. This result is obtained uniquely from standard quantum mechanics without any particular interpretation. The natural hint seems to be that particles and waves cohexist objectively, unless one is willing to accept the idea that atomic reality is different according to whether the screen in fig.9 is present or not.

Another indication of the fact that the division of the particle and wave aspects is less sharp than expected from the usual "complementarity" interpretation of quantum mechanics has been produced by Pierre Noyes[38]. In his operational analysis of the double slit experiment Noyes introduces a particular device, a kind of "semi-transparent" counter which is supposed to be able to detect a particle crossing it with a certain probability p and instead not to detect it with probability 1-p. In the latter case one assumes naturally that no random phase-change of the wavefunction takes place. This counter is supposed to be incorporated in both slits in such a way that a particle crossing a slit must always cross also the counter.

Noyes is thus able to predict that three different statistical distributions will be detected in an array of (small) particle detectors placed at sufficient distance behind the two slits and playing the role of the "second screen" of the usual double-slit apparatus. In this way a statistical distribution showing an interference pattern arises which is composed of events in which neither of the slit detectors fires, while other two distributions arise (without interference) out of the events in which one or the other of the slit detectors fires in addition to one of the detectors in the final array.

Noyes' interesting apparatus seems to contradict some well known formulations of the complementarity principle, like Fock's idea of relativity with respect to the observational means.

Complementarity states that an experiment allowing one to determine which slit the particle passes through is complementary to (i.e. excludes) an experiment yielding the double slit pattern. Noyes shows instead that the same apparatus can lead to complementary outcomes like (a) revealing which slit the electron passed through (particle behaviour), and, (b) revealing an interference pattern (wave behaviour).

Similar considerations can be made about an interpretation of the Schrödinger equation which was originally proposed by de Broglie[39] and Bohm[40].

Considering Schrödinger's equation for a particle with mass m in an external potential V

$$ih \frac{\partial \psi}{\partial t} = - \frac{h^2}{2m} \nabla^2 \psi + V\psi \tag{7}$$

and putting

$$\psi = R \exp(iS/h) \tag{8}$$

one can easily show that the complex equation (7) is equivalent to the two following real equations

$$\frac{\partial R^2}{\partial t} + \vec{\nabla}.(R^2 \frac{\vec{\nabla} S}{m}) = 0 \tag{9}$$

$$\frac{\partial S}{\partial t} + \frac{1}{2m}(\vec{\nabla} S)^2 + V - \frac{h}{2m} \frac{\nabla^2 R}{R} = 0 \tag{10}$$

Planck's constant does not enter at all in eq.(9) while in eq.(4) it enters only in the last term

$$Q = - \frac{h^2}{2m} \frac{\nabla^2 R}{R}$$

Obviously Q has dimensions of energy : it is called "the quantum potential". Eq.(9) and (apart from Q) eq.(10) are well known in classical physics : the first one is a continuity equation for a fluid of density $\rho(xyzt)$ and velocity $\vec{v}(xyzt)$ given by

$$p(xyzt) = [R(xyzt)]^2 \; ; \; \vec{v}(xyzt) = \frac{1}{m} \vec{\nabla} S(xyzt)$$

In fact, since $R^2 = |\psi|^2$, (9) is precisely the well known continuity equation.

It is usually interpreted in terms of conservation of total probability, with $R^2$ assumed to be the probability density and $\vec{v}$ the local velocity of the probabilistic fluid.

For a given wave-function $\psi$ one can then consider a statistical ensemble of particles randomly distributed within their all identical wave functions with a density (over the statistical ensemble) given by $|\psi|^2 = R^2$ and with a local velocity of the probabilistic fluid given by $\vec{v} = \frac{1}{m}\vec{\nabla}S$. In this way $\vec{v}(xyzt)$ becomes very naturally the velocity of the particle located in the point xyz at atime t. Obviously then the momentum of the same particle is $m\vec{v} = \vec{\nabla}S$ : the second and third terms of eq.(10) are therefore the kinetic and the potential energy of the particle located in the point xyz at time t. But this equation has then exactly the structure of the Hamilton-Jacobi equation where S(xyzt) in Jacobi's function and gives the first term of (10) with its partial derivative with respect to time and the particle momentum with its gradient changed of sign. The only difference with respect to the classical Hamilton-Jacobi equation is represented by the quantum potential Q which is absent in classical physics. The quantum interpretation of this term is, according to de Broglie, the following. If the particle is embedded in a plane wave

$$\psi_0(xyzt) = R_0 \exp\{2\pi i(\vec{u}.\vec{x}-\nu t)\}$$

with $R_0$ constant, the quantum potential obviously vanishes. No quantum force is therefore acting on the particle and this can propagate on a straight line accordingly to the principle of inertia. The situation changes however if the plane wave hits on a screen with a slit : the well known phenomenon of diffraction takes place and beyond the slit one has to do with a more or less conical wave. The amplitude of such a wave is not a constant in space anymore and this means that one has now in general $Q \neq 0$. This implies that the wave exerts a force on the particle and bends its path even in those cases in which it passes through the slit : but this is a welcome feature since energy (associated to the particles) is found on a second screen to have a diffraction -like distribution with maxima and minima which in all cases occupies a surface larger than that covered by the slit on the first screen.

In this way one can assume that the quantum potential Q represents in all case the action of the wave on the particle. One can show that the diffraction -like distribution of the particles is indeed generated by Q for a uniform distribution of particles on the slit.

The de Broglie-Bohm approach to quantum physics has been recently reconsidered by Philippidis, Dewdney and Hiley[41] who showed in particular how a detailed calculation of the quantum potential Q for the two-slit experiment can give rise to interference without the need to abandon the notion of a well defined particle trajectory. Their result is that every single particle follows a well defined trajectory and that, in spite of this, if the initial particle distribution over the two slits is quantum mechanical the presence of the quantum potential in (10) ensures that the particle probability density is equal to $|\psi|^2$ for all subsequent times.

The numerical calculations were performed using an electron's energy of 45 keV, the separation between the slits $10^{-4}$ cm and the slit width $10^{-5}$ cm. In Fig.10 we reproduce the particle trajectories calculated by Philippidis, Dewdney and Hiley. Initially the trajectories from each slit fan out in a manner consistent with diffraction from a single slit. The subsequent "kinks" in the trajectories coincide with the troughs in the quantum potential. They arise because, when a particle enters the region of a trough, it experiences a force in the direction perpendicular to the symmetry line of the slits which accelerates the particle rapidly through the trough into a plateau region where the forces are again weak. In consequence most of the trajectories run along the plateau regions giving rise to the bright fringes, while the troughs coincide with the dark fringes.

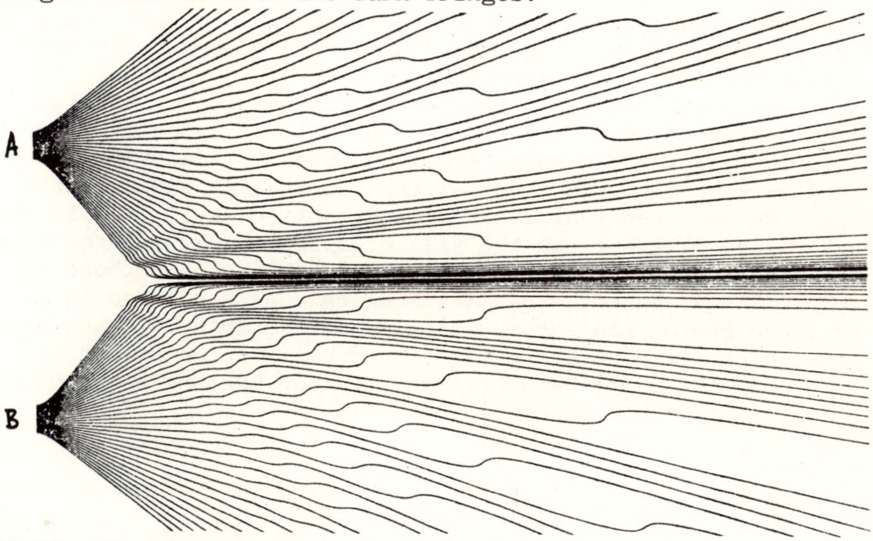


FIG.10 "Particle trajectories" from slits A and B as calculated by Philippidis, Dewdney and Hiley.

An interesting feature of this result is that no trajectory crosses the symmetry plane : this means that all particles observed on the lower (upper) half of the second screen come from the lower (upper) slit, a result charged of profound philosophical implications for those who believed that the double-slit experiment did not allow any simple interpretation.

The picture of the double-slit experiment is very simple for Philippidis, Dewdney and Hiley :

"The approach through the quantum potential retains a pointlike particle and each particle in the original ensemble follows a well defined trajectory which passes through one or the other of the slits. This ensemble produces the required interference pattern and, at the same time, shows that the final position of the particle on the screen allows us to deduce through which slit it actually passes. Thus it is possible to retain the trajectory concept and, at the same time, account for the interference. There is no longer a mystery as to how a single particle passing through one slit "knows" the other slit is open. The information is carried by the quantum potential so that we no longer have a conceptual difficulty in understanding the results obtained in very low intensity interference experiments".

A further interesting feature of this calculation is that the quantum potential turns out to contribute extremely little to the total energy of the electrons. Its absolute maximum value is only about $10^{-4}$ eV, whereas the kinetic energy of the electrons is about 45 keV. While the average value of $|Q|$ is probably at least two orders of magnitude below its maximum value it should be remembered that even a wave energy fourteen orders of magnitude below the particle energy would have been observed in recent neutron interferometric experiments.

A highly interesting feature of these results is the fact that no particle trajectory crosses the axis of symmetry between the slits : exactly the same result had been obtained in 1976 by Richard Prosser[42] in his study of the double-slit experiment with photons. He observed that lines of constant phase of the electromagnetic wave always cross the axis of symmetry between the slits at right angles and hence the flow lines, which are orthogonal to the phase lines, cannot cross it. From this he concluded that there is no photon trajectory which can cross this axis and that photons which illuminate the right-hand part of the screen always pass through the right-hand slit and that no photons pass into this region via the left-hand slit.

REFERENCES

1. A. Einstein ; Ann. der Phys. 18, 639 (1905)
2. A. Einstein ; Ann. der Phys. 17, 132 (1905)
3. Letter from A. Einstein to L. de Broglie (1954). Quoted in : L. de Broglie ; Ann. Fond. L. de Broglie 4, 13 (1979)
4. L. de Broglie ; Ann. Fond. L. de Broglie 2, 1 (1977)
5. R.L. Pfleegor and L. Mandel ; Phys. Rev. 159, 1084 (1967)
6. U. Bonse and H. Rauch, editors ; Neutron Interferometry ; Clarendon Press, Oxford (1979)
7. L. Janossy and Zs. Naray ; Suppl.Nuovo Cim. 9, 588 (1958)
8. H. Rauch, W. Treimer and U. Bonse ; Phys. Lett. 47A, 369 (1974)
9. A. Adam, L. Janossy and P. Varga ; Ann. der Phys. 16, 408 (1955)
   F. Davidson and L. Mandel ; Phys. Lett. 25A, 700 (1967)
   J.F. Clauser ; Phys. Rev. D9, 853 (1974)
10. L. de Broglie ; Comptes rendus 177, 548 (1923). Reprinted in

Annales de la Fond. L. de Broglie 7, 101 (1982)
11. N. Bohr, H.A. Kramers and J.C. Slater ; Phil. Mag. 47, 785 (1924)
Reprinted in SQM : the quoted sentence is at page 166
12. SQM, p. 167
13. G.D. Blake and D. Scarl ; Phys. Rev. A19, 1948 (1979)
14. P.W. Milonni and P.L. Knight ; Opt. Comm. 9, 119 (1973)
15. H. Kuhn ; Jour. Chem. Phys. 53, 101 (1970)
16. E.T. Jaines ; in Proc. Third Rochester Conference on Coherence and Qunatum Optics, ed by L. Mandel and E. Wolf ; Plenum Press, New York (1973), p. 35
17. R. Hanbury Brown and T.Q. Twiss ; Nature 177, 27 (1956)
18. L. Mandel in Progress in Optics XIII, E. Wold, ed., North-Holland (1976), p. 27
19. A. Landé ; Quantum Mechanics in a New Key ; Exposition Press, New York (1973), p. 4
20. Albert Einstein : Philosopher-Scientist ; P.A. Schilpp, editor, Library of Living Philosophers (1970), p. 666
21. Ibid., p. 206
22. F. Selleri ; Ann. Fond. L. de Broglie ; 7, 45 (1982). See also : On the Direct Observability of Quantum Waves, Found. Phys., in print (1982)
23. R.H. Dicke ; Phys. Rev. 93, 99 (1954)
24. E.g. see Q.H.F. Vrehen, in Proceedings of the Fourth Int. Conf. Rottach-Egern, June 11-15, 1979 ; H. Walther and K.W. Rothe, editors ; Springer-Verlag, Heidelberg (1979)
25. A. Garuccio, V. Rapisarda and J.P. Vigier ; Phys. Lett. 90A, 17 (1982)
26. See ref. 5
27. See ref. 23
28. A. Gozzini ; Communication at the Symposium on Wave-Particle Dualism (in honor of Louis de Broglie's 90th birthday), Perugia, April 22-30 (1982)
29. L. S. Vasilenko, V. P. Chebotayev and A. V. Shishaev ; JETP, Lett. 12, 113 (1970)
30. B. Cagnac, G. Grynberg and F. Biraben ; Jour. de Phys. 34, 845 (1973)
31. G. Grynberg, F. Biraben, M. Bassini and B. Cagnac ; Phys. Rev. Lett. 37, 283 (1976)
32. J. and M. Andrade e Silva ; Comptes rendus 290, 501 (1980)
34. N. Herbert ; FLASH-A superluminal communicator based upon a new kind of quantum measurement ; preprint, Boulder Creek (1981)
33. E.g. see : M. Sargent III, M.O. Scully and W.E. Lamb, Jr. ; Laser Physics, Addison-Wesley, Reading (1974)
35. J.A. Wheeler ; in Proceedings of the New Orleans Conference on the Mathematical Foundations of Quantum Theory ; edited by A.R. Marlow ; Academic, New York (1978)
36. L.S. Bartell ; Phys. Rev. D21, 1698 (1980)
37. W.K. Wooters and W.H. Zurek ; Phys. Rev. D19, 473 (1979)

38. P. Noyes ; Communication at the Symposium on Wave-Particle Dualism (in honor of Louis de Broglie's 90th birthday), Perugia, April 22-30 (1982)
39. L. de Broglie ; J. Physique 6, 225 (1927)
40. D. Bohm ; Phys. Rev. 85, 166 and 180 (1952)
41. C. Philippidis, C. Dewdney and B.J. Hiley ; Nuovo Cimento, 52B, 15 (1979).

# ON THE POSSIBILITY OF REALISING A LOW INTENSITY INTERFERENCE EXPERIMENT WITH A DETERMINATION OF THE PARTICLE TRAJECTORY

Adriano Gozzini

Istituto di Fisica - Università di Pisa - Italy

It has been proposed to perform a single photon interference experiment in a two-arm interferometer so as to determine along which arm the particle has passed and to ascertain whether the interference persists under such conditions (ref. 1).

To this end it has been suggested that one could use an LGT (Laser Gain Tube), that is to say a system that emits a pair of identical photons from a single incident photon, by means of stimulated emission. One of the photons is used to indicate the path chosen by the photon, and the other for the interference.

Since the incident photon and each of the emitted photons are physically indistinguishible, each photon of the pair has the same interference properties as the incident one. The proposed experiment is illustrated schematically in fig. 1, in which a mirror of a Mach Zender interferometer is substituted by the beam splitter $BS_2$. Another beam splitter $BS_4$ is inserted in the arm 2 of the interferometer in order to equalize the intensities of the beams recombined at the exit beam splitter $BS_3$. All of the beam splitters have R=T=1/2. A, B and C are photomultipliers; B and C are placed in the constructive and destructive interference regions, respectively, so that the probability that a photon incident on $BS_1$ be received by A, B or C is 1/4, 1/2 and 0, respectively.



*S. Diner et al. (eds.), The Wave-Particle Dualism, 129–137.*

Let us now insert in arm 1 an "ideal duplicator", i.e. a light amplifier such that in the absence of light, at its entrance no light is emitted at its output, and which emits a pair of identical photons when a single photon crosses it. Under these conditions, if an incident photon is reflected from $BS_1$ it is received by the photomultiplier B, whereas if it is transmitted, then a pair reaches $BS_2$. In one third of the events both photons of the pair are reflected towards $BS_3$ and no signal is furnished by A; in a third of the events both photons are transmitted and no signal is furnished by B or C, in one third of the events one photon of the pair is transmitted and the other is reflected. In the absence of noise, a coincidence AB or AC can only be produced by a pair created by the LGT that has been subdivided by $BS_2$ in a reflected and a transmitted photon. The signals furnished by A feed two coincidence circuits, together with the respective signals furnished by B and by C.

If one observes coincidences AB but not AC, then interference occurs also when the trajectory of the particle inside the interferometer is known, whereas if an equal number of coincidences AB and AC occur, then we have a proof that under these conditions the interference is destroyed.

The feasibility of the experiment described above is subject to the possibility of realizing an LGT having features which are as close as possible to those of the ideal duplicator.

A light amplifier is usually made from a laser tube maintained below the threshold, that is to say by a system of atoms maintained in a steady state such that the population of the excited level is greater than that of the lower state of the transition between which the amplifier works. Clearly such a system emits light, even in the absence of light at its entrance, because of the spontaneous emission of the excited atoms.

If the predicted amplification is $1+\varepsilon$ (with $\varepsilon \ll 1$ so as to avoid the emission of more than two photons corresponding to one incident photon) and N is the stationary population of the excited state, then a beam of n photons per second with an angular spread $\Omega$ which is incident on the instrument gives rise to a beam having the same angular spread, composed of $n(1-\varepsilon)$ photons which have not undergone interaction, $n\varepsilon$ pairs and $NA\Omega/4\pi$ photons resulting from spontaneous decay to the lower state. Such photons produce coincidences AC or AB with each other or with the non duplicated photons of the interfering beam. These constitute noise [1]. Since $A \sim 10^8$ such noise is eccessive if n is composed of a few pairs per second. It therefore appears to be very difficult to realize such an experiment using a normal LGT amplifier as a duplicator.

However, in two experiments recently reported in the literature, it appear that a photon duplicator having acceptable characteristic can indeed be realized (refs.2-4)

In such experiments a non-linear process involving three photons has been investigated, by means of which an atom passes from the ground to an excited state. The process is sketched in figs. 2 and 3. A two level system g , e whose energetic separation is $\hbar\omega_0$ is irradiated with two laser beams having frequencies $\omega_1=\omega_0+\delta$ and $\omega_2=\omega_0+2\delta$, respectively, so as to satisfy the resonance condition $2\omega_1-\omega_2=\omega_0$ , with $\delta \gtrless 0$ and $|\delta| \gg \Gamma$ , where $\Gamma$ is the natural width of the level e .

Rare processes are produced in which the atom absorbs a photon $\hbar\omega_1$ , thus passing to a virtual state B, from which it decays by stimulated emission of a photon $h\omega_2$ to a second virtual state C, and then passes to the excited level e absorbing a second photon $\hbar\omega_1$ . Finally, it decays from the level e by spontaneous emission to the ground state, emitting a photon $\hbar\omega_0$ . In the process, there is therefore absorption of the beam $\omega_1$ and amplification of the beam $\omega_2$ .

The probability of the process is, per atom, proportional to $I_1^2 I_2/\delta^4$ where $I_1$ and $I_2$ are the intensities of the two beams in the interaction region.

A light amplifier utilizing such a process presents, with respect to a normal LGT, important advantages:

a) the possibility of focussing the beams in a very small volume ( $\sim 10^{-8} cm^3$) which therefore contains a small number N of atoms excited to the intermediate level B, all of which are useful for the amplification;
b) a low noise level, due essentially to the photons of the pumping beam, scattered from the walls of the container and diffused by the Rayleigh scattering,

and which are received by A. These can produce coincidences with the photons received by the other phototubes;

c) the possibility, by means of an appropriate detuning $\delta$, of filtering the frequency signal $\omega_2$ from the noise at the frequencies $\omega_1$ and $\omega_0$ ;

d) the noise consists mainly of coincidences between the n $(1 - \mathcal{E})/2$ unamplified photons per second transmitted to A and the photons $\omega_1$ resulting from Rayleigh scattering. The number of photons of frequency $\omega_0$ that are emitted in the whole solid angle of $4\pi$ sterad-units is equal to the total number of pairs produced. They can be collected and utilized to discriminate the coincidences due to the pairs from the background.

Using the above process Cagnac et al. (ref.3) have studied the hyperfine structure of the $D_1$ line of Na, with a geometry in which the total momentum exchanged by the three photons with the atom is zero (fig.2b). Under such conditions the process is Doppler free. The hyperfine structure has been studied by varying the detuning $\delta$ so as to obtain the resonance condition successively for the different hyperfine components of the transition $^2S_{1/2} \rightarrow {}^2P_{1/2}$ . The detection is achieved through the fluorescence at the frequency $\omega_0$ due to the spontaneous decay from the $^2P_{1/2}$ level (see fig.2a).

Cohen-Tannoudji et al. (ref.4), on the other hand, have studied the process by irradiating the system with a single beam of frequency $\omega_L = \omega_1 = \omega_0 + \delta$. In this case the three photon process occurs via the absorption of two photons $\hbar\omega_1$ of the exciting beam and the spontaneous emission of a photon $\hbar\omega = \hbar(\omega_0 + 2\delta)$. The analysis of this experiment shows that, using the three photon process to realize the LGT amplifier, the single photon interference experiment previously described is surely feasible.

In the latter experiment the time correlation between the two photons $\hbar\omega, \hbar\omega_0$ spontaneously emitted in a single process has been investigated. A strontium atomic beam is excited by the off resonance laser. The spectrum of the fluorescence light consists of a triplet, whose central line has the laser frequency $\omega_L = \omega_0 + \delta$ , and whose intensity is $10^4$ times that of the two sidebands, of frequencies $\omega_0$ and $\omega_0 + 2\delta$, respectively. These sidebands originate from the three photon process where the transition between the two virtual levels B, C (fig. 3) occurs via spontaneous emission, whereas the central line is due to Rayleigh scattering.

The values of the parameters of interest in this experiment are as follows.

Resonance frequency of the transition ($^1S_0 \rightarrow {}^1P_1$):

$$\omega_0/2\pi = 6.5 \times 10^{14}\,Hz \quad (\lambda_0 = 4607\,\text{Å})$$

Detuning of the exciting light from the resonance:

$$(\omega_L - \omega_0)/2\pi = \delta/2\pi = 4 \times 10^{12}\,Hz \quad (\Delta\lambda = 28\,\text{Å})$$

Natural width of the resonance:

$$\Gamma = 2 \times 10^{8}\,Hz$$

Intensity of the exciting light: 2w, focalized into a $2.10^{-8}$ volume region of the atomic beam enclosing $2.10^4$ atoms

Corresponding value of the Rabi mutation frequency:

$$\omega_R/2\pi = \mu E/\hbar = 8.10^{10}\,Hz$$

Under these conditions the number of processes has been 2.10 /sec i.e. the probability that an atom jump to the excited state absorbing two photons $\hbar\omega_L$ and spontaneously emitting one photon $\hbar\omega$ is 1 per atom per second. The number of the fluorescence photons at the laser frequency $\omega_L$ due to the Rayleigh scattering has been $2.10^8$ /sec, i.e. $10^4$ per atom per second.

Let us now consider the interferometer of fig. 1, in which the LGT duplicator is obtained by using the interaction region of this experiment. In the absence of any light at the interferometer entrance and any filter interposed, the total number of the fluorescence photons received by the photomultipliers A, B and C is $\mathcal{N}_0 = 2.10^4\,\Omega/4\pi$ per second, at each of the frequencies $\omega$ and $\omega_0$, and $10^4 \mathcal{N}_0$ at the laser frequency $\omega_L$. Here $\Omega$ is the solid angle, which is assumed be the same for the three photomultipliers, from which the sensitive surfaces $S$ of the photomultipliers are seen from the interaction region. With S = 0.1 cm$^2$ and the distance of the photomultipliers from the LGT equal to R = 2 m the phototubes A B C will receive, respectively, 20,10 and 10 photons/second.

Let us now consider the interferometer fed by a beam of intensity 2n photons/sec., of frequency $\omega = \omega_0 + 2\delta$. The n photons/sec. beam transmitted by B cross the interaction region and is amplified.

The probability of the three photon process has been calculated (ref.2) and results to be

$$\gamma = \frac{2}{\Gamma} \frac{1}{\varepsilon_0^3} n_1^2 n_2 \frac{\omega_1^2 \omega_2}{\delta^4} \langle\mu\rangle^5 / \hbar^3 ,$$

where $n_1$, $n_2$ are the number of photons per unit volume at the two frequencies respectively.

Inserting the appropriate values given in ref. 4 one gets $\gamma = 8\times 10^{-7} n_2$. Therefore the number of pairs produced per second results to be

$$\mathcal{N}_p = N\gamma = 1.6\times 10^{-3}\, n / \Sigma c ,$$

where $\Sigma$ is the cross section of the interaction region in the direction of the n photons/second beam. Assuming this region, whose volume is $2\times 10^{-8}$ cm$^3$, to be of spherical shape we have $\Sigma c = 10^4$ and $\mathcal{N}_p = 1.6\times 10^{-7}$ n.

The coincidence signals are therefore.

$$\mathcal{N} = \frac{1}{3} \mathcal{N}_p \eta^2 = 5\cdot 10^{-8} n \eta^2 / \mathrm{sec}$$

whereas the number of noise coincidences $\mathcal{N}_{AC}$ per second results to be

$$\mathcal{N}_{AC} = 10 \times \frac{n}{2} \eta^2 \Delta$$

where $\eta$ is the photomultiplier efficiency and $\Delta$ is the coincidence width.

The ratio of the signal to noise coincidences does not depend on the intensity n and is greater than 1 if $\Delta < 10^{-8}$ sec. Putting $n = 10^7$ /sec, $\eta = 0.1$ and $\Delta = 10^{-8}$ sec let us measure the number of coincidences $A \wedge C$ which occur during T seconds. Their number is $5\times 10^{-2}$ T or $7.5\times 10^{-2}$ T according as to whether the interference persists or is destroyed under these conditions. Of course the signal to noise ratio can be easily improved by a large factor by reducing the detuning $\delta$ or by increasing the exciting power, because the noise due to the Rayleigh scattering varies as $1/\delta^2$ whereas the signal varies as ( $I_1^2/\delta^4$ ). In conclusion, we have wished to demonstrate that the experiment suggested by Garuccio, Rapisarda and Vigier is feasible by simply repeating the experiment of ref. 4.

If in the experiment the interference is destroyed, by commutating the beam transmitted by $BS_2$ to the photomultiplier A with that, of the same intensity, reflected by $BS_4$, the information is erased and the interference must reappear. This perhaps will be the most astonishing result of an experiment of this kind.

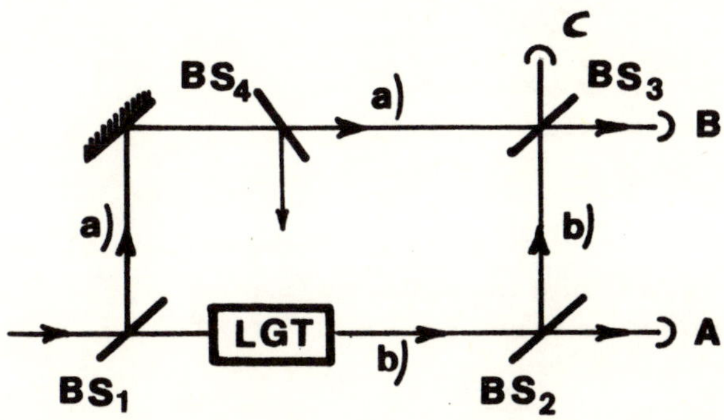


fig. 1

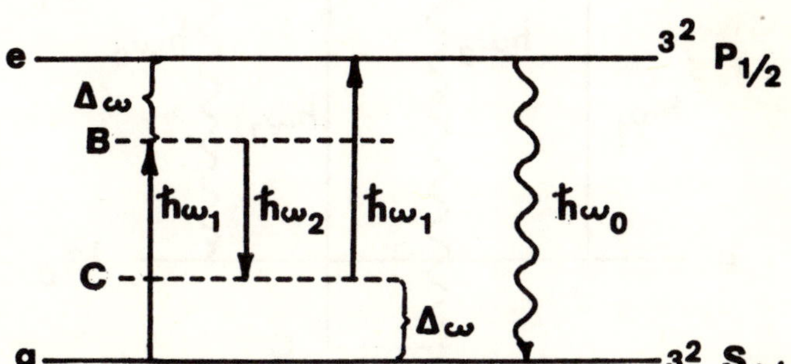


fig. 2a)

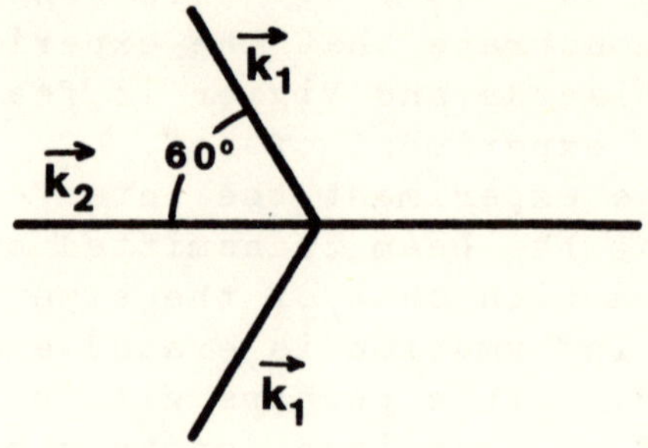
$\vec{k}_1$
$\vec{k}_2$
60°
$\vec{k}_1$

fig. 2b)

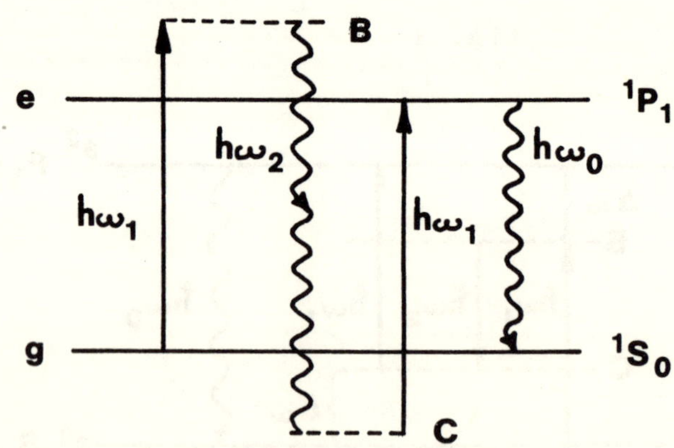
B
e
$^1P_1$
$h\omega_2$
$h\omega_0$
$h\omega_1$
$h\omega_1$
g
$^1S_0$
C

fig. 3

Note:

1 - The non duplicated photons interfere and do not produce noise. This originates from the fluorescence photons of the LGT. The required amplification can therefore be very small if the LGT fluorescence is small.

References:

(1) Garuccio, A., Rapisarda, V., and Vigier, J.P., Phys. Lett., to be published.
Andrade e Silva, J., Selleri, F., and Vigier,J.P., preprint (Paris 1982).
This and other experiments connected with the problem of duality which can be realised with LGT, are also discussed in the contribution to this volume by F. Selleri.

(2) Cagnac, B., Grynberg, G., and Biraben, F.,: 1973, Journal de Physique 34, p. 845.

(3) Cagnac, B., Grynberg, G., Biraben, F.,and Bassini,M.: 1976, Phys. Rev. Lett. 37, p. 283.

(4) Aspect, A., Roger, G., Raynaud, S., Dalibard, J., and Cohen-Tannoudji, C.,: 1980, Phys.Rev. Lett., 45, p. 617.

# FROM GHOST TO REAL WAVES: A PROPOSED SOLUTION TO THE WAVE-PARTICLE DILEMMA

G. Tarozzi

Istituto Matematico dell'Università - Bologna
Istituto di Filosofia dell'Università - Perugia

The most controversial epistemological problem of the foundations of quantum mechanics arises in connection with the interpretation of the fundamental theoretical term of quantal formalism, i.e. the wave-function or state-vector $|\Psi>$, which is at the origin of the problem of the wave-particle duality.

This essential property of microscopic phenomena finds its most significant manifestation in the well-known double-slit experiment.

In this peculiar physical situation one considers a source of photons (so weak that at a given time only one photon can be found in the experimental device) impinging on a diaphragm, endowed with two slits A and B, which are detected on a second screen X.

As it can be easily shown the description of the physical situation given starting from hypotheses which seem all absolutely reasonable leads to a result in open contradiction with experimental evidence. Let us consider the event $X_p$ "the photon has arrived at some point on the detecting screen X" which will be given by

$$(1) \qquad X_p = X_p \wedge (A_p \vee B_p)$$

where $A_p$ and $B_p$ indicate respectively the passage of the photon for the slit A or B respectively. This means that the probability $P(X_p)$ that the photon arrives at X is



S. Diner et al. (eds.), The Wave-Particle Dualism, 139–148.

(2) $$P(X_p) = P\left[X_p \wedge (A_p \vee B_p)\right]$$

which for the principle of distributivity becomes

(3) $$P(X_p) = P\left[(X_p \wedge A_p) \vee (X_p \wedge B_p)\right]$$

Moreover as $A_p$ and $B_p$ are disjunctive propositions, so that the event $(A_p \wedge B_p)$ is never true, one can write (3) as

(4) $$P(X_p) = P(X_p \wedge A_p) + P(X_p \wedge B_p)$$

The relation (4) is however contradicted by quantum mechanics. According to this last theory the wave-function $\Psi(\underline{x})$ which describes the photon in the point $\underline{x}$ of the detecting screen X is given by

(5) $$\Psi(\underline{x}) = \Psi_A(\underline{x}) + \Psi_B(\underline{x})$$

where $\Psi_A(\underline{x})$ and $\Psi_B(\underline{x})$ are complex quantities whose square modulus corresponds respectively to

(6) $$|\Psi_A(\underline{x})|^2 = P(X_p \wedge A_p)\ ;\ |\Psi_B(\underline{x})|^2 = P(X_p \wedge B_p)$$

In a similar way $|\Psi(\underline{x})|^2$ represents the probability $P(X_p)$ which will therefore in this case correspond to

(7) $$P(X_p) = |\Psi_A(\underline{x}) + \Psi_B(\underline{x})|^2 = |\Psi_A(\underline{x})|^2 + |\Psi_B(\underline{x})|^2 + \Psi_A^*(\underline{x})\Psi_B(\underline{x}) + \Psi_B^*(\underline{x})\Psi_A(\underline{x}) = P(X_p \wedge A_p) + P(X_p \wedge B_p) + I_t$$

where

(8) $$I_t = \Psi_A^*(\underline{x})\Psi_B(\underline{x}) + \Psi_B^*(\underline{x})\Psi_A(\underline{x})$$

indicates the interference terms of this sort of strange wave-like probability calculus.

As is well known the experimental evidence confirms (7) and contradicts (4). However when one locates behind the slits two measuring instruments allowing to establish <u>through which slit</u> the photon passes, the interference term (8) are destroyed and the right description is proved by (4). The physical situation appears of an even more difficult comprehensibility when one considers all the cases in which one has only one measuring apparatus located behind one slit which <u>has not revealed anything</u> and the photon has been anyhow detected in some point of the second screen.

In this physical situation the disappearance of the interference terms (8), which is a well-confirmed experimental result, can be justified only assuming that a measuring apparatus acting only as a source of information which allows to establish through which slit the photon has passed, can modify the empirical result of the experiment. But this appears of course a very paradoxical result as Renninger (1) was the first to stress.

This incapability to provide a reasonable explanation to the behaviour of the photon in the double slit experiment, incapability which can be extended to the more general problem of the wave-particle duality, conceived in the light of the complementary interpretation of the Cöpenhagen school as an insuperable dilemma has been emphatized also by Feynam:
"One might still like to ask: How does it work? What is the machinery behind the law? No one has found any machinery behind the law. No one can explain any more than we have just explained. No one will give you any deeper representation of the situation. We have no ideas about a more basic mechanism from which these results can be deduced. It is all quite mysterious. And the more you look at it, the more mysterious it seems (...)." (2)

To this completely renouncing epistemological attitude is opposed the realist interpretation of the wave-particle dualism which was historically put forward by Einstein and de Broglie. According to this interpretation the wave-function $\Psi(\underline{x}, t)$ associated to the particle would represent an objectively existing wave and not only a complex mathematical expression. In particular the particle would be associated to the wave in such a way as to be found in different points of the wave itself with a probability proportional to its square amplitude.

In 1978 Selleri and myself proposed a realist interpretation of the double- slit experiment (3), in conformity with the Einstein-de Broglie conceptions of wave-particle duality (4), according to which the events $A_p$ and $B_p$ would have to correspond to the more complex events:

(9) $$A_p = A_1 \wedge \gamma ,$$

$$B_p = B_1 \wedge \gamma ,$$

where $A_1$ ,$B_1$ and $\gamma$ mean respectively,

(10) $A_1$ ($B_1$ ) :the particle passed through slit A(B)

(11) $\gamma = A_2 \quad B_2$ : the associated quantum wave passed through both A and B.

In this maintaining the same mathematical description of quantum mechanics both the presence of the interference terms (8) as a consequence of the event and their destruction, provoked by the interaction of the measuring apparata in correspondence with the slits with the Einstein-de Broglie wave described by (11), can be explained. The realist interpretation would be able to account also for the paradox of Renninger, since even the insertion of one apparatus behind one slit, although not revealing the particle, would have destroyed the cohérence of the wave passing through the two slits and wiped out the interference pattern. In this case one would be dealing of course with the modified events

$$A'_p = A_1 \wedge (A_2 \wedge B'_2) \quad ; \quad B'_p = B_1 \wedge (A'_2 \; B_2)$$

where $A'_2$ ($B'_2$) indicates the absorption of the wave by the apparatus located behind A(B).

There is however a very strong objection against the possibility of writing relations (11) which regards the inobservability of the Einstein-de Broglie wave. In fact on which basis can we maintain that the apparatus located behind B(A) has absorbed, or at least interacted with the wave when it has not revealed any kind of physical effect? This inobservable reality of the wave associated to the particle has been of course object of violent criticism on the part of the supporters of the orthodox interpretation and Einstein himself used to talk in these cases about "ghost fields" (Gespensterfelder).

Anyhow this impossibility of determining any kind of predictable attributes for these "objects" implied that they could not be considered as sensible carriers of predicates in the language of quantum mechanical events. In a similar way the result of an interaction of this wave with other physical objects could be paradoxically compared to the one produced by other complex mathematical terms like the square root of a negative number.

One is faced therefore with a situation which seems without a way out since if on one hand the assumption of the reality of the Einstein-de Broglie wave can explain what in the orthodox interpretation appears completely inexplicable, on the other hand one must pay the price of postulating the existence of an object which cannot be characterized through any kind of observable quantity like energy momentum charge mass as in the case of every other proper physical object.

An answer to this paradoxical situation is very likely contained in a proposal of experimental test which has been discussed for the first time by F. Selleri (5), and which is now carried out by Gozzini (6), in Pisa.

According to Selleri's proposal one considers a source of photons S whose intensity is reduced by an absorber in such a way that only one photon is present at any given instant in the experimental device (Fig.1). Each of these individual photons falls on a halfsilvered mirror SRM which reflects half its intensity towards the counter $PH_1$ and transmits the other half to the counter $PH_2$, which are linked together to the extent of performing coincidence measurements. In the path of the reflected "photon" is placed a laser gain tube LGT (i.e. a piece of matter composed of unstable physical entities whose excitation energy corresponds to the one of the photons emitted by S).

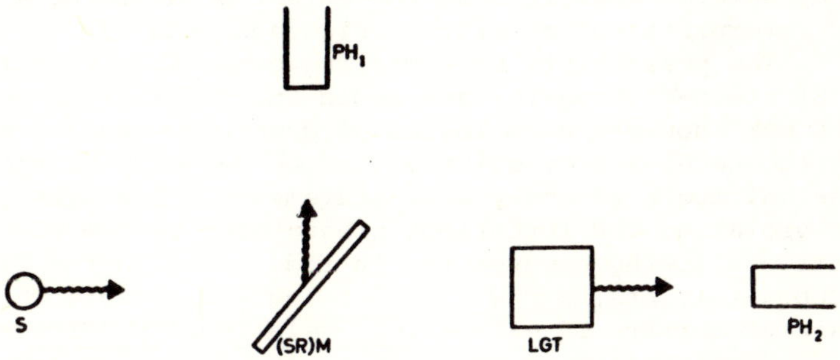


Fig.1 - Selleri's proposal for testing the properties of quantum waves.

Let us consider the number of the coincidences of the results of the measurements performed by $PH_1$ and $PH_2$. If this number will be found higher than the one of the random coincidences this positive result would mean that "something" different from the particle, which has been detected by the counter $P_2$, has produced a transition of probability on the lifetime of the atoms of LGT. We assume this "something" to be nothing but the Einstein-de Broglie wave. In this way a positive result would confirm the possibility to characterize the wave via the following predictable property: the property of inducing transition of probability on the lifetime of the excited physical systems whose excitation energy is the same one of the associated particle. One would be dealing however with a property of a particular nature which appears weaker than the ones which can be attributed to the particles.

This impossibility of characterizing by means of intrinsic or monadic properties, as happens for other ordinary physical objects, with the opportunity of determining a relational or better a diadic property between waves and some particular classes of particles would lead in turn to weaker notion of physical reality. And such a deepening of the conception of realism will be perhaps able to provide a satisfactory explanation of the nature of the wave particle duality.

A positive result from the previous experiment would also involve very important consequences for the logic of empirical theories since it would produce an automatic resolution of all the difficulties connected with the "weakening" of classical logic - i.e. through the renouncing of the principle of tertium non datur, or to the one of distributivity which has previously assumed in passing from (2) to (3) in the derivation of (4) - required by several logicians to the extent of avoiding a contradictory interpretation of microscopic phenomena (7).

The possibility of revealing observable effects induced by the Einstein-de Broglie wave would imply a radical refutation of the "weak" non-standard logics of quantum mechanics and consequently the necessity of maintaining the validity of the principles of classical logic as analytical statements which cannot be denied without violating the inventions presupposed by the very procedure of negation allowing however the introduction, in the language of microscopical events, a new class of logical objects which could be sensible carriers of exclusively relational predicates.

If the standard double-slit and Selleri's one, in which the screen with the two holes A and B has been substituted by the beam splitter M (or alternatively by a semi-reflecting mirror SRM), are compared, one can easily realize that the events $A_p$ to "the photon has been transmitted through SRM" and "the photon has been reflected by SRM". Such an equivalence can be extended of course to all events described by relations (9), (10) and (11).

On the basis of the previously introduced definitions, Selleri's experiment appears as a test of the event

(13) $B_1 \wedge A_2$

(or with a trivial modification of the geometry of the experiment $A_1 \wedge B_2$) corresponding to the statement "the particle has been reflected by and the wave has been transmitted through SRM".

We are not dealing, therefore with a complete test of the event $y$, that is of the fact that, in conformity with the Einstein-de Broglie realist interpretation of the double-slit experiment, the quantum wave must have been both transmitted and reflected (passed through both A and B according to the language of the standard double slit).

There is, moreover, another point connected with the testability of the properties of quantum waves, that must be stressed. As a matter of fact, if one found that quantum waves have in addition to the property of producing interference pattern, as Janossy and Naray showed in a conclusive way, also the one of inducing stimulated emission, then these physical objects would result characterizable through two different but mutually exclusive empirical properties, since interference and stimulated emission would manifest themselves with different experimental devices in two incompatible physical situations. One could, therefore, try to apply the notion of complementarity and in this case it would be perhaps more correct to speak of "meta-complementarity", since we could not be dealing with a dualism between the ondulatory and corpuscular properties of physical objects but with a dualism between the two ondulatory properties of microobjects - to the properties of quantum waves of producing either interference or alternatively stimulated emission.

With the aim of bypassing the previous problems I should like to propose two generalizations of Selleri's experiment.

My first variant represents a test of the complete event y corresponding to the statement "the quantum wave has been both transmitted and reflected" while the second experiment is a unique test of both the empirical properties of quantum waves, i.e. the design of a device able to test at the same time either the stimulated emission and the interference pattern produced by this physical object.

My first proposal is shown in fig. 2, where S is a very weak source of photons, SRM and $SRM^{I}$ semireflecting mirrors, LGT and $LGT^{I}$ laser gain tubes, $PH_1$ $PH_2$ and $PH_3$ photomultipliers.

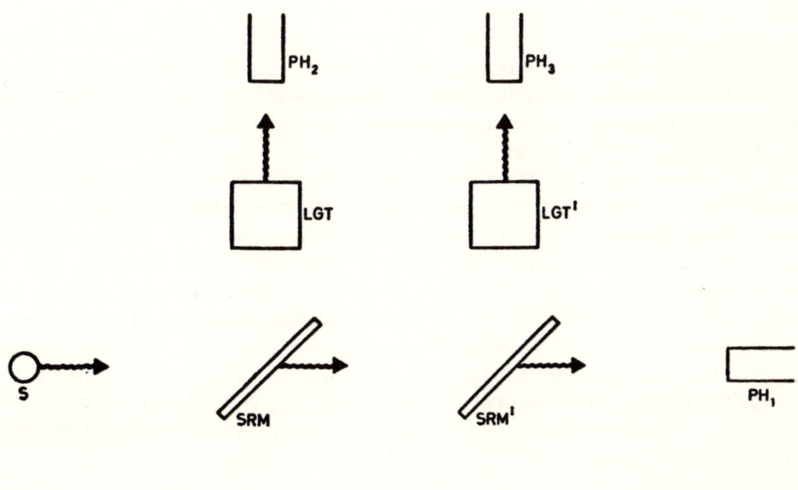


Fig.2 - Test of the event $A_1 \Lambda$ corresponding to "the particle has been transmitted through and the quantum wave has been both transmitted through e reflected by SRM".

If we denote by $P_i(t)$ the event,

(14) $P_i(t)$: "a photon has been detected by $PH_i$ at time t"

it appears clear that each coincidence of the detections of the three counters at any given time t, corresponding to

(15) $P_1(t) \wedge P_2(t) \wedge P_3(t)$

would represent a confirmation of the event

(16) $A_1 \wedge \gamma$.

As a matter of fact

(17) $|\overline{P_2(t) \wedge P_2(t)}|$ $(A_1 \wedge B_2)$,

that is, if we have a coincidence in the detection of PH and PH this means that, since the particle has been transmitted through SRM (and SRM ) as is confirmed by the arrival of a photon on PH , the second contemporary arrival of another photon on PH is a consequence of the stimulated emission produced by the wave reflected by SRM (Selleri's experiment).

Moreover,

(18) $|\overline{P_1(t) \wedge P_2(t) \wedge P_3(t)}| \rightarrow (A_1 \wedge A_2 \wedge B_3)$

that is, if $PH_1$, $PH_2$ and $PH_3$ detect at the same time a photon, we have the previous situation expressed by (17) plus the event corresponding to the reflection of the quantum wave by SRM which means in turn that such a wave has been not only reflected but also transmitted by SRM.

Our second variant for testing the empirical properties of quantum waves is shown in Fig. 3 where SRM and $SRM^I$ are semireflecting mirror, LGT is a laser gain tube, PH is a photomultiplier and IR a detector devised by Fleegor and Mandel (9) which is built with a stack of thin glass plates each of which has a thickness corresponding to a half-fringe width. The plates are cut and arranged in such a way that photon falling on the odd plates are fed to one photomultiplier $PH_o$ while photons falling on the even plates are fed to the other one $PH_e$. Each of these photomultipliers is connected to PH for the purpose of performing coincidence measurements.

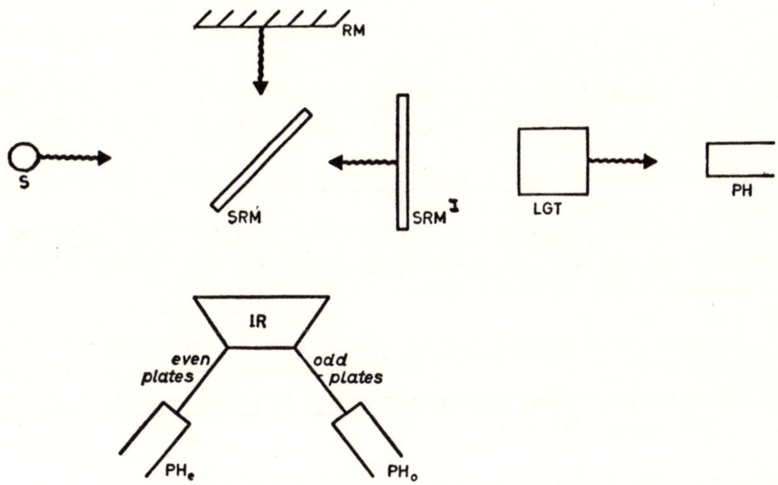


Fig. 3. - Experimental device for testing both the hypothesis that quantum waves induce stimulated emission and their property of producing an interference pattern in a unique physical situation.

We can perform in this way a unique test of both the empirical properties of the event $\gamma$.

As a matter of fact the coincidences between the detections of PH $PH_o$ and PH $PH_e$, that is with the notations previously introduced

(19) $$|\overline{P}(t) \wedge P_o(t)| \vee |\overline{P}(t) \wedge P_e(t)|$$

would put in evidence the property of quantum wave inducing stimulated emission while a difference between the previous coincidence rates, that is

(20) $$N_o \neq N_e,$$

when $N_o$ and $N_e$ indicate the number of the events $P(t) \wedge P_o(t)$ and $P(t) \wedge P_e(t)$, respectively, would show the other property of quantum wave of producing interference.

(1) M. Renninger, Zeitschrift für Physik 158, 417 (1960)

(2) R. Feynman, Lectures on Physics, III (Quantum Mechanics) pp.8 and 13, Addison Wesley (New York, 1972)

(3) F. Selleri and G. Tarozzi, Il Nuovo Cimento, 43 B, 31 (1978)

(4) A. Einstein, Annalen der Physik 17, 132 (1905); L. de Broglie, Annals de la Fondation L. de Broglie, 2, s (1977)

(5) F. Selleri, Nuovo Cimento Lettere, 1, 908 (1969)

F. Selleri, Realism and the Wave-Function of Quantum Mechanics in Foundations of Quantum Mechanics, ed. by B. d'Espagnat, Academic Press (New York, 1971)

F. Selleri, Annals de la Fondation L. de Broglie, 7, 45 (1982)

(6) G. Gozzini, Some experiments on the wave-particle duality of light preprint (1982)

(7) M.L. Dalla Chiara, Logica, ISEDI (Milano, 1974)

(8) L. Janossy and Zs. Naray, Suppl. Nuovo Cimento, 9, 588 (1958)

(9) R.F. Pfleegor and L. Mandel, Phys. Rev., 159, 1084 (1967). The use of such an interference detector in a similar tests of quantum waves has been proposed for the first time by A. Garuccio, V. Rapisarda and J.P. Vigier, Physics Letters, to be published .

# PERMANENCE OF THE CORPUSCULAR APPEARANCE AND NON LINEARITY OF THE WAVE EQUATION

D. FARGUE

Ecole des Mines de Paris



## I. INTRODUCTION

The two fold character of matter, undulatory and corpuscular, sets problems of mathematical representation which are not yet really solved. The easier to picture is certainly the wave : there are numerous partial differential equations which can be used and are well studied, at least in the linear domain.

It remains to account for the corpuscle and, above all, to connect it in some way with the wave. A first solution, the one which is at present favoured by most physicists, consists in



*S. Diner et al. (eds.), The Wave-Particle Dualism, 149–172.*

localizing the corpuscle when ever it is impossible to avoid it in view of the experimental results, the link between corpuscle and wave being the weakest conceivable, that is of a statistical nature only. In every individual case it is well known that quantum mechanics gives no answer at all, one notes the result without attempting to understand it [the situation has not changed since Schrödinger ([1]) noticed this in the foreword to the collection of his famous 1926 memoirs].

A second position, the one adopted by the founders of wave mechanics, Einstein, de Broglie, Schrödinger, consists in trying to incorporate in every circumstance the corpuscle in the wave, which in turn becomes a physical wave. This is done by representing the particle as a small region of large amplitude, or of large concentration of energy, a limiting case being a mathematical singularity. This is Einstein's, and later de Broglie's, "bunch field" which would now be called a soliton. Clearly one can then speak of a permanence of the corpuscle, as the mathematical model allows not to give up the particle concept, even when the undulatory aspect appears obvious.

Such a theory must fulfill a number of requirements ; in the following we shall examine the three main ones :

1. The permanence of the corpuscle must be ascertained : the bump in the field must not disappear, at least as long as the particle is not acted upon by too large force gradients.

2. A dynamics must be recovered, that is a law of motion for the corpuscle, which is in good agreement with experiment, or, for lack of it, with the former theories (classical or quantum) in their domain of validity.

3. One must also recover the results of the statistical experiments, the description of which is claimed to be one of the great successes of quantum theory, as it is commonly used in practice.

After a brief historical sketch of such attempts, these three points will be successively studied, particularly for those models which fall under the general framework of de Broglie's double solution theory.

## 2. EARLY THEORIES

Two broad classes of theories have been proposed since the beginning of quantum theory, which intend to account for the association between wave and corpuscle. The first ones are those which attempt to reduce the description of wave and corpuscle to some physical fields, previously known. For instance, this is the case for modified electromagnetisms such as Mie or Born-Infeld theories (although the latter had been designed to be later quantified). This is essentially the case for Einstein's general relativity theory.

2.1. The corpuscle in general relativity

One of the goals of this theory, at least in the mind of its creator, was to help or even to be sufficient to describe the corpuscles. For instance in a 1919 article ([2]) he modifies slightly the equations of general relativity and defines his objective : "The above reflexions show the possibility of a theoretical construction of matter out of gravitational field and electromagnetic field alone, without the introduction of hypothetical supplementary terms on the lines of Mie's theory". The line which is drawn is to construct in general relativity the corpuscle with as few as possible : it is a monist attitude, since then adopted in some so-called unitary theories. But in this first attempt, an equation is missing so that there are infinitely many spherically symmetric solutions and Einstein's conclusion is : "Thus the problem of the constitution of the elementary quanta cannot yet be solved on the immediate basis of the given field equations ".

Carrying on this same idea, he publishes in 1927 ([3]), in collaboration with J. Grommer, a famous article on the dynamics of neutral particles. There they tackle again this problem in a way which they explain as follows : "Alle Versuche der letzten Jahre, die Elementarteilchen der Materie durch kontinuierliche Felder zu erklären, sind fehlgeschlagen... Man wird so zu dem Wege gedrängt, die Elementarteilchen als singuläre Punkte, bzw., singuläre Weltlinien auf zufassen". As they can give no answer to the problem of the composition of the particles, and although they find that admitting their existence outside the theory constitutes a "provoking dualism", they show that they can nevertheless give an answer to the second problem mentioned above : the problem of finding a dynamics. For it is not necessary to know the exact structure of the particles to show that their law of motion is a consequence of the field equations : they recover indeed the right point dynamics, that is the principle of geodesics, when they only assume that the corpuscle is a singularity (in the mathematical sense) in the wave. One has thus gone over from a theory in which two sets of equations were needed (one set for the field and another for the motion of the particles in the field) to another in which the equations of the field alone are already sufficient .Besides, this result had at the same time been achieved in a different way by G. Darmois ([4]) who assumed the corpuscle enclosed in a thin 4-dimensional tube, and imposed only conditions for the solutions inside and outside the tube to agree on its surface. This makes the characteristics appear, which include precisely the geodesics. These two methods are clearly linked together (see A. Lichnerowicz ([5])).

Yet, in spite of this success, achieved thanks to the singularities, Einstein did not leave the idea of describing physical reality with the help of fields free of singularities and wrote with N. Rosen ([6]) that "a singularity brings so much arbitrariness into the theory that it actually nullifies its laws". But the only bright result of all this struggle remains the principle of geodesics. The whole of these steps can be summed up in the following chart (fig. 1).

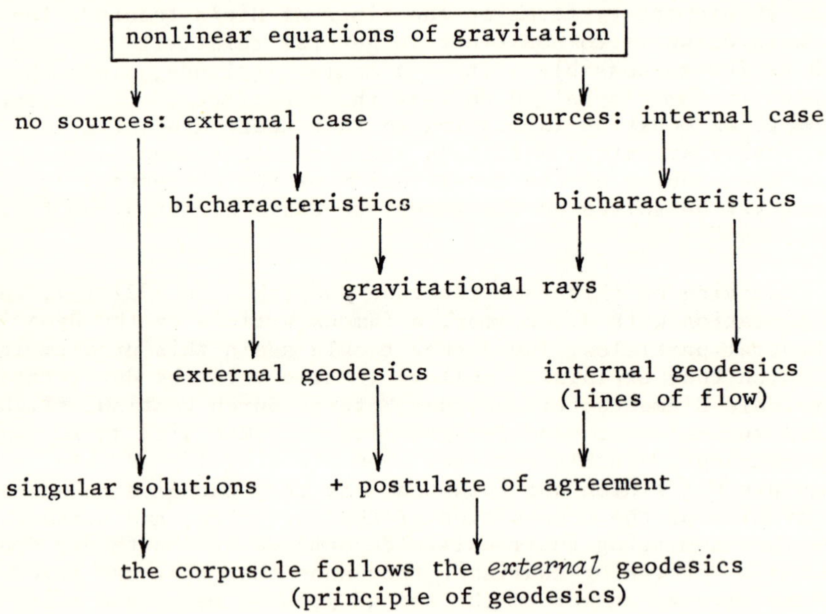


- fig. 1 -

One notices on this chart that, if the singular solutions are considered only as a makeshift, the principle of geodesics can only be recovered, for the time being, by means of a supplementary postulate of agreement between the internal and external solutions. We shall find a similar situation in wave mechanics. This seems to come from the fact that true continuous bump-like solutions of the field equations are not known, analogous to the ones already known a long time ago in hydrodynamics, and called now solitons.

Within this field theory, which general relativity is, the first question asked above has up to now no definite answer, the third one has not been asked, and only the second one has a satisfying solution. This leads one to try a theory less unitary than Einstein's, and introduce from the very beginning an additional field typical of quantum phenomena, the natural choice of which is de Broglie's matter wave.

2.2. Early attempts in wave mechanics

One knows that, guided by Einstein's work on the photon, de Broglie associated a wave with every corpuscle, but for him, contrary to quantum mechanics later development, the particle always remained localized in the wave as is shown by the following definition of his program or research ([7]) : "Mais toute la theorie ne deviendra vraiment claire que si l'on parvient à définir la structure de l'onde lumineuse et la nature de la singularité constituée par le quantum dont le mouvement devrait pouvoir être prévu en se plaçant uniquement au point de vue ondulatoire". The analogy with the question raised in general relativity is obvious.

If one adopts the mathematical description of the wave given by Schrödinger's equation, there are two possibilities to picture the corpuscle : the mathematical singularity, and the wave packet. The latter was immediately considered by Schrödinger ([8]), and soon abandoned, as is well-known, because the linearity of the equation involves the spreading of the wave packet in free motion ([9]), so that it is incapable of ensuring the permanence of the corpuscle in the wave. Besides, linearity entails a superposition principle which is contrary to the very idea of distinct and well-defined particles ([10]).

De Broglie ([11]) notices also another drawback which is that the spatial dimension of the wave packet must be larger than a wavelength, and an electron inside an atom cannot be considered as having at every moment a well defined position. He was thus led to use the singular solution of the wave equation, an idea which he attempted to develop until 1927, giving first ([11]) an example of such a solution in the case of the rectilinear motion, and afterwards ([12]) the main formulas, particularly the guidance formula (eq (18) hereafter) which links the motion of the singularity with the phase of the wave and thus plays a part analogous to the principle of geodesics in general relativity. But this wave is a physical wave, written $u(\vec{x},t)$ which cannot coincide, as it possesses a singularity, with the continuous solution $\psi$ of quantum mechanics, the statistical meaning of which had just been recognized. It remaind then to link these two waves, so as to find the position probability density $|\psi|^2$ ; this can be achieved with the help of the principle of phase harmony which, in that case, says that the phases of both waves $u$ and $\psi$ coincide outside the singular region of $u$ .

But the theory encountered a serious difficulty. It was obviously obligatory to show that the wave equation actually possessed solutions with a singularity (of pole type as a matter of fact). Although this is the case, as we shall see later (§ 3.5), the lack of such a proof made de Broglie submit at the 1927 congress a simplified version, the pilot-wave theory. This theory removes the use of the physical wave $u$ and sets as a principle the guidance formula and the position probability density, written with the help of the statistical wave $\psi$. But if this attitude avoids difficult mathematical problems it encounters conceptual difficulties even more serious ([13]). Indeed, the guidance formula written with the $\psi$ wave implies that it is this wave which actually guides the corpuscle. Now, the statistical meaning of $\psi$ is well established and involves that it depends on the information which the observer has about the system. This information would then have consequences on the motion of the particle and this leads to deviate further from classical conceptions rather than come closer to them.

These considerations, and some others, led de Broglie to abandon these attempts until 1952 when the pilot-wave theory was resumed and developped by D. Bohm ([14]). In conjunction with a careful study of the probabilistic interpretation of quantum mechanics ([15]), this led to a new development of the double solution theory, improved essentially by the idea of the necessity of non linear equations.

In all that follows, we opt for a classical-like description of matter with the help of one (or more) function $u(\vec{x},t)$ which may be called matter-wave, fluid... We shall see later how this is linked with the probabilistic wave $\psi$ and why the reasonings (not to speak of the paradoxes) held in quantum mechanics have always wavered between these two conceptions of the wave.
De Broglie's "double solution" takes into account this pair of functions $u$, $\psi$, the former describing a physical wave, the latter a statistical one.

## 3. NON LINEAR WAVE EQUATIONS AND THE DYNAMICS OF PARTICLES

### 3.1. Non linear wave equations

The use of non linear wave equations seems to be unavoidable for several reasons. If one does not want to give up every precise description of the corpuscle and its internal structure, while schematizing it in the extreme by a singularity, we saw that the evolution equation must be non linear. Another important reason is that linear dynamics, as is well-known, although it can supply directly the stationary motions, is incapable of explaining their birth and their stability. This is of course true of quantum mechanics and has even led to set as a principle something which is but a flaw in the theory, the inobservability of quantum jumps. Schrödinger ([16]) has even written : "while describing so minutely the so-called stationary states which the atom had normally, i.e. in the comparatively uninteresting periods when nothing happens, the theory was silent about the periods of transition or quantum jumps". Every attempt to study these problems imply using non linear theories ([17], [18],[19]).

It remains then to show the mathematical feasability of carrying out such a theory. In hydrodynamics it is well known([20]), as remarked already de Broglie ([21]), that some non linear partial differential equations have the property of ensuring in course of time the permanence of a bump, region of large amplitude. Such solutions are nowadays called solitons and have drawn the attention of mathematicians. Unfortunately, in view of the extreme diversity of partial differential equations able to exhibit such solutions, no general theory is up to now available, which would be comparable with the theory of linear equations, but only a range of more or less particular studies, for a long time focused on the search for exact solutions ([22], [23]).

The basic property of solitons is to guarantee the existence of a bump of amplitude when it encounters various obstacles ; this implies of course the existence of static bump-like solutions. In the quantum domain, several non linear equations having this property have already been proposed (see § 3.5 below). But most of them offer only a classical model of particles designed to be later quantified. The non linearity appears thus not as an explanation of the existence of particles, contrary to de Broglie's attempt. Besides few of them have been sufficiently developped to test the permanence of the bump, for instance in a collision between two such solutions, as the problem is generally extremely difficult (see e.g ([24]) on Born-Infeld equation, or ([25]) for sinus-Gordon equation).

After a brief sketch of some of these theories, we will only bring out their common general features, without insisting on any particular model. We shall see that nevertheless important results can be achieved, for instance the dynamics of corpuscles is determined in a way similar to the one we recalled in general relativity.

3.2. Madelung's hydrodynamical interpretation

It is necessary that, at least in some cases (low amplitude, statistical mean) the matter wave $u$ satisfies with a very good approximation the linear equation of wave mechanics . This leads us to recall the material interpretation of this equation given by Madelung ([26]) who called it the hydrodynamical interpretation . Although the corpuscle is missing, owing to the linearity, this formalism will be a convenient starting point to the extensions to non linear equations .

Let us consider Schrödinger equation for an electron in an external electrostatic field, written for the physical wave $u$ , such that $\int |u|^2 = m$, particle mass.

$$i \hbar \partial_t u = -\frac{\hbar^2}{2m} \Delta u + eUu \tag{1}$$

Define $f$ and $\phi$ , real functions of $\vec{x}$ and $t$ by

$$u = f \exp(\frac{i}{\hbar} \phi) \tag{2}$$

From (1) and (2), by separation of the real and imaginary parts, one gets

$$\text{(real)} \quad \partial_t \phi + \frac{1}{2m} \partial_i \phi \, \partial_i \phi + e\,U = \frac{\hbar^2}{2m} \frac{\Delta f}{f} \tag{3}$$

$$\text{(imaginary)} \quad \partial_t (f^2) + \frac{1}{m} \partial_i (f^2 \, \partial_i \phi) = 0 \tag{4}$$

One recognizes easily that these equations are respectively the equation for the velocity potential $\phi/m$, and the continuity equation of a fluid of mass density $\mu$, charge density $\rho$ and velocity $\vec{w}$ given by

$$\mu = f^2 \quad , \quad \rho = \frac{e}{m} f^2 \quad , \quad w_i = \frac{1}{m} \partial_i \phi \tag{5}$$

Equation (1) describes the irrotational evolution of this fluid under the action of the electrostatic potential U and a new internal one - as it depends only on $\mu$(or $f$) which is called ([12]) the quantum potential, Q

$$Q = \frac{\hbar^2}{2m^2} \frac{\Delta f}{f} \tag{6}$$

This quantum potential expresses the existence of a new,"quantum", law of behavior and corresponds to a specific density of force $\partial_i Q$ which can be derived from a stress tensor :

$$\mu\ \partial_j Q = \partial_i [\ \frac{\hbar^2}{2m^2}\ \mu\ \partial_i \partial_j\ \ell n \sqrt{\mu}\ ] \tag{7}$$

It is worth mentioning that, just as the stress tensor of a macroscopic fluid expresses the effect of short-ranged forces between the molecules of this fluid, so can the quantum potential be deduced for instance from a short-ranged interaction between the différent points in the medium, corresponding to a repulsive potential in $q(r)$ having an energy of the form

$$\frac{1}{2} \iint f(\vec{x}) q(|\vec{x}-\vec{y}|) f(\vec{y}) d^3x\ d^3y \tag{8}$$

We shall call (4) the continuity equation and (3) the Jacobi equation, although the latter is not strictly speaking the Jacobi equation of this fluid (which is a more general functional equation, admitting the solutions of (3) as particular solutions), for it coincides with the Jacobi equation for a point when the quantum potential is negligible.

Such a fluid interpretation of the usual formalism of wave mechanics can be extended to the other equations if one takes as starting point the fact that the fluid velocity is simply, as it is almost obvious in (5), the velocity which would be deduced from the probability current of quantum mechanics, but with a different meaning as the wave in question is the physical wave $u$ , and not the probabilistic one $\psi$ .

3.3. Some non linear quantum wave equations

The theory discussed above is obviously not sufficient : the corpuscle is missing (although its necessity could be questioned ([16])) . To compensate the unavoidable spreading of the wave packet, it is necessary to modify the evolution equation and add some new terms able to ensure the cohesion and thus thepermanence of the corpuscle, but the action of which is only perceptible inside the domain actually filled by it, i.e. where the wave amplitude is large, and this is clearly possible only for non linear terms. They can be introduced with the help of two different principles of generalization : one starts from the mathematical expression of the wave equation, the other from physical considerations, for instance suggested by what is already known in physics.

In the first case one adds one (or more) supplementary term as simple as possible, and nevertheless producing the desired effect, i.e. the existence of solitons, while disturbing as little as possible the other properties .For instance, adding a term proportional to $(\psi\psi^*)^2$ in the lagrangian ([28,29]) leads to the cubic Schrödinger (or Klein-Gordon) equation

$$i\hbar\, \partial_t \psi = -\frac{\hbar^2}{2m}\,\Delta\psi + eU\psi - g\,\psi^2\psi^* \tag{9}$$

Its energy, taken as the value of the hamiltonian, is :

$$H = \int \left[\frac{\hbar^2}{2m}\;\partial_i\psi^*\partial_i\psi + eU\psi^*\psi - \frac{g}{2}\,(\psi^*\psi)^2\right] d^3x \tag{10}$$

One understands that the additional negative energy (if $g > 0$) of the last term allows the advent of solitons, as has been shown ([28,29]).

But there is no reason to stop here, and other equations have been studied, of a similar type

$$i\hbar\, \partial_t \psi = -\frac{\hbar^2}{2m}\,\Delta\psi + eU\psi + F(\psi\psi^*)\psi \tag{11}$$

Where F is a real function, so as not to disturb the continuity equation. For instance F can be a logarithm ([30]) ; the evolution of the corresponding solitons has been studied in various circumstances ([31]), which, by the way, suggest that the cohesion is then not strong enough to maintain the unity of the particle. More generally, when one wants to determine this function F in a reasonable way, one can start from particular solutions of the linear wave equation and impose that they must not be much perturbed, as they are supposed to possess a profound physical meaning . This leads ([32]) once more to the cubic Klein-Gordon equation, as the simplest solution.

In a similar way, a particular form of soliton can be given, which seems a priori acceptable, and then work backwards to find an equation of the preceding form (11), which has this soliton as a solution ([33]). Equation (11) then appears as typical of a broad class of non linear equations, including a closely related one, as the sinus-Gordon equation ([10,25]) .

The other point of view can be traced back to Lorentz electron model .But an electrically charged globule is unstable under the action of its own electrostatic forces . To solve this problem, the first idea was to suppose it to be rigid, i.e. to use infinitely large forces .This is at best a first approximation, even in macroscopic physics, a fortiori if relativity has to be taken into account. This latter requirement led Poincaré ([34]) to introduce a particular force : "[elle] peut être assimilée à une pression extérieure constante, agissant sur l'électron déformable et compressible" ; the electron becomes "une capacité creuse soumise à une pression interne constante" (a negative pressure of course).

A number of classical-like theories for particles have since been proposed, mainly to account for the structure of elementary particles, and designed to be later quantified on the usual way ([35,36]). But, once more, quantum phenomena do not appear from the beginning.

It is nevertheless possible to retain the idea of internal forces, analogous to Poincaré's ones, and at the same time adopt Madelung's fluid interpretation. This is all the more necessary, as, if $e\, f^2 = e\, u\, u^*$ is interpreted as an electric charge density - and this is necessary as the electric force density in the dynamical equation deduced from (5) is $-e\, f^2\, \partial_i U$, and as radiation theory must also be recovered - then this electrical density has no reason for non acting on itself. Now it is well-known ([1,26,37]) that this leads to unacceptable conclusions : on the one hand the energy levels of the stationary states are then wrong, on the other hand, the spreading of the wave packet becomes all the worse, for internal electrostatic forces lead, as does also the quantum potential, to the bursting of the electron.

Macroscopic physics suggests then two kinds of attractive forces (as gravitation seems much too weak to play a significant part). Surface forces can be used, analogous to surface tension ([35]), but this obviously implies that the particle has a strictly finite size and leads to discontinuities which are awkward to handle and foreign to a true undulatory theory.

The simplest hypothesis then consists in introducing ([18,38,39]) an attractive potential of cohesion derived from a law of distance g(r), and which can be written as :

$$G(\vec{x},t) = \int \mu(\vec{a},t)\; g(|\vec{x}-\vec{a}|)\, d^3a \tag{12}$$

The evolution equation of the fluid then becomes

$$\mu \frac{dw^i}{dt} = \mu\partial_i Q - \mu\, \partial_i G - \rho(\partial_i U + \partial_i V) \tag{13}$$

where V is the internal electrostatic potential. This term and the cohesion term could naturally be put together and this would lead to modifying electromagnetism at short distances.

Such an evolution is hamiltonian, the value of the energy being

$$H = \int \left(\mu \frac{w^2}{2} - \mu\, Q + \frac{1}{2}\mu\, G + \frac{1}{2}\rho\, V + \rho\, U\right) d^3x \tag{14}$$

where the continuity equation (4) has been taken into account. With the help of Madelung's method, a wave function $u$ can be used, which is linked with $\mu$ and $\vec{w}$ by (2) and (5), and the evolution of which reads :

$$i\hbar\, \partial_t u = -\frac{\hbar^2}{2m} \Delta u + eUu + \frac{e}{m}\frac{1}{K_o} u \int \frac{u(a)u^*(a)}{|\vec{x}-\vec{a}|} d^3a$$

$$+ u \int u(a)u^*(a)\, g(|\vec{x}-\vec{a}|) d^3a \tag{15}$$

If the self-action electrostatic term is omitted and $g$ taken as $-g\,\delta(x-a)$, one sees that (15) goes over into (9), i.e. the cubic Schrödinger equation.

Equation (15) possesses static, spherically symmetric solutions which are every where finite ([40]), for H is non convex but has a lower bound. We presented here a simplified version, sufficient for slow motions, but if the propagation delay of electromagnetic actions is taken into account, such a theory also provides a description of quantum transitions and an explanation for the stability of stationary states ([38]).

We have thus seen that a rather broad class of non linear wave equations, which have been proposed to account for the existence of corpuscles as regions of large amplitude in an outspread wave (solitons) can be written in the form of a pair of real equations :

$$\partial_t\phi + \frac{1}{2m}\, \partial_j\phi\, \partial_j\phi + eU = \frac{\hbar^2}{2m}\frac{\Delta f}{f} + \text{non linear selfaction terms} \tag{16}$$

$$\partial_t(f^2) + \partial_i \left(f^2 \frac{\partial_i\phi}{m}\right) = 0 \tag{17}$$

in the non-relativiste case. Whatever the final formulation may be, it seems that this separation in two will remain valid : one equation describes the evolution of a velocity potential, analogous to a classical Jacobi equation but completed with the quantum potential and a non linear term, the other is simply the continuity equation.

## 3.4. The corpuscle and its dynamics. The guidance formula

The previous equations, as different as they are, have in common the existence of bump-like solutions. It remains to check if these bumps, supposed to picture the corpuscle, do actually propagate in a way corresponding to what we already know about particles.

To this end, we will use the fact that the continuity equation (17) is common to all of them. The assumption (which is strongly supported by the results mentioned above) of the permanence of the bump allows then to show that it must follow, with a very good approximation, the lines of flow of Madelung's fluid, as soon as the size of the bump is small enough for the phase of the wave to vary much more slowly than the amplitude in the vicinity of the corpuscle. This in turn leads to the guidance formula ([12,21]) relating the velocity $\vec{v}$ of the particle to the phase of the wave :

$$\vec{v} = \frac{1}{m} \nabla\phi \tag{18}$$

This equation coincides formally with the last of (5), but has a different meaning. Equation (18) expresses indeed the velocity of the corpuscle, and the value of $\nabla\phi$ must be taken at the place actually taken up by the particle, when equation (5) expressed the local velocity of the flow, everywhere in space.

This is one of the main formulas of the double solution theory, and also, as it seems, of every theory attempting to describe matter with the help of a physical wave. Therefore we will give briefly one proof of it ([21]), and this will allow us to specify the exact limits of its applicability.

We shall call guidance lines the flow lines $\vec{\xi}(\vec{x}_o,t)$ of the fluid, $\vec{x}_o$ being the initial position of the point considered (Lagrange coordinates) ; it follows that

$$\partial_t \, \xi^i(x_o,t) = w^i = \frac{1}{m} \partial_i \phi(\xi^j(x_o,t),t) \tag{19}$$

This lines are thus the characteristics of the continuity equation (17), thought of as a partial differential equation in $f^2$. Integrating this equation along a guidance line gives

$$f^2(\xi(x_o,t),t) = f^2(x_o,0)\exp\left[-\int_o^t \frac{\Delta\phi(\xi(x_o,\tau),\tau)}{m} d\tau\right] \tag{20}$$

which expresses the conservation of $f^2$ (i.e. the mass in the hydrodynamical interpretation). This shows that, on condition that $\Delta\phi$ is not too large, or more precisely that $\phi$ varies not much in the domain where $f^2$ increases rapidly, the exponential is close to 1, and the points where $f^2$ is large at time $t$ are the evolved through $\vec{\xi}(\vec{x}_o,t)$ of those where $f^2$ had a large value at time 0. The bump moves thus with the local velocity $\vec{w}$ given by (19), whence (18).

This result would be strictly valid for an incompressible fluid, as in that case $\text{div}\,\vec{w} = \Delta\phi/m = 0$, this would also be the case for the motion of a mathematical singularity, as suggested by (20) with $f \to \infty$ as one comes closer to the singularity, and as can be seen directly from the singular solutions of the linear equation ([41]). Finally it would also be strictly valid if $f^2$ possessed a discontinuity on a surface enclosing the corpuscle ; this is the fundamental property of characteristics which is imperative only in that case. When, more generally the bump is smooth, it can steal out of the guidance tube if $\Delta\phi$ is large enough, i.e. is the phase is rapidly varying. But the size of the corpuscle, assumed to exist, is about $10^{-15}$m when in low-energy experiments (excluding the experiments performed in elementary particle physics, where precisely the linear Schrödinger equation is obviously inadequate) the phase changes over a distance of the order of the wavelength, say $10^{-10}$m ; the guidance formula is thus certainly valid.

It is worth noting that the known solitons if the one-dimensional cubic Schrödinger equation move with the velocity given by the guidance formula, as is immediately seen on their expression for a free propagation :

$$\psi = \exp\{\frac{i}{\hbar}[mvx-(\frac{mv^2}{2}-\alpha)t]\}(\frac{2\alpha}{g})^{1/2}\operatorname{sech}\left[\frac{(2m\alpha)^{1/2}}{\hbar}(x-vt)\right] \quad (21)$$

The "gaussons", solitons of the logarithmic Schrödinger equation ([30]) have also the same property.

At this stage, we have thus obtained, in non linear wave mechanics, the analogue of the internal case in general relativity : we know that the particle must move along the guidance lines of the associated wave u , exactly as it follows the internal geodesics in general relativity. But in this theory, as the structure of the corpuscle was not known, its motion had to be connected to the solution in the external case, as we have seen above (§ 1). In wave mechanics the situation is similar : as the exact non linear wave equation is unknown, we must link the motion of the corpuscle to the solution, called v by de Broglie, of the linear equation.

Let then u be the bump-like solution of the non linear equation and v the solution of the linear equation, for the same initial and boundary conditions. As in general relativity, we need a postulate of agreement between these two solutions. The basic idea of de Broglie's thesis gives the answer : it is the postulate of phase harmony : the phase of the soliton u and the phase of the "linear" wave v must be equal, at least in the vicinity of the corpuscle. This appears the more plausible as we know that usually the phase changes much more slowly than the amplitude .

The guidance formula remains then expressed by (18) but $\phi$ must now be taken as the phase of the wave v , solution of the linear equation. This gives an answer to the second question asked in the introduction : the dynamics of the corpuscle is determined and, as we shall see below, in good agreement with both the experimental data, and usual quantum mechanics. Naturally the postulate of phase harmony is only provisional : as soon as a non linear equation is selected, it must result from the solution of this equation.

3.5 . The singular solutions of the linear equation

We mentioned several times above the singular solutions of the wave equation. They are interesting in two respects : firstly knowing their properties corroborates the assumptions made above, concerning the bump-like solutions, secondly they supply an approximate representation of the true solution, which is perfectly valid, as soon as the size of the bump remains much smaller than the other characteristic length of the physical problem.

Such solutions are known to exist as well for the Klein-Gordon equation ([41]) as for the Dirac and higher spin equations ([42]). More precisely, for such solutions, the singularity moves along any given space-time trajectory, which is then a guidance trajectory for the solution. This line is determined by the phase of the solution itself for, if the amplitude $f$ possesses a pole-singularity, the phase is defined everywhere just as its first derivatives. It is also possible to make two solutions, one singular, the other regular, agree along a 3-surface made up with guidance lines, in such a way that their amplitudes are proportional and that the phase harmony principle holds up to the second derivatives of $\phi$. This justifies the principle in this case, and also the dynamics of varying proper mass which results from it([43]).

Finally the approximation given by the singular solutions is already good enough for them to possess some interesting properties ([41]) : in an experiment of diffraction by a periodic medium, the singularities are necessarily placed on the directions given by Bragg's law. This is a first example of a typically undulatory phenomenon which is nevertheless compatible with the permanence of the corpuscle, pictured by the singularity in the wave.

3.6. General scheme of a non linear theory in wave mechanics

The previous connections between the (as yet hypothetical) non linear equation of wave mechanics in the double solution theory, and the linear equation, as well as their respective solutions, can be summed up in the following diagram where the connections with classical dynamics have also been given.

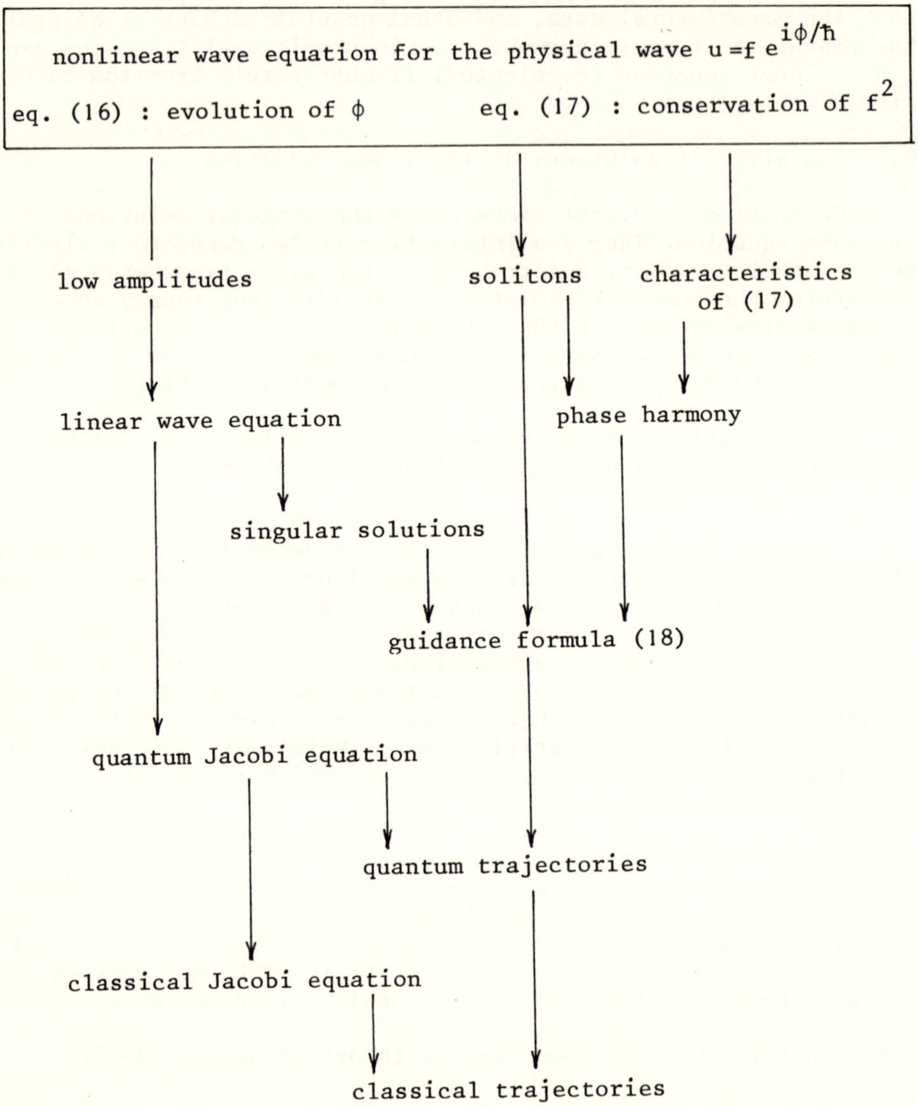
nonlinear wave equation for the physical wave u = f e^{iϕ/ħ}
eq. (16) : evolution of ϕ
eq. (17) : conservation of f^2
low amplitudes
solitons
characteristics of (17)
linear wave equation
phase harmony
singular solutions
guidance formula (18)
quantum Jacobi equation
quantum trajectories
classical Jacobi equation
classical trajectories

- fig. 2 -

## 4. CONNECTIONS WITH THE PROBABILISTIC FORM OF QUANTUM MECHANICS

The question is now to see how the theories developped above satisfy the third condition formulated in the introduction : explain the statistical results of quantum mechanical experiments; we shall also see to what extent the usual formalism of quantum mechanics can be recovered, as well as the probabilistic wave $\psi$.

In the case of interference-like experiments, in which the position probability density $|\psi|^2$ is tested, it has been known for a long time that both theories coincide ([12]), at least within the framework of the singular solutions, and we shall try in the following to state precisely the validity of this approximation. The measure of other microscopic quantities for free states (i.e. when a beam of particles is sent through the measuring apparatus) can then be reduced to the former one by a careful analysis of the part played by the measuring device ([14,44,45]), at least in every practically performed experiment (it appears not necessary to recover the whole axiomatical measure theory of quantum mechanics, the usefulness of which may be questioned ([46])).

It remains the case of the bound states, for which the problem seems much more difficult, the exact meaning of $|\psi|^2$, even in quantum mechanics, appearing not so clearly ([47]). We shall only give two kinds of possible attitudes, having complementary advantages and drawbacks, the question seeming up to now impossible to settle.

### 4.1. Experiments with an incident beam of particles

We shall take, as a typical example, interference experiments performed on electrons. The source of the particles is a heated filament and its associated accelerating and focusing electrodes. The aim of this device is to obtain a coherent beam, with electronic trajectories as well defined and stable as possible. Thus, before falling on the interference apparatus, every particle is, according to the description given above, in translational motion together with its wave, which is a soliton $u(\vec{x}-\vec{v}t)$, $u(.)$ being the function describing the free particle. In the most general case, the initial conditions are Cauchy's conditions for $u$, thus belonging to some space of functions ; but in the particular case of these experiments, we see that they are sufficiently specified by the coordinates of the barycentre of $f^2$, say $\vec{x}_o$, $\vec{v}_o$.

Now, as the trajectories behind the condenser are well defined in every point, knowing $\vec{x}_o$ is actually sufficient to determine the evolution of the (bump)-wave, for $\vec{v}_o$ is known as soon as $\vec{x}_o$ is. We are thus able to associate to every corpuscle (quasi point-like as we saw) a trajectory belonging to a 3-parameter set, $\{\vec{x}_o\}$, which corresponds to a class of classical motions associated with a solution $\phi$ of the corresponding classical Jacobi equation. As the evolution equation (16) coincides then, outside the region taken up by the particle, with the classical Jacobi equation, the guidance formula (18) coincides also with the value of the momentum given by Jacobi's theory : $m\vec{v} = \nabla\phi$. The wave associated with every corpuscle is thus the same outside the large amplitude domain and can be written

$$v = a \exp(\frac{i}{\hbar}\phi) \tag{22}$$

Where $\phi$ the previous Jacobi function, and a is an amplitude slowly varying, and of the size of the associated wave train (much larger than the corpuscle).

As the electrons have different origins on the filament, $\vec{x}_o$ changes and only probability density of initial positions, $\rho_o(\vec{x}_o)$ is known .The joint probability distribution of position and velocity is thus [14,44].

$$R_o(\vec{x}_o,\vec{v}_o) = \rho_o(\vec{x}_o)\delta[\vec{v}_o - \frac{1}{m}\nabla\phi(\vec{x}_o)] \tag{23}$$

and, in particular, the probability distribution of velocity is

$$\rho_o(\vec{v}_o) = \rho_o(\vec{x}_o) \left(\frac{D(\frac{1}{m}\nabla\phi)}{D(\vec{x}_o)}\right)^{-1} \tag{24}$$

This is generally not the momentum probability distribution given by quantum mechanics, for (24) is a "hidden" probability distribution valid before the interaction with the measuring apparatus [48]. Nevertheless, according to the common use in quantum mechanics, we set

$$\psi_o = \rho_o^{1/2} \exp(\frac{i\phi}{\hbar}) \tag{25}$$

but this will act as a <u>definition</u> of the initial probabilistic wave $\psi_o$.

When the corpuscle comes later near the interface device, the classical motion becomes insufficient, and $\phi$ evolves according to equation (16) where the quantum potential plays an important part. Outside the singular region, as u satisfies the Schrödinger equation, the boundary conditions are essential. One can also think (but this is not de Broglie's point of view) that in some cases the corpuscle may lose its particle-like appearance and spread out. In the case of Young's experiment for instance, the wave associated with a corpuscle (the dimensions of which are of the order of magnitude of the associated wave train) covers both holes at the same time and, even if the corpuscle passes through only one hole [49,50], as it is guided by the phase of the wave, its motion depends on the presence of the second hole.

Let then $\psi(\vec{x},t)$ be the wave function resulting from $\psi$ through Schrödinger equation, such then as it is done in quantum mechanics. As the wave $u(\vec{x},t)$ complies with this equation outside the bump, and as $u(\vec{x},0)$ and $\psi_o$ have approximately the same phase and proportional amplitudes outside the corpuscle, at least in the vicinity of the holes, one sees that after leaving the interference device, the agreement of the phase of the low-amplitude part of u, and of $\psi$ is conserved (this is once more the "principle of the double solution") and the trajectories of the corpuscles are not only the guidance lines of u but also of $\psi$ : this justifies the pilot-wave theory, which is a very good approximation in this kind of experiments.

It is then easy to show that, behind the interference device, the position probability density is always equal to $|\psi|^2$. For behind this device, the corpuscles are indeed certainly point-like. They thus have well defined positions and velocities, $\vec{x}$, $\vec{v}$. As the trajectories are defined by the differential equation

$$\frac{d\vec{x}}{dt} = \frac{1}{m}\nabla\phi \tag{26}$$

there corresponds to every initial position $\vec{x}_o$ (which, in turn, determines $\vec{v}_o$) one and only one position $\vec{x}(\vec{x}_o,t)$ at time t . The conservation of probability entails then that the position probability density at time t is $\rho(\vec{x},t)$ such that

$$\rho_o(\vec{x}_o,t)d^3x_o = \rho(\vec{x},t)d^3x \tag{27}$$

This result is general. If one assumes now, as de Broglie's does, that in these experiments the corpuscle is always point-like, then (27) holds $\forall t$ and, together with (26) this gives

$$\partial_t\rho + \mathrm{div}(\rho\,\frac{1}{m}\nabla\phi) = 0 \tag{28}$$

Now, $\psi$ is a solution of Schrödinger equation, thus

$$\frac{\partial|\psi|^2}{\partial t} + \mathrm{div}\,(|\psi|^2\ \frac{1}{m}\nabla\phi) = 0 \tag{29}$$

with the same phase $\phi$ as in (28). As $\rho_o = |\psi_o|^2$, this gives $\rho = |\psi|^2$ $\forall t$ and the interference principle (position probability density equal to $|\psi|^2$) is demonstrated.

In the pilot-wave theory, this result is evident, but only because this theory is but an interpretation of quantum mechanics, without attempting to give a more profound description of microphysics. In non linear theories such as the double solution the preceding result is only approximate for, rigorously, $|u|$ and $|\psi|$ are only approximately proportional outside the bump, and their phases not quite identical, because the actual quantum potential is proportional to $\Delta|u|/|u|$ and not to $\Delta|\psi|/|\psi|$. But in all experiments similar to those studied here, these differences are inobservable. It then seems that it is possible to describe every experiment of this kind, performed up to now, while retaining a description steadily using a localized corpuscle and thus without resorting to the complementarity notion. Besides, it seems that the above reasoning is much closer to the one actually held by physicists than the reasoning which should be held according to the axiomatics of theoretical quantum mechanics, and which is never perfectly applied in practice (and perhaps impossible to use ([46])).

4.2. Bound states

In the previous analysis, one can easily adopt the picture of minute-size corpuscles, almost point-like. This is not the case for bound states.

Let us indeed consider the simple case of the stationary states of the hydrogen atom. In spherical coordinates r, θ, φ they are

$$\psi_{n\ell m} = F_{n\ell}(r,\theta) \exp\left(\frac{i}{\hbar} E_n t + im\phi\right) \tag{30}$$

for the states with energy $E_n$, with the usual notations. The phase is thus

$$\Phi = E_n t - \hbar m\phi \tag{32}$$

(in this paragraph the mass will be noted $m_o$ and the phase $\Phi$ to avoid any confusion).

The guidance formula (18) assigns to the corpuscle, if it remains well-localized even in this case, a velocity $v_\phi$ tangential to a parallel with axis Oz :

$$v_\phi = -\frac{1}{m_o}\frac{1}{r}\frac{\partial\Phi}{\partial\phi} = -\frac{1}{m_o r} m\,\hbar \tag{32}$$

By the way, this gives for the angular momentum the usual expression $M_z = m\ r\ v_\phi = m\ \hbar$. The motion of the electron, in phase with its wave is thus a uniformly circular motion round Oz. In the case of the ground state, the electron is even motionless, $v = 0$, since the attraction due to the nucleus is counterbalanced by the repulsion caused by the quantum potential.

Nevertheless, these results encounter a number of difficulties. In the ground state, as the electron is motionless, the model is very far from being spherically symmetric, contrary to the quantum solution, and there is an analogous problem for every other state. Thus, how is it possible to account for the probability distribution $|\psi|^2$ ? It seems that some stochastic element must be introduced, concerning either the whole set of electrons on which the experiment is performed, or the single electron in question, on condition that it performs, about the regular motion given by the guidance formula, a kind of brownian motion caused by fluctuating interactions with its surroundings ([51,43]). But this could also result from the action of the zero-point electromagnetic field, as is assumed in stochastic electrodynamics ([52]). This last hypothesis seems the more adequate as it permits the solution of another important problem set by guidance trajectories : the states cannot remain stationary for they should continuously radiate electromagnetic energy, if the electric charge is assumed proportional to $|u|^2$, since the electron is almost point-like and performs accelerated motions, except in the ground state. The vacuum field could help to overcome this difficulty.

Another solution is to consider that the electron is no more point-like as soon as it is in a strongly bound state ([38]). As a matter of fact it can be assumed that the non linearity (cohesive potential for instance) which is responsible for the point-like character in the free states, is no more strong enough in the bound states to prevent the u wave from spreading. The wave should then have, at least for not too much excited states, approximately the same shape as the $\psi$ wave calculated according to Schrödinger's equation. When in interference experiments it is a statistical effect upon the whole set of particles which allows us to compare u outside the bump with $\psi$, in the bound states it would be a compensation between the internal electrostatic potential and the centrifugal forces on the one hand, and the cohesive force on the other hand, which would allow for the success of the linear equation. The additional assumption of a standing stochastic element would then become superfluous. Besides, the problems connected with radiation would be solved since the stationary states correspond then to steady motions which do not radiate.

These two conceptions appear a priori equally interesting, only their subsequent development and its comparison with experiment will allow us to determine which one is the best.

## REFERENCES

(1) E. Schrödinger, Mémoires sur la mécanique ondulatoire, trad. Al. Proca, Alcan, Paris, 1933.

(2) A. Einstein, Sitz. Pr. Ak., 1919.

(3) A. Einstein, J. Grommer, Sitz. Pr. Ak., 1927, p. 2

([3]) A. Einstein, ibid., 1928, p. 235.

([4]) G. Darmois, Les équations de la gravitation einsteinienne, Mémorial des Sc. Math. 25, G.V., Paris, 1927.

([5]) A. Lichnerowicz, Théories relativistes de la gravitation et de l'électromagnétisme, Masson, Paris, 1955.

([6]) A. Einstein, N. Rosen, Phys. Rev. 48, 1935, p. 73.

([7]) L. de Broglie, C.R. Ac. Sc. 179, 1924, p. 1039.

([8]) E. Schrödinger, Naturwiss. 28, 1926, P. 664.

([9]) E. Prugovecki, Quantum mechanics in Hilbert space, Ac. Press, 1971 ; (for a proof valid in the whole of $L^2$).

([10]) U. Enz, Phys. Rev. 131, 1963, p. 1392.

([11]) L. de Broglie, J. Phys. 7, 1926, p. 321.

([12]) L. de Broglie, J. Phys. 8, 1927, p. 225.

([13]) L. de Broglie, C.R. Acad. Sc. 233, 1951, p. 641.

([14]) D. Bohm, Phys. Rev. 85, 1952, p. 166 and p. 180.

([15]) L. de Broglie, Les incertitudes d'Heisenberg..., G.V., Paris, 1982 ; (these lecture notes from 1951-1952 had never been published before).

([16]) E. Schrödinger, Brit. J. Phil. Sc. 3, n° 11, 1952.

([17]) J. Andrade e Silva, F. Fer, P. Leruste, G. Lochak, Cahiers Phys. 15, 1961, p. 210 and 16, 1962, p. 1.

([18]) F. Fer, L'irréversibilité fondement de la stabilité du monde physique, G.V., Paris, 1977.

([19]) N. Gisin, Ann. Fond. L. de Broglie 7, 1982, p. 275.

([20]) H. Lamb, Hydrodynamics, 1932, Dover reprint 1945, § 252.

([21]) L. de Broglie, Une tentative d'interprétation causale et non linéaire..., G.V., Paris, 1956.

([22]) A.C. Scott, F.Y.F. Chu, D.W. Mac Laughlin, Proc. IEEE 61, 1973, p. 1443.

([23]) G.B. Whitham, Linear and non linear waves, Wiley, N.Y., 1974.

([24]) G. Lochak, C.R. Ac. Sc. 250, 1960, p. 1986 and 2146.

([25]) J.K. Perring, T.H.R. Skyrme, Nucl. Phys. 31, 1962, p. 550.

([26]) E. Madelung, Z. Phys. 40, 1926, p. 322.

([27]) See ([21]) p. 194, and also ([43]).

([28]) R. Finkelstein, R. Lelevier, M. Ruderman, Phys. Rev. 83, 1951, p. 326.

([29]) N. Rosen, H.B. Rosenstock, Phys. Rev. 85, 1952, p. 257.

([30]) I. Bialynicki-Birula, J. Mycielski, Ann. Phys. (N.Y.) 100, 1976, p. 62.

([31]) T.A. Minelli, A. Pascolini, in Dynamical systems and microphysics, CISM n° 261, Springer, Wien, 1980.

([32]) G. Petiau, Suppl. Nuovo Cimento 9, 1958, p. 542.

([33]) A. Kumar, Ya.P. Terletskii, Ann. Fond. L. de Broglie 3, 1978, p. 35.

([34]) H. Poincaré, C.R. Ac. Sc. 140, 1905, p. 1504. and Rend. Circ. Mat. Palermo 21, 1906.

([35]) P.A.M. Dirac, Proc. Roy. Soc. 268 A, 1962, p. 57.

([36]) L. de Broglie, D. Bohm, P. Hillion, F. Halbwachs, T. Takabayashi, J.P. Vigier, Phys. Rev. 129, 1963, p. 438.

([37]) E. Schrödinger, Congrès Solvay 1927.

([38]) F. Fer, C.R. Acad. Sc. 258, 1964, pp. 2983, 3215, 3435 ; 262, 1966, p. 1417 ; 263, 1966, p. 103.

([39]) D. Fargue, F. Fer, Ann. Fond. L. de Broglie 1, 1976, p.30.

([40]) D. Fargue, F. Fer, C.R. Acad. Sc. 288 A, 1979, p. 323.

([41]) F. Fer, Thèse, Paris, 1956 ; and also in, Louis de Broglie, sa conception du monde physique, G.V., Paris, 1973.

([42]) M. Thiounn, thèse, Paris, 1963.

([43]) L. de Broglie, La réinterprétation de la mécanique ondulatoire, G.V., Paris, 1971.

([44]) L. de Broglie, La théorie de la mesure en mécanique ondulatoire, G.V., Paris, 1957.

([45]) L. de Broglie, G. Lochak, A. Beswick, J. Vassalo-Pereira, Found. Phys. 6, 1976, p. 3.

([46]) F. Fer, this volume.

([47]) P. Claverie, S. Diner, Israel J. Chem. 19, 1980, p. 54.

([48]) G. Lochak, this volume, p. 1.

([49]) R.D. Prosser, Int. J. Theor. Phys. 15, 1976, p. 169.

([50]) C. Philippidis, C. Dewdney, B.J. Hiley, Nuovo Cimento 52 B, 1979, p. 15.

([51]) D. Bohm, J.P. Vigier, Phys. Rev. 96, 1954, p. 208.

([52]) S. Diner, this volume, p. 215.

# COULD SOLITONS BE ADIABATIC INVARIANTS ATTACHED TO CERTAIN NON LINEAR EQUATIONS ?

Pierre LOCHAK

Centre de mathématiques de l'E.N.S. Ecole Normale Supérieure 45, rue d'Ulm 75230 PARIS CEDEX 05

In the first three parts, arguments are given to support the claim that solitons should be the adiabatic invariants associated to certain non linear P.D.E. ; a precise mathematical form of this conjecture is then stated. In the last part, a particular case of the conjecture is further studied.

## 1. INTRODUCTION

The recent progress in the mathematical study of certain non linear partial differential equations (P.D.E.) has made it possible to revive the idea of a "non linear quantum physics", an idea which stems from Louis de Broglie's work in the midfifties, since the "onde à bosse" of the "théorie de la double solution" represented nothing but that which is now termed a "soliton".

As a working definition of the word soliton, we shall adopt a slight generalization of that given in the very useful review paper of Scott, Chu and Mc Laughlin (1). It has two aspects :

i) From a geometrical point of view, a soliton is a solitary wave, represented by a function with a (more or less) sharp unique maximum and damped at infinity.

ii) From a dynamical point of view, solitons should behave particle-like under collisions, i.e. retain their shape or be glued together etc.

As (ii) in this definition already stresses, if solitons are really entitled to eventually "embody" particles in a non linear quantum field theory, they should certainly possess the stability properties which common sense and to a certain extent physics, attach to the word "particle" ; stability theory, however, has many aspects depending both upon the particular definition of the



*S. Diner et al. (eds.), The Wave-Particle Dualism, 173–182.*

term "stability" which is under consideration and which determines the physical aspects of the problem, and upon the mathematical methods employed.

In this context, (asymptotic) stability with respect to the initial conditions is certainly the less involved from a mathematical point of view and it has already been extensively studied although the results are far from being complete (see (2) for a beautiful analysis by elementary means). In this paper, we shall be concerned with a different kind of stability, which should allow one to view solitons as thermodynamically stable quantized states : namely adiabatic stability.

But this kind of stability is certainly the most difficult to verify ; for instance, the adiabatic stability of the eigenvectors of the hamiltonian operator under the Schrödinger equation was not established rigorously -by T. Kato (3)- before 1950, although at that time the validity of the result had long been beyond doubt for physicists and many "partial" demonstrations had already been given. Also, one was dealing with one, well-defined, linear equation... whereas since physicists have not yet been friendly enough to provide us whith the "right" equation, we shall have to consider an arbitrary non linear equation. In fact, it is probably better to view our tentative theorem as a criterion which the "right" equation should satisfy.

## 2. A LIST OF FACTS

Why are solitons susceptible of yielding adiabatic invariants ?

### i) The break-up phenomenon

In some equations containing solitons, one can prove the following statement :

"Given an arbitrary function (satisfying such and such regularity and growth properties) as the initial condition, the solution eventually breaks up into a definite number of soliton waves, travelling at certain predictable speeds with certain predictable phases".

This can be easily established for the Korteweg- de Vries (KdV) equation using the inverse scattering method (see (4) for a neat proof of what follows), the equation is :

$$u_t + u\,u_x + u_{xxx} = 0 \qquad (1)$$

We shall sometimes use subscripts to denote partial derivatives ; equation (1) contains the solitons :

$$S(x - ct + \delta, c) = 3c\ \mathrm{sch}^2\{\frac{\sqrt{c}}{2}(x - ct + \delta)\} \qquad (2)$$

then, if $u(x) \equiv u(x,0)$ satisfies :

$$u \in \mathcal{S}^{\infty}(\mathbb{R}) \; ; \int_{-\infty}^{+\infty} (1 + |x|).|u(x)|.dx < \infty \tag{3}$$

there exist N and $(\eta_j, \delta_j^{\pm})_{j=1}^{j=N}$ such that

$$\forall \varepsilon > 0, \; \limsup_{t\to\pm\infty \, |x|>\varepsilon|t|} \left| u(x,t) - \sum_{j=1}^{j=N} S(x - 4\eta_j^2 t - \delta_j^{\pm}, 4\eta_j^2) \right| = 0 \tag{4}$$

This property of "breaking-up" seems reminiscent of the transition from a "pure state" to a "mixed state" in quantum mechanics, with an eventual independence of the solitons when they are very far from each other ; alternatively, energy could be said to concentrate irreversibly on stable states of motion.

ii) A family of operators associated with solitons

Consider a general non linear equation of the form :

$$\frac{\partial u}{\partial t} = \phi(u,\ldots,u^{(n)}) \tag{5}$$

$\phi$ may be a polynomial expression for instance, and we shall assume that there is only one space dimension ; we further suppose that the Cauchy problem is solvable (forward or in both time directions) for $u(x) \equiv u(x,0)$ in a certain Banach space of functions . One can then look at the operators :

$$c \in \mathbb{R}^+ \qquad \Pi_c : \mathcal{B} \to \mathcal{B} \qquad (\Pi_c u)_{(x)} \equiv \lim_{t\to\infty} u(x + ct, t) \tag{6}$$

(one puts a "+" sign for waves travelling to the right).

The well-defined operator $\Pi_c$ extracts the part of the solution travelling at speed c to the right. If we further suppose (as it is certainly the case when the break-up phenomenon occurs) that this part is always a solitary wave (or zero), we claim that the operators $\Pi_c$ share many properties with the spectral projection operators of quantum mechanics, which, we recall, are the adiabatical invariants of the Schrödinger equation.

- ($\alpha$) $\Pi_c$ is a first integral of (5)
- ($\beta$) It is idempotent ($\Pi_c^2 = \Pi_c$).

Now we suppose further that (5) preserves $E(u) \equiv \int u^2$ (and $\mathcal{B} \subset \mathcal{H} \equiv L^2(\mathbb{R})$) or any other "energy function" invariant under translation, giving rise to a norm ; then :

- ($\gamma$) $\Pi_c$ is a contraction operator in $\mathcal{H}$
- ($\delta$) The $\Pi_c$ 's are in a sense orthogonal (for different c 's) ; in fact, the speed spectrum is continuous but a given u will contain only finitely many solitons of energy greater than $\varepsilon$ ( $\varepsilon > 0$) because energies of solitons add up in the initial function. One has :

$$\forall u \in \mathcal{B}, \quad E(u) \leqslant \sum_{(c)} E(\Pi_c u) \tag{7}$$

the difference between the two sides being provided by the continuous spectrum (radiative part).

. (ε) In the case of the KdV equation, using the inverse scattering method, one sees that the operators $\Pi_c$ coincide with the spectral projections of the operator $L = -\frac{\partial^2}{\partial x^2} + u$ acting on $L^2_{(\mathbb{R})}$ and that they are therefore linear operators when viewed on the space of scattering data for L. Another way of studying the quasi-linear behaviour of solitons is via the Bäcklund transform and especially the Biankin superposition principle (see (5) on this subject).

We now move on to a precise mathematical statement of a tentative adiabatic theorem for solitons.

## 3. AN ADIABATIC THEOREM FOR SOLITONS

Let us consider an equation of the form :

$$\frac{\partial u}{\partial t} = \phi_t(u,\dots,u^{(n)}) \tag{8}$$

where the subscript t means that the coefficients of the equation are time-dependent (in the second part, we shall consider the special case : $\frac{\partial u}{\partial t} = -\alpha(t)u\,u_x - u_{xxx}$).

Since we want to consider an increasingly slow variation, we put $t = \tau S$ ($S \in [0,1]$) and let τ tend to infinity, supposing the coefficients are in fact functions of S only, which allows us to transform (8) into :

$$\frac{\partial u}{\partial S} = \tau\ \phi_S(u,\dots,u^{(n)}) \tag{9}$$

The reader will easily be able to check that we are thus really looking at an equation with more and more slowly varying coefficients. We make two assumptions :

i) The energy (or norm) $E(u) = \int u^2$ is conserved under (9) ; we emphasize that one could think of different first integrals and the choice of E is arbitrary but proves the "physical" one, at least in the case studied in part 4. However, this assumption could certainly be modified according to the equation under study.

ii) For any $S_0 \in [0,1]$ and $c > 0$, the equation

$$\frac{\partial u}{\partial t} = \phi_{S_0}(u,\dots,u^{(n)}) \qquad (10)_{S_0}$$

contains a soliton travelling at speed c (in part 4, (10) will read : $u_t + \alpha(t_0)u\,u_x + u_{xxx} = 0$ with $\alpha(t_0) \equiv \alpha$, fixed ; this is the KdV equation) and there is a dispersion relation relating the energy and speed of solitons in a monotonous way (in part 4, we shall have $E_{(c)} \sim c^{3/2}$). This monotonicity condition translates the principle that "big waves tend to travel faster than small ones". It enables us to invert the relation in the form of a function $c = c_{S_0}(E)$ giving the speed c as a function of E in the equation $(10)_{S_0}$.

We write the soliton in (10) in the form

$$\sigma = \sigma(S_0, c, x - ct + \delta) \tag{11}$$

A small difficulty has to be pointed out ; it is not possible to relate the phases $\delta$ for different equations (10) i.e. for different $S_0$ since the determination of $\delta$ involves an arbitrary origin for each equation. In the following tentative theorem, however, we use a function $\delta(S)$ ; this can be done by relating the different origins e.g. chosing "the top of the wave" or the center of mass, or anything else as a marker.

We now state the conjecture in a mathematical form :

We consider the equation :

$$\frac{\partial u}{\partial S} = \tau\, \phi_S(u,\dots,u^{(n)}),\ S \in [0,1],\quad \tau > 0 \tag{9}$$

satisfying the conditions stated above. If $u(x) \equiv \sigma(0, c, x)$ is a fixed soliton initial condition and $u_\tau(X,S)$ is the solution of (9) for the fixed value $\tau$ of this parameter, with $u_\tau(X,0) = u(x)$ $(\forall\tau)$, one has :

$$\forall S \in [0,1],\ \lim_{\tau\ \infty} u_\tau(X,S) = \sigma(S, C_S(E), x - \tau\int_0^S C_u(E)du + \delta(S)) \tag{12}$$

where E is the energy of u (at time zero, hence at any time $t = \tau S$).

We call the function on the right hand side of (12), the "drifting soliton" associated to equation (9). In words, we can state an <u>adiabatic principle for solitons</u> :

"Drifting solitons are asymptotic solutions of non linear equations containing solitons, with slowly varying parameters".

I insist that this is certainly very difficult to prove in the general case (in fact, it may be false ! ) and should rather be viewed as a criterion to select physically relevant equations.

Finally, we make a few mathematical remarks :

i) The limit in (12) is taken in some function space, but we cannot make it more precise in the general case (of course, one thinks of $L^2(\mathbb{R})$ but it may be too simple). At this point, it should be noticed that a constant feature of the recent progress in the field is the replacement of function space estimates by algebraic relations ; this will be illustrated in the next part.

ii) The most important statement concerns the <u>shape</u> of the solution ; its <u>phase</u> is only a reasonnable guess in which a function $\delta(S)$ appears. This function can be determined a little more, using the two following observations :

- $u(X,S)$ is regular (usually $C^\infty$), which determines the regularity of $\delta(S)$.

- A more subtle idea is that there should exist, by analogy with the adiabatic theorems in classical and quantum mechanics, a principle of "kinematical simplicity" expressed in terms of $\frac{\partial\sigma}{\partial S}(S, C_S, x + \delta(S))$ (independenly of $\tau$), and more precisely in terms of the action of $\Pi_C^{(S)}$ (the operators corresponding to the equations $(10)_S$) on this expression.

## 4. A PARTICULAR CASE ; THE KdV EQUATION

We now study a particular case of this conjecture, first stating the mathematical result, and then analysing the connection with the above discussion. A first task is to find sufficiently simple one-parameter families of time-independent non linear P.D.E. containing solitons. Equation (13) is certainly one of the simplest :

$$u_t + \alpha(t)\, u\, u_x + u_{xxx} = 0 \tag{13}$$

In the sequel, $\alpha$ will be a positive nowhere vanishing function on the real line, with the inessential restriction that $\alpha(0) = 1$. When $\alpha$ is a constant, equation (13) is denoted by $(\mathrm{KdV})_\alpha$ ; all these equations are equivalent up to scaling, and for convenience, we refer to the KdV equation as $(\mathrm{KdV})_1$. We make the physical remark that equation (13) can be "realized" by heating the medium (or performing any operation which continuously modifies the viscosity coefficient) in which waves obeying equations of the KdV type are travelling.

We will not discuss the Cauchy problem for equation (13), so that our result will be purely algebraic. Now, it is well-known that the equations $(\mathrm{KdV})_\alpha$ have a whole series of first integrals (see (6) for instance) $H^\alpha_{n,\alpha}$, of the form :

$$H_{n,\alpha}[u] = \int_{-\infty}^{+\infty} T_{n,\alpha}[u]\,dx \tag{14}$$

where $T_{n,\alpha}$ is a polynomial in $u, u', \ldots, u^{(n)}$, homogeneous of degree n with respect to the grading given by :

$$\deg u \equiv 1 \quad ; \quad \deg \frac{\partial}{\partial x} \equiv \frac{1}{2}$$

To normalize $T_{n,\alpha}$, we shall say that it is obtained from $T_{n,1}$ by changing u into $\alpha u$ and then factoring out a power of $\alpha$, so that the resulting $T_{n,\alpha}$ is not divisible by $\alpha$ ; for example :

$$T_{1,\alpha}[u] = u \quad ; \quad T_{2,\alpha}[u] = u^2 \quad ; \quad T_{3,\alpha}[u] = \frac{1}{3}\alpha u^3 - u_x^2 \tag{15}$$

We notice that the two "physical" integrals $H_1 \equiv H_{1,\alpha}$ (total momentum) and $H_2 \equiv H_{2,\alpha}$ (total energy) are still invariant under equation (13), whereas for general n, one has the following :

Proposition :

Let u be a solution of (13), then :

$$\forall n \geqslant 1,\ \forall t \in \mathbb{R},\ \frac{d}{dt}H_{n,\alpha}[u] = \alpha'_{(t)} \cdot \int_{-\infty}^{+\infty} \frac{\partial}{\partial\alpha} T_{n,\alpha}[u]dx \tag{16}$$

Proof : Straightforward ; the algebraic relations entailing the constancy of $H_{n,\alpha}$ for constant $\alpha$, make it possible to eliminate the time derivatives arising from the variation of u.

Next, we construct the drifting soliton for equation (13), starting from the soliton of $(KdV)_\alpha$ travelling at speed c :

$$S(x,t,\alpha,C) \equiv 3C\alpha^{-1} \operatorname{sech}^2\left(\frac{1}{2} C^{1/2}(x - ct)\right) \tag{17}$$

An elementary computation furnishes the momentum and the energy of this soliton :

$$H_1[S] = \int S = 12.C^{1/2}.\alpha^{-1} \ ;\ H_2[S] = \int S^2 = 24.C^{3/2}.\alpha^{-2} \tag{18}$$

By definition, the drifting soliton is given by :

$$S(x,t,\alpha) \equiv 3C(t)\alpha^{-1}_{(t)}.\operatorname{sech}^2\left(\frac{1}{2}C(t)(x - \int_0^t C(s)ds)\right) \tag{19}$$

where C(t) and $\alpha(t)$ are related by :

$$C(t) \equiv C_0\ \alpha^{4/3}_{(t)} \qquad (C_0 > 0) \tag{20}$$

That is, by (20), the drifting soliton is a travelling wave with constant energy. We emphasize that :

i) The phase in (19) is tentative since no information will be obtained by means of the study of the integrals to support this guess.

ii) The choice to preserve the energy rather than momentum appears as arbitrary (but will prove essential).

One has :

$$H_1[S(x,t,\alpha)] = 12C_0^{1/2}.\alpha^{-1/3}_{(t)} \quad ; \quad H_2[S(x,t,\alpha)] = 24\ C_0^{3/2} \tag{21}$$

Now the theorem is :

Let (13) be the equation :

$$u_t + \alpha(t)u\,u_x + u_{xxx} = 0 \qquad (\alpha(t) > 0\ \forall t) \tag{13}$$

and put :

$$S(x,t,\alpha) \equiv 3C\alpha^{-1}.\operatorname{sech}^2\left(\frac{1}{2}C^{1/2}(x - \int_0^t C(S)ds)\right) \tag{19}$$

with (20)

$$C(t) \equiv C_0\ \alpha^{4/3}_{(t)}$$

that is (19 bis)

$$S(x,t,\alpha) = 3C_0\alpha^{1/3}.\mathrm{sech}^2\left[\frac{1}{2}C_0^{1/2}.\alpha^{2/3}(x-C_0\int_0^t \alpha^{4/3}_{(S)}\, dS)\right]$$

Then, for n = 2,3,4 :

$$\forall t \in \mathbb{R} \; ; \; \frac{d}{dt} H_{n,\alpha}[S] = \alpha'_{(t)}\cdot\int_{-\infty}^{+\infty}\frac{\partial}{\partial\alpha} T_{n,\alpha}(S)dS \tag{21}$$

<u>Conjecture</u> : The relation (21) extends to all n, n $\geq$ 2.

We make two remarks :

i) (21) means that the drifting soliton satisfies <u>the same relations</u> (16) as does the actual solution of (13). This string of relations will determine the spectral data associated to the operator $L_t = -\frac{\partial^2}{\partial x^2} + u_{(x,t)}$ at time t and the conjecture is that they are all satisfied by drifting soliton (19) <u>except for the first</u>.

ii) This should generalize in some way to the higher order equations of the KdV hierarchy.

We now prove the theorem by inspection :

(I) (21) is satisfied for n = 2 by construction.

(II) n = 3 ; here (21) reads :

$$\frac{d}{dt}\int (\frac{1}{3}\,\alpha S^3 - S_x^2) = \frac{1}{3}\,\alpha'(t)\int S^3 \tag*{(21)$_3$}$$

To check this identity, one uses the algebraic relations satisfied by (19) because when $\alpha$ is constant S is the soliton of $(KdV)_\alpha$ travelling at speed C :

$$\begin{aligned} &\text{i)}\; -C\,S_x + \alpha\,S\,S_x + S_{xxx} = 0\\ &\text{ii)}\; -C\,S + \frac{1}{2}\,\alpha\,S^2 + S_{xx} = 0\\ &\text{iii)}\; -C\,S^2 + \frac{1}{3}\,\alpha\,S^3 + S_x^2 = 0 \end{aligned} \tag{22}$$

In this way, $(21)_3$ is transformed into :

$$\frac{d}{dt}\int (\frac{2}{3}\,\alpha\,S^3 - C\,S^2) = \frac{1}{3}\,\alpha'\int S^3 \tag*{(21)$_3$}$$

This identity is checked directly (see the table below).

(III) The case n = 4 is dealt with similarly, using relations (22), and the elementary identity :

$$\int_{-\infty}^{+\infty}\mathrm{sech}^{2n}x.dx = 2^{2n-1}\int_1^{\infty}(y-1)^{n-1}.y^{-2n}.dy \tag{23}$$

One has :

$$T_{4,\alpha}[S] = \frac{1}{4}\alpha^2S^4 - 3\alpha S\,S_x^2 + \frac{9}{5}S_{xx}^2 = \frac{17}{10}\alpha^2S^4 - \frac{24}{5}\alpha CS^3 + \frac{9}{5}C^2S^2 \tag{23}$$

and (21) reads :

$$\frac{d}{dt}\int T_{4,\alpha} \doteq \frac{3}{2}\,\alpha.\alpha'.\int S^4 - 3\,C\,\alpha'\int S^3 \qquad (21)_4$$

bearing in mind $C = C_0\alpha^{4/3}$ ; the right hand side has been transformed, using (22), after the derivation is performed.

The values of the integrals and their derivatives are tabulated here :

| n | $H_{n,\alpha}[S,t]$ | $\frac{d}{dt}H_{n,\alpha}[S,t]$ |
|---|---|---|
| 1 | $2^2.3\;C_0^{1/2}.\alpha^{-1/3}$ | $-2^2C_0^{1/2}.\alpha^{-4/3}.\alpha'$ |
| 2 | $2^3.3.C_0^{3/2}$ | 0 |
| 3 | $\frac{2^3.3^2}{5}.C_0^{5/2}.\alpha^{4/3}$ | $\frac{2^5.3}{5}\;C_0^{5/2}.\alpha^{1/3}.\alpha'$ |
| 4 | $\frac{2^3.3^4}{5.7}\;C_0^{7/2}.\alpha^{8/3}$ | $\frac{2^6.3^3}{5.7}\;C_0^{7/2}\;\alpha^{5/3}.\alpha'$ |

After these purely mathematical considerations, we return to the connection of this result with the conjectured adiabatic theorem. As we already pointed out, the result above is purely algebraic and does not provide any inequality in a function space bounding the difference between the drifting soliton and the actual solution. However, one can argue as follows : suppose $\alpha$ is slowly varying, say $\alpha(t) = \tilde{\alpha}(\frac{t}{\tau})$ with $\tilde{\alpha}$ a fixed function and $\tau$ large. Now (16) can be considered as an asymptotic series in the inverse powers of $\tau$ since :

$$\frac{d\alpha}{dt} = \frac{1}{\tau}\frac{d\tilde{\alpha}}{dS}\,, \qquad t = \tau S \qquad (25)$$

By the adiabatic theorem, one would expect the matching of the respective first terms (coefficients of $\tau^{-1}$) of the series giving $\frac{dH_{n,\alpha}}{dt}(u)$ and $\frac{dH_{n,\alpha}}{dt}(S)$ (where u is the actual solution and S is the drifting soliton). But these series have only one term, so that they in fact coincide and one does not have to consider a "slowly varying $\alpha$". This is how algebra works. The fact that the identity fails to occur for the momentum may indicate that one should not chose $L^1(\mathbb{R})$ as a function space to work in, when trying to derive inequalities related to the adiabatic conjecture.

REFERENCES

(1) *Scott, A.C., Chu, F.Y.F., and Mc Laughlin, D.W. ; The soliton: a new concept in applied science* ; Proceedings of the I.E.E.E., *51 (1973)*

(2) *Benjamin, T.B. ; The stability of solitary waves*, Proc. of the Roy. Soc. of London, *A 328 (1972)*

(3) *Kato, T.* ; Journal of the Phys. Soc.of Japan *(1950)*

(4) *Tanaka, S., ; KdV : Asymptotic behavior of solutions*, Publ. R.I.M.S. *Kyoto Univ.* 10 *(1975)*
(5) Lecture Note in Maths n° 515 ; *edited by R.M. Miura*, Springer Verlag
(6) *Gardner, C., Kruskal, M., and Miura, M. ; KdV equation and generalizations II*, Journal of Math. Phys. *9 (1968)*.

Note added in proof : The result stated as a "conjecture" in part 4 has now been demonstrated and extended to other equations ; an account of this work will appear shortly.

# INTRODUCTION OF THE CHRONON IN THE THEORY OF ELECTRON AND THE WAVE-PARTICLE DUALITY

P. Caldirola

Istituto di Scienze Fisiche dell'Università
Via Celoria, 16 - 20133 MILANO - Italy



## 1 - INTRODUCTION

Immediately after the discovery of the electron many attempts have been made at understanding its existence on the basis of the classical theory of electromagnetic field. To this goal many hypotheses have been put forward, like some models of extended electron (1-5). However, such models were not completely satisfactory, mainly because they could not be easily put in a relativistically invariant form. In connection with these difficulties, the opportunity of considering the electron as a point particle was suggested by Frenkel (6) and subsequently developed by several authors and particularly by Dirac (7). Some mathematical divergences are introduced by point like model and, once they are (rather artificially) eliminated, in some problems we are faced with a motion of the electron not only of a pathological character but also in disagreement with experimental evidence.



S. Diner et al. (eds.), The Wave-Particle Dualism, 183–213.

In order to overcome these difficulties, a theory of the classical electron has been developed by us (8) many years ago, based on the introduction of a fundamental interval of time (the so called chronon). Such a theory has recently been reconsidered and partially improved (9).

One of the most interesting results of our theory appears to be the fact that, while in its external (or macroscopical) motion the electron may be considered a point particle, in its internal (or microscopical) motion the electron results to be associated with a microuniverse and this gives an explanation of the intrinsic properties of the electron, like the intrinsic energy related to the rest mass, the intrinsic "anomalous" magnetic moment, etc.

It has been also possible (10) to derive quantal equations (describing in particular the wave-like behaviour of the particle) based on the introduction of the chronon. Such a formulation has allowed us, without the introduction of any new additional hypothesis, to reach important results, like the interpretation of the muon as an excited mass state of the electron. Let us recall that a general hystorical survey of the introduction of the chronon in physics and in particular in the theory of the electron proposed by us has been given by Arzeliès (11), Casagrande (12) and Lacroix (13); the epistemological implications of such a theory of the electron have been discussed by Bruzzaniti (14) and its physical meaning in the framework of Relativity has been stressed by Ivanenko (15), one of the pioneers in the introduction of the chronon.

In the present paper we are going to summarize the more important results obtained in the electron theory based on the chronon and to stress some peculiarities of the wave-particle duality directly connected with the introduction of the chronon itself

## 2 - SUMMARY OF THE CLASSICAL MOTION OF THE ELECTRON

The classical theory of the electron based on the introduction of the chronon leads to equations of motion of finite-difference type in the velocity variable. To this equation the so-called "transmission law" is associated, that is the relation between position and velocity inside any fundamental interval of time typical of the theory.

The fundamental relativistic equations of the theory for the classical electron may be written as follows:

a) for the radiating electron

$$\frac{m_\circ}{\tau_\circ}\{u_\alpha(\tau)-u_\alpha(\tau-\tau_\circ)+\frac{u_\alpha(\tau)u_\beta(\tau)}{c^2}[u_\beta(\tau)-u_\beta(\tau-\tau_\circ)]\} = \frac{e}{c}F_{\alpha\beta}(\tau)u_\beta(\tau) \qquad (1)$$

$$x_\alpha(\tau)-x_\alpha(\tau-\tau_\circ) = \frac{\tau_\circ}{2}\left[u_\alpha(\tau)+u_\alpha(\tau-\tau_\circ)\right] \tag{1}$$

b) <u>for the absorbing electron</u>

$$\frac{m_\circ}{\tau_\circ}\left\{u_\alpha(\tau+\tau_\circ)-u_\alpha(\tau)+\frac{u_\alpha(\tau)u_\beta(\tau)}{c^2}\left[u_\beta(\tau+\tau_\circ)-u_\beta(\tau)\right]\right\} = \frac{e}{c}F_{\alpha\beta}(\tau)u_\beta(\tau)$$

$$x_\alpha(\tau+\tau_\circ)-x_\alpha(\tau) = \frac{\tau_\circ}{2}\left[u_\alpha(\tau+\tau_\circ)+u_\alpha(\tau)\right] \tag{2}$$

c) <u>for the non-radiating non-absorbing electron</u>

$$\frac{m_\circ}{2\theta_\circ}\left\{u_\alpha(\tau+\theta_\circ)-u_\alpha(\tau-\theta_\circ)+\frac{u_\alpha(\tau)u_\beta(\tau)}{c^2}\cdot\left[u_\beta(\tau+\theta_\circ)-u_\beta(\tau-\theta_\circ)\right]\right\} = \frac{e}{c}F_{\alpha\beta}(\tau)u_\beta(\tau) \tag{3}$$

$$x_\alpha(\tau+\theta_\circ)-x_\alpha(\tau-\theta_\circ) = 2\theta_\circ u_\alpha(\tau)$$

In the equations written above

$$\theta_\circ = \tau_\circ/2 = 2/3 \cdot e^2/m_\circ c^3 \tag{4}$$

is the chronon associated to the electron; $u_\alpha(\tau)$ is the four-vector velocity satisfying the condition

$$u_\alpha(\tau)u_\alpha(\tau) = -c^2 \qquad \text{for} \quad \tau = n\tau_\circ \tag{5}$$

(where $n = 0,1,2,\ldots$ and $\alpha = 1,2,3,4$) and $F_{\alpha\beta}$ is the electromagnetic field tensor [1].

The univocally determined solutions in the points $\tau = n\tau_\circ$ of the different systems considered give the so-called "macroscopical motion" of the electron defined at the instants $\tau = n\tau_\circ$. To this "macroscopical" motion an arbitrary "internal motion", vanishing at the extreme points $\tau=n\tau_\circ$ of every elementary interval of time, may be superimposed. Such a motion is connected with some intrinsic properties of the particle.

It is noticeable the fact that the theory allows also the

description of the positon, considered as the antiparticle of the electron. In fact the so-called reinterpretation principle introduced in the ordinary classical electrodynamics by Stückelberg (16), adopted by Feynman (17) and systematically embodied into the theory of relativity by Recami and co-workers (18), allows us to verify that equation (2) for an electron (charge e) moving backwards in time under the action of the external advanced electromagnetic field is equivalent to equation (1) for a positon (charge -e) going forward in time, acted upon by the ordinary retarded fields.

The corresponding equations in the non-relativistic approximation can be written as follows:

a) for the radiating electron

$$\frac{m_\circ}{\tau_\circ}\left[\underline{\underline{v}}(t)-\underline{\underline{v}}(t-\tau_\circ)\right] = \underline{\underline{F}}(t)$$

$$\underline{\underline{r}}(t)-\underline{\underline{r}}(t-\tau_\circ) = \frac{\tau_\circ}{2}\left[\underline{\underline{v}}(t)+\underline{\underline{v}}(t-\tau_\circ)\right] \qquad (1')$$

with

$$\underline{\underline{F}}(t) = e\left(\underline{\underline{E}}(t) + \frac{\underline{\underline{v}}(t)}{c} \times \underline{\underline{H}}(t)\right)$$

b) for the absorbing electron

$$\frac{m_\circ}{\tau_\circ}\left[\underline{\underline{v}}(t+\tau_\circ)-\underline{\underline{v}}(t)\right] = \underline{\underline{F}}(t)$$

$$\underline{\underline{r}}(t+\tau_\circ)-\underline{\underline{r}}(t) = \frac{\tau_\circ}{2}\left[\underline{\underline{v}}(t+\tau_\circ)+\underline{\underline{v}}(t)\right] \qquad (2')$$

c) for the non-radiating non-absorbing electron

$$\frac{m_\circ}{2\theta_\circ}\left[\underline{\underline{v}}(t+\theta_\circ)-\underline{\underline{v}}(t-\theta_\circ)\right] = \underline{\underline{F}}(t)$$

$$\underline{\underline{r}}(t+\theta_\circ)-\underline{\underline{r}}(t-\theta_\circ) = 2\theta_\circ\underline{\underline{v}}(t) \qquad (3')$$

The closed system of the equations of motion give a solution which is univocally determined in a sequence of discrete instants of time, separated from each other by the interval $\tau_\circ$, that is 0, $\tau_\circ, 2\tau_\circ \ldots r\tau_\circ \ldots$. Such a motion will be called the macroscopic motion of the electron.

Let us recall that the second equation of each system (1), (2) or (3) is called the transmission law: such a law assumes a constant value of the velocity in the interval $(n-1)\cdot\tau_\circ \leqslant \tau \leqslant n\tau_\circ$. But the macroscopical motion remains inaltered if, instead of a constant value of $u_\alpha(\tau)$ within each elementary of time, we assume any function varying in such a way that its average value is equal to

$$\frac{1}{2}\left[u_\alpha((n+1)\tau_\circ)+u_\alpha(n\tau_\circ)\right]$$

This means in particular that the particle's macroscopical motion is unaltered by the superposition on its uniform internal motion involved in the transmission law, of any other internal or microscopic motion, provided the average velocity of the latter is zero in an elementary interval of time $\tau_\circ$ .

In any case, even if we are interested only in the macroscopic motion, condition (5) is still verified for any $\tau = n\tau_\circ$.

The solutions of system (1) (or (2) or (3)), or of the corresponding non-relativistic equations, are explicity obtained for many problems. As we can see the worst difficulty of the classic Dirac equation is thus overcome: the macroscopic motion of the electron is completely determined once the initial position and velocity are given. In this connection we point out that higher derivatives of the velocity appearing in the classical Dirac's equation of motion can be explained since this equations may be obtained from our finite-difference equation by cutting off the expansion in $\tau_\circ$ powers of the function $u_\alpha(\tau-\tau_\circ)$ after a finite number of terms. In this approximation we have no more internal solutions, but a number of arbitrary constants higher than two is now required for the determination of the macroscopical motion.

Let us remark that, in the theory proposed by us, time flows continuously as in ordinary theory; however under the action of an external force a sequence of sudden variations ("discontinuities") takes place, along the electron's world-line, concerning the particle's dynamical state. Those discontinuities are separated from each other by the same interval of proper time $\tau_\circ$. More precisely by a discontinuity we mean a sudden variation (taking place in an interval of time $\tau_\circ$) of the velocity vector due to the action of an external force: the electron velocity jumps from the value $u_\alpha(\tau-\tau_\circ)$ to the value $u_\alpha(\tau)$ , where $\tau = s/c$ is the proper time of the electron.

From this fact it follows (19), according to a suggestion made by Caianiello (20), that the acceleration cannot overcome in the relativistic theory a maximal value represented by

$$a_\circ = c/\tau_\circ = 3/4 \cdot m_\circ c^4/e^2$$

## 3 - THE MOTION OF THE RADIATING ELECTRON

We will now discuss the equations (1) or (1') for the radiating electron and give a list of solved problems.

First at all let us point out here that the finite-difference equation by putting

$$\Delta Q_\alpha = m_\circ \left[u_\alpha(\tau) - u_\alpha(\tau-\tau_\circ)\right] \tag{6}$$

$$R_\alpha(\tau) = -\frac{m_\circ}{2\tau_\circ}\left\{\frac{u_\alpha(\tau)u_\beta(\tau)}{c^2}\left[u_\beta(\tau+\tau_\circ)+u_\beta(\tau-\tau_\circ)-2u_\beta(\tau)\right]\right\} \tag{7}$$

$$S_\alpha(\tau) = \frac{m_\circ}{2\tau_\circ}\left\{\frac{u_\alpha(\tau)u_\beta(\tau)}{c^2}\left[u_\beta(\tau+\tau_\circ)-u_\beta(\tau-\tau_\circ)\right]\right\} \tag{8}$$

may be written in the form

$$\frac{\Delta Q_\alpha(\tau)}{\tau_\circ} + R_\alpha(\tau) + S_\alpha(\tau) = \frac{e}{c} F_{\alpha\beta}(\tau)u_\beta(\tau) \quad . \tag{9}$$

The first term $\Delta Q_\alpha(\tau)/\tau_\circ$ represents the (mean) variation for unit of proper time, in the interval between $\tau-\tau_\circ$ and $\tau_\circ$, of the momentum-energy vector of the particle.

The second term $R_\alpha(\tau)$ is a dissipative one, since it contains only even derivatives of the velocity, as one can prove with an expansion of $u_\beta(\tau+\tau_\circ)$ and $u_\beta(\tau-\tau_\circ)$ in terms of $\tau_\circ$. Furthermore, it can be proved that it is never negative and, therefore, it represents the momentum-energy radiated by the electron in the unit of proper time.

The third term $S_\alpha(\tau)$ is a conservative one and it represents the rate of change of the reaction momentum-energy of the electron in the proper time.

We note that, by writing the equation corresponding to $\alpha = 4$, we have

$$\frac{T(\tau)-T(\tau-\tau_\circ)}{\tau_\circ} + R_\circ(\tau) + S_\circ(\tau) = P_{ext}(\tau) \tag{10}$$

where, by using the formalism of finite-difference electrodynamics, the first term is the analogous of the proper-time derivative of the kinetic energy.

$$T(\tau) = m_\circ c^2 \left\{\frac{1}{\sqrt{1-\beta^2}} - 1\right\}$$

the second term $R_\circ(\tau)$ is the rate of the radiated energy, $S_\circ(\tau)$ is the change of the reaction energy of the electron in the unity of proper time (radiative correction) and $P_{ext}(\tau)$ is the rate of work done by external forces in the proper time.

With equation (10) we are able to explain the origin of

the so-called acceleration energy appearing in the energy conservation of Dirac equation. Expanding all the terms on the left-hand side of equation (10) in a power series of the parameter $\tau_\circ$ and keeping only terms of the first order, we get

$$\frac{T(\tau)-T(\tau-\tau_\circ)}{\tau_\circ} \simeq \frac{dT}{d\tau} - \frac{2}{3}\frac{e^2}{m_\circ c^3}\frac{d^2T}{d\tau^2} = \frac{dT}{d\tau} - \frac{2}{3}\frac{e^2}{c^3}\frac{da_\circ}{d\tau} \tag{11}$$

$$R_\circ(\tau) \simeq 1/\sqrt{1-\beta^2} \cdot 2/3 \cdot e^2/c^3 \cdot a_\mu a_\mu \tag{12}$$

$$(a_\mu = du_\mu/d\tau)$$

$$S_\circ(\tau) \simeq 0 \tag{13}$$

To the first order in $\tau_\circ$ we can, therefore, write equation (10) in the form

$$\frac{dT}{d\tau} - \frac{2}{3}\frac{e^2}{c^2}\frac{da_\circ}{d\tau} + \frac{1}{\sqrt{1-\beta^2}}\frac{2}{3}\frac{e^2}{c^3} a_\mu a_\mu = P_{ext}(\tau) \tag{14}$$

or, going from the proper time $\tau$ to the observer time t,

$$\frac{dT}{dt} - \frac{2}{3}\frac{e^2}{c^2}\frac{da_\circ}{dt} + \frac{2}{3}\frac{e^2}{c^3} a_\mu a_\mu = \frac{d\tau}{dt} P_{ext}(\tau) \tag{15}$$

In this relation, which is identical with the equation of the energy conservation law deduced by Fulton and Rohrlich (21) for the Dirac equation, we have the derivative of the term (2/3) $(e^2/c^2)$ $a_\circ$. This term is just the acceleration energy, also known as Schott energy. Going back to our deduction of equation (14), we clearly see now that Schott energy arises by expanding in a power series of $\tau_\circ$ the change of the kinetic energy during the fundamental interval $\tau_\circ$ of proper time. The acceleration energy formally disappears if we use our finite-difference equation instead of the classical Dirac equation.

The solution of the relativistic equation (1) for the radiating electron, and also of the corresponding non-relativistic equations (1') is generally rather difficult. Here we limit ourselves to a list of solved problems.

a) Exact relativistic solutions
   1) Free electron
   2) Electron under the action of an instantaneous electromagnetic field (Cirelli (22))
   3) Hyperbolic motion (Lanz (23))

b) Non-relativistic approximation solutions
   1) Electron under the action of time-dependent forces (9)
   2) Electron in a constant uniform magnetic field (Prosperet-

ti (24))

3) Electron on a straight line under the action of an elasticrestoring force (Caldirola, Casati, Prosperetti (25)). Some other solutions have been obtained by means of numerical calculations and Petzwold and Sorg (26) have studied from the mathematical point of view a class of equations generalizing equation (1).

4 - INTERNAL MOTIONS AND THE ELECTRON'S STRUCTURE

As we have already pointed out, the finite-difference equations for the electron motion allow arbitrary solutions that have to be added to the ones describing the particle macroscopic motion in the point $\tau = n\tau_0$ in a univocal way. These solutions correspond to the so-called internal motions, that is to the motions taking place inside the fundamental interval of proper time $(n-1)\tau_0 \leqslant \tau \leqslant n\tau_0$ being connected in a certain way to the "structure" of the particle.

We would like to point out here that the internal motions we considered are, as already observed by Bohm and Weinstein (27) for the non-relativistic equation (1'), of non-radiating kind. In the framework of the general formalism of finite-difference electrodynamics, this follows immediately from the fact that the vectors $R_\alpha(n\tau_0)$ and $S_\alpha(n\tau_0)$ describing the electron's radiation and self-reaction, which we considered earlier in sect. 3, are not influenced at all by the existence of such motions.

A more general way of dealing with the problem of the "internal" behaviour and, therefore, of the "structure" of the electron is given by interpreting the internal solutions of the finite-difference equations of motions in terms of a general geometrical description based on relativity (28) (29).

From this point of view, the more interesting case is the study of the free electron ($F_{\alpha\beta} = 0$). The relativistic equation of motion becomes

$$u_\alpha(n\tau_0) - u_\alpha[(n-1)\tau_0] = 0 \qquad (16)$$

The "internal" solutions are the solutions of this equation that can be added to $u_\alpha(\tau) = \text{const.}$

With the substitution $\tau \to \tau + \theta_0$, we can, without any loss of generality, limit ourselves to the consideration of the proper-time interval $-\theta_0 \leqslant \tau \leqslant \theta_0$.

The four chronotopical coordinates x, y, z, ct involved in the internal motion of the electron satisfy during this interval the inequality

$$c^2t^2 - x^2 - y^2 - z^2 \leqslant c^2\theta_0^2$$

The points $P \equiv (x,y,z,ct)$, corresponding to the internal motions, are then the points external to the three-dimensional quadric ("absolute of Cayley-Klein"), whose equation is

$$c^2 t^2 - x^2 - y^2 - z^2 = c^2 \theta_0^2 \tag{17}$$

It is easy to prove that a consistent geometrical interpretation of this fact can be obtained by assuming that the internal motions of the point-like electron take place in a de Sitter (absolute) microuniverse $V_4$ with positive constant curvature $1/c^2\theta_0^2$, associated to the electron itself, but they are actually observed in a (relative) microuniverse (also called Castelnuovo microuniverse) $V_4^*$ obtained by a mapping of $V_4$ on a flat manifold (extension of Beltrami theorem): in such a mapping the images of the geodesics of $V_4$ reduce to straight lines.

In the relative microuniverse the "internal motions of the electron happen to arise in the interior part of the elementary time interval $0 \leqslant \tau \leqslant 2\theta_0$.

Such internal motions are superimposed to the "macroscopic" motion of the electron, which instead is defined only on the discrete instants $\tau = 0, 2\theta_0, 4\theta_0, \ldots$

As a matter of fact, in this model of the electron one actually considers two kinds of time: the one, called "macroscopic" or "extrinsic" $\tau$, associated to the macroscopic motion of the electron, the other, called "microscopic" or "intrinsic" $t^*$, associated to the electron internal motion. There exists a definite connection between these two kinds of time, since to a variation of the external time $\tau$ it corresponds (in the relative description) a variation of the intrinsic time $t^*$, as shown in figure 1.

By considering the "simple rotations" of the Fantappiè group (30)[2], it is also possible to show (23) that, in the "relative description, to the pointlike electron it is associated

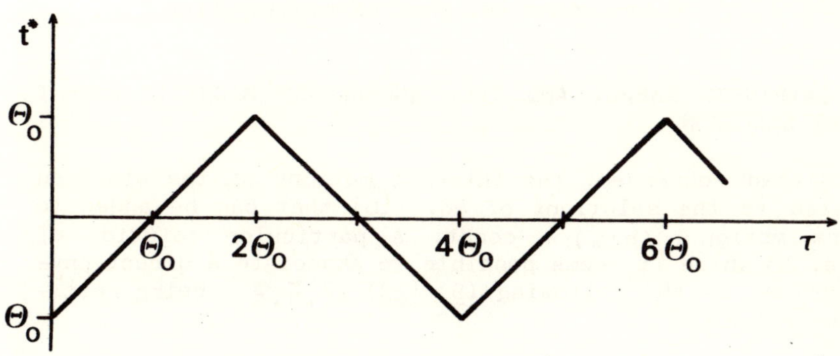


Fig. 1 - Connection between "intrinsic" time t* and "extrinsic" time τ

an oscillating spherical microcosmos in the ordinary 3-dimensional space: the variation of the radius $r^*$ of this microcosmos is represented in Fig. 2 as a function of the "extrinsic" time $\tau$.

At the extreme of any elementary interval of the proper time, i.e. at the instants at which the macroscopic motion is defined, the electron microcosmos reduces to a point: in other words, as far as the description of macroscopic motion is concerned, the electron may actually be considered as a point particle moving in the usual Minkowski space-time of ordinary special relativity: such a motion is described by a discrete sequence of points corresponding to the positions of the electron in the instants of proper time $\tau = 0, 2\theta_\circ, 4\theta_\circ, \ldots$.

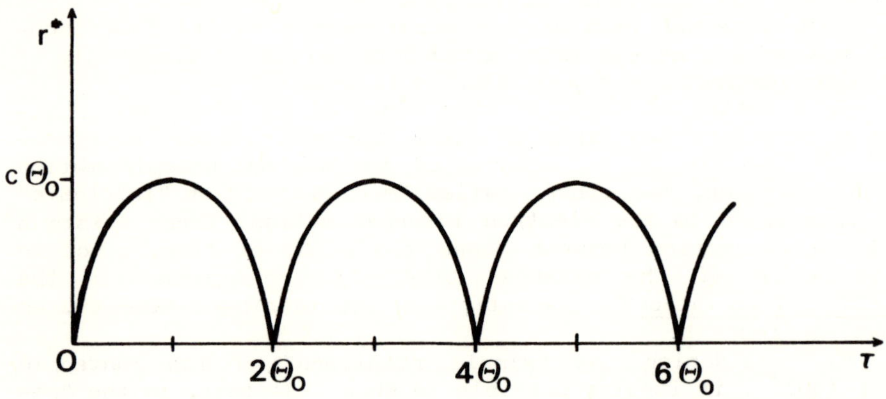


Fig. 2 - The radius $r^*$ of the relative electronic microcosmos vs. the "extrinsic" time $\tau$

## 5 - THE INTRINSIC ENERGY AND THE "ANOMALOUS" MAGNETIC MOMENT OF THE ELECTRON

As already observed, the internal motions of the electron are related to the solutions of eq. (16) that can be added to a uniform motion $u_\alpha(n\tau_\circ) = \text{const.}$ A particular solution of this type, to which it seems possible to associate a direct physical meaning, is the following (9) (28) ($\bar{\zeta}, \bar{\eta}, \bar{\xi}$ being rectilinear co-ordinates):

$$\frac{d\bar{\zeta}}{dt} = \dot{\bar{\zeta}} = -\beta_\circ c \sin(2\pi \frac{\bar{t}}{\tau_\circ}) \tag{18}$$

$$\frac{d\bar{\eta}}{dt} = \dot{\bar{\eta}} = \beta_\circ c \cos(2\pi \frac{\bar{t}}{\tau_\circ}) \tag{18}$$

$$\frac{d\bar{\xi}}{dt} = \dot{\bar{\xi}} = 0$$

from which we get

$$\bar{\zeta} = \frac{\beta_\circ c \theta_\circ}{\pi} \cos(\pi \frac{\bar{t}}{\theta_\circ})$$

$$\bar{\eta} = \frac{\beta_\circ c \theta_\circ}{\pi} \sin(\pi \frac{\bar{t}}{\theta_\circ})$$

$$\bar{\xi} = 0$$

Obviously the solution here considered describes in the absolute microuniverse a geodesic giving a uniform circular motion of radius $a_\circ = \beta_\circ c \theta_\circ / \pi$, velocity $\beta_\circ c$ and frequency $\nu_\circ = 1/2\theta_\circ$ .

Let us now observe that the kinetic energy of such an internal motion is given by

$$E_{kin,int} = m_\circ c^2 \{\frac{1}{\sqrt{1 - \beta_\circ^2}} - 1\}$$

In agreement with the electron model here adopted, this energy shall be equal to the intrinsic energy of the electron. It is then necessary to assume $\beta_\circ^2 = 3/4$.

For a correct understanding of this point, i.e. of the fact that <u>the intrinsic energy of the electron is originated entirely by its internal motion</u>, some comments are necessary. First of all let us recall that in relatively theory the quantity $W = m_\circ c^2 / \sqrt{1-\beta^2}$ is usually considered as the <u>complete</u> energy of the particle moving with velocity $\beta c$. But it is not very clear what this really means. In fact according to such an assumption, a particle at rest has an energy $W_\circ = m_\circ c^2$. If the particle is a composite one, the bond of the intrinsic energy of that particle with the energy of the composing ones is clear. But what is the meaning of the inner or proper energy of an electron conceived as a point particle? According to Frenkel (6) the relation $W_\circ = m_\circ c^2$ for such a particle is <u>simply a definition</u> and is not connected with its electromagnetic self-field. On the other hand, in our model such a relationship expresses the kinetic energy of the free point electron in its internal motion. In order to connect the two points of view, let us quote the exact words of Frenkel: "We can observe only <u>external</u> forces which are exerted on one electron by others. The <u>self-force</u> represents a metaphysical fiction, apart from the fact that it is epistemologically and physically meaningless to divide an electron up into further elements .... It seems not natural and simple to regard the electrons as unextended force centres ..... Hence it becomes necessary to define the mass more accurately, the rest mass of an electron and its magnetic moment as primary properties independent of the charge.... These properties cannot be reduced to the electromagnetic field of the electron, but must be regarded as its own mechanical attributes which are not directly connected with this field".

Coming back to our model, the total energy of a moving electron is given by

$$W = E_{kin,macr} + E_{kin,int}$$

that is the sum of the kinetic energy of the macroscopic motion and the internal one. Therefore, we have

$$W = m_\circ c^2 \left\{\frac{1}{\sqrt{1-\beta^2}} - 1\right\} + m_\circ c^2 \left\{\frac{1}{\sqrt{1-\beta_\circ^2}} - 1\right\} = \frac{m_\circ c^2}{\sqrt{1-\beta^2}}$$

where $\beta c$ is the velocity of the macroscopic motion.
For a particle at rest in its macroscopic motion ($\beta=0$) we get

$$W = m_\circ c^2 \left\{\frac{1}{\sqrt{1-\beta_\circ^2}} - 1\right\} = m_\circ c^2 \qquad (19)$$

that is the ordinary expression of the intrinsic energy of the electron.
Let us observe that the internal motion gives rise to a magnetic moment of the electron directed along the $\xi$ axis having a value

$$\mu_a = \frac{1}{c}\,\pi a_\circ^2 e \nu_\circ = \frac{1}{3\pi}\,\frac{e^3}{m_\circ c^2}\,\beta_\circ^2 = \frac{1}{4\pi}\,\frac{e^3}{m_\circ c^2} \qquad (20)$$

This value coincides with the first approximation for the "anomalous" magnetic moment predicted by Schwinger as an electrodynamic effect.

One needs not be surprised by the fact that it is possible to deduce, on a simply classical basis, the existence of the electron anomalous magnetic moment, generally considered as a typical result of quantum electrodynamics, since formula (20) does not contain Planck's constant $\hbar$ : in other words, the first-order contribution to the theoretical expression of the anomalous magnetic moment of the electron is not essentially a quantum effect. The higher-order contributions to such a moment are on the contrary a real quantum effect, for which it would be absurd to try a classical explanation.

In the relative description the macroscopic motion inside any fundamental interval of time is, according to the transmission law, the uniform motion of the point corresponding to the centre of the circular motion in the absolute description. This point ("guiding centre") is the centre of the ordinary 3-dimensional microcosmos associated to the electron and then, in a certain sense, it can be considered as the centre of mass of the electron itself. Furthermore, the macroscopic motion (defined in the instants $\tau = 0, \tau_o, 2\tau_o ...$) is represented in the Minkowski space-time by the world-line of a point particle having rest mass $m_o$, intrinsic energy $m_o c^2$, electric charge e and magnetic moment $\mu_a$.

Coming back to equations (18), describing the internal motion of the electron, we observe that the electron model deriving from the introduction of the chronon in the classical equations of electron motion has the peculiarity of combining some advantages of both the usual pointlike electron and the extended electron.

Moreover, it seems to us very interesting to notice that the model of electron resulting from our theory is completely in agreement with the requirements that, following the already recalled opinion of Frenkel, a correct theory concerning the nature of the electron must be satisfied: i.e. the fact that the energy as well as the linear and angular momentum of the electron cannot be reduced to its electromagnetic field and that, according to the Einstein's principle of relativity, the rest mass $m_o$ and the (classical) magnetic moment

$$\mu_a = \frac{1}{4\pi} \frac{e^3}{m_o c^2} = \frac{1}{4\pi} \left(\frac{3}{2}\right)^{3/2} m_o^{1/2} \, c \, \theta_o^{3/2}$$

must be considered as <u>primary</u> properties independent of the charge. Let us remark that to the internal motion, inside the de Sitter microuniverse associated with the particle, an elementary angular momentum of the order of

$$A_o = \frac{m_o \beta_o c}{\sqrt{1 - \beta_o^2}} \simeq \frac{e^2}{c} \qquad (21)$$

is related.

Keeping in mind that in quantum theory, to the "macroscopical motion" of the electron (e.g. the motion of the electron in the hydrogen atom) is, on the other hand, associated the quantal elementary momentum $A_q \simeq \hbar$ we have the following relation:

$$\frac{A_o}{A_q} \simeq \frac{e^2}{\hbar c} = \alpha$$

This result shows that the electron model here suggested might well give a promising start for the understanding of the fine-structure constant $\alpha$.

Finally the existence at a microscopical level of quantities, like (19), (20) and (21), associated to the electron non depending from the quantum constant $\hbar$ seems to impose a solution of the wave-particle duality different from that assumed in the so-called orthodox or "Kopenhagener" interpretation of quantum mechanics.

6 - HEAVY CHARGED LEPTONS AS EXCITED STATES OF ELECTRON

One of the more interesting results of the present theory of the electron based on the introduction of the chronon in the motion equations is given by the interpretation of muon and other heavy leptons as excited mass states of the electron. This point has been examined in different versions, but we will here follow the simple one based on the application of suitable Bohr-Sommerfeld quantization conditions. To this aim we recall that the de Sitter universe can be thought of as a 4-dimensional pseudesphere of equation

$$x_1^2 + x_2^2 + x_3^2 - x_4^2 + x_5^2 = R^2$$

embedded in a 5-dimensional metric

$$-ds^2 = dx_1^2 + dx_2^2 + dx_3^2 - dx_4^2 + dx_5^2$$

Taking into account the momentum-energy relation

$$p_1^2 + p_2^2 + p_3^2 - E^2/c^2 + M_\circ^2 c^2 = 0$$

we can consider the momentum $p_5 = M_\circ c$ as conjugate to the fifth coordinate $x_5 = w$. As a consequence the corresponding Bohr-Sommerfeld condition gives

$$\oint M_\circ c \, dw = h \tag{22}$$

By assuming the existence of a set of possible de Sitter micro-universes of radius $a^{(n)}$ in correspondence to different excited mass states $M_\circ^{(n)}$ we obtain a set of relations

$$M_\circ^{(n)} c \cdot 2\pi a^{(n)} = h$$

from which

$$M_\circ^{(n)} a^{(n)} = \hbar/c$$

Putting $a^{(1)} = c\theta_\circ$ and $M_\circ^{(0)} = m_\circ$, we immediately obtain the following formula for mass spectrum of our family of charged leptons:

$$M_o^{(n)} = \left(\frac{3}{2}\frac{\hbar c}{e^2}\right)^n m_o \tag{23}$$

The first excited lepton's mass turns out to be

$$M_o^{(1)} = \frac{3}{2}\frac{\hbar c}{e^2} m_o \simeq 206\ m_o \tag{24}$$

that is practically coincident with the muon's rest mass $m_\mu$. Therefore we will interprete the muon as the first excited state of the electron. This is in agreement with some other different evaluations that we will consider later.
Let us observe that, as a consequence of Bohr-Sommerfeld conditions, the chronon to be associated to the muon results

$$\theta_o^{(1)} = \frac{2}{3}\frac{e^2}{m_\mu c^3} = \theta_o \frac{m_o}{m_\mu} \tag{25}$$

and the radius of its corresponding de Sitter microuniverse $a_\mu = c\theta_o^{(1)} = c\theta_o\, m_o/m_\mu$.

In general the condition (22) points to the existence of a set of possible charged leptons $e^{(n)}$ $(n = 0,1 \ldots.)$ having the same charge e and masses $M_o^{(n)}$.

As we will see later a refined form of equation (23) is to be written

$$M^{(n)} = M^{(n-1)} + M_o^{(n)} \tag{23'}$$

and correspondingly we have for the muon mass

$$m_\mu = M^{(1)} = M^{(0)} + M_o^{(1)} = \left(1 + \frac{3}{2}\frac{\hbar c}{e^2}\right) m_o \tag{24'}$$

and so on.

Limiting ourselves to the consideration of the first three mass states, we have

| particle | mass | de Sitter radius ($10^{-13}$ cm) |
|---|---|---|
| $e^{(o)}$: electron | $M^{(o)} = M_o^{(o)} = m_o = .511$ MeV | $a_o \simeq 1.87$ |
| $e^{(1)}$: muon $\mu$ | $M^{(1)} = (1+\frac{3}{2}\frac{\hbar c}{e^2})m_o \simeq 105.55$ MeV | $a_1 \simeq .9\ 10^{-2}$ |
| $e^{(2)}$: heavy electron | $M^{(2)} = M^{(1)} + (\frac{3}{2}\frac{\hbar c}{e^2})^2 m_o \simeq 21.58$ GeV | $a_2 \simeq .44\ 10^{-4}$ |

Following previous considerations we have that the region in the de Sitter universe, where the trajectories of the normal electron take place, is given by the shell between the surface of radius $a_\circ = c\theta_\circ$ and the surface of radius $a_1 = c\theta_\circ^{(1)}$, that of the muon by the shell between the surface of radius $a_1 = c\theta_\circ^{(1)}$ and the surface of radius $a_2 = c\theta_\circ^{(2)}$, that of the heavy electron $e^{(2)}$ by the shell between the surface of radius $a_2 = c\ \theta_\circ^{(2)}$ and the surface of radius $a_3 = c\ \theta_\circ^{(3)}$, etc.

We will now consider (31) the question wether in the different regions of the de Sitter microuniverses related to the various leptons $e^{(0)}$, $e^{(1)}$, $e^{(2)}$ ... new excited motions of the particles can exist, deriving from possible self-considerations inside a single particle $e^{(n)}$.

To this purpose let us recall that in our theory the lepton $e^{(n)}$ moves in the associated microuniverse along a circular trajectory with velocity $\beta_\circ = \sqrt{3}/2$, to which corresponds an internal kinetic energy (observed as intrinsic energy) $M^{(n)}c^{(2)}$ and an "intrinsic" magnetic moment

$$\mu_a^{(n)} = 1/4\pi \cdot e^2/M^{(n)}c^2$$

We recall now a Barut's suggestion (32) according to which excited states could be derived for the electron from the coupling of its intrinsic magnetic moment with its self-field.

Starting from such a suggestion we will assume, as a mere heuristic hypothesis, that to the particle $e^{(n)}$ moving in the de Sitter microuniverse of radius $a_n = c\theta_\circ^{(n)}$ and having therefore a mass $M^{(n)}$, an extra interaction interval self-energy $\varepsilon^{(n,p)}$ is associated given by [3]

$$\varepsilon^{(n,p)} = (2p)^4 \left(\frac{3}{2}\frac{\hbar c}{e^2}\right)^n m_\circ c^2 = (2p)^4\ M_\circ^{(n)} c^2 \qquad (26)$$

being p an integer positive number.

Let us now observe that calling $M^{(n,p)}$ the mass of the internal excited state corresponding to the quantum number p, that is $M^{(n,p)} = M^{(n)} + \varepsilon^{(n,p)}/c^2$, a condition must be imposed to the values of p in order to assure that the trajectory of the particle $e^{(n)}$ takes place in the n-th shell of the de Sitter microuniverse. Such a condition is expressed by

$$\varepsilon^{(n,p)} < (M^{(n+1)} - M^{(n)})\ c^2 \qquad (27)$$

Consequently, as it is very easy to verify, the quantum number p has only the value p = 0 for n = 0 and the values p = 0,1 for $n \neq 0$.

Therefore for n = 0 we have the singlet $M^{(0,0)}$ and for $n \neq 0$ a doublet of masses

$$M^{(n,p)} = M^{(n)} + (2p)^4 M_\circ^{(n)} \quad (\text{with} \quad p = 0,\ 1) \qquad (28)$$

We obtain in such a way the following set of mass values for the charged leptons (for n = 0,1,2):

a) lepton $e^{(0)}$
   p = 0: ordinary electron of mass $M^{(0)}=M_\circ^{(0)}=m_\circ \simeq 0.511$ MeV

b) lepton $e^{(1)}$
   p = 0: ordinary muon of mass $M^{(1)}=m_\circ+M_\circ^{(1)} \simeq 105.55$ MeV.

   p = 1: "internal" excited state of muon of mass

   $M^{(1)*}=M^{(1)}+2^4\ M_\circ^{(1)} \simeq 17M_\circ^{(1)} \simeq 1794$ MeV

c) heavy electron $e^{(2)}$
   p = 0: ordinary heavy electron of mass $M^{(2)}=M_\circ^{(2)} \simeq 21.58$ GeV
   p = 1: "internal" excited state $e^{(2)*}$ of mass

   $M^{(2)*}=M^{(2)}+2^4 M_\circ^{(2)} \simeq 17M_\circ^{(2)} \simeq 367$ GeV.

In particular we see that the electron admits no "internal excited state and the muon an "internal" excited state of mass $M \simeq 1794$ MeV. This last state can be identified with the lepton $\tau$ . In fact let us recall that the fairly precise measurement of the $\tau$ mass has been performed by means of the DESY - Heidelberg experiment (33): the obtained value is 1790(+7;-10) MeV is in very good agreement with our theoretical result. We can therefore conclude that completing the theory of the electron proposed by us and based on the hypothesis of the chronon by taking into account the existence of further "internal" states for the different leptons $e^{(n)}$ due to a self-interaction, we reach a unitary self-consistent description of all the charged leptons until now discovered by experiment. Furthermore, as we have seen, the theory suggests the existence of a heavy lepton having a mass equal to about 23 protonic mass that seems undergo (if its mean life is long enough) the possibility of an experimental test: the production of such a particle should be possible by means of the reaction $\mu^+ + \mu^- \rightarrow e^{(2)+} + e^{(2)-}$ at sufficiently high energy.

Let us remember (10) that the introduction of the chronon can be extended to the description of the neutrinos associated to the charged leptons. The neutrinos result having a finite mass and moving on the same geodesics in the de Sitter microuniverses as the corresponding charged leptons.

In particular the mass of the electronic neutrino results to be of the order $m_{\nu_e} \simeq 0.01$—$0.1$ eV and that of the muonic neutrino $m_{\nu_\mu} = m_\mu / m_e \cdot m_{\nu_e} \simeq 2$——$20$ eV

Finally it is interesting to stress some others consequences of the quantized theory of leptons founded on the introduction of the chronon. Keeping in mind the existence of electromagnetically non radiating classical internal motions (see Sect. 4), we can easily understand, on the basis of the correspondence principle, why the muon's electromagnetic decay into an electron and a $\gamma$ quantum is forbidden and consequently we can give an explanation of the so-called "muon-puzzle". Furthermore the "internal" field (corresponding to a pure weak interaction) results associated to a quantum, the mass of which, evaluated by means of a suitable application of the Heisenberg uncertainty relations, seems to be of the order of the Weinberg-Salam heavy photon $Z_o$.

## 7 - WAVE EQUATION OF THE ELECTRON WITH THE INTRODUCTION OF THE CHRONON

A few years ago we succeded in extending the introduction of the chronon to quantum mechanics (10) (34). As a first step we suppose the electron described by the non relativistic Schrödinger equation: the introduction of the chronon brings about the substitution of the differential equation with a finite-difference equation in the time variable. This procedure appears to be justified by the fact that in the non-relativistic approximation the local time coincides with the proper time $\tau$ .

Recalling that in quantum mechanics the radiation of the electron is due to the interaction with the external electromagnetic field and that, in the absence of this interaction, the particle is in a stationary state, we shall consider the <u>symmetrical</u> finite-difference Schrödinger equation

$$i\hbar \frac{\Psi(t+\theta_o)-\Psi(t-\theta_o)}{2\theta_o} = \hat{H}\Psi(t) \tag{29}$$

where $\hat{H}(\hat{p};\ \hat{x};\ t)$ is the Hamiltonian operator which appears in the ordinary Schrödinger differential equation

$$i\hbar \frac{\partial \Psi}{\partial t} = \hat{H}\Psi \tag{30}$$

In some sense, equation (29) corresponds to the classical symmetrical non-relativistic equation (3') for a non-radiating electron, that is

$$\frac{m_o}{2\theta_o} \left|\underline{\underline{v}}(t+\theta_o)-\underline{\underline{v}}(t-\theta_o)\right| = \underline{\underline{F}}(t)$$

If we suppose the Hamiltonian H to be time-independent, the solution of equation (29) satisfying the initial condition

$$\Psi(0) = \Psi_o \tag{31}$$

can be written as

$$\Psi(t) = \Sigma_n c_n \exp[\lambda_n t] u_n \quad (32)$$

where $\hat{H} u_n = W_n u_n$ and $c_n = (u_n, \Psi_\circ)$.
The "generalized complex eigenvalues" $\lambda_n$ are given by the equation

$$\exp[\lambda_n \theta_\circ] - \exp[-\lambda_n \theta_\circ] = -2iE_n \quad (33)$$

where $E_n$ are adimensional eigenvalues of H, that is $E_n = W_n \theta_\circ / \hbar$.
From equation (33) we obtain

$$\exp[\lambda_n \theta_\circ] = -iE_n \pm \sqrt{1 - E_n^2} \quad (34)$$

The behaviour of the solutions is very different in correspondence of the two cases $|E| < 1$ or $|E| > 1$ (35):

a) case $|E| < 1$.
Taking $\lambda_n = k_n - i\nu_n$ we obtain

$$k_n = 0 \quad , \quad \sin \nu_n \theta_\circ = E_n \quad (35)$$

and therefore

$$\nu_n = 1/\theta_\circ \cdot \arcsin E_n \quad (35')$$

Here we have chosen the solution of the trascendental many valued equation (34) which gives back for $|E_n| \ll 1$ the ordinary Schrödinger eigenvalues $\nu_n = W_n/\hbar$.
The probabilities $P_n(t) = |c_n|^2$ are constant, therefore the states are stationary as for the corresponding usual differential Schrödinger equation.

The restriction $|E_n| \leqslant 1$ has been interpreted (8) (10) assuming that the critical value

$$W_M = \hbar/\theta_\circ = \frac{3}{2} \frac{\hbar c}{e^2} m_\circ c^2 \simeq \frac{3}{2} \text{"137"} m_\circ c^2$$

corresponds to the creation of a muon, whose rest mass will be consequently given by

$$m_\mu = m_\circ + W_M/c^2 \simeq 206.5 \; m_\circ \quad (36)$$

As observed before (28) for $|E| < 1$ the equation (29) is equivalent to the following

$$i\hbar \frac{\partial \Psi}{\partial t} = \tilde{H}\Psi \quad (37)$$

where $\tilde{H} = \hbar/\theta_\circ \arcsin \theta_\circ \hat{H}/\hbar$.

The operator $\tilde{H}$ is densely defined and self-adjoint in the subspace $S < L^2(\underline{\underline{R}}^3)$ spanned by the eigenvectors $u_W$ with $|W| < \hbar/\theta_0$. Consequently the observables (for the ordinary electron) must be restricted to self-adjoint operators leaving the subspace $S$ invariant.

b) case $|E_n| > 1$.
Imposing the continuity to the solution (33) we obtain

$$\nu_n = \pi/2\theta_0$$
$$k_n = \pm 1/\theta_0 \; \ln\left[\, |E_n| + E_n^2 - 1 \,\right] \tag{38}$$

We conclude that the eigenstates are unstable, according to the fact the coefficients of $u_n$ in the expression (32) are such that the probability for the occupation of the n.th state is given by

$$p_n(t) = |c_n|^2 \; \exp\left[2k_n t\right] \tag{39}$$

and therefore it decays or diverges in time.
The dependence of $k_n$ from $|E_n|$ is illustrated in Fig. 3, which shows clearly the transition from stationary to non-stationary states in correspondence to the threshold $|E_n| = 1$. Obviously positive eigenvalues $k_n$ must be excluded in order to prevent a diverging behaviour to the wave function. Furthermore from equation (39) it derives that the probability of occupation of a state with $|E_n| > 1$ decreases with time and vanishes for $t \to \infty$.

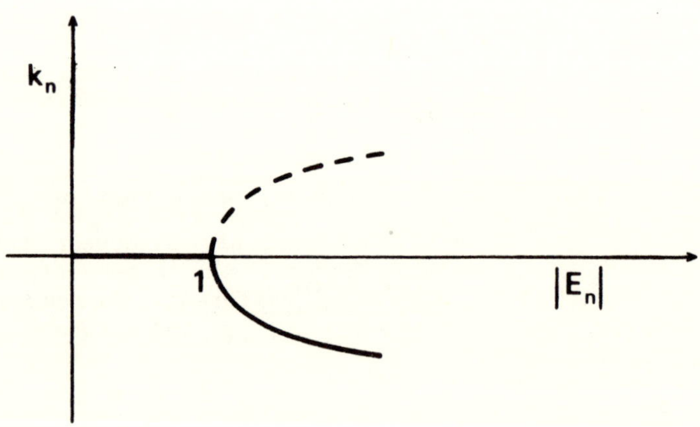


Fig.3 - Instability of finite-difference Schrödinger equation's solutions.

We remark that if one includes the spectrum with $|E_n| \gtrsim 1$, equation (1) can still be written in the form (37) with $\tilde{H}=\tilde{\nu}+i\tilde{k}$. Here $\tilde{\nu}$ and $\tilde{k}$ are Hermitian operators which are functions of $\hat{H}$ defined for the whole spectrum by their eigenvalues $\nu_n(E_n)$ and $k_n(E_n)$ given by equations (35) and (38). This operator $\tilde{H}$ coincides with the one defined in equation (37) only in the subspace $S(|E_n| < 1)$ where $k_n = 0$.

## 8 - THE PROBABILISTIC SOLUTION OF WAVE-PARTICLE DUALITY

Coming back now to the general solution (39), due to the completeness of the eigenvectors $u_n$, we have

$$||\Psi_t||^2 = \Sigma_n |c_n|^2 \exp[2k_n t] =$$

$$= \Sigma_{|E_n|\leq 1} |c_n|^2 + \Sigma_{|E_n|>1} |c_n|^2 \exp\left[-\frac{2t}{\theta_\circ} \ln\left(|E_n| + \sqrt{E_n^2-1}\right)\right] \qquad (40)$$

Clearly the norm is constant only if $|c_n| = 0$ for $|E_n| > 1$. In other words the norm is conserved only for the vectors belonging to the Hilbert subspace $S$ spanned by the eigenvectors of $\hat{H}$ having <u>no component</u> on the vectors $u_n$ with $|E_n| > 1$. However it is clear that the restriction to subspace $S$ introduces the artificial assumption that it is not possible to prepare a particle in whatever eigenstate of the energy. As a consequence the usual probabilistic interpretation of Schrödinger equation, that is considered the cornerstone of quantum mechanics, seems to fall down. However we will show that, by modifying in a proper way the epistemological background of the "Kopenhagener Geist", it is still possible to restore a satisfactory interpretation.

In fact it is possible in our opinion to reach this goal by supposing that formula (40) has to be interpreted in the following way.

A particle can be "prepared", at the instant $t = 0$, in whatever state corresponding to $|E_n| \leqslant 1$ or to $|E_n| > 1$. However when the particle is prepared in a state with $|E_n| > 1$, owing to its instability, it will spontaneously disappear from the original Hilbert space spanned by all the eigenvectors of $\hat{H}$ transforming itself into <u>something different</u>. Correspondingly we will suppose that $||\Psi||_t^2 = P_{ex}(t)$ is the probability of existence of the particle at the instant $t$ without being transformed into something different, while $|\Psi(\mathbf{x};t)|^2$ is the probability density of finding the particle in $\mathbf{x}$ at instant $t$. Obviously in the case of a vector $\Psi\in\{u_n\}$ with $|E_n| < 1$, we have, as in ordinary quantum mechanics, $||\Psi||_t^2 = P_{ex}(t) = P_{ex}(0) = 1$ according to the particle's stability. On the contrary for an unstable state ($|E_n| > 1$) we have

$$\|\Psi\|_t^2 = P_{ex}(t) = \exp\left[-\frac{2t}{\theta_\circ} \ln(|E_n| + \sqrt{E_n^2 - 1})\right] \quad (41)$$

showing clearly how the probability for the existence of the particle decays with a time constant that increases according to a logaritmic law with its energy. The "mean life" $\tau_D$ is given by

$$\tau_D = \frac{\theta_\circ}{2} \frac{1}{\ln[|E_n| + \sqrt{E_n^2 - 1}]} \quad (42)$$

If the particle has an energy slightly above the threshold value $|E_n| = 1$, i.e. for $|E_n| = 1 + \Delta E$ with $\Delta E \ll 1$, equation (42) reduces to

$$\tau_D \simeq \frac{1}{2\sqrt{2\ \Delta E}}\ \theta_\circ \quad (43)$$

The mean life can thus be much larger than the chronon $\theta_\circ$. For example in the case of a harmonic oscillator with frequency $\omega \ll 1/\theta_\circ$ prepared in a state $W_n = \hbar/\theta_\circ + \hbar\omega$ (or $E_n = 1 + \omega\theta_\circ$), equation (43) becomes

$$\tau_D = \frac{\theta_\circ}{2\sqrt{2\omega\ \theta_\circ}} >> \theta_\circ \quad (44)$$

Let us emphasize that in general $\tau_D$ goes to $\infty$ approaching the threshold for instability as it really ought and, above the threshold, decreases as the square root of $\Delta E$. In general for an arbitrary initial state having components with $|E_n| > 1$ and $|E_n| < 1$, the asymptotic value of $P_{ex}(t)$ reduces, according to equation (40), to

$$P_{ex}(\infty) = \Sigma_{|E_n|<1} |c_n|^2 < 1$$

This value should be proportional to the number of particles which have not been transformed, whereas $1 - P_{ex}(\infty)$ is proportional to the initial number of unstable particles.

In conclusion the symmetrical finite-difference Schrödinger equation (40) can describe a stable particle for $|W_n| \leqslant \hbar/\theta_\circ$ and a desappearing particle (transforming itself in something different) for $|W_n| > \hbar/\theta_\circ$.

This result gives a strong support to the assumption made by us according to which a bound electron with an energy $|W| > \hbar/\theta_\circ$ transforms itself in a muon with a rest mass $m_\mu$ given by (36). Furthermore by associating a chronon $\theta_\mu$ to the muon given by

$$\theta_\mu = \frac{2}{3} \frac{e^2}{m_\mu c^3} = \theta_\circ \frac{m_\circ}{m_\mu} \quad (45)$$

we obtain the following finite-difference equation for the muon

$$i\hbar \frac{\Psi(t+\theta_\mu)-\Psi(t-\theta_\mu)}{2\theta_\mu} = \hat{H}\Psi(t) \tag{46}$$

This implies the existence of the same family on heavy leptons considered in Sect. 6.

## 9 - CONSIDERATIONS ON A FREE PARTICLE MOTION

Let us now consider the case of unbounded states, in order to show that the limitations on the eigenvalues of the operator $\hat{H}$ do not imply any restriction on the values of the kinetic energy of the particle.

For the sake of simplicity we shall examine the problem of the free particle. A general solution of equation (29) is given for this problem by

$$\Psi(\mathbf{x},t)=\frac{1}{(2\pi\hbar)^{2/3}} \int c(\mathbf{p})\exp\left[-i\alpha(\mathbf{p})t+i\frac{\mathbf{p}\cdot\mathbf{x}}{\hbar}\right] d^3\mathbf{p} \tag{47}$$

where the function $\alpha(\mathbf{p})$ is determined by a condition similar to (35'):

$$\alpha(\mathbf{p}) = \frac{1}{\theta_\circ} \arcsin \frac{\theta_\circ}{\hbar} \frac{p^2}{2m_\circ} \tag{48}$$

For $\theta_\circ \to 0$ this condition reduces to $\alpha(\mathbf{p}) = (p^2/2m_\circ)/\hbar$. Hence, in this limit, $\alpha(\mathbf{p})\hbar$ coincides with the kinetic energy of the particle $W = p^2/2m_\circ$.

The coefficients $c(\mathbf{p})$ are determined by the initial condition $\Psi(\mathbf{x},0)=\Psi_\circ(\mathbf{x})$ as

$$c(\mathbf{p}) = \frac{1}{(2\pi\hbar)^{3/2}} \int \exp\left[-i\frac{\mathbf{p}\cdot\mathbf{x}}{\hbar}\right] \Psi_\circ(\mathbf{x})\, d^3\mathbf{x} \tag{49}$$

Let us emphasize that equations (47) and (49) are defined only if $c(\mathbf{p})=0$ for $p \geqslant p_M=\sqrt{2\hbar m_\circ/\theta_\circ}$ , otherwise equation (47) is meaningless.

This condition ensures that the normalization condition is preserved at all times and, via equation (49), it implies the existence of a minimum spread for the wave function which we shall discuss later on.

In order to find the behaviour of the wave packet, we may apply the condition of the stationary phase

$$\frac{d}{d\mathbf{p}}\left[\frac{\mathbf{p}\cdot\mathbf{x}}{\hbar} - \frac{t}{\theta_\circ}\arcsin\frac{\theta_\circ}{\hbar}\frac{p^2}{2m_\circ}\right] = 0$$

We find

$$\underline{\underline{x}} - \frac{\underline{\underline{p}}}{m_\circ} \frac{t}{\sqrt{1 - (\theta_\circ^2/\hbar^2)(p^4/4m_\circ^2)}} = 0 \tag{50}$$

If we assume that the function $c(\underline{\underline{p}})$ corresponds to a probability distribution for $\underline{\underline{p}}$ having a peak for $\underline{\underline{p}} = \underline{\underline{p}}_\circ$ , then the wave packet motion will be uniform along the direction of $\underline{\underline{p}}_\circ$ with the velocity

$$\underline{\underline{v}} = \frac{\underline{\underline{p}}_\circ}{m_\circ} \frac{1}{\sqrt{1 - (\theta_\circ^2/\hbar^2)(p_\circ^4/4m_\circ^2)}} \tag{51}$$

which coincides with the group velocity of the wave packet. The quantity p must satisfy the condition $W = p^2/2m_\circ \leqslant \hbar/\theta_\circ$ and therefore the maximum value $p_M$ of p for which

$$p_M^2/2m_\circ = \hbar/\theta_\circ \tag{52}$$

corresponds, as a consequence of equation (51), to $v = \infty$ . From this it follows that <u>the introduction of the chronon $\theta_\circ$ does not impose any restriction on the particle translation velocity</u>.

Equation (51) suggests that it is convenient to identify the kinetic momentum of the particle with the vector

$$\underline{\underline{p}}^* = \frac{\underline{\underline{p}}}{\sqrt{1 - (\theta_\circ^2/\hbar^2)(p^4/4m_\circ^2)}} \tag{53}$$

Under this assumption the <u>kinetic energy</u> of the particle becomes

$$E_{kin} = \frac{1}{2} m_\circ v^2 = \frac{p^{*2}}{2m_\circ} = \frac{p^2}{2m_\circ [1 - (\theta_\circ^2/\hbar^2)(p^4/4m_\circ^2)]} \tag{54}$$

which for $p \to p_M$ (and consequently $v \to \infty$ ) becomes infinite.

Let us observe that the restriction

$$p \leqslant \sqrt{\frac{2 m_\circ \hbar}{\theta_\circ}} = p_M \tag{55}$$

implies that no wave packet exists associated to a particle with

$$\Delta p_x \geqslant \Delta p_\circ \simeq \sqrt{2\hbar m_\circ/\theta_\circ}$$

and correspondently with a position spread

$$\Delta x \leqslant \Delta x_\circ \simeq \hbar/\Delta p_\circ \simeq \sqrt{\hbar\theta_\circ/2m_\circ} \simeq \sqrt{r_\circ \lambda_\circ/2} \tag{56}$$

where $r_\circ$ is the so-called "classical" radius and $\lambda_\circ$ the Compton

lenght of the electron. This new type of limitation, pertaining to the measurement of a single variable rather than two variables simultaneously, has no counterpart in the mathematical formulation of current quantum mechanics. Correspondently let us remark how the existence of a maximum value of p implies a restriction on the possible values of the de Broglie wave length, the fundamental concept of whole wave mechanics.
In fact we have

$$\lambdabar = \hbar/p \geqslant \hbar/p_M = \hbar/\sqrt{2m_\circ\hbar/\theta_\circ} = \tag{57}$$
$$= \sqrt{\hbar\theta_\circ/2m_\circ} = \sqrt{\tfrac{1}{2}\lambdabar_\circ r_\circ}$$

that is the existence of a minimum value $\lambdabar = \sqrt{\frac{1}{2}\lambdabar_\circ r_\circ}$ of the de Broglie wavelength. This result deserves in our opinion a deep consideration for the correct understanding of the present theory and in particular of the problem of wave-particle duality.

## 10 - INTRODUCTION OF THE CHRONON IN RELATIVISTIC QUANTUM THEORY OF ELECTRON

The extension of the procedure described in the previous sections to the case of the relativistic Dirac quantum equation for the electron raises some difficulties. In fact in this equation the time evolution of the state of the particle, contrary to what occurs in the relativistic classical equations of motion, is not expressed by means of the proper time $\tau$, but <u>via</u> the four chronotopical local coordinates $x_k \equiv (x,y,z,ct)$.

This fact, however, suggest (<u>10</u>) (<u>36</u>) to adopt the following finite-difference quantum Dirac equations for the electron (for the sake of simplicity we limit ourselves to the case of a free particle):

$$\gamma_k \frac{\Psi(x_k+\Delta x_k) - \Psi(x_k-\Delta x_k)}{2\Delta x_k} + \chi\Psi(x_k) = 0$$

$$(k=1,2,3,4), \tag{58}$$

$$\gamma_k \frac{\Psi^\dagger(x_k+\Delta x_k) - \Psi^\dagger(x_k-\Delta x_k)}{2\Delta x_k} - \chi\Psi^\dagger(x_k) = 0$$

where $\chi = m_\circ c/\hbar$ and the quantities $\Delta x_k$ satisfy the relation

$$-\Delta x_k \Delta x_k \equiv (c\Delta t)^2 - (\Delta r)^2 = c^2\theta_\circ^2 \tag{59}$$

Let us remark that equations (58) do not imply the introduction of the discreteness into space-time, in the sense that the points in a space-time manifold do not form a continuum, but represent some discrete, denumerable set represented by points on a lattice, as supposed by many authors (37). In our theory the space-time is still a continuous manifold, but the variation of the dynamical state of the electron takes place in a minimum interval of proper time $\theta_o$ by means of the relation (59) introduced many years ago by Landè (38) and by Born (39). Due to the complexity of equations (58), in our attempt for the introduction of the chronon in the electron's relativistic wave equations we have assumed, as a very rough approximation, instead of equations (56), the differential Dirac equations as extended by Fock (40), by Nambu (41) and by Corben (42) for the five-dimensional case

$$\begin{aligned} &(\gamma_k \frac{\partial}{\partial x_k} + \chi)\ \phi = \gamma_5 \frac{\partial \phi}{\partial x^5} \\ &(\gamma_k \frac{\partial}{\partial x_k} - \chi)\ \phi^\dagger = \gamma_5 \frac{\partial \phi^\dagger}{\partial x^5} \end{aligned} \qquad (k=1,2,3,4) \qquad (60)$$

and imposing to the function $\phi$ to be a periodic function of $x^5$ with period $l_o = 2\pi c \theta_o$ . One obtains as a first excited mass state of the electron a particle with a mass

$$m^{(1)} = m_o \sqrt{\left(\frac{3}{2}\frac{\hbar c}{e^2}\right)^2 + 1} \simeq \frac{3}{2}\frac{\hbar c}{e^2}\, m_o \simeq m_\mu \qquad (61)$$

that can be identified with the muon.

Obviously this procedure doesn't provide a satisfactory solution to the problem of the mass of the lepton family.

We believe that a more proper way to attack the problem consists in describing in a unitary way both the "macroscopical" and the "internal" motions considered in Sect. 5 for the classical electron, by means of a field defined on a fiber bundle, the structural group of which is taken to be the $L_{4,1}$ de Sitter group, according with a formalism developed from a general point of view by Drechsler (43). Such a program, though in a preliminary stage (44), seems to lead as a first approximation to an equation for the electron having a structure (u is the wave-function associated to the particle)

$$L(u) + \chi_o^2\, u + N(u) = 0 \qquad (62)$$

where $L(u) + \chi_o^2 u = 0$ will be a usual linear equation of the particle in the ordinary Minkowski space-time while N(u) satisfies a non linear equation defined inside a de Sitter microuniverse associated to the particle and describes the internal properties of the particle itself.

It is at present very difficult to say that our attempt will lead or will not lead to a positive result. What seems important to us is the very close connection of equation (62) with some ideas proposed and developed a quarter of a century ago by Louis de Broglie (45).

11 - APPENDIX:

The solution of the finite-difference Schrödinger equation

The solution of the equation

$$i\hbar \frac{\Psi(t+\theta_\circ)-\Psi(t-\theta_\circ)}{2\theta_\circ} = \hat{H}\Psi(t) \tag{1a}$$

with a time indipendent $\hat{H}$ and satisfying the initial condition

$$\Psi(0) = \Psi_\circ \tag{2a}$$

can be written as

$$\Psi(t) = \Sigma_n \; c_n \; \exp[\lambda_n t] \quad u_n \tag{3a}$$

where $\hat{H}\, u_n = W_n u_n$ and $c_n = (u_n, \Psi_\circ)$.
The complex eigenvalues $\lambda_n$ are determined by the condition

$$\exp[\lambda_n \theta_\circ] - \exp[-\lambda_n \theta_\circ] = -2iE_n \tag{4a}$$

with

$$E_n = |W_n| \theta_\circ / \hbar \tag{5a}$$

By putting $z = \exp(\lambda_n \theta_\circ)$ we have $z^2 - 2iE_n z - 1 = 0$ from which

$$z = iE_n \pm \sqrt{1 - E_n^2}$$

and

$$\lambda_n = \frac{1}{\theta_\circ} \ln\left[iE_n \pm \sqrt{1 - E_n^2}\right]$$

where ln is defined in the field of complex numbers. We can consider two cases:

case a: $E_n < 1$.

Since $\left| iE_n \pm \sqrt{1-E_n^2} \right| = 1$ we have

$$iE_n \pm \sqrt{1 - E_n^2} = \exp[i\phi_\pm]$$

with $\phi_+ = \pi - \phi_-$.
Consequently $\phi_+ = \arcsin E_n$ and $\phi_- = \pi - \arcsin E_n$.
The corresponding values of $\lambda_n$ are

$$\lambda_{n-}^{(m)} = \frac{i}{\theta_\circ} (\pi - \arcsin E_n + 2\pi m)$$

$$(m = 0, \pm 1, \ldots)$$

$$\lambda_{n+}^{(m)} = \frac{i}{\theta_\circ} (\arcsin E_n + 2\pi m) \qquad (m = 0, \pm 1, \ldots)$$

We conclude that equation (4a) admits infinite solutions as a consequence of the polidromic character of the functions square root and logarithm. However the physical solution is $\lambda_{n+}^{o} = i/\theta_\circ \arcsin E_n$ because is the only one that for $\theta_\circ \to 0$ reduces to the solution of the ordinary Schrödinger equation.

case b: $E_n > 1$

We have again two classes of solutions

$$\lambda_{n+}^{(m)} = \frac{i}{\theta_\circ}(\frac{\pi}{2} + 2\pi m) + \frac{1}{\theta_\circ} \ln^{(R)} [E_n + \sqrt{E_n^2 - 1}]$$

$$\lambda_{n-}^{(m)} = \frac{i}{\theta_\circ}(\frac{\pi}{2} + 2\pi m) - \frac{1}{\theta_\circ} \ln^{(R)} [E_n + \sqrt{E_n^2 - 1}]$$

We observe that the solution $\lambda_{n-}^{(m)}$ decay to zero for $t \to \infty$, whereas the solutions $\lambda_{n+}^{(m)}$ diverge.
The physical solution is selected by adopting the same criteria used for the case $E_n < 1$. This implis to take m = 0 and the value of the square root which corresponds to first "folium" of the Rieman surface, i.e. $+\sqrt{1-E_n^2}$ for $E_n < 1$ and $+i\sqrt{E_n^2-1}$ for $E_n > 1$. In this way the divergent solution is eliminated.

Finally let us observe that the divergent solution corresponding to the second "folium" of the Riemann surface leads to time reversed behaviours. In fact (taking m = 0) for the decaying solution we have:

$$\left.\begin{array}{ll} \text{for } E_n<1: & \exp\{-it/\theta_\circ \cdot \arcsin E_n\} \\ \text{for } E_n>1: & \exp\{-t/\theta_\circ \cdot (i\frac{\pi}{2} + \ln^{(R)} [E_n^2 + \sqrt{E_n^2 - 1}])\} \end{array}\right\} \quad (I)$$

on the other hand for the amplified solutions we have

$$\left.\begin{array}{ll} \text{for } E_n < 1: & \exp\{-i\pi t/\theta_\circ\} \exp\{+it/\theta_\circ \cdot \arcsin E_n\} \\ \text{for } E_n > 1: & \exp[-i\pi t/\theta_\circ] \exp\{t/\theta_\circ \cdot (i\frac{\pi}{2} + \\ & \quad + \ln^{(R)} [E_n^2 + \sqrt{E_n^2 - 1}])\} \end{array}\right\} \quad (II)$$

Neglecting the unessential common factor $\exp[-i\pi t/\theta_\circ]$ we see that the solution (II) can be obtained from the solution (I) by changing t into -t. Obviously, as the original equation (1a) is invariant under time reversal, we obtain both amplified and damped solutions (and the second is the time reversed of the first).

The solutions having physical meaning in the forward time evolution are the damped ones, those having physical meaning in the backwards evolution are the amplified ones. Let us observe that the time reversal corresponds to the passage from one "folium" of the Riemann surface associated to the square root function to the other.

NOTES

([1]) More exactly $F_{\alpha\beta}$ is in equation (1) the retarded field, in equation (2) the advanced field and in equation (3) the half advanced and half retarded field respectively

([2]) The Fantappiè group is the group representing in $V_4^{4*}$ the group of motions of a de Sitter manifold $V^4$, i.e. the projectivity which changes in itself the Cayley-Klein absolute (17)

([3]) The expression of the extra interaction internal self-energy assumed here is different from the one given by Barut; in particular it is connected with the description of the different leptons $e^{(n)}$ deriving from our theory of the electron based on the introduction of the chronon.

([4]) The appendix is due to L.A. Lugiato.

REFERENCES

( 1) M.Abraham: Ann. der Phys., 10, 105 (1903)
( 2) H.A. Lorentz: Amsterdam Proceedings, 12, 986 (1904)
( 3) H.Poincarè: Rend. Palermo, 21, 129 (1906)
( 4) G.A.Schott: Electromagnetic Radiation (Cambridge, (1912)
( 5) L.Page: Phys. Rev. 18, 58 (1921)
( 6) J.Frenkel: Lehrbuch der Elektrodynamik (Berlin, 1926-28), p. 284
( 7) P.A.M.Dirac: Proc. Roy. Soc. 161A, 148 (1938)
( 8) P.Caldirola: Suppl. Nuovo Cimento, 3, 297 (1956), and references therein
( 9) P.Caldirola: Riv. Nuovo Cimento, 2, N. 13 (1979), and references therein
(10) P.Caldirola: Nuovo Cimento, 45A, 549 (1978) and references

therein
(11) H.Arzeliès: "Rayonnement du corpuscule chargè fortement accéleré" - (Paris, 1966) (in part. III$^{me}$: Chap.XVIII)
(12) F.Casagrande: Scientia, 112, 416 (1977)
(13) R.Lacroix: Can. Journ. Phys. 56, 1109 (1978)
(14) G.Bruzzaniti: Epistemologia IV, N.2 (1981).
(15) V.Ivanenko: "Perennial modernity of Einstein's theory of gravitation" - chapt. 9, in: Albert Einstein 1879-1979 - "Relativity, Quanta and Cosmology in the Development of the Scientific Thought of A.Einstein", Vol. I (New York, N.Y. 1979) p. 285
(16) E.C.P.Stückelberg: Helv. Phys. Acta, 14, 589 (1941)
(17) R.P.Feynman: Phys. Rev. 76, 746, 769 (1949)
(18) E.Recami and M.Mignani: Riv. Nuovo Cimento, 4, 209 (1974); Lett. Nuovo Cimento 18, 5 (1977); E.Recami and G.Ziino, Nuovo Cimento, 33A, 205 (1976)
(19) P.Caldirola: Lett. Nuovo Cimento, 32, 264 (1981).
(20) E.R.Caianello: Lett. Nuovo Cimento,32, 65 (1981)
(21) N.Fulton and F.Rohrlich: Ann. of Physics, 9, 499 (1960)
(22) R.Cirelli: Nuovo Cimento, 1, 260 (1955)
(23) L.Lanz: Nuovo Cimento, 23, 195 (1962)
(24) A.Prosperetti: Nuovo Cimento, 57B, 253 (1980)
(25) P.Caldirola, G.Casati and A.Prosperetti: Nuovo Cimento, 43A, 127 (1978)
(26) J.Petzold and M.Sorg: Zeits Phys. 283 a, 207 (1977)
(27) D.Bohm and W.Weinstein: Phys. Rev., 74, 1789 (1948)
(28) P.Caldirola: Nuovo Cimento, 49A, 497 (1979)
(29) V.Benza and P.Caldirola: Nuovo Cimento, 62A, 175 (1981)
(30) L.Fantappié: Rend.Acc.Naz.Lincei, 8, 873 (1954)
(31) P.Caldirola: Lett.Nuovo Cimento, 27, 225 (1980)
(32) A.O.Barut: Phys.Lett.B, 73, 310 (1978); Phys.Rev.Lett. 42, 1251 (1979)
(33) G.J.Feldmann: Proc. of the Nineteenth International Conference on High-Energy Physics, edited by S.Homma, M.Kawaguchi and H.Migazawa (Tokio, 1978) (Physical Society of Japan, Tokio, 1979)
(34) P.Caldirola: Lett.Nuovo Cimento 16, 151 (1976)
(35) R.Bonifacio and P.Caldirola: Lett. Nuovo Cimento 33, 197 (1982)
(36) P.Caldirola: Lett.Nuovo Cimento, 18, 465 (1977)
(37) V.Ambarzumian and D.Ivanenko: Zeits.Phys., 64, 563 (1930); H.T.Flint: Phil.Mag. 7, 29 (1940); Phys.Rev., 71, 38 (1974); A.Schild: Phys.Rev., 79, 414 (1948); Canad. Journ. Math., 1, 29 (1949); B.T..Darling: Phys. Rev., 80, 460 (1950); A.Das: Nuovo Cimento 8, 482 (1960); E.A.B.Cole: Nuovo Cimento, 65A, 645 (1970); K.Hasche: Progr. Theor. Phys., 48, 1742 (1972); A.Messeen: Ann.Soc.Sci.Bruxelles, 86, 89, (1972); Nuovo Cimento, 12A, 491 (1972); L.G.Welch: Nuovo Cimento, 31B, 279 (1976)
(38) A.Landé: Journ.Franklin Inst., 229, 767 (1940)

(39) M.Born: Rev. Mod. Phys., 71, 38 (1947)
(40) V.Fock: Phys. Zeits. Sov. Un., 12, 404 (1937)
(41) Y.Nambu: Progr. Theor. Phys., 5, 82 (1950)
(42) H.C.Corben: Nuovo Cimento, 9, 580 (1952)
(43) W.Drechsler: Fortschritte der Phys. 23, 607 (1975)
(44) V.Benza and P.Caldirola: Preliminary unpublished results
(45) L.de Broglie: C.R. Acad. France 246, 2077 (1958)

# THE WAVE-PARTICLE DUALITY AS AN INTERPLAY BETWEEN ORDER AND CHAOS

Simon Diner

Institut de Biologie Physico-Chimique
13 rue P. et M. Curie
75005 PARIS

Der Begriff eines Noumenon ist also nicht der Begriff eines Objekts, sondern die unvermeidlich mit der Einschränkung unserer Sinnlichkeit zusammenhängende Aufgabe, ob es nicht von jener ihrer Anschauung ganz entbundene Gegenstände geben möge.

E. KANT
Kritik der reinen Vernunft

## QUANTUM THEORY : A UNITARY THEORY FOR LIGHT AND MATTER

De Broglie's fundamental idea was to extend to particles the laws of the quantum theory of light. It was the idea of a Unitary Theory, "une grande loi de la Nature" as he said himself.

Such a Unitary Theory is a general formulation of the energy and momentum concepts, independently of the physical system (light or matter).

This requirement of unity seems a guarantee for success, opening the way to the concept of wave properties of matter, asserting its effectiveness in the general concept of quantization. This lead in return to the developement of quantum electrodynamics (and quantum field theory) along formal rules identical to those of quantum mechanics.

The great conceptions of Nature are those which are free from specifical properties of objects, at the price of an abstract formu-



*S. Diner et al. (eds.), The Wave-Particle Dualism, 215–229.*

lation which offers dreadful problems of interpretation. This is the case for the fundamental law of dynamics ($\vec{F} = m\vec{\gamma}$), for the law of gravitation ($\vec{F} = g \frac{mm'}{2}$), for the law of conservation of energy, for the second principle of thermodynamics.

This is also the case for the renormalization techniques and the associated universal properties in the theory of critical and chaotic phenomena.

Such a unitary strategy is still at work with success in quantum field theory where for example a conception as supersymmetry allows to handle bosons and fermions together.

Quantum Theory discloses such universal principles which are at first sight unexpected.

The Kirchhoff theorem on the independance of the law of black-body radiation upon the material nature of the cavity, is some kind of "miracle" which has to be included in every theory, even if some explanation for it is not given. This is indeed the gateway through which Planck proceeded when he introduced abstract oscillators whose physical nature was left undefined.
These oscillators are in some way present in Quantum Mechanics which appears in many circumstances as a general model of abstract harmonic oscillator. A striking example being given by the universality of the response theory of quantum systems : there is a unique general form for the fluctuation-dissipation theorem. This simplicity of the quantum case is in sharp contrast with the high specificities of the classical case.
The universality of the wave-particle duality is another universal abstract principle. It works surprisingly in physical systems as different as electrons or neutrons, helium atoms or hydrogen molecules, which exhibit typical wave phenomena.
This universality is a challenge to the fundamental specificities of physical systems. As such it must be the manifestation of a kind of universal reality. The nature of this reality has to be discovered.

## ACCESSIBILITY OF THE VARIOUS LEVELS OF REALITY

The debate upon the possibility and conditions of knowledge is as old as Science and Philosophy. The dispute over "occult" qualities, characterized as insensible, as opposed to "manifest" qualities, goes back to Aristotle and the Christian Aristotelianism, and has been a major epistemological impasse not surmounted until the seventeenth century. At that time occult qualities became fully and consciously accepted in natural philosophy, just as it became recognized that no qualities were ever directly perceived.

This evolution is a major component of the Scientific Revolution and has culminated in Kant's conception of the *noumenon* and the *phenomenon*.
A new aspect of this debate, typical of contemporary Science, is the systematic inclusion of limitations to knowledge in the scientific doctrines themselves. One has to distinguish between experimentally accessible and non accessible elements of reality. A philosophy expressed in the "weak anthropic principle" : "What we can expect to observe must be restricted by the conditions necessary for our presence as observers".
The Carnot theorem, the second principle of thermodynamics and the conception of available energy where the first clear manifestations of such an operational limitation to knowledge. In the 20th century such manifestations appear everywhere : Special Relativity is built on a limitation of the speed of light, Quantum Mechanics is built on limitations upon certain types of measurement. A new type of limitation appears with Gödel's theorem, where it is shown how there exist properties which are non-provable in the framework of certain formal systems.
To prove that certain questions are unsolvable has become a mood, even a fashion.
*The theory of algorithmic complexity provides a precise definition of randomness (or chaos) through limitations in computability* [1].
Theoretical physics describes many elements of reality which experimental physics is unable to observe, and in many cases one witnesses the appearance of a theory of the non-observability.
How many phantoms of that kind ?
Magnetic monopole, advanced action, gravity waves, quarks, black holes, multiple worlds ...
Myths or reality ? What is reality ?
To the naïve conception of physics as an experimental science is opposed a time long neo-platonic tradition which accompanies to day a true metaphysics : *the physics of the inaccessible.*
But while neo-platonism invaded physics by following the use (and abuse) of formalistic mathematics (represented at best by the Bourbaki group), among mathematics itself there is a renewal of empiricism (constructive mathematics, non standard analysis) which is also manifest in the concerns of information theory and computing theory [2].

## VACUUM : AN EVERYWHERE PRESENT INACCESSIBLE REALITY

The conception of the world which was shaped in the XVIIth century, in conclusion to the Renaissance period, is being deeply modified nowadays.
The rising modern atomism of Boyle and Gassendi was linked with a precise conception of the vacuum, supported by the experimental work of Pascal and Torricelli. The vacuum was considered as a

perfectly empty region with no matter and no shape at all.
Such a conception of the vacuum, faithful to Democrite but opposed to the ideas of Aristotle or Leibniz, is naturally linked to a conception of infinite space and in someway of continuum. The concept of chance is rooted in such an interpretation of the vacuum. Chance as emptiness, as absence of determinations and causes.
*The Mechanical Philosophy is a subtle intertwining of Atomism, Empty Space and Chance.*
*The contemporary revolution comes from the fact that the discovery of chaotic (or stochastic) motion destroys the traditional antagonism between determinism and chance* [3] *and that atomism only survives if one introduces a new conception of the vacuum.* A vacuum not empty at all. A new kind of aether.
As soon as relativity theory discarded the idea of an elastic motionless ether, the vacuum was populated again.
*The quantum vacuum is no more a medium but a state of a quantized system. It is the state of lowest energy of a quantum field.* Due to the interaction between quantum fields, a unique Vacuum must exist, as the state of lowest energy of the Universe.
Energy appears as one of the most fundamental attribute of reality. As such it is submitted to precise limitations.
All energy is not dynamically available according to the second law of thermodynamics.
The third law of thermodynamics (Nernst principle) says that the state corresponding to a zero temperature cannot be reached. At that temperature matter is annihilated. So that there appears an energetic reality in a situation where no more "normal matter" exists. There is some reality beyond matter. This is the vacuum, an inaccessible reality, which nevertheless manifests itself in the behaviour of matter. *The roots of the quantum behaviour seem to be in the very existence of the vacuum.*
The relationship between zero-temperature, vacuum and quanta appeared first in the work of Planck in 1911 and 1912.
That Planck in his second theory of black-body radiation obtains the exact formula with the quantum term $\frac{1}{2}\hbar\nu$ must be looked upon as purely coincidental [4]. But one must think much more about Planck's work on the Nernst theorem in 1912 [5]. He gives the new formulation showing the decrease of entropy to zero with temperature, and establishes the link between this theorem and the Planck constant $\hbar$. This constant determines the size of cells in phase space.
Planck says : "the entropy of a state has a quite definite, positive value, which, as a minimum becomes zero, while in contrast therewith the entropy may, according to classical thermodynamics, decrease without limit to minus infinity. For the present, I would consider this proposition as the very quintessence of the hypothesis of quanta".
But the champion of the zero point energy has been Nernst, who considered it has an "ordered energy" of the radiation field or of the ether. A champion even not challenged by Einstein and Stern

who disagree with themselves at the Congrès Solvay of 1913, considering that the arguments they have given for the existence of a zero point energy are not right [6].

In fact Stern was not really convinced of that, as he testifies in a letter written to C.P. Enz in 1960 [7]

> "Pauli and I have continually discussed the question of the zero-point energy in Hamburg in the early twenties. We always went together from the Institute to the house for lunch, and thereby there were two problems which we talked over time and again ... I for my part always tried to convert Pauli to the zero-point energy against which he had the gravest hesitations".

Pauli was never convinced, and wrote in Wellenmechanik in 1933 [8]:

> "At this point it should be noted immediately that it is more consistent here, in contrast to the material oscillator, not to introduce a zero point energy of $\frac{1}{2} h\nu$ per degree of freedom. For, on the one hand, the latter would give rise to an infinitely large energy per unit volume due to the infinite number of degrees of freedom, on the other hand, it would be principally unobservable since nor can it be emitted, absorbed or scattered and hence, cannot be contained within walls and, as is evident from experience neither does it produce any gravitational field".

Nevertheless in 1962, Stern wrote : "Wave mechanics is not only compatible with the third law of thermodynamics but I believe that it should be possible under quite general assumptions, to derive the content and formalism of wave mechanics with the help of the third law" [9].

"Electromagnetic vacuum" is an electromagnetic field without real photons.

The requirements of a unitary theory, where the field becomes the fundamental physical reality and the particles are just excitations of the field, leads to associate a vacuum field to any microphysical particle.

The "particle vacuum" supports in some way the particle. It creates the particle, as a singularity. But, as any singularity, the particle organizes the "mother field". *One must without doubt consider the existence of a wave in the vacuum, resulting from the action of the moving particle on the supporting vacuum. The particle appears as a "dressed particle" or "quasiparticle", remembering the properties of the inaccessible underlying vacuum.*

In absence of any measurement for localizing the particle, it is lost in the vacuum. The measurement breaks the umbilical cord between the particle and the vacuum. It gives a reality to a potentiality. It creates the observable reality.

This idea of a quantum vacuum, out of reach aether, has been reasserted by Dirac in his famous small paper in Nature in 1951 [10] :

> "We can now see that we may very well have an aether, subject to quantum mechanics and conforming to relativity, provided we are willing to consider the perfect vacuum as an idealized

state, not attainable in practice. From the experimental point of view there does not seem to be any objection to this. We must make some profound alterations in our theoretical ideas of the vacuum".

In fact Dirac modifies drastically the relationship between vacuum and chance, because he remarks that a complex random medium is not suitable as a reference frame, the same for an absolute vacuum. The non existence of an absolute reference frame implied by the Relativity theory leads either to an absolute vacuum or to a chaotic vacuum. In 1920 Einstein remarked that from the point of view of General Relativity empty space must have physical properties, so that an aether must exist, but the concept of motion cannot be applied to it [11]. A chaotic aether fulfills such a requirement. Recovering the ancient myths, contemporary science is lead to consider that *Vacuum is Chaos*. A modern version of the Tohu-Bohu. But the microphysical chaos must be described by the quantum formalism which is, as we shall see later, a theory of some category of chaotic behaviour. In that way *the chaotic vacuum must be described as a quantum state, a state of a quantum field.*

## AN ELECTROMAGNETIC FIELD IN THE VACUUM. VIRTUAL PHOTONS.

According to classical electrodynamics systems of charged particules emit radiation in a continuous way. But it is not logical to consider radiation from material systems and not to consider the presence of radiation in the space. One is lead to suppose that a universal electromagnetic field is present in space even in region where there is no matter or detectable field, even in the Vacuum. But this electromagnetic field of the vacuum cannot be detected directly, due to the impossibility of a permanent transfer of energy from the vacuum.

In quantum electrodynamics this vacuum field is the lowest state of the quantized electromagnetic field. It must be a random field. The direct inaccessibility of this field corresponds to the absence of actual associated photons. The photon is experimental evidence of the presence of the electromagnetic field.

But if one considers this vacuum field as an actual electromagnetic field, one must describe it as any other electromagnetic field, on a quantum basis, introducing the conception of photon, but in the present case a virtual photon.

To speak of a virtual photon means that the associated field is not detectable, or that one cannot transfer energy from that field to a material detector in a permanent way.

*The concept of a virtual particle extends the wave-particle duality to the quantum vacuum itself.*

## THE FAILURE OF A CLASSICAL STOCHASTICAL PICTURE OF THE VACUUM : STOCHASTIC ELECTRODYNAMICS.

It has been thought very often that one can do without a quantum model of the electromagnetic field. This is the conception of all "semi-classical theories of radiation-matter interaction" where matter is described by the quantum theory whereas radiation is a pure classical electromagnetic field. It seems largely proved to day that such theories obtain only limited successes and that electromagnetic radiation has actually non classical properties, mainly as statistical properties are involved [12][13].
*Stochastic Electrodynamics* belongs to such attempt of using a classical picture of the electromagnetic field. It *describes the motion of classical charged particles, obeying the laws of classical electrodynamics* (Abraham-Lorentz-Dirac theory), *subjected to the effect of a fluctuating classical electromagnetic field in the vacuum.* The concepts and the results of Stochastic Electrodynamics (S.E.D.) have been largely reviewed elsewhere, so that we need only here recall and comment briefly the failure of S.E.D. [14].
S.E.D. consider a stochastic electromagnetic field in the vacuum with spectral statiscal properties corresponding to the ("quantum mechanical") second Planck law for black-body radiation. The energy density is given by

$$\rho_E(\omega) = \frac{\omega^2}{\pi^2 c^3}\left\{\frac{1}{2}\hbar\omega + \frac{\hbar\omega}{e^{\frac{\hbar\omega}{kT}} - 1}\right\}$$

where the first term is the zero-point energy, the only remaining term at T = 0.
Spectral density, and correlation function for the stochastic field are easily deduced according to :

$$\rho_E(\omega) = \frac{3}{4\pi^2}\mathcal{S}_E(\omega)$$

$$\mathcal{S}_E(\omega) = \int_{-\infty}^{+\infty} e^{-i\omega\theta}\mathcal{C}_E(\theta)d\theta$$

$$\mathcal{C}_E(\theta) = \langle E(t)E(t+\theta)\rangle \ .$$

This zero point radiation spectrum is the unique spectrum of stochastic electromagnetic radiation which is Lorentz invariant. This invariance makes the field in the vacuum not detectable (by a Milchelson and Morley experiment for example).
Let us remark that using a spectral density deduced from Planck law does not mean introduction of any quantization. We remain in a strictly classical Kolmogorovian probability calculus, where it is always possible to build stochastic processes reproducing any probability law (even of quantum origin) [15].

Recall that in quantum mechanics the zero point energy is a consequence of the commutation relations.
The fundamental equation of S.E.D. for a charged particle (charge e, mass m) in an external force field $\vec{F}$ is the Braffort-Marshall equation :

$$m\ddot{\vec{r}} = \vec{F}(\vec{r}) + m\psi\vec{r} + e\,\vec{E}(t)$$

where $\tau = \frac{2e^2}{3mc^3}$.
One eliminates the runamway solutions by using an integro differential form

$$m\ddot{\vec{r}} = \int_0^\infty \vec{F}[\vec{r}(t+\tau s)e^{-s}ds + e\underbrace{\int_0^\infty \vec{E}(t+\tau s)e^{-s}ds}_{\vec{E}^{(\tau)}(t)}$$

where $\vec{E}^{(\tau)}(t)$ is a new stochastic field, with a spectral density

$$\mathcal{S}_E^{(\tau)} = \mathcal{S}_E(\omega)/1 + \tau^2\omega^2$$

A simple development gives

$$m\ddot{\vec{r}} = \vec{F}(\vec{r}) + \tau(\overline{\nabla}\vec{F})\dot{\vec{r}} + e\,\vec{E}^{(\tau)}(t)$$

Intead of the mechanical moment $\vec{p} = m\dot{\vec{r}}$, which is a too irregular stochastic process, one uses the more regular elektro kinetic momentum

$$\vec{p}^{ek} = \int \vec{F}[\vec{r}(t')]dt'$$

to give a phase space representation

$$\frac{d\vec{p}^{ek}}{dt} = \vec{F}(\vec{r})$$

$$m\,\frac{d\vec{r}}{dt} = \vec{p}^{ek} + \tau\,\vec{F}(\vec{r}) + e\vec{\mathfrak{A}}^{(\tau)}(t)$$

where $\vec{\mathfrak{A}}^{(\tau)}(t)$ is a new field : $\vec{\mathfrak{A}}^{(\tau)}(t) = \int \vec{E}^{(\tau)}(t')dt'$ with a spectral density

$$\mathcal{S}_{\mathfrak{A}}^{(\tau)} = \mathcal{S}_E(\omega)/\omega^2(1 + \tau^2\omega^2).$$

The ultra-violet catastrophy has disappeared.
Among the successes of S.E.D. one can list the following problems: harmonic oscillator, black body radiation law (Planck law, "photon" term in the fluctuations of the electromagnetic field), absorption of light by the plane and the spatial rigid dipolar rotator, Lamb shift, spontaneous emission, Vand der Waals forces. Some of these successes are superficial : if one uses non linear oscillators as model of the black body cavity one gets the Rayleigh Jeans law instead of Planck's, the Kirchoff law is not satisfied, absorption

curves for the rotator are too broad. These are forewarning signs of deeper failures.

The three major failures of S.E.D. are :

1) *No waves appear associated to a particle.* No wave properties are to be deduced from the fundamental equations. *Schrödinger equation is certainly not deducible from the Braffort Marshall equation.*
2) *The absorption of light gives rise to broad absorption bands.* There is nothing like narrow line spectra.
3) *There are major discrepancies between S.E.D. and Quantum Mechanics for non linear systems.*

The study of non linear systems is made possible through an approximation of markovian type. This is due to the presence in the equations of the very small factor $\tau \sim 1/c^3$. Using the Kharsminski-Lax procedure one is able in principle to write a Fokker-Planck-Kolmogorov (FPK) equation for a markov process which is a very good approximation of the true solution.

For *the hydrogen atom* (Kepler problem) where the external force field acting on the electron is defined by a potentiel $V = -K/r$, the F.P.K. equation can be reduced to only two variables choosen among the constants of motion. The probability density W is a function of these two variables. If we choose the energy E and the total angular momentum M as variables, the reduced FPK equation reads :

$$8\pi^2 MT \frac{\partial W}{\partial t} = \text{div}\left\{W\vec{C} + \mathbb{C}\,\text{grad } W\right\}$$

where T, the orbital period, equal $2MK\sqrt{\frac{m}{|2E|^3}}$ and the drift vector $\vec{C}$ and the diffusion matrix $\mathbb{C}$ are

$$\vec{C} = \begin{vmatrix} C^E \\ C^M \end{vmatrix} \qquad \mathbb{C} = \begin{vmatrix} C^{EE} & C^{EM} \\ C^{ME} & C^{MM} \end{vmatrix}$$

where

$$C^E = 16\pi^3 \tau m K^2 \left\{\frac{E}{M^2} + \frac{3}{2}\frac{mK^2}{M}\right\}$$

$$C^M = 16\pi^3 m K^2 \frac{1}{M}$$

$$C^{EE} = 16\pi^3 \tau\hbar M \frac{|2E|^4}{mK^2} \Phi_3(\varepsilon)$$

$$C^{MM} = 16\pi^3 \tau\hbar M(2E)\ \Phi_1(\varepsilon)$$

$$C^{ME} = C^{EM} = 16\pi^3 \tau\hbar M \frac{|2E|^{5/2}}{(mK^2)^{1/2}} \Phi_2(\varepsilon)$$

and $\Phi_1(\varepsilon)$, $\Phi_2(\varepsilon)$, $\Phi_3(\varepsilon)$ are Kapteyn series expressed in term of Bessel functions, where $\varepsilon$ is the eccentricity

$$\varepsilon = \sqrt{1 + \frac{2EM^2}{mK^2}}$$

As Claverie, Pesquera and Soto have shown [16] a stationary solution $W_0$ = Constant is possible because the stationarity condition is

$$\mathrm{div}\left\{W_0\vec{C} + \mathbb{C}\,\mathrm{grad}\ W_0\right\} = 0$$

and it is easy to show that div $\vec{C} = 0$.
They don't know if there is any other stationary solution but if any, it must be non recurrent, which implies non ergodic.
In any case this is in deep disagreement with the experimental stability and symetry of the hydrogen atom, both properties reproduced by quantum mechanics.
The origin of this instability is due to the damping force (Lorentz force) on which the drift coefficients depend exclusively. They do not depend on the stochastic force exerted by the vacuum. This is due mainly to the fact that we have not introduced any dependency of the stochastic force on the point of the phase space.
The stochastic force is not position or velocity dependant.
That means that *having neglected any back reaction of the charged particle on the vacuum, we are not able to prevent a continuous transfer of energy from the vacuum to the particle. This is in principle strictly forbidden.* Hence the failure of the model. We see that it must be necessary to study this problem further *to introduce some kind of non linear interaction between the particle and the vacuum.* According to modern conceptions this could open the road to wave phenomena in the vacuum (not directly detectable) associated to an intrinsic stochastic behaviour of the particle.
But one has for the time being no physical arguments to write down this non linear interaction.

## QUANTIZATION AND WAVE-PARTICLE DUALISM AS A MANIFESTATION OF SIMULTANEOUS APPEARANCE OF ORDER AND CHAOS

Quantization is a mathematical procedure for defining a model satisfying some fundamental requirements of "physical observables".
One thinks usually that the main virtue of this model is to allow a transition from a continuum to a discrete physics. Just look at the broad use of the word quantization instead of discretization.
The initial purpose of quantization for classical electrodynamical systems was to provide a model with two fundamental properties :

1) stability of atomic and molecular systems
2) discrete properties of spectral responses (line spectra for absorption and emission of light).

In both cases one faces strong manifestations of Order. That a concept of probability, randomness, Chaos, slips into the quantized model is considered as an unavoidable parasitic phenomenon.
Probability looks as a by product of quantization, which is not clearly accounted for. One has really to make the best of it, all the more so since those are strange probabilities, not ruled by

the classical Kolmogorovian calculus of probability [17].
The contemporary progress in the field of classical dynamical systems on the mechanism of generation of randomness [3] sets a new intellectual demand for the justification of chaotic phenomena. These phenomena appear in precise conditions as complex dynamical behaviour induced by some kind of interaction (in principle nonlinear interactions). It is no more possible just to say that quantum systems show an "irreducible randomness". One day or another it will be unavoidable to give the reasons for the appearance of the "microphysical chaos".
In that spirit *it seems essential to consider quantization as the assertion of the very existence of a microphysical chaos, and of the probabilistic strategy which gives a regularized description of this chaos.*
Quantization is above all a way of mathematical correspondence between a classical non chaotic (or chaotic) dynamical system and some new kind of chaotic system. In any case a new kind of chaos appears. But in this chaos it appears simultaneously some kind of order, of organization, of auto-organization. Order and chaos are interwoven. *Quantum phenomena set the problem of "structure of chaos"* [18]. Historically the wave-particle duality is born in such a problematic ; it expresses a structure of the statistical properties of the electromagnetic field in black-body radiation. The quantum formalism is well adapted to this problematic introducing both order and chaos through the mathematical properties of operators. For instance the discrete spectrum of operators expresses organization while eigenfunctions give a probabilistic measure (regularized measure) of the underlying chaos.
The wave-particle structure of fluctuations arises from the non-commutation of operators, as the appearance of the zero-point energy in the harmonic oscillator [12].
One is forced to evoke those phenomena where order and chaos coexist, as turbulence or more generally auto-organisation phenomena. More precisely auto-organisation in turbulence is produced by non linear equations but the auto-organized state is described by equations which can be linear. This remembers also of the situation for the soliton where the original non-linear equation can be solved through linear equations. From this point of view the wave function is just a mathematical device.
As for de Broglie wave, the wave of the wave-particle dualism, it is a peculiar wave, both order and chaos. A random wave ? Such the same for the corpuscle which is at the same time an expression of the statistical properties of the field (the wave) and of its organized properties (exchange of energy through quanta).
*Chaos is organized and organization is not just order. A duality which gives rise to Dualism.*

## NON KOLMOGOROVIAN CALCULUS OF PROBABILITY AS A REGULARIZED DESCRIPTION OF THE MICROPHYSICAL CHAOS

We have already seen that quantum theory must be considered as a probability calculus which gives a regularized description of chaotic phenomena called "the microphysical chaos".
It has been often stressed that the quantum probability calculus is not all equivalent to the classical Kolmogorovian probability calculus. In fact the latter considers implicitely a restricted class of chaotic phenomena, where all the possible observables are endowed with regularized properties (existence of measures). That chaotic phenomena exists which are not giving rise to Kolmogorov calculus, is clear from the theory of pseudo-random functions due to J. Bass [19]. In that theory it is possible to show the existence of chaotic phenomena which though happening simultaneously and being individually described by asymptotic measures, do not allow any regularization for their simple product.
Just the same situation as in quantum mechanics. We think that requiring a quantum description is fundamentally requiring the use of a non Kolmogorovian probability calculus. Quantum physics is the manifestation of a microphysical chaos which must be described probabilistically in a different way than a billiard or a game of dice. The actual nature of this chaos is completely unknown.

## IS MICROPHYSICAL CHAOS BROUGHT ABOUT BY THE OBSERVATION ?

There is a wide agreement to say that Quantum Mechanics does not describe a system in itself but only deals with the results on actual observation on it. Quantum mechanics is a probability calculus for the results of measurement on microphysical systems. These affirmations imply in fact that observation plays an active role in the appearance of results. G. Lochak writes for example : "C'est pourquoi Piron (28.0 livraison 19. Lettres épistémologiques de l'Association Gonseth) a parfaitement raison de dire "C'est la mesure qui, perturbant le système, crée les valeurs trouvées". Cette phrase n'a aucun rapport avec le subjectivisme, l'indéterminisme, la non objectivité des lois de la nature, ni avec d'autres problèmes philosophiques. Elle ne fait que résumer un état de chose expérimental qui provient de ce que le seul procédé de mesure que nous connaissions en microphysique consiste à étudier la réponse d'un système à un champ perturbateur et que cette réponse est extrêmement compliquée en raison des propriétés ondulatoires de la matière" [20].
In fact one can ask if the perturbation introduced by the measurement is not responsible for the appearance of the microphysical chaos (or a part of it). Such a concept of chaotic phenomena generated by perturbation of non-linear systems is becoming a current concept for classical dynamical systems and plays some role in the problem of "quantum chaos" [21][22].

## TEN THESIS AS CONCLUSION

1. There is a Vacuum, inaccessible by direct experiments.

2. Vacuum is chaotic.

3. There is a great diversity of chaotic phenomena.

4. Kolmogorov probability calculus is not the only possible regularized description of chaotic phenomena.

5. Quantum mechanics is a non-kolmogorovian probability calculus.

6. All "microphysical chaos" including phenomena in the vacuum must be described by the quantum formalism on a probabilistic level.

7. Wave-particle dualism is a manifestation of an interplay between order and chaos in a physical system.

8. The actual physical nature of the microphysical chaos is unknown.

9. The observation could play a role in the generation of the microphysical chaos.

10. Be patient ! Some Order will appear in all that Chaos.

## REFERENCES

(1) The best accounts of this subject are :
Zwonkin, A.K., Levin, L.A., Usp. Mat. Naouk 25, 85, 1970
Dellacherie, C., Gazette des mathématiciens n°11, 1978
Alekseev, V.M., Yakobson, M.V., Phys. Reports, 75, 287, 1981
Popular accounts are given in Ford, J., Physics Today, 40, April 1983

(2) See for example the special issues of International J. of Theor. Physics (1982) devoted to the proceedings of a conference on "Physics of Computation"

(3) Berry, M.V., in Jorna, S., ed. Topics in non linear dynamics Am. Inst. of Physics; 1978
Helleman, R.H.G., in Cohen, E.G.D., ed. Fundamental problems in statistical mechanics vol.5, North-Holland 1980

(4) Planck, M., Verhandl. Dtsch. phys. Ges. 13, 138, 1911

(5) Planck, M., Uber neuere thermodynamische Theorien (Nernstsches Wärme theorem und Quantenhypothese) Leipzig 1912, Ber. Deutsch. Chem. Ges. 45, 5, 1912. Phys. Z. 13, 165, 1912

(6) Einstein, A., Stern, O., Ann. Phys. 40, 551, 1913

(7) Enz, C.P., in Enz, C.P., Mehra, J., eds Physical Reality and Mathematical description p. 125

(8) Pauli, W., "Die Állgemeinen Prinzipien der Wellenmechanik" in Handbuch der Physik vol. 24, part 1, 83, Springer 1933

(9) Stern, O., Helv. Phys. Acta 367, 1962

(10) Dirac, P.A.M., Nature 168, 906, 1951

(11) Einstein, A., Ather und Relativitätstheorie, Julius Springer, Berlin 1920

(12) See the article by Milonni in this book

(13) Mandel, L., in Progress in Optics, Wolf, E., ed. vol.XIII p. 27, North-Holland 1976
Loudon, R., Rep. Progr. Phys. 43, 911, 1980

(14) Milonni, P.W., Physics Reports 25C, 1, 1976
Claverie, P., Diner, S., Int. J. Quant. Chem. 12, Supp. 1, 23, 1978
Boyer, T.H. in Barut, A.O., ed., Foundations of radiation theory and quantum electrodynamics, Plenum 1980, p. 49
Claverie, P., in Blaquière, A., Fer, F., Marzollo, A., eds. Dynamical systems and microphysics, Springer 1980, p. 111

(15) For such a simulation of quantum mechanical probability distributions, cf :
Nelson, E., Dynamical theories of brownian motion, Princeton U.P. 1967
Erber, T., Rynne, T.M., Sklar, A., Acta Physica Austriaca 53, 145, 1981

(16) Claverie, P., Pesquera, L., Soto, F., Physics Letters 80A, 113, 1980

(17) See the article by Accardi in this book, p. 297.

(18) The problem of "structure of chaos" is the main recurrent thema of the self-organization phenomenon in chaotic or turbulent media.
Cf. Haken, H., Synergetics, Springer, 1978
Hasegawa, A., Europhysics News, 13, 1, 1982

(19) Bass, J., J. of Math. analysis and applications 47, 354, 1974
Bass, J., Ann. Inst. H. Poincaré, A, XXXIII, 301, 1980
Bass, J., in Diner, S., Fargue, D., Lochak, G., "La Pensée Physique Contemporaine" p. 391, A. Fresnel, 1982

(20) Lochak, G., Epistemological letters, Association F. Gonseth, 33 livr.

(21) Tomita, K., Physics Reports, 86, 113, 1982

(22) Zaslavsky, G.M., Physics Reports 80, 157, 1981.

# STRUCTURES IN SEMICLASSICAL SPECTRA: A QUESTION OF SCALE

M.V. Berry

H.H. Wills Physics Laboratory
Tyndall Avenue, Bristol BS8 1TL, U.K.

## ABSTRACT

Theories of semiclassical bound state spectra for systems with N freedoms are reviewed, emphasizing the different features occurring on successively finer scales of energy E, measured in terms of $\hbar$, and attempting to correlate these with whether the underlying classical motion is regular or irregular.

## 1. INTRODUCTION

Substantial efforts are now being made [1-7] to unravel the semiclassical energy-level structure of bound Hamiltonian systems, motivated for example by the need to understand the high-lying states of vibrating molecules. The difficulty of the quantal problem reflects the complexity of the underlying classical motion [8-11]: for a system with N freedoms (i.e. with a 2N-dimensional phase space), where $N \geqslant 2$, motion may be <u>regular</u> (ordered, predictable), i.e. orbits wind smoothly round N-dimensional tori in phase space; or it may be <u>irregular</u> (chaotic, unpredictable), i.e. exploring 2N-1 dimensional regions in phase space with neighbouring orbits separating exponentially. What corresponds, quantum-mechanically, to these classical distinctions? My intention here is to give a brief discussion of this question, concentrating on the distribution of energy levels; a more substantial review, including an account of the morphologies of wave functions, will be published elsewhere [12].

Obviously one expects the nature of the classical motion to influence the quantal spectrum under <u>semiclassical</u> conditions,



*S. Diner et al. (eds.), The Wave-Particle Dualism, 231–252.*

i.e. when $\hbar$ is in some sense small (in comparison with classical quantities having the same dimensions). In fact when $\hbar \to 0$ the spectrum displays a richness of different types of behaviour on clearly separated energy scales. I shall describe this picture of the spectrum by starting in § 2 with the coarsest energy scales (larger than $O(\hbar)$), on which the spectrum is described by the mean level density $\bar{d}(E)$. In § 3 I shall explain how oscillatory corrections to $\bar{d}$, corresponding to level clustering on scales of order $\hbar$, are associated with classical closed orbits. § 4 is devoted to the 'quantization conditions' giving individual levels, whose spacing is of order $\hbar^N$. Then, in § 5 the texture of the spectrum on scales smaller than $\hbar^N$ will be discussed in terms of the distributions of spacings between neighbouring levels. Finally, in § 6 the most delicate features of all, on the finest scale-degeneracies between energy levels - will be considered.

It will be helpful to have in mind certain extreme types of classical motion. The extreme of regularity is displayed by integrable systems, where there are N global constants of motion (including the energy) and all orbits are confined to phase-space tori. Systems displaying extreme irregularity are ergodic: only the energy is conserved and almost all orbits (i.e. all except a set of zero measure) explore almost all of the 2N-1 dimensional 'energy surface'. But most Hamiltonians (including the anharmonically coupled oscillators representing vibrating molecules) are neither integrable nor ergodic: a finite volume of phase space is filled with N-tori, and a finite volume with 2N-1-dimensional chaotic orbits; often the proportions change as the energy increases, being nearly integrable ('quasi-integrable') at low energies and nearly ergodic at high energies.

## 2. SCALES LARGER THAN $\hbar$: THE MEAN LEVEL DENSITY

Consider a bound system with coordinates $\underset{\sim}{q} = \{q_1 \ldots q_N\}$ and momenta $\underset{\sim}{p} = \{p_1 \ldots p_N\}$, with classical Hamiltonian $H(\underset{\sim}{q},\underset{\sim}{p})$. Suppose the corresponding Hamiltonian $\hat{H} = H(\hat{\underset{\sim}{q}},\hat{\underset{\sim}{p}})$ generates the discrete spectrum of energies $E = E_1, E_2 \ldots = \{E_j\}$, labelled in order of increasing E. A complete description of this spectrum would be obtained from knowing the level density d(E), defined as

$$d(E) \equiv \sum_{j=1}^{\infty} \delta(E-E_j) = \mathrm{Tr}\,\delta(E-\hat{H}) \qquad (1)$$

There is a beautiful semiclassical theory for d(E), whose principal architects were Gutzwiller [13-17] and Balian and Bloch [18-21]. This is based on representing d(E) in the form

$$d(E) = \bar{d}(E) + d_{osc}(E), \qquad (2)$$

where $\bar{d}$ is the mean level density and $d_{osc}$ is a series of oscillatory corrections. The terms on the right side of (2) correspond to successive smoothings of the singular function d(E). On the coarsest scale, that is after smoothing over energy ranges ΔE large enough to obliterate all traces of individual levels and all scales of level clustering, only $\bar{d}(E)$ survives. As ΔE is made smaller, more and more terms in $d_{osc}(E)$ contribute, with faster oscillations, until eventually they sum to give a series of delta functions at the energies of the levels. In simple terms, the representation (2) is a generalization of the following 'Poisson' expansion for a series of equally spaced delta functions:

$$\sum_{n=-\infty}^{\infty} \delta(E-n) = 1 + 2 \sum_{n=1}^{\infty} \cos 2\pi mE \qquad (3)$$

We begin by studying the mean level density $\bar{d}(E)$. This is given by the simple semiclassical rule [e.g. 22] that each quantum state is associated with a phase-space volume $h^N$, so that it predicts

$$\bar{d}(E) \approx \frac{1}{h^N} \int d\underset{\sim}{q} \int d\underset{\sim}{p}\ \delta(E-H(\underset{\sim}{q},\underset{\sim}{p})), \qquad (4)$$

i.e. the level density is proportional to the 'size' of the energy surface.

One way to obtain this formula - which is convenient in that it also leads to a theory for $d_{osc}$ - is to write (2) in terms of the outgoing time-independent Green function:

$$G^+(\underset{\sim}{q},\underset{\sim}{q}';E) \equiv \left\langle q' \right| \frac{1}{E+i\varepsilon-\hat{H}} \left| q \right\rangle \qquad (5)$$

so that

$$d(E) = -\frac{1}{\pi} \operatorname{Im} \int d\underset{\sim}{q}\ G^+(\underset{\sim}{q},\underset{\sim}{q}';E)_{\underset{\sim}{q}' \to \underset{\sim}{q}} \qquad (6)$$

In view of the fact that $G^+$ is the probability amplitude for observing at $\underset{\sim}{q}$ particles emitted from $\underset{\sim}{q}'$ with energy E, it is not surprising that the semiclassical theory [13-21; see also the review 23] gives $G^+$ as a sum over all classical trajectories leading from $\underset{\sim}{q}'$ to $\underset{\sim}{q}$ with energy E. As $\underset{\sim}{q}' \to \underset{\sim}{q}$ this sum is dominated by the direct path going straight from $\underset{\sim}{q}'$ to $\underset{\sim}{q}$, and the contribution to (6) from this path alone gives precisely $\bar{d}$ as given by (4).

In the important special case where H takes the form

$$H(\underset{\sim}{q},\underset{\sim}{p}) = \frac{|\underset{\sim}{p}|^2}{2\mu} + V(\underset{\sim}{q}), \qquad (7)$$

(4) gives $\bar{d}$ as an integral over the classically accessible space, namely

$$\bar{d}(E) \approx \left[\frac{\mu}{2\pi\hbar^2}\right]^{\frac{N}{2}} \frac{1}{\Gamma(N/2)} \int d\underset{\sim}{q} \left[E-V(\underset{\sim}{q})\right]^{\frac{N}{2}-1} \Theta\left[E-V(\underset{\sim}{q})\right], \qquad (8)$$

where $\Theta$ denotes the unit step function. If N=2 this reduces to the simple formula

$$\bar{d}(E) = \frac{\mathcal{A}(E)\mu}{2\pi\hbar^2} \qquad (9)$$

where $\mathcal{A}(E)$ is the area of the classically accessible $\underset{\sim}{q}$-space. In the context of 'billiard' systems, where $V(\underset{\sim}{q})$ vanishes within a curve B and is infinite outside B, so that classical trajectories are straight lines reflecting specularly at B and quantal wave functions vanish on B, (9) is called the Weyl formula (for a review see [24]).

For numerical tests of (4) it is convenient to work not with the spiked function d(E) but with its integral, namely the <u>mode number</u> $\mathcal{N}(E)$ defined by

$$\mathcal{N}(E) \equiv \int_{-\infty}^{E} d(E')dE' = \sum_{j=1}^{\infty} \Theta(E-E_j). \qquad (10)$$

The procedure is to compute the levels $E_j$, construct the exact stepped curve $\mathcal{N}(E)$ and compare its trend $\bar{\mathcal{N}}(E)$ with what (4) would give. In one such test, Marcus and his coworkers [25,52] obtain very good agreement. Here I show another test, for the 'desymmetrized Sinai billiard', whose boundary B is shown on fig. 1. This system (which I studied in detail in [26]) was chosen because its classical trajectories were proved to be ergodic (Sinai [27]) whenever R > O (when R = O the system is trivially integrable). Fig. 2 shows $\mathcal{N}(E)$ (stepped curve) and $\bar{\mathcal{N}}(E)$ as given by (9) and (10) (full curve) for five values of R. Evidently the agreement is poor. The reason is that (4) is an <u>asymptotic</u> formula which is here being tested on <u>low-lying states.</u> It is necessary to include correction terms. These are not contributions to $d_{osc}$ (which will be discussed in § 3), but are smooth terms of lower order in $\hbar$, depending on aspects of billiard geometry [24, 18] other than the area $\mathcal{A}$. When these corrections are included, the agreement improves spectacularly, as the dashed curves in fig. 2 show.

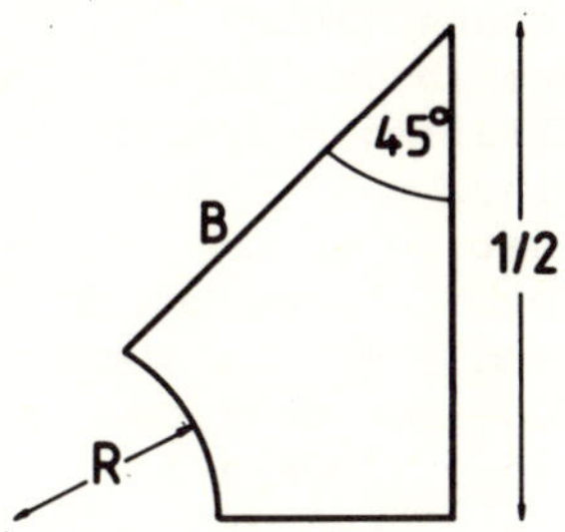


Figure 1. Desymmetrized Sinai billiard.

We draw two conclusions from the formula (4) for $\bar{d}$. The first is that the mean spacing of the levels, $\bar{d}^{-1}$, is proportional to $\hbar^N$. The second is that on these coarsest scales, embodied in $\bar{d}$, the spectrum is so smoothed as to no longer display any evidence of the regularity or irregularity of the underlying classical motion. To find such evidence, we must take an 'energy microscope' and inspect the spectrum more closely.

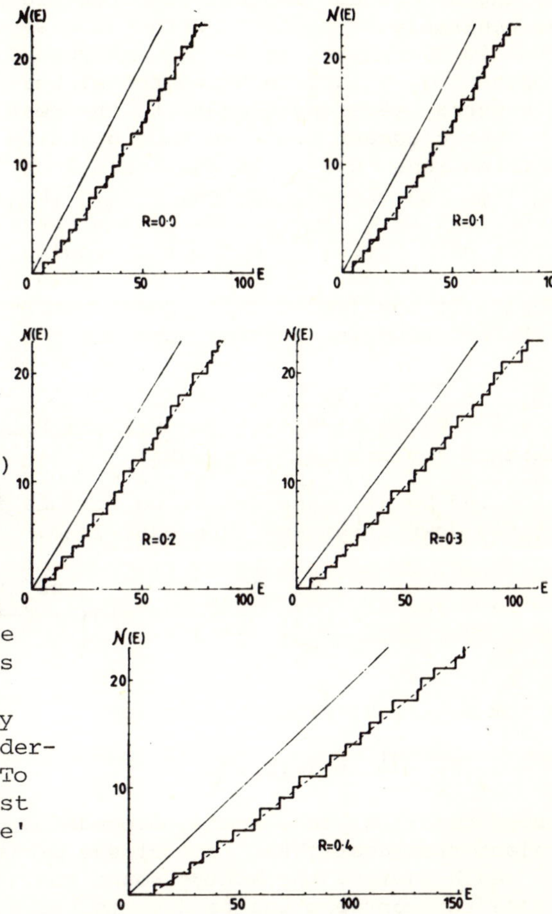


Figure 2. Mode number (stepped curve) for Sinai billiard, with uncorrected mean mode number (full curve) and corrected mean mode number (dashed curve).

## 3. SCALES OF ORDER $\hbar$: OSCILLATORY LEVEL CLUSTERING

The fact that the mean level spacing is of order $\hbar^N$ might lead us to expect that the mean level density $\bar{d}(E)$ would be obtained simply by smoothing the delta-spikes of the exact $d(E)$ over an energy range of order $\hbar^N$. But such an expectation is mistaken, because as we shall see the corrections $d_{osc}$ oscillate with energy 'wavelength' of order $\hbar$, which when $N > 1$ (i.e. in nontrivial cases) is infinitely larger than the mean spacing as $\hbar \to 0$. In this important respect the spectrum in the general case contrasts with those special cases (which actually correspond to N=1 as analyzed for example in [23]) represented by (3), in which the slowest-oscillating corrections have energy wavelength equal to the mean spacing.

The clearest route to understanding the nature of the contributions to $d_{osc}$ is the Green function method based on (5) and (6). As already stated, $G^+(\underset{\sim}{q},\underset{\sim}{q}';E)$ is a sum over classical paths linking $\underset{\sim}{q}'$ and $\underset{\sim}{q}$. As $\underset{\sim}{q}' \to \underset{\sim}{q}$ these include not only the direct path giving $\bar{d}$, but paths looping back to $\underset{\sim}{q}$ after a finite excursion. A looping path need not be a closed orbit, because it may (and usually does) return to $\underset{\sim}{q}$ with momentum $\underset{\sim}{p}$ different from its initial momentum $\underset{\sim}{p}'$. But in the integration over $\underset{\sim}{q}$ in (6) such non-closed looping paths can be shown [16, 12, 23] to give negligible contributions.

The conclusion is that <u>only closed classical orbits with energy E contribute to $d_{osc}(E)$;</u> these include repetitions (labelled by p) of primitive (i.e. unrepeated) closed orbits (labelled by j). The resulting formula for $d_{osc}$ will be written down first and then explained:

$$d_{osc}(E) = \sum_j \sum_{p=1}^{\infty} \frac{A_{j,p}(E)}{\hbar^{1+\ell_j/2}} \sin\left\{ \frac{pS_j(E)}{\hbar} + p\alpha_j \right\} \qquad (11)$$

In the phase, $S_j(E)$ is the action

$$S_j(E) = \oint \underset{\sim}{p}_j(\underset{\sim}{q};E).d\underset{\sim}{q} \qquad (12)$$

around the j'th path, and $\alpha_j$ depends on the focusing of trajectories near the closed orbit.

As E varies, the action round the j'th path changes, and causes its contribution to $d_{osc}$ to oscillate with energy wavelength $\Delta E$ given by

$$\frac{p}{\hbar}\frac{dS_j}{dE}\Delta E = 2\pi \quad \text{i.e. } \Delta E = \frac{h}{pT_j} \qquad (13)$$

where $T_j$ is the period of the j'th primitive orbit. Therefore the oscillations are indeed on the scale $\hbar$ (much longer than $\hbar^N$ as $\hbar \to 0$) as asserted. Longer paths give faster oscillations.

The amplitude in (11) depends on whether the j'th path is isolated or nonisolated, and whether it is stable or unstable. In (11) the integer $\ell_j$ measures the degree of isolation: the j'th orbit is embedded in an $\ell_j$-parameter family of closed orbits. $\ell_j$ may vary from zero (for an isolated orbit) to N-1 (for the torus-filling closed orbits of an integrable system - taking account of the fact that each closed orbit occupies one dimension). As an example, the stadium billiard (fig. 3) which Bunimovich [28] proved to be ergodic, has both isolated closed orbits (e.g. the 'long-diameter' and 'rectangular' orbits in fig. 3) and non-isolated closed orbits (the 'short-diameter' orbits in fig. 3).

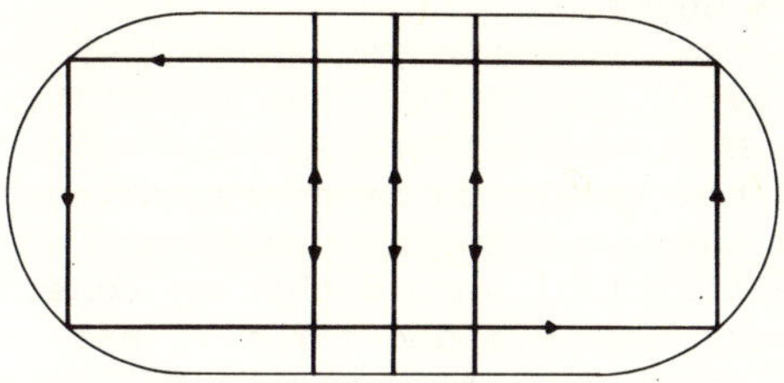

Figure 3. Stadium billiard, with two isolated closed orbits and three members of a family of nonisolated closed orbits.

In (11) the behaviour of the factor $A_{j,p}$ as a function of repetitions p depends on the stability of the j'th orbit. For isolated orbits, $A_{j,p}$ oscillates with p if the orbit is stable, and decays exponentially if the orbit is unstable - as expected in view of the repeated focusing or continued de-focusing of beams of trajectories in the two cases. For integrable systems, where orbits are embedded in N-1 parameter families, Berry and Tabor [29, 30] showed that $A_{j,p}$ decreases as $p^{-(N-1)/2}$.

The Green function theory leading to (11) is reviewed by Rajaraman [31] and deWitt-Morette et al. [32]. In the original papers, Gutzwiller [16] emphasizes the question of stability, and Balian and Bloch [20] emphasize the question of isolation.

Now I want to dispose of a fallacy based on a mis-interpretation of (11). Consider the terms with given j, i.e. those corresponding to all repetitions of a single closed orbit. The terms will interfere constructively if

$$S_j(E_m) = (2\pi m - \alpha_j)\hbar \qquad (14)$$

defining a series of energies $E_m$ corresponding to integers m, at which the sum over p gives a contribution to $d_{osc}$ with some sort of singularity whose nature depends on $A_{j,p}$. The fallacy is to suppose that $E_m$ must be eigenvalues of $\hat{H}$, i.e. that (14) is a semiclassical quantization condition associating individual quantum states with repetitions of individual closed orbits.

Why is this a fallacy? For a start, (14) gives energies with separations of order $\hbar$, whereas the true levels have separation $\hbar^N$. So there are too few levels. But why not superpose the level sequencies obtained from (14) with all topologically different orbits? Because this would give too many levels! An instructive example is motion in a rectangle, where the true levels are labelled by two quantum numbers (see § 4) and the false levels given by (14) are labelled by three quantum numbers and moreover have the 'infrared catastrophe' of existing with arbitrarily low energies.

Nevertheless, there are two circumstances where (14) does give semiclassical quantal levels correctly. The first (trivial) case is potential wells with N=1, where there is only one topology of closed orbit and the levels do have separation $\hbar$ (see e.g. [23]). The second case is when the orbits are isolated and stable. Then Miller [33] showed that by considering lowest-order fluctuations about the periodic orbit it was possible to generalize (14) into a condition with a full set of quantum numbers. But Voros [34] pointed out that isolated stable orbits are always surrounded by tori, and explained how this quantum condition is really an approximate version of the 'torus quantization' to be discussed in § 4.

In general, though, a single closed orbit gives not individual levels but a collective property of the spectrum, namely an oscillatory clustering with scale $\Delta E$ given by (13). Conversely, the determination of individual levels from (11) involves the close orbits collectively, and would require the summation over sufficiently many closed orbits for individual delta functions in d(E) to emerge as the result of interference. Is this a feasible procedure for calculating individual levels? I shall argue that it is not.

To begin to see delta functions emerging from (11), it is necessary to include at least all orbits giving oscillations whose energy wavelength (13) exceeds the mean spacing $d^{-1}$. Since the longer orbits give faster oscillations, it is necessary to sum over at least all closed orbits with period less than $T_{max}$, which is easily calculated to be

$$T_{max} = \frac{\int d\underset{\sim}{q} \int d\underset{\sim}{p} \;\; \delta(E-H(\underset{\sim}{q},\underset{\sim}{p}))}{h^{N-1}} \qquad (15)$$

As $\hbar \to 0$, $T_{max}$ increases and it is necessary to include ever more closed orbits. How many? As discussed for example in [12] and [26], the number is of order $\hbar^{-N(N-1)}$ for integrable systems, and of order $\exp\{+\mathrm{const}/\hbar\}$ for chaotic systems where closed orbits are vastly more numerous. This means that the number of operations required to determine levels by summing over closed orbits is much greater than in competing methods (see § 4).

Nevertheless, it is instructive to see delta functions emerging as more topologies are included, and I illustrate this with an integrable system with N=2, whose Hamiltonian involves a Morse potential:

$$\left.\begin{array}{c} H = \dfrac{p_x^2+p_y^2}{2\mu} + V_o\left[e^{-2\delta(r-r_o)} - 2e^{-\delta(r-r_o)}\right] \\ (\mu=1 \text{ proton mass, } V_o = 0.2 \text{ eV, } r_o = 0.25 \text{ nm, } \delta = 10 \text{ nm}^{-1} \\ r \equiv (x^2+y^2)^{1/2} \end{array}\right\}; \qquad (16)$$

there are 166 bound levels. Fig. 4 shows some simple topologies of closed orbit (classified by the number of rotations and librations before closure), and fig. 5 shows the effect on d(E) of including increasingly many topologies. In the last frame, which includes several hundred closed orbits, delta functions are beginning to emerge very clearly (arrows mark exact levels, the chain curve is $\bar{d}(E)$). For more details see [29].

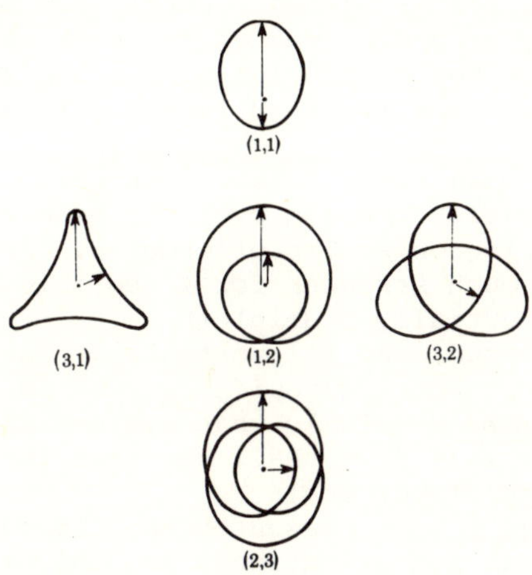


Figure 4. Some closed orbits in a central potential.

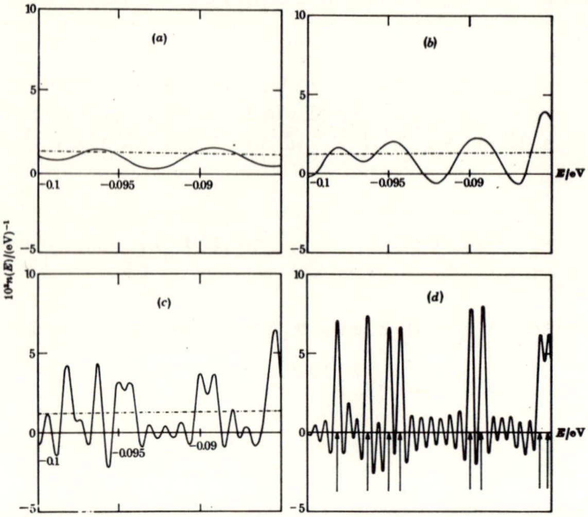


Figure 5. Emergence of δ functions in d(E) as more oscillatory corrections, from closed orbits, are added to $\bar{d}(E)$.

For ergodic systems, direct summation over closed orbits appears hopeless as a means of locating eigenvalues. However, in a tour de force, Gutzwiller [35] has recently evaluated by an indirect (and not very economic) method the path sum for the (almost certainly ergodic) Kepler problem with anisotropic kinetic energy. He obtains the first 18 levels with an accuracy of a few percent.

It is therefore the case that, rather than being a means of calculating individual levels, the sum over closed orbits gives information about clustering on scales of order $\hbar$. According to (11), this clustering is much stronger for integrable systems (orbits nonisolated and neutrally stable) than for ergodic systems (most orbits isolated, and all unstable). This is the first evidence that systems whose classical motion is irregular have levels distributed <u>more</u> regularly than systems with regular classical motion; in § 5 we shall see that the same property holds down to the finest scales.

And finally, although I have presented the closed orbit expansions as semiclassical approximations, and this is how they arose in the Green function theories [13-20], it is encouraging that an <u>exact</u> correction is known between the spectrum and the closed orbits for nonintegrable systems. For the wave equation on a smooth compact Riemannian manifold (which need not have constant curvature), if the eigenwavenumbers are

$k_j\,(=2\mu E_j)^{1/2}/\hbar)$, then the function

$$Q(\lambda) = \sum_{j=1}^{\infty} e^{ik_j\lambda} \tag{17}$$

was proved by Chazarain [36] to have singularities for each λ equal to the length of a closed geodesic ($Q(\lambda)$ is the Fourier transform of the k density of states). Balian and Bloch [21] obtained a similar exact result for the Schrödinger equation: the transform with respect to $\hbar^{-1}$ is singular at the action values of closed classical orbits.

## 4. SCALES OF ORDER $\hbar^N$: QUANTIZATION CONDITIONS

Now we step down from the scale $\hbar$ to the scale $\hbar^N$, and seek a semiclassical quantization formula giving the positions of individual energy levels with an accuracy that increases as $\hbar \to 0$. Such a formula exists for regular motion, that is for orbits lying on phase-space tori; it is known by various names: Bohr, Sommerfeld, Einstein [37], Brillouin, Keller [38], Maslov [39] ... I shall refer to the procedure, now to be described, simply as 'torus quantization'.

Let tori be labelled by the values $I_i$ of the actions round their i'th irreducible circuits $\gamma_i$:

$$I_i \equiv \frac{1}{2\pi}\oint_{\gamma_i} \underset{\sim}{p}(\underset{\sim}{q}).d\underset{\sim}{q} \quad (1 \leqslant i \leqslant N). \tag{18}$$

Then the eigenstate with quantum numbers $\underset{\sim}{m} \equiv \{m_1 \ldots m_N\}$ is associated with the torus whose actions $\underset{\sim}{I} \equiv \{I_i \ldots I_N\}$ are quantized by

$$\underset{\sim}{I}_{\underset{\sim}{m}} = (\underset{\sim}{m} + \underset{\sim}{\alpha}/4)\hbar \tag{19}$$

where $\underset{\sim}{\alpha} \equiv \{\alpha_1 \ldots \alpha_N\}$ is a set of N integers giving the number of real-space caustics encountered during each cycle $\gamma_i$ of the torus. The energy levels $E_{\underset{\sim}{m}}$ are simply the energies of these quantized tori, i.e.

$$E_{\underset{\sim}{m}} = H(\underset{\sim}{I}_{\underset{\sim}{m}}) \tag{20}$$

where H denotes the Hamiltonian in action representation (for elementary derivations of this torus quantization formula see [40] and [12]).

The tori picked out by (19), associated with individual

quantum states, almost never have commensurable frequencies and so do not contain closed orbits. On the other hand, we saw in § 3 that closed orbits can be made to generate the whole spectrum. This is apparently contradictory but actually no more than the fact that an irrational number can be approximated by a series of rationals.

The explicit formula (20) is exact as $\hbar \to 0$. For fixed E this corresponds to high states, but (20) often gives results accurate to a few percent even for the ground state (e.g. when N=1, in which case it is the familiar WKB level formula [23]); in a few cases (harmonic oscillator, Coulomb potential, rectangular boxes) it gives all levels exactly. For integrable systems, the whole phase space is filled with tori and (20) approximates all the levels. For quasi-integrable systems, the Kolmogorov-Arnol'd-Moser (KAM) theorem [8-11] guarantees the existence of tori filling part of phase space, and so (20) can be employed to find a finite proportion of the levels, provided the actions and energies of the tori can be determined. This is a difficult problem of classical mechanics, which has been tackled analytically by perturbation and iteration methods, and numerically by studying caustics and the Poincare surface of section. By these techniques, Marcus and his colleagues [1-4], Percival and Pomphrey [5], Jaffe and Reinhardt [6] and Chapman, Garrett and Miller [7] obtained eigenvalues for chemically interesting quasi-integrable systems in very good agreement with 'exact' computations.

What determines the levels if large regions of phase space are filled with irregular trajectories? The extreme is an ergodic system. Then torus quantization must fail, for there are no tori, no actions and so no quantum numbers. Nobody has so far succeeded in finding an explicit quantum condition to put in its place, i.e. to give the levels in what Percival [41] called the 'irregular spectrum'. For billiard systems with N=2, however, and in particular for the ergodic Sinai billiard (fig. 1), it is possible to devise an implicit quantum condition [26] in the form of a determinant, obtained by a Green function procedure, which is much more rapidly convergent than those obtained from conventional basis-set expansions. The determinant has effective size MxM, where

$$M = \frac{\text{perimeter of billiard boundary}}{\text{de Broglie wavelength } \lambda \text{ of state being studied}} \qquad (21)$$

Because $\lambda$ is proportional to $\hbar$, the work of finding levels increases as $\hbar \to 0$, but less rapidly than in any other method known to me.

It is possible by (nontrivial) analytical manipulation of the explicit torus quantization condition (20), or the implicit determinantal condition for billiard eigenvalues, to give alternative derivations [29,26] of the 'large-scale' formulae for $\bar{d}$ and $d_{osc}$ already discussed in §§ 2 and 3.

## 5. SCALES SMALLER THAN $\hbar^N$: LEVEL SPACINGS

Now that we are down on the scale $\hbar^N$ corresponding to individual levels, the next step is a further application of the energy microscope, to study the fine-scale texture of the spectrum as embodied in the distribution P(S) of spacings S between neighbouring levels. S is measured as a fraction of the mean level spacing $[\bar{d}(E)]^{-1}$, and P(S) is defined by

P(S) dS = probability that the spacing of a randomly chosen pair of neighbouring levels lies between S and S+dS. (22)

The ensemble over which the probability is taken is defined semiclassically, as the infinitely many levels near any given E as $\hbar \to 0$. The function P(S) was introduced [42] in a non-semiclassical context, to describe the many-body spectrum of nuclei.

We shall study P(S) as $S \to 0$, because this gives information about the finest scales of level clustering. If $P(S) \to 0$ as $S \to 0$, neighbouring levels can be considered to 'repel' each other, leading to a degree of regularity in the arrangement of levels, which can be quantified by the manner in which P(S) vanishes. If on the other hand $P(S) \to$ constant as $S \to 0$, neighbouring levels cluster rather than repel.

In § 6 I shall outline an argument strongly suggesting that for generic (i.e. 'almost all') systems,

$$P(S) \sim \text{const} \times S \text{ as } S \to 0 \qquad (23)$$

(the same result holds for eigenvalues of the ensembles of random matrices [42] considered in nuclear physics). Therefore generic systems are expected to display linear level repulsion. Fig. 6 shows a test of this prediction, obtained by calculating several hundred quantal eigenvalues [26] of the classically ergodic desymmetrized Sinai billiard (fig. 1), for a range of R between .20 and .44. It is clear that the levels do repel, and that the linear law gives a good fit. McDonald and Kaufman [43], and Casati et al. [44], in computations of P(S) for the desymmetrized stadium billiard (one quarter of the shape in fig. 3), also obtain level repulsion.

On the other hand, for systems with torus quantization Berry and Tabor [45] showed from (20) that in almost all cases (and if $N > 1$), P(S) has the universal form

$$P(S) = e^{-S} \qquad (24)$$

which is finite as $S \to 0$, in contrast to (23), and corresponds

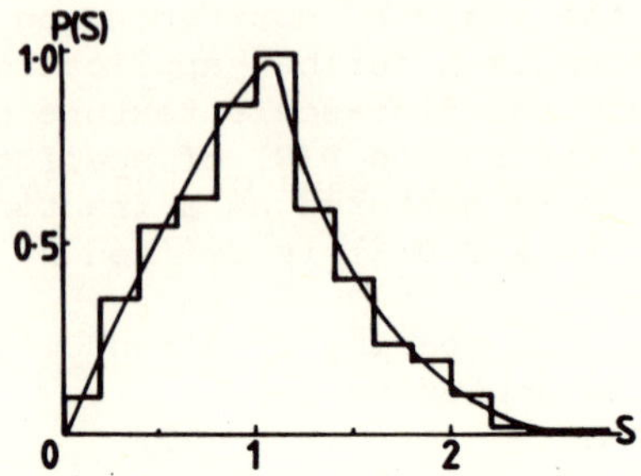


Figure 6. Level spacings distribution for desymmetrized Sinai billiard.

to levels arriving irregularly with a degree of clustering. Fig. 7 shows tests of this prediction for two integrable systems with N=2: (a) corresponds to a rectangle with side ratio 2, and (b) corresponds to a potential which is a square well in the x direction and a harmonic oscillator in the y direction. Evidently the negative exponential is a very good fit to the computed histograms.

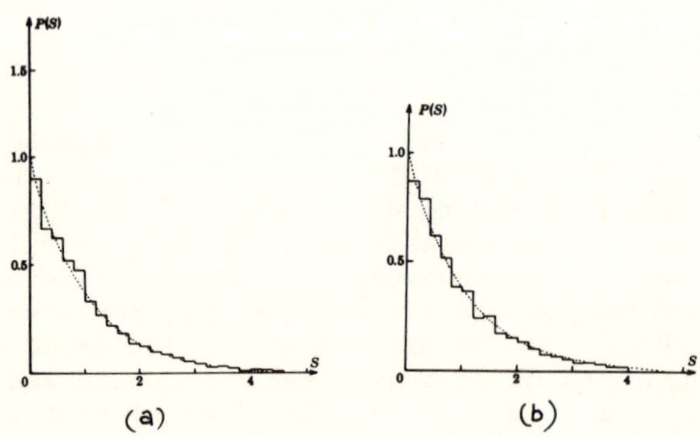


Figure 7. Level spacings for (a) rectangle with side ratio 1/2, and (b) square well along x and harmonic oscillator along y.

It is tempting on the basis of (23) and (24) to use P(S) to discriminate between a spectrum associated mainly with irregular motion and a spectrum associated mainly with regular motion. Indeed the implication

classically irregular motion in Hamiltonian without symmetry → level repulsion given by (23) (25)

is almost surely correct. But the reverse implication is wrong; the existence of level repulsion does not indicate classically irregular motion.

To see this, consider the delicate case of the 'desymmetrized square torus billiard' (fig. 8), which is one of a class of two-dimensional systems that Richens and Berry [46] showed to possess the 'pseudointegrable' property of having two constants of motion confining orbits not to tori but to phase-space surfaces with the topology of multiply-handled spheres (two handles in the case of fig. 8, as opposed to one for tori). If there is any chaos here, it is of a very rudimentary nature - not 2N-1 dimension filling, and with no exponential orbit separation. Nevertheless, P(S) shows clear level repulsion, as fig. 9 indicates.

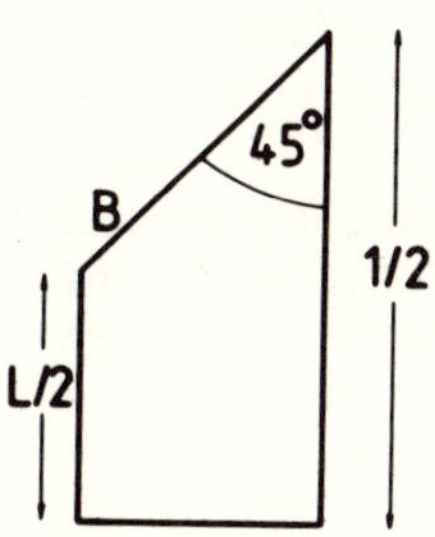


Figure 8. Desymmetrized square torus billiard.

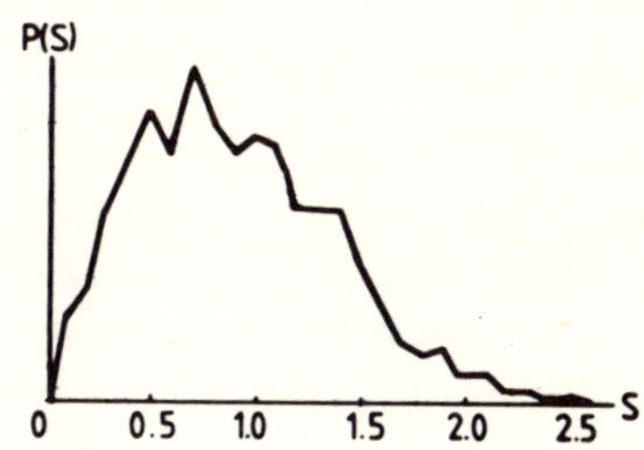


Figure 9. Level spacings distribution for desymmetrized square torus billiard.

Moreover, even for systems with tori the quantization condition (20) forming the basis of the level clustering law (24) is approximate rather than exact. For most systems with tori, and indeed for most integrable systems (where phase space is entirely filled by tori) I expect ([12] and § 6) that for very small values of S, of order $\exp\{-\text{const}/\hbar\}$, P(S) will be given not by (24) but will fall to zero like (23). So when does clustering occur in the exact (rather than semiclassical) spectrum? For the very special case where Schrödinger's equation is separable in the coordinates (as opposed to more general phase-space separability under the canonical transformation to action-angle variables, whose existence is guaranteed by integrability).

It should also be pointed out that there are certain very special classes of system (all integrable) for which P(S) does not exist. For example, consider two equal harmonic oscillators with Hamiltonian

$$H = \frac{p_x^2 + p_y^2}{2\mu} + \tfrac{1}{2}\mu\omega^2\left(q_1^2 + q_2^2\right) \tag{26}$$

The exact levels are

$$E_{m_1 m_2} = \hbar\omega(m_1+m_2+1) \qquad (0 \leqslant m_1, m_2 < \infty) \tag{27}$$

and occur at energies

$$E = \hbar\omega p \qquad (1 \leqslant p < \infty) \tag{28}$$

with degeneracies p. The mean level density is, from (4) (or directly)

$$\bar{d}(E) = \frac{E}{\hbar^2\omega^2} \tag{29}$$

so that the mean spacing of states is of order $\hbar^2$ as expected for N=2. But the actual levels (28) have spacings $\hbar$ and arrive in ever-more-degenerate groups, so that P(S) does not exist - the clustering structure continually changes as $\hbar \to 0$.

## 6. THE FINEST SCALE: DEGENERACIES

The most delicate question we can ask concerning neighbouring levels is: under what circumstances do these coincide? In other words: when do degeneracies occur? We shall see that the answers to these questions will enable us to understand the spacings distributions P(S) just discussed.

If the Hamiltonian $\hat{H}$ has any symmetry, this may produce degeneracies, whose nature can be studied using group theory. I shall not consider degeneracies of this type, and so when $\hat{H}$ does have symmetry I shall consider only states which all have the same symmetry class. This procedure is equivalent to considering all the states in a suitably 'desymmetrized' Hamiltonian, and that is why I have used this term in connection with the billiards in figs. 1 and 8.

For a typical (generic) such Hamiltonian it seems clear that degeneracy is infinitely improbable. But it might be expected that in a one-parameter family of Hamiltonians,

$$\hat{H} = H(\hat{\underset{\sim}{q}}, \hat{\underset{\sim}{p}}; A) \qquad (30)$$

the eigenvalue curves E(A) could cross, and so two levels could degenerate, for isolated values of the parameter A. But the surprising fact is that this picture is not correct: for typical systems with real eigenfunctions (the only ones considered here) it is necessary to vary two parameters, not one, in order to make two levels degenerate. This is the content of a theorem due originally to von Neumann and Wigner [47] and Teller [48] and later generalized by Arnol'd (ref. [9], appendix 10). The proof is based on a simple argument [12] employing degenerate perturbation theory.

Let the two parameters be A and B. Then the theorem just mentioned implies that the connection of eigenvalue surfaces $E = E_{\pm}(A,B)$ in E, A, B space takes the form of a double cone (diabolo) (fig. 10) with sheets joined at the 'diabolical point' $E^*$, $A^*$, $B^*$, where $A^*$, $B^*$ are the parameters for which the degeneracy occurs. If only one parameter A is varied, the diabolical point will almost surely be missed and the curves $E_{\pm}(A)$ will avoid each other like branches of a hyperbola obtained by slicing the cone, rather than crossing.

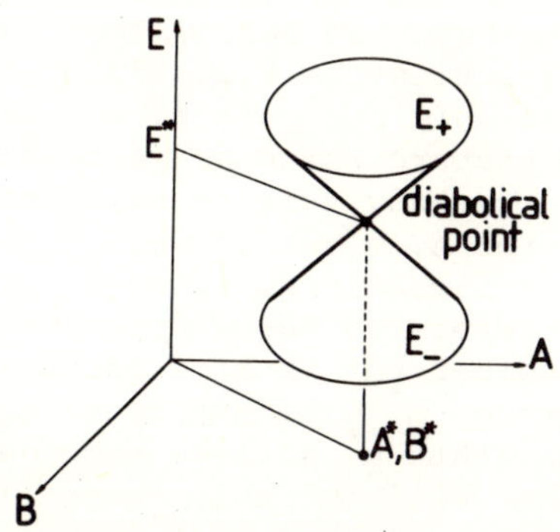


Figure 10. Diabolical degeneracy structure in space of energy E and parameters A, B.

In testing the resulting picture of the spectrum, it is plausible to assume that a one-parameter family of non-symmetric classically ergodic systems will be typical in the quantal sense, and so will produce no level crossings. Such a family is the desymmetrized Sinai billiard (fig. 1) with the radius R acting as parameter. The levels $E_j(R)$ were computed 'exactly' [26], and the resulting spectrum is shown in fig. 11. There are many near-degeneracies, but close examination shows that no two levels actually cross, so this test is successful.

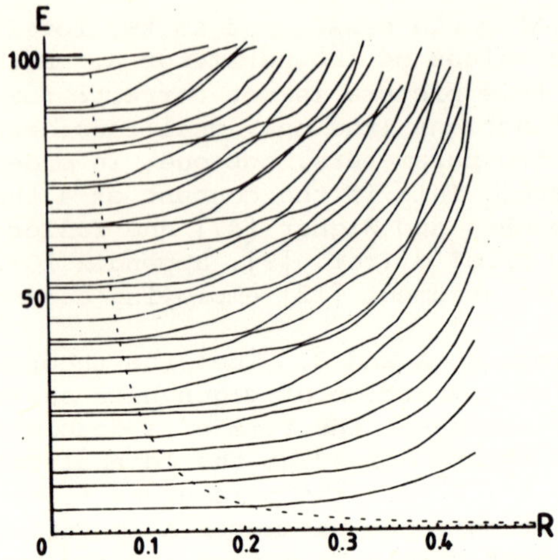


Figure 11. Spectrum of an ergodic system: levels E of desymmetrized Sinai billiard as functions of R.

The two-parameter rule does not hold for systems whose levels are given by the torus quantization rule (20). In these cases level crossings occur when only one parameter A is varied, thus reinstating the intuitive picture which as we have seen is wrong for the general case. To see why this is so, regard (20) as defining, for a given parameter A, a set of N-1 dimensional hypersurfaces $E=H(\underset{\sim}{I}_{\underset{\sim}{m}};A)$ in the N-dimensional space of quantum numbers $\underset{\sim}{m}$. The levels occur at energies whose hypersurfaces intersect a lattice point. Consider the lattice point $\underset{\sim}{m}^*$. Typically, the hypersurface through $\underset{\sim}{m}^*$ will not contain any other lattice point, and so the state $\underset{\sim}{m}^*$ will be nondegenerate. But on varying A the hypersurface will smoothly change its orientation and there will typically be values $A^*$ where it cuts another lattice point, so that $A^*$ corresponds to a pair of degenerate states. A simple example of this behaviour is the (integrable) family of rectangular boxes with sides 1, A, whose eigenvalues are

$$E_{m,n} = \frac{\pi^2\hbar^2}{2\mu}\,(m^2+A^2n^2) \qquad (31)$$

Degeneracies occur whenever $A^2$ is rational; for example, the states (2,1) and (1,3) degenerate when $A^2 = 3/8$.

One-parameter crossings will occur in the exact spectrum in the rare cases when torus quantization is exact. They will occur in separable systems for which torus quantization is not exact, because the basis for the conclusion still holds, namely that the energy can be written as a continuous function in the space of quantum numbers. But for nonseparable systems with tori I expect (20) to be modified by some multidimensional analogue of barrier penetration, causing the one-parameter degeneracies it predicts to be split by energies of order $\exp\{-1/\hbar^2\}$, so that the exact levels will show the typical property of two-parameter crossings.

There are some special integrable systems whose levels are all proportional to integers, and which are therefore not naturally embedded in continuous families, whose degeneracy structure is very strange. We have already mentioned the two equal harmonic oscillators (26), whose spectrum (28) shows increasing degeneracies. Another example is the $45^\circ$ right triangle (desymmetrized square), whose levels are proportional to $m^2+n^2$. It can be shown [26] that as $\hbar \to 0$ the levels are increasingly multiply degenerate, with multiplicity proportional to $\ln(\hbar^{-1})$, and are separated by energies of the order $\hbar^2 \ln(\hbar^{-1})$ rather than $\hbar^2$.

Marcus and his coworkers [25, 52] suggest that the presence of many near-degeneracies (overlapping avoided crossings) in curves of energy levels as functions of a single parameter, is an indication that the corresponding quantum states are associated with classically chaotic motion. This is unlikely to be correct, because two-parameter degeneration is the generic case, i.e. exceptions such as classes of systems with one-parameter degeneracies (or some other number not equal to two) occur with measure zero and so are fragile (like nodal-line intersections of wave functions when N=2 [12, 49, 50]). On the other hand, regular motion such as occurs on tori is robust as guaranteed by the KAM theorem and does not occur with measure zero. Therefore most torus-associated spectra should show two-parameter degeneration, and hence avoided crossings, albeit on very fine scales; only separable systems should show true one-parameter crossings. Support for this idea that avoided crossings need not imply classical chaos is given by computations of Richens and Berry [46] on the 'pseudointegrable' billiard of fig. 8, where orbits are confined to a two-handled (two-dimensional) sphere in phase space: energy levels as a function of L show avoided crossings very similar to those in fig. 11. The further suggestion [25, 52] that the reverse implication holds, i.e. that the energy levels of quantum states associated with classical chaos will show many avoided crossings, is almost certainly correct.

To conclude, I shall now as promised outline the connection between degeneracy structure and the level spacings distribution P(S); for details and a more formal argument, see [12] and [26].

Consider first the case of a generic Hamiltonian, which can be embedded in a family with two parameters A, B and having degeneracies at diabolical points in E, A, B space as in fig. 10. Let the actual Hamiltonian under study have parameters $A_o$, $B_o$. Then the line in E, A, B space with $A = A_o$, $B = B_o$ will thread its way among cones, which if $\hbar$ is small will presumably be thickly distributed (according to an unknown, and in this context unimportant, law).

Now, P(S) is defined according to (22) as an average over spacings for fixed A, B, i.e. as an 'energy average'. But by hypothesis there is nothing special about the parameters $A_o$, $B_o$ and so the energy average can be augmented by an ensemble average over a region of A, B near $A_o$, $B_o$. Thus can P(S) be expressed as an integral over a volume V in E, A, B space. For small S, the only contributions come from the neighbourhoods of diabolical points in V, and are proportional to the measure of the intersection of a diabolo by two parallel sheets separated by S. But this measure (length of intersection) is always proportional to S, whatever the orientation, ellipticity or cone angle of the diabolo, and this implies directly the linear level repulsion (23).

A similar argument for nongeneric systems which require m parameters to produce a degeneracy leads to

$$P(S) \sim \text{const} \times S^{m-1} \quad \text{as } S \to 0 \qquad (32)$$

We have seen that m=1 for systems with torus quantization (as opposed to 2 generically), and this leads to the 'clustering' prediction $P \to$ const as $S \to 0$, consistent with (24). For systems where degeneracies are strictly forbidden, such as finite one-dimensional systems, and two-dimensional harmonic oscillators with irrationally related frequencies, we can take $m = \infty$ and predict, in accord with observation [51, 12, 45], that P(S) vanishes faster than any power of S.

## ACKNOWLEDGEMENTS

I thank the University of Utrecht for hospitality whilst this paper was written. The work was not supported by any military agency.

## REFERENCES

[1] Noid, D.W., and Marcus, R.A.: 1977, J. Chem. Phys. 67, pp. 559-567.
[2] Eastes, W., and Marcus, R.A.: 1974, J. Chem. Phys. 61, pp. 4301-4306.

[3] Noid, D.W., and Marcus, R.A.: 1975, J. Chem. Phys. 62, pp. 2119-2124.
[4] Noid, D.W., Koszykowski and Marcus, R.A.: 1979, J. Chem. Phys. 71, pp. 2864-2873.
[5] Percival, I.C., and Pomphrey, N.: 1976, Mol. Phys. 31, pp. 97-114.
[6] Jaffe, C., and Reinhardt, W.P.: 1979, J. Chem. Phys. 71, pp. 1862-1869.
[7] Chapman, S., Garrett, B., and Miller, W.H.: 1976, J. Chem. Phys. 64, pp. 502-509.
[8] Ford, J.: 1975,in 'Fundamental Problems in Statistical Mechanics' (ed. E.G.D. Cohen) Vol. 111, pp. 215-255. North-Holland.
[9] Arnol'd, V.I., 'Mathematical Methods of Classical Dynamics', (Springer, New York 1978).
[10] Berry, M.V.: 1978, in 'Topics in Nonlinear Dynamics' (ed. S. Jorna), Am. Inst. Phys. Conf. Proc. 46, pp. 16-120.
[11] Helleman, R.: 1980, in 'Fundamental Problems in Statistical Mechanics (ed. E.G.D. Cohen), Vol. V, pp. 165-233, North-Holland.
[12] Berry, M.V.: 1981, 'Semiclassical Mechanics of Regular and Irregular Motion', Les Houches Summer School Lectures, to be published by North-Holland).
[13] Gutzwiller, M.C.: 1967, J. Math. Phys. 8, pp. 1979-2000.
[14] Gutzwiller, M.C.: 1969, J. Math. Phys. 10, pp. 1004-1020.
[15] Gutzwiller, M.C.: 1970, J. Math. Phys. 11, pp. 1791-1806.
[16] Gutzwiller, M.C.: 1971, J. Math.Phys., 12, pp. 343-358.
[17] Gutzwiller, M.C.: 1978, in 'Path Integrals and their Application in Quantum, Statistical and Solid State Physics' (eds.: G.J. Papadopoulos and J.T. Devreese), pp. 163-200, Plenum, N.Y.
[18] Balian, R., and Bloch, C.: 1970, Ann. Phys. N.Y. 60, pp. 401-447.
[19] Balian, R., and Bloch, C: 1971, Ann. Phys. N.Y. 64, pp. 271-307.
[20] Balian,R., and Bloch,C.:1972, Ann.Phys.N.Y.69,pp. 76-160.
[21] Balian, R., and Bloch, C: 1974, Ann. Phys. N.Y. 85, pp. 514-545.
[22] Landau, L.D., and Lifshitz, E.M.: 1965, in 'Quantum Mechanics (Nonrelativistic Theory)' (Oxford: Pergamon)
[23] Berry, M.V., and Mount, K.E.: 1972, Reps. Prog. Phys. 35, pp. 315-397.
[24] Baltes, H.P., and Hilf, E.R. 1978, 'Spectra of Finite Systems'(B-I Wissenschaftsverlag, Mannheim).
[25] Marcus, R.A.: 1980, in 'Nonlinear Dynamics', Ann. N.Y. Acad. Sci. 357, pp.169-182 (ed. R.H.G. Helleman).
[26] Berry, M.V.: 1981, Ann. Phys. N.Y. 131, pp. 163-216.
[27] Sinai, Ya. G.: 1970, Russ. Math. Surv. 25 No. 2, pp. 137-189.
[28] Bunimovich, L.A.: 1974, Funct. Anal. Appl. 8, pp. 254-255,

1979, Commun. Math. Phys. 65, pp. 295-312.
[29] Berry, M.V., and Tabor, M.: 1976, Proc. Roy. Soc. A349, pp. 101-123.
[30] Berry, M.V., and Tabor, M.: 1977, J. Phys. A.10, pp. 371-379.
[31] Rajaraman, R.: 1975, Phys. Rept. 21, pp. 227-313.
[32] DeWitt, C., Maheshwari, A., and Nelson, B.: 1979, Phys. Rept. 50, pp. 255-372.
[33] Miller, W.H.: 1975, J. Chem. Phys. 63, pp. 996-999.
[34] Voros, A.: 1976, Annls. Inst. H. Poincare XXIV, pp. 31-90.
[35] Gutzwiller, M.C.: 1980, Phys. Rev. Lett. 45, pp. 150-153.
[36] Chazarain, J.: 1974, Inv. Math. 24, pp. 65-82.
[37] Einstein, A.: 1917, Ver. Deut. Phys. Ges. 19, pp. 82-92.
[38] Keller, J.B.: 1958, Ann. Phys. N.Y. 4, pp. 180-188.
[39] Maslov, V.P., and Fedoriuk, M.V., in 'Semi-classical approximation in quantum mechanics' (D. Reidel: Dordrecht 1981) (original Russian edition 1965).
[40] Percival, I.C.: 1977, Advan. Chem. Phys. 36, pp. 1-61.
[41] Percival, I.C.: 1973, J. Phys. B6, pp. L229-232.
[42] Porter, C.F. (ed.): 1965, in 'Statistical Theory of Spectra: Fluctuations' (Academic Press, N.Y. 1965).
[43] McDonald, S.W., and Kaufman, A.N.:1979, Phys. Rev. Lett. 42, pp. 1189-1191.
[44] Casati, G., Valz-Gris, F., and Guarneri, I.: 1980, Nuovo Cimento Lett. 28, pp. 279-282.
[45] Berry, M.V., and Tabor, M.: 1977, Proc. Roy. Soc. A356, pp. 375-394.
[46] Richens, P.J., and Berry, M.V.: 1981, Physica 2D, pp. 495-512.
[47] Neumann, J. von, and Wigner, E.P.: 1929, Physik, Z.30, pp. 467-470.
[48] Teller, E.: 1937, J. Phys. Chem. 41, pp. 109-116.
[49] Pechukas, P.: 1972, J. Chem. Phys. 57, pp. 5577-5594.
[50] Uhlenbeck, K.: 1976, Amer. J. Math. 98, pp. 1059-1078.
[51] Pokrovskii, V.L.: 1966, JETP Lett. 4, pp. 96-99.
[52] Noid, D.W., Koszykowski, M.L., Tabor, M., and Marcus, R.A.: 1980, J. Chem. Phys. 72, pp. 6169-75.

# RAY OPTICS FOR DIFFRACTION: A USEFUL PARADOX IN A PATH INTEGRAL CONTEXT

L.S. Schulman

Physics Department
Technion-Israel Institute of Technology
Haifa, Israel

ABSTRACT

Geometrical diffraction theory uses ray tracing techniques to calculate diffraction and other properties of the electromagnetic field generally considered characteristically wave like. We here study this dualism of the classical electromagnetic field so as to distinguish those aspects of quantum dualism that arise simply as properties of oscillatory integrals and those that may have deeper origins. By a series of transformations the solutions of certain optics problems are reduced to the evaluation of a Feynman path integral and the known semiclassical approximations for the path integral provide a justification for the geometrical diffraction theory. Particular attention is paid to the problem of edge diffraction and for a half plane barrier a closed form solution is obtained. A classical variational principle for barrier penetration is also presented.

## I. INTRODUCTION

This article is about what wave-particle dualism isn't. That is, I will discuss a theory of little relevance to the fundamental problems of quantum mechanics and show that it has many characteristics of wave-particle dualism. This is being done not only out of general orneriness, but is offered in the spirit of the counterexample in mathematics: any philosophical conclusions one would wish to draw from wave-particle dualism ought not to be based on those aspects of dualism that I shall discuss below.



*S. Diner et al. (eds.), The Wave-Particle Dualism, 253–272.*

The propagation of light is historically the first example of people being torn between wave and particle interpretations. The characteristic property of light leading to a particle interpretation was its propagation along straight lines or rays, while the wave property was later demonstrated through diffraction and interference. The theory I shall discuss manages to encompass in particle language - or more precisely, in terms of rays - many of the characteristic wave features of light: diffraction, tunneling and the illumination of "forbidden" regions near a caustic. It is known by the paradoxical juxtaposition: "Geometrical Diffraction Theory" and is due to J.B. Keller (1). Since the introduction of these ideas in the 1950s they have been developed not only by mathematicians interested in asymptotic expansions but for practical technological reasons as well, for example in the interpretation of radar reflections or in the construction of antennas (2).

As we shall see below, what makes the whole thing work is an oscillatory integral with a large parameter $k$, the wave number. The same is true of the semiclassical approximation in quantum mechanics and therein lies the close resemblance. But the quantum parameter $\hbar$ carries more philosophical baggage and it is my purpose here to allow one to distinguish between the deep features of quantum mechanics and between those that arise merely because of the asymptotic behavior of an oscillatory integral. It may also be worth mentioning that geometrical diffraction theory has nothing to do with photons. The only scale parameter around is the (large) ratio of characteristic distances to the wavelength of light; there would be no change if Planck's constant were ten orders of magnitude smaller or indeed zero.

In this article I will develop the geometrical diffraction theory from the Feynman path integral. Then I will show that many familiar semiclassical techniques lie behind the geometrical diffraction theory. But first I will say a few words about geometrical diffraction theory as well as about path integration and its relation to optics.

## II. GEOMETRICAL DIFFRACTION THEORY

Consider an electromagnetic plane wave impinging on a perfectly conducting half plane (Fig. 1). Geometrical optics predicts a direct and reflected ray but there is also a diffracted wave. In fact, Sommerfeld solved this boundary value problem for Maxwell's equation in closed form (3). He found that if the incoming wave is $\exp(i\vec{k}\cdot\vec{x})$ then to leading order in $k = |\vec{k}|$ the field in the shadow region is $D\, e^{ik\rho}/(k\rho)^{1/2}$ where $\rho$ is the distance from the edge of the knife edge and $D$ is an angle

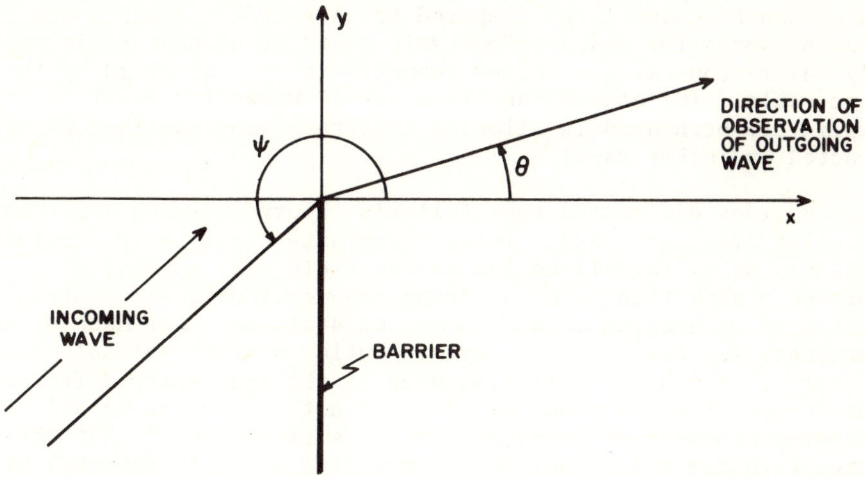


Fig. 1. Geometry of waves impinging upon and scattered off an infinite half plane perfectly conducting barrier. All angles are measured positively from the x-axis.

dependent factor of order one. Keller used this formula to suggest a kind of Huygens' principle for diffraction. He postulated rays emanating from the edge and spreading into both the illuminated and shadow regions. As in the usual optics these rays carry a phase; this is the factor $\exp(ik\rho)$. The dropoff $\rho^{-1/2}$ is appropriate to a cylindrical wave so that in all respects, once launched from the knife edge this ray is perfectly respectable. Its special character as a diffracted ray is expressed in the factor $k^{-1/2}$, which is a term that disappears in the $k\to\infty$ limit. It may be a perfectly respectable ray - once launched - but it has a reduced amplitude due to its origin in a breakdown in the conventional laws of geometrical optics.

This example is canonical, that is there is rigorous information on the diffracted rays. Working from a number of such examples Keller develops a scheme for calculating the diffracted field in geometries not amenable to exact solution. Namely, when the medium is continuous do the usual ray tracing, but when encountering a discontinuity look for a solved canonical problem which is locally of the same form and use it to determine the properties of the resulting diffracted rays. Then once these rays leave the neighborhood of the discontinuity they are treated like any other rays. The key work above is "locally". The wave nature of the phenomenon is suppressed - asymptotically speaking - on distance scales large compared to $\lambda = 2\pi/k$. This is why ray tracing works for media effectively constant on that scale and why the matchup with a solved geometry is also governed by the same scale. (Of course the linearity of Maxwell's equations has implicitly been used in allowing arbitrary superposition of diffracted and other rays).

Besides diffracted rays Keller's theory includes something he calls imaginary rays. These illuminate classically forbidden regions, as in tunnelling and in the field near a caustic. Barrier penetration seems a strong argument for a wave interpretation of quantum mechanics but we shall see that much of the machinery for dealing with rays in optics or with particle trajectories in mechanics is also used in the apparently different barrier penetration problem. I shall not review the details of Keller's handling of imaginary rays; suffice to say that they agree with the WKB formulas obtained from the path integral below.

## III. PATH INTEGRAL FOR OPTICS

Let $u(x,t)$ be some component of the electromagnetic field or its potential. In a medium with a slowly varying index of refraction $n(x)$, $u$ satisfies

$$\left(\frac{n(x)}{c}\right)^2 \frac{\partial^2 u}{\partial t^2} - \nabla^2 u = 0 \tag{3.1}$$

For waves of definite frequency $\omega$, $u(x,t) = \psi(x) \exp(-i\omega t)$, and we suppose that for any direction $n(x) \to \text{const}$ as $|x| \to \infty$. Call this constant $\sqrt{E}$; for free space $E = 1$. By substituting and rescaling the spatial coordinate as follows

$$x = \beta y c/\omega \tag{3.2}$$

Eq. (3.1) becomes

$$-\frac{1}{\beta^2} \nabla_y^2 \psi + V(y)\psi = E\psi \tag{3.3}$$

where $V(y) \equiv E-(n(x))^2$. Thus a stationary solution of the wave equation satisfies a Laplace equation which in turn can be made to look like Schroedinger's equation by choosing the right letters of the alphabet. Of interest is the one letter for which I have desisted from full alphabetic identification. $\beta$ will play the role of $\hbar^{-1}$ but I draw attention to its independent physical nature. In equation (3.2), $c/\omega$ is essentially the wavelength of the radiation, so that if $y$ has order unity, then $\beta$ is the scale factor measuring $x$ in units of wavelengths. Geometrical optics is the large $\beta$ limit. But then (3.3) tells us that our knowledge of the $h \to 0$ behavior of the Schroedinger equation can be used to gain information about geometrical optics.

A convenient way to study semiclassical properties of (3.3) is by the path integral (4). Introduce an artificial "time" variable $\tau$ such that $\psi(y,\tau) = \exp(-iE\tau\beta)\psi(y)$. We invoke the path integral to write down a propagator for $\psi(y,\tau)$:

$$\psi(y,\tau) = \int dy' \, G(y,\tau;y')\psi(y',0) \tag{3.4}$$

$$G(y,\tau;y') = \int_{(y',0)}^{(y,\tau)} dy(\cdot)\exp\{i\beta S[y(\cdot)]\} \tag{3.5}$$

$$S[y(\tau')] = \int_0^{\tau} d\tau' \left[\frac{1}{4}\left(\frac{dy}{d\tau'}\right)^2 - V(y)\right] \tag{3.6}$$

$S$ is the classical action corresponding to (3.3) and the sum in (3.5) is over all paths starting at "time" zero at $y'$ and arriving at "time" $\tau$ at $y$. $\tau$ is eliminated by Fourier transformation. To summarize, the solution to the electromagnetic problem (with fixed $\omega$) is provided by the following quantity

$$G(b,a;E) = \int_0^{\infty} d\tau \exp(i\beta E\tau) \int_{(a,o)}^{(b,\tau)} dy(\cdot)\exp\{i\beta S[y(\cdot)]\} \tag{3.7}$$

As discussed in Ref. 4, Fermat's principle and the usual geometrical optics are obtained by a stationary phase approximation ($\beta\to\infty$) for the $\infty+1$ dimensional integral above. In particular, the requirement that S be stationary with respect to variation of $y(\cdot)$ leads to the Euler-Lagrange equations. To get the semiclassical approximation to (3.7) we use the well-known form of this approximation for (3.5):

$$G(b,\tau;a) = \sum_{\alpha} \left[\det\left(\frac{i\beta}{2\pi}\frac{\partial^2 S_\alpha}{\partial b \partial a}\right)\right]^{1/2} \exp(i\beta S_\alpha) \tag{3.8}$$

where $\alpha$ runs over classical paths only, $S_\alpha$ is the action evaluated along path $\alpha$ and $\partial^2 S_\alpha/\partial a \partial b$ is the derivative of $S_\alpha$ with respect to the endpoints. Such a derivative is meaningful only when the classical paths do not form a continuous family and indeed such continuous families are characteristic of focusing where $\partial^2 S_\alpha/\partial b \partial \alpha$ is singular. The determinant allows for $b$ and $a$ to be vectors. Eq. (3.8) is a major simplification of (3.5) since only paths satisfying the classical equations of motion appear. The object in the square root is the van Vleck determinant and was introduced to path integration by Cecile DeWitt (5). In optics it is deduced from the conservation of power in tubes of rays.

It is remarkable that in nearly all problems in which the time dependent Green's function is known in closed form it is given exactly by (3.8). Since we shall use the propagator for a free particle in two dimensions I give its explicit form

$$G(\vec{b},\tau;\vec{a}) = \frac{\beta}{4\pi i\tau} \exp\left[\frac{i\beta}{4\tau}(\vec{b}-\vec{a})^2\right] \tag{3.9}$$

## IV. EDGE DIFFRACTION

To study knife edge diffraction, or the infinite half plane barrier, as illustrated in Fig. 1, it is sufficient to restrict attention to a two dimensional cross section. Obliquely arriving plane waves require consideration of the third dimension, but it turns out that there is a simple transformation that eliminates this complication. As mentioned this was solved by Sommerfeld and we give the solution in the form provided by Lewis and Boersma (6). The geometry and angles are defined in Fig. 1. The incident plane wave is

$$u_o^i = \exp(-ik(x\cos\psi + y\sin\psi)) = \exp(-ik\rho\cos(\psi-\theta)) \tag{4.1}$$

where $(\rho,\theta)$ are the polar coordinates of the observation point. Define the function

$$h(t) = \pi^{-1/2}\, e^{-i\pi/4} \int_{-\infty}^{t} \exp(it'^2)dt' \tag{4.2}$$

with $h(+\infty) = +1$. Then at the point $(\rho,\theta)$ the total field is

$$u = \exp[-ik\rho\, \cos(\psi-\theta)]\; h[\sqrt{2k\rho}\cos \tfrac{1}{2}(\psi-\theta)]$$

$$\mp \exp[ik\rho\, \cos(\psi+\theta)]\; h[-\sqrt{2k\rho}\, \sin \tfrac{1}{2}(\psi+\theta)] \tag{4.3}$$

where the upper sign (here and in the sequel) is for the boundary condition $u=0$ on the half plane and the lower sign corresponds to $\partial u/\partial n = 0$. Comparing (4.3) and (4.1) we see that when variation of $h$ can be ignored the first term in $u$ is essentially the incident plane wave. Similarly we can define a reflected wave by $u_o^r = \exp(ik\rho\, \cos(\psi+\theta))$ and this too appears in (4.3).

For large $t$ $h$ is issentially constant; 0 or 1, depending on the sign of $t$. The asymptotic expansion for $h$ gives finer corrections to $u$, namely the diffracted wave. For large $t$

$$h(t) = \theta(t) - \frac{1}{2}\pi^{-1/2} e^{i\pi/4} e^{it^2} \frac{1}{t} \sum_{n=o}^{\infty} a_n (it^2)^{-n} \tag{4.4}$$

where $\theta(t)$ is the step function, $a_o=1$ and $a_n = (n - 1/2)a_{n-1}$. A bit of algebra yields

$$u \sim \theta(\cos\tfrac{1}{2}(\psi-\theta))u_o^i \mp \theta(\cos\tfrac{1}{2}(\psi+\theta))u_o^r + \frac{De^{ik\rho}}{\sqrt{k\rho}} + O(k^{-3/2}) \tag{4.5}$$

where

$$D = e^{i\pi/4} [\sec\tfrac{1}{2}(\psi-\theta) \mp \sec\tfrac{1}{2}(\psi+\theta)]/2\sqrt{2\pi} \tag{4.6}$$

For the "shadow boundary", namely $\psi=\theta+\pi$, the form (4.5) is invalid ($t=0$) and going back to (4.3) we see that the incident wave amplitude is down by exactly a factor 1/2 whereas the reflected contribution leads only to a diffracted wave given by the second term in $D$.

The salient features of the diffracted wave are its overall $k^{-1/2}$ amplitude and its form as a cylindrical wave (thinking 3-dimensionally). As mentioned in the introduction, this justifies thinking of the diffracted wave as comprised of rays emanating from the edge of the barrier.

## V. SEMICLASSICAL PROPAGATOR FOR EDGE DIFFRACTION

A semiclassical calculation of the propagator for the knife edge may seem pointless given knowledge of the exact solution. But here our attitude is the same as Keller's: if your approximations can be verified against known solutions then you can have confidence in them when exact solutions are not available. Even more: out approach makes no use of special features of the geometry. It would be valid for a curved edge, or for any geometry whose scale is large compared to $\lambda$. Thus the verification of our approximation against the leading terms in (4.5) is a justification of Keller's method.

First we calculate the propagator for fixed initial and final points and "time" $\tau$ and then take appropriate limits and transformations to obtain the electromagnetic field. The geometry is shown in Fig. 2. The propagator is given by the path integral

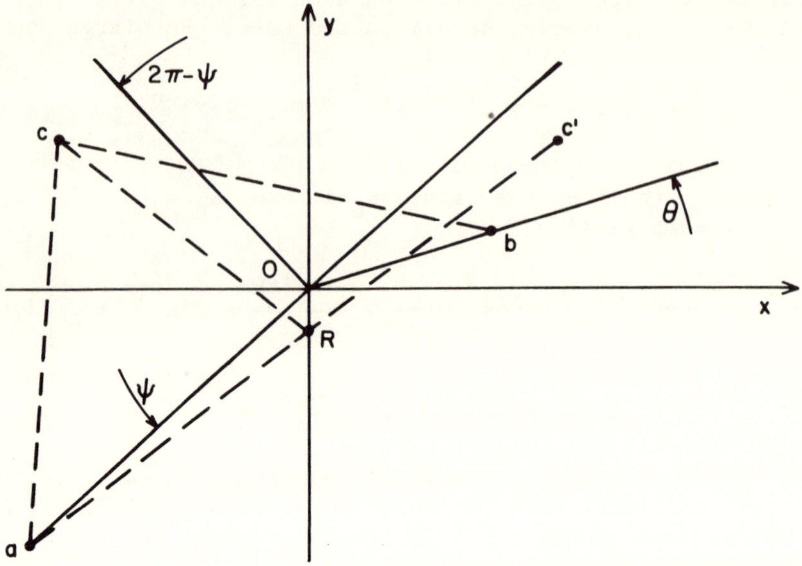


Fig. 2. The initial point is a and the final point b, having polar coordinates $(\rho_a,\psi)$ and $(\rho_b,\theta)$ respectively; all angles are measured counterclockwise from the positive x axis. The intermediate point of integration is c. The dotted line ac is the direct path from a to c; similarly for the dotted line cb. The pair of dotted lines aR-Rc form the (specularly) reflected path from a to c and together have the same length as the dotted line ac' where c' is the reflection of c in the y axis. The polar coordinates of c are $(\rho,\phi)$ and the direct-direct path is not blocked by the barrier for $\theta+\pi\geqslant\phi\geqslant\psi-\pi$. Similarly reflection off the barrier can occur for $\phi\geqslant2\pi-\psi$ and the direct path cb is not blocked for $\phi<\theta+\pi$.

$$G(\vec{b},\tau;\vec{a}) = \int_{(\vec{a},o)}^{(\vec{b},\tau)} dy(\cdot)\ \exp\{i\beta \int_0^\tau d\tau'\ [\tfrac{1}{4}(\frac{dy}{d\tau'})^2]\} \qquad (5.1)$$

summed over all paths that avoid the barrier. Except for the barrier, n(x)=1 so that $V(\vec{y})=0$ and E=1 (notation of Sec. II). We first pick an intermediate time $\tau_c$ and use the general relation

$$G(\vec{b},\tau;\vec{a}) = \int d^2c\ G(\vec{b},\tau-\tau_c;\vec{c})\ G(\vec{c},\tau_c;\vec{a}) \tag{5.2}$$

As polar coordinates for $\vec{a}$, $\vec{b}$ and $\vec{c}$ take $(\rho_a,\psi)$, $(\rho_b,\theta)$ and $(\rho,\phi)$ respectively. Choose $\tau_c$ such that $\rho_a/\tau_c = \rho_b/(\tau-\tau_c)$. Now we make our major approximation. Instead of summing over all paths from $\vec{a}$ to $\vec{c}$ and $\vec{c}$ to $\vec{b}$ we sum only over the classical paths using the semiclassical propagator given in Eq. (3.9).

For the geometry of Fig. 2, specifically for $0<\theta<\frac{\pi}{2}$ and $\pi<\psi<3\pi/2$, there is but a single classical path from $\vec{c}$ to $\vec{b}$ and this only for $\phi<\theta+\pi$. The propagator on this leg of the journey is therefore approximated by

$$G_{\text{Direct}}(\vec{b},\tau-\tau_c;\vec{c}) = \frac{\beta}{4\pi i(\tau-\tau_c)} \exp\left[\frac{i\beta(\vec{b}-\vec{c})^2}{4(\tau-\tau_c)}\right] \tag{5.3}$$

To get from $\vec{a}$ to $\vec{c}$ there are two possibilities: (1) a direct path with classical action $(\vec{c}-\vec{a})^2/(4\tau_c)$ provided $\phi\geq\psi-\pi$ and (2) a reflected path, bouncing specularly off the barrier and which can exist provided $\phi\geq2\pi-\psi$. The action for the reflected path is $(|\vec{aR}| + |\vec{Rc}|)^2/4\tau_c$ (vertical bars denote distance), which is easy to calculate using $\vec{c}'$, the reflection of $\vec{c}$ in the $y$ axis. The semiclassical propagator form $\vec{a}$ to $\vec{c}$ is therefore

$$G_{\text{Direct}} \mp G_{\text{Reflected}} = \frac{\beta}{4\pi i\tau_c}\left\{\exp\left[\frac{i\beta}{4\tau_c}(\vec{c}-\vec{a})^2\right] \mp \exp\left[\frac{i\beta}{4\tau_c}(\vec{c}'-\vec{a})^2\right]\right\} \tag{5.4}$$

with the proviso that each term appears only in its classically allowed region. Eqs. (5.3) and (5.4) are inserted in (5.2) to give

$$G(\vec{b},\tau;\vec{a}) = G_o \frac{\beta\gamma}{4\pi i}\int_o^\infty \rho d\rho \left[\int_{\psi-\pi}^{\theta+\pi} d\phi e^{i\beta A_{DD}} \mp \int_{2\pi-\psi}^{\theta+\pi} d\phi\ e^{i\beta A_{RD}}\right]$$

$$\equiv G_{DD} \mp G_{RD} \tag{5.5}$$

where we have used the notation

$$\gamma = \frac{1}{\tau_c} + \frac{1}{\tau-\tau_c} \tag{5.6}$$

$$V = \frac{\rho_a}{\tau_c} = \frac{\rho_b}{\tau - \tau_c} \tag{5.7}$$

$$G_o = \frac{\beta}{4\pi i \tau} \exp\left[\frac{i\beta}{4\tau}(\rho_a + \rho_b)^2\right] \tag{5.8}$$

$$\left.\begin{matrix} A_{DD} \\ A_{RD} \end{matrix}\right\} = \frac{1}{4}\gamma\rho^2 \mp \frac{1}{2}\rho v\, [\cos(\phi \mp \psi) \pm \cos(\phi - \theta)] \tag{5.9}$$

$v$ is the velocity a classical particle would have going from $\vec{a}$ to 0 to $\vec{b}$ and the argument of the exponent in $G_o$ is the classical action (time $i\beta$) for that path. It is instructive to examine the quantity $A_{DD}$. For the special case of the shadow boundary, namely $\psi = \theta + \pi$, we have identically $A_{DD} = \gamma\rho^2/4$. $G_{DD}$ is now exactly 1/2 $G_o$. This should be compared to the results of Sec. IV. At the shadow boundary (4.6) does not apply and $u$ is given by $1/2\ u_o^1$ + [diffracted term]. The diffracted term is as in (4.5) except that D includes only the second of its contributions in (4.6), namely that arising from the "reflected" wave. There is an earlier path integral calculation for the edge diffraction problem (7) in which only the shadow boundary case is considered and which gets the factor 1/2 but not the additional diffracted term. The latter term for the shadow boundary case in fact arises from $G_{RD}$. In general both $G_{DD}$ and $G_{RD}$ contribute to diffraction.

Returning to the general case, we rewrite $A_{DD}$ in terms of

$$\omega_1 = \frac{1}{2}(\psi + \theta) \quad \omega_2 = \frac{1}{2}(\psi - \theta) - \frac{\pi}{2} \quad \alpha = \phi - \omega_1 \tag{5.10}$$

For $\alpha$ the limits of integration in $G_{DD}$ are $|\alpha| < \pi/2 - \omega_2$ and $A_{DD}$ becomes $A_{DD} = 1/4\ \gamma\rho^2 + \rho v \cos\alpha \sin\omega_2$. The behavior of $G_{DD}$ for large $\beta$ is given by the stationary phase approximation for the integral. It is clear that $\alpha = 0$ is the only stationary point in $\alpha$. Then the existence of nonexistence of a stationary point for the $\rho$ integration depends on the sign of $\sin\omega_2$. For $\omega_2 > 0$ the stationary point in $\rho$ lies on the negative $\rho$ axis, outside the range of integration. Examination of Fig. 2 shows this to be the situation in which the barrier blocks the straight line path from $\vec{a}$ to $\vec{b}$. This forces the leading asymptotic term to be of smaller order in $\beta$. Specifically, when the stationary point falls within the range of integration the two integrals give a factor $1/\beta$. Having the point fall outside will introduce an additional $1/\sqrt{\beta}$. This is the price of having a broken line path with a single bend.

For $\omega_2 < 0$ the stationary point has $\rho > 0$; however, although the stationary point lies within the range of integration there is still a diffraction term due to cutting off the integral and this contribution is again $1/\sqrt{\beta}$ relative to the leading term.

To do the integral use orthogonal coordinates $\xi,\eta$ with the $\xi$ axis at an angle $\omega_1$ to the x axis. For the case $\omega_2 > 0$ $G_{DD}$ becomes

$$G_{DD}=G_o\frac{\beta}{4\pi i}\gamma \int_0^\infty d\xi \int_0^{\xi/\tan\omega_2} 2d\eta \exp\{\frac{i\beta}{4}[\gamma\xi^2+\gamma\eta^2+4\xi v \sin\omega_2]\} \quad (5.11)$$

Extending the $\eta$ integration to $\infty$ brings $G_{DD}$ to the form

$$G_{DD} = G_o \frac{e^{-i\pi/4}}{\sqrt{\pi}} e^{-im^2} \int_m^\infty e^{it^2} dt = G_o e^{-im^2} h(-m) \quad (5.12)$$

with

$$m = \sqrt{\frac{\beta}{\gamma}}\, v \sin\omega_2 \quad (5.13)$$

Consider now the error introduced by extending the upper limits of integration for $\eta$ to $\infty$. This is problematic for small $\xi$ and from Eq. (4.4) we see that the error is $0(1/\xi\sqrt{\beta})$. Therefore if it were possible to discard from the $\xi$ integration a range of $\xi$ such that for all remaining $\xi$ $\xi\sqrt{\beta} >> 1$, then the replacement of $\xi/\tan\omega_2$ by $\infty$ would not affect the asymptotic behavior with $\beta$. Take the range to be discarded to be $0<\xi<\xi_{max} = g(\beta)/\sqrt{\beta}$ where $g(\beta)$ is a function that goes to $\infty$ with $\beta$ more slowly than $\sqrt{\beta}$, for example $\beta^{1/4}$. The effect of this is to change the limits of integration in (5.13) to a new value of $m'$ where $m' = m + 1/2\sqrt{\beta\gamma}\,\xi_{max}$. Since $m$ is growing with $\beta$, to find the relative effect on the value of $G_{DD}$ we look at $(m'-m)/m$ which is $0(g(\beta)/\sqrt{\beta})$ and therefore goes to zero. Our arguments do not justify (5.12) as a uniform approximation (for $\omega_2$ small); however, one notices that for $m=0$ $G_{DD}$ collapses to $1/2\, G_o$, the correct value. In fact (5.13) is uniform since for small $\omega_2$ the error from extending the limits of integration is $0(\omega_2/\xi\sqrt{\beta})$. Hence $\xi_{max}$ can be taken to be $\omega_2\, g(\beta)/\sqrt{\beta}$. This cancels the $\omega_2$ appearing in $m$ when one considers the ratio $(m'-m)/m$.

For the case $\omega_2 < 0$ we write $G_{DD}$ as the integral over the entire plane minus an integral very much like (5.11). Thus

$$\int_0^\infty \rho d\rho \int_{\psi-\pi}^{\theta+\pi} d\phi = \int_{-\infty}^\infty d\xi \int_{-\infty}^\infty d\eta - \int_0^\infty d\xi 2 \int_0^{\xi/|\tan\omega_2|} d\eta \quad (5.14)$$

The first term is the free propagator from $\vec{a}$ to $\vec{b}$. The second term has the same structure as (5.11) and therefore as (5.12) with $m'' = -\sqrt{\beta/\gamma}\ v \sin\omega_2$. Thus

$$G_{DD} = G_o e^{-im^2} - G_o e^{-im''^2} \frac{e^{-i\pi/4}}{\sqrt{\pi}} \int_{m''}^{\infty} e^{it^2} dt$$

$$= G_o e^{-im^2} h(-m) \tag{5.15}$$

exactly as in (5.12). It follows that our solution for the direct-direct part of the Green's function has all the simplicity of Sommerfeld's direct ray part of his solution. It did not have to work out this way since our calculation is only valid to leading order in $\beta$ and to the next lower order, or relative size $1/\sqrt{\beta}$. It should be borne in mind that we work with the "time" representation of the Green's function whereas Sommerfeld deals with the Fourier transform. Our spatial variable a is also transformed away in his solution. It is quite possible to have WKB give exact answers in one representation while the asymptotic approximation of the Fourier transform is no longer exact. (The hydrogen atom provides an example (4,8).)

Next consider $G_{RD}$, the term that includes a reflected path. Defining $\alpha = \phi - [\pi-\omega_2]$ $G_{RD}$ is brought to a form resembling (5.11) and we find that

$$G_{RD} = G_o e^{-in^2} h(-n) \tag{5.16}$$

with

$$n = \frac{\beta}{\gamma} v \sin\omega_1 \tag{5.17}$$

Eqs. (5.16) and (5.17) can also be derived by noting that $G_{DD} \to G_{RD}$ under the transformation $\psi \to 3\pi - \psi$. This also provides an easy check that $G_{DD} - G_{RD} = 0$ for $\psi = 3\pi/2$.

To summarize

$$G(\vec{b},\tau;\vec{a}) = G_o[e^{-im^2} h(-m) \mp e^{-in^2} h(-n)] \tag{5.18}$$

with m and n given respectively by (5.13) and (5.17), v, $\gamma$ and $G_o$ by (5.6) (5.7) and (5.8) and $\omega_1$ and $\omega_2$ by (5.10). The function h is given in (4.2). (Formula (5.18) has turned out to be an exact solution for the time dependent propagator (12)).

To compare this to (4.3)-(4.6) we must get plane wave incoming boundary conditions, Fourier transform from $\tau$ to E and set E = 1. Instead of using Fourier transformation on $\vec{a}$ it is easiest to set $\vec{a} = \rho_a \hat{a}$. Then let $\rho_a \to \infty$ and multiply G by an appropriate factor to match (4.1). Thus

$$u = N\int_0^\infty d\tau\, e^{i\beta\tau} \frac{\beta}{4\pi i\tau} \exp[\tfrac{i\beta}{4\tau}(\rho_a+\rho_b)^2]\{e^{-im^2} h(-m) \mp e^{-in^2} h(-n)\} \tag{5.19}$$

and

$$m = \frac{\rho_a \beta \rho_b}{\tau} \sin \omega_2 \qquad n = \frac{\rho_a \beta \rho_b}{\tau} \sin \omega_1$$

We consider the first term in (5.19) (deriving from $G_{DD}$), as the second term follows by the replacement $m \leftrightarrow n$. Call this term $u_{DD}$. It is convenient to go back to (15.11) but still allow the integral to run to $+\infty$. Then

$$u_{DD} = N \int_0^\infty d\tau \int_0^\infty d\xi \frac{\beta}{4\pi i\tau} \frac{\beta\gamma}{4\pi i}$$

$$\times \exp\{i\beta[\tau + \frac{\gamma\xi^2}{4} + \xi\, v \sin\omega_2 + \frac{1}{4\tau}(\rho_a+\rho_b)^2]\} \tag{5.20}$$

with appropriately substituted values for $v$ and $\gamma$. For the $\tau$ integral we do a stationary phase approximation, with large parameter $\rho_a$. Since $\rho_a \to \infty$ this is an exact evaluation. The stationary value of $\tau$ is

$$\bar{\tau} = \frac{1}{2}\rho_a + \frac{1}{2}\rho_b + \frac{\xi^2}{4\rho_b} + \xi \sin \omega_2 + O(1/\rho_a) \tag{5.21}$$

and after a bit of tedium and with an appropriate N one obtains

$$u_{DD} = \exp[-i\beta\rho_b \cos(\psi-\theta)]h\, (\sqrt{2\beta\rho_b} \cos \tfrac{1}{2}(\psi-\theta)) \tag{5.22}$$

Recall that $\rho_b$ is related to physical coordinates by Eq. (3.2) which is equivalent to $\beta\rho_b = k|\vec{x}|$. With this correspondence the expression (5.22) is seen to be exactly the same as Sommerfeld's first term.

The reflected term presents no new challenge and we find complete agreement with the exact solution. In fact we have the remarkable result that the semiclassical approximation to the path integral used at a single intermediate time ("$\tau_c$") is enough to recover the exact solution.

Since we have recovered Sommerfeld's exact solution we have a fortiori obtained the asymptotic form (4.5)-(4.6). The same asymptotic expansion used there when applied to the time dependent

G of Eq. (5.18) shows it to be composed of direct and reflected rays and of diffracted terms that can be considered to have arisen from trajectories from $\vec{a}$ to 0 to $\vec{b}$. The latter terms are smaller by a factor $\sqrt{\beta}$ relative to the former. In our calculation we did not get them from a particular path from $\vec{a}$ to 0 to $\vec{b}$ but rather by summing over paths that were defective at one point. That is, from $\vec{a}$ to the integration point $\vec{c}$ they followed a path of stationary action (a straight line or a straight line reflected off the barrier). The same is true from $\vec{c}$ to $\vec{b}$. But at $\vec{c}$ they violated the stationary action condition with a single bend. The price of this deviation (when all these paths are summed) is $1/\sqrt{\beta}$.

## VI. INFINITE DIMENSIONAL PERSPECTIVE

In the last section we were involved in the details of a long calculation. Now I'd like to take a more global view. The propagator is the sum over all paths from $\vec{a}$ to $\vec{b}$ (Fig. 2) of $\exp(i\beta S)$. The usual stationary phase approximation for G is given in (3.10) and consists of a sum over paths $y(\cdot)$ for which $\delta S=0$. But for the case in which $\vec{a}$ and $\vec{b}$ have the barrier between them ($\omega_2 > 0$ in the notation of Sec. V) there is no such path. In the 2 dimensional integrals of Sec. V this was manifested in that the stationary point lay outside the range of integration. We can look at the infinite dimensional integral (3.5) in the same way. The impenetrable barrier restricts the range of integration in function space, eliminating paths that pass through the barrier and in particular eliminating the path for which $\delta S=0$.

We review some features of the stationary phase approximation. Let

$$f(\beta) = \int_0^\infty dx\, e^{i\beta(x-x_o)^2}$$

For $x_o > 0$ and $x_o\sqrt{\beta} >> 1$, we know that $f(\beta) \sim \sqrt{i\pi/\beta}$. On the other hand, for $x_o < 0$ and $\beta|x_o| >> 1$ we have $f(\beta) = 0(1/\beta)$. To see this integrate by parts:

$$f(\beta)=\exp(i\beta x_o^2)\int_0^\infty dx[-2i\beta(x_o-x)]^{-1} \frac{\partial}{\partial x} \exp[-2i\beta x(x_o-x/2)]$$

$$= \frac{e^{i\beta x_o^2}}{2i\beta x_o} + \frac{e^{i\beta x_o^2}}{2i\beta} \int_0^\infty dx \exp[-2i\beta x(x_o - \frac{x}{2})] \frac{\partial}{\partial x} \frac{1}{x_o-x}$$

The first term above is $0(1/\beta)$ while the second term is $(1/\beta)$ times an integral of the same form as f itself. Another integration by parts on this second integral reduces it by another power of $\beta$ so that to leading order $f$ is given by the first term alone.

There is no substantial change in the foregoing results when $f$ is a multidimensional integral. Consider

$$f(\beta) = \int_{\Omega} dx \exp[i\beta\phi(x)]$$

where $x \in R^n$ and $\Omega \subset R^n$ is a convex domain of integration. If $\nabla\phi$ vanishes at $x_o$ within $\Omega$, then $f = 0(\beta^{-n/2})$. This follows by diagonalizing the matrix of second derivatives of $\phi$ around the point where its first derivative vanishes (we assume this matrix is nonsingular). Now suppose $\nabla\phi \neq 0$ in $\Omega$. Then we can integrate by parts using the divergence theorem of vector calculus (see (9))

$$\frac{1}{i\beta} \nabla\cdot(\vec{\omega} e^{i\beta\phi}) = e^{i\beta\phi} + \frac{e^{\beta\phi}}{i\beta} \nabla\cdot\vec{\omega} \tag{6.1}$$

with $\vec{\omega} = \nabla\phi/|\nabla\phi|^2$. Therefore $f$ is written as a volume integral plus a surface integral with a factor $1/\beta$ multiplying both. The volume integral can be subjected to another integration by parts (i.e. the divergence theorem) losing another factor of $\beta$. The surface integral is of one lower dimension and for a closed volume $\Omega$ of integration and smooth function $\phi$, it will be of order $\beta^{-(n-1)/2}$. This is because $\nabla\phi$ when restricted to $\partial\Omega$ (the surface of $\Omega$) must vanish somewhere since $\phi$ must assume its maximum on the boundary. Assembling the various powers of $\beta$, we see that just as for the one dimensional case the absence of a stationary point within the volume of integration costs a factor $\sqrt{\beta}$.

Application of the foregoing considerations to the functional integral is not immediate since the volume of integration (all paths that do not pass through the barrier) is not closed. However, we know by more prosaic calculations that the diffracted ray contribution is down by one factor of $\sqrt{\beta}$, so there is reason to believe that we are on the right track. In fact from the previous sections we know that when the Green's function is reduced to a one dimensional integral a factor $\sqrt{\beta}$ is lost because the stationary point is outside the range of integration. When we are not too deep in the diffracted ray regime, i.e. when the bend in the line $\vec{a}0\vec{b}$ is not too great, then in function space there is a point where $\delta S$, the gradient of the classical action, vanishes just outside the domain of integration. Under these circumstances it is possible to show in linear approximation that there must be a point on the surface at which $\delta S$ - when restricted to the surface - does vanish. This is what is needed to get the extra $\sqrt{\beta}$ dropoff.

It is also of interest to look at $\delta S$ along the path $y(t)$ which goes in a straight line from $\vec{a}$ to 0 and from 0 to $\vec{b}$ with constant velocity except for its discontinuity (in direction) at 0. This path is on the surface of the domain of functional integration and it is the classical action evaluated along this path that gives the diffracted ray contribution to the Green's function. This leads us to believe that $\delta S$ when restricted to the surface should vanish. To evaluate $\delta S$ we note that $\ddot{y}(t) = (|\vec{a}|+|\vec{b}|)(\hat{a}+\hat{b})\delta(t-\tau_c)$. Since for a general variation $\eta$ about $y$ we have

$$S(y+\eta) = \frac{\tau}{4}\,(|\vec{a}|+|\vec{b}|)^2 + \int_0^\tau (-\ddot{y}(t))\eta(t)dt+\delta^2 S + \ldots$$

it follows that $\delta S$ is proportional to a delta function. This can also be put in a discrete form in which case $\delta S$ is the vector with components

$$(\delta S)_j = -(|\vec{a}|+|\vec{b}|)\;(\hat{a}+\hat{b})\delta_{j\,j_c}$$

where $j_c$ is the index of the discrete time at which $y$ makes its bend.

At present I have not brought everything together by showing $\delta S$ to be perpendicular to $\partial\Omega$ at $y$. What is clear is that trivial application of (6.1) will not always work and for some cases the gradient vanishes nowhere on the surface. (Consider two separated barriers where one expects a factor $1/\beta = (1/\sqrt{\beta}^2)$.)

## VII. TUNNELLING AND CAUSTICS

Tunnelling occurs in optics as well as in quantum mechanics. In fact R. Dicke when teaching quantum mechanics brought in a pair of prisms to demonstrate that total internal reflection within one of the prisms could break down if the other prism is placed close enough (13). One can think of total internal reflection as due to a potential barrier (cf. Eq. (3.3)) and bringing the second prism close means that an allowed region is not far away and tunnelling through the forbidden region is possible.

The path integral propagator $G(\vec{b},\tau;\vec{a})$ is not a suitable object for study of barrier penetration since there is a path through any finite potential. So we Fourier transform, as is required for the optics problem anyway. Consider therefore the quantity $G(b,a;E)$ as given in (3.7). Inserting the semiclassical expression (3.8) for the time dependent propagator yields

$$G(b,a;E) = \sum_{\alpha} \int_{o}^{\infty} d\tau \left[\det\left(\frac{i\beta}{2\pi} \frac{\partial^2 S_\alpha}{\partial b \partial a}\right)\right]^{1/2} \exp\{i\beta\tau E + i\beta S_\alpha(b,a;\tau)\} \tag{7.1}$$

where we explicitly indicate the dependence of $S_\alpha$ on the endpoints. That is, for a particular classical path $y_\alpha(\cdot)$, with $y_\alpha(0) = a$, $y_\alpha(\tau) = b$, $S_\alpha(b,a;\tau) = S[y_\alpha(\cdot)]$ with $S[\cdot]$ given by (3.6). For fixed $E$ (7.1) is evaluated by the stationary phase approximation and the important value of $\tau$ is that for which

$$\frac{\partial S_\alpha(b,a;\tau)}{\partial \tau} + E = 0 \tag{7.2}$$

From classical mechanics we know that this $\tau$ is given by

$$\tau = \frac{1}{2} \int_a^b dy \, [E - V(y)]^{-1/2} \tag{7.3}$$

and that for this path

$$S_\alpha(b,a;E) = \int_a^b \sqrt{E-V(y)} \, dy - E\tau \tag{7.4}$$

We combine these formulas and for convenience ignore the determinant in (7.1) (see Sec. 18, Ref. 4, for proper treatment of this factor) to get

$$G(b,a;E) \sim \exp[i\beta \int_a^b \sqrt{E - V(y)} \, dy] \tag{7.5}$$

The integral in (7.5) is one dimensional and we neglect the possibility of there being several paths (these extra bounces off walls are unimportant for barrier penetration).

Now suppose that somewhere on the path of integration $V$ exceeds $E$. Then there is no classical path from $a$ to $b$ and (7.1) breaks down. But consider the expression (7.5). It is an analytic function of $E$ in the upper half $E$ plane even if $V(y)$ is not particularly smooth. The formula for quantum mechanical barrier penetration comes from just this analytic continuation. Suppose $E > V(y)$ for $a<y<y_1$ and for $y_2<y<b$, while $E < V(y)$ for $y_1<y<y_2$. Then taking the branch of $G$ that leaves it bounded for

$E \to i\infty$ we have

$$G(b,a;E) = \exp[i\beta(\int_a^{y_1} + \int_{y_2}^{b})\sqrt{E-V}]\ \exp[-\beta \int_{y_1}^{y_2} \sqrt{V(y)-E}\, dy] \quad (7.6)$$

the usual barrier penetration formula.

My purpose in presenting this material is to comment on the extent to which we can ascribe a classical path to the quantities in (7.6). From (7.3) note that the $\tau$ that gives rise to (7.6) is complex and has an imaginary contribution from the range $y_1<y<y_2$. Thus for the $\tau$ integral in (7.1) there is no stationary phase point on the contour of integration. However, one can deform the contour and bring this contribution to G. Is this an "imaginary ray"? I lean to McLaughlin's view (10) of this as an analytic continuation in $\tau$ or, in the present description, in E. In particular, I am conservative about letting y be complex. My reason is that the barrier penetration formula holds with no particular analyticity assumptions on V(y) so there is no a priori way to extend V into the complex y plane - nor is there any need to.

Nevertheless, one can still think of the G of (7.6) as arising from the solution of a variational problem. For a pair of points a and b pick a real number E such that $E > V(a)$ and $E > V(b)$ but there is an interval between a and b for which $V(y) > E$. Let $\tau$ be the complex number given in (7.3). Let y(t) be a real valued function of the complex parameter t with $y(0) = a$ and $y(\tau) = b$, and $S = \int_0^\tau [\frac{1}{4}(dy/dt)^2 - V(y)]dt$. We want S to be stationary relative to variations in y. First imagine that the contour of integration for t in the complex t plane (from 0 to $\tau$) is specified. Vary y in the usual way to get the Euler-Lagrange equations. Thus any y(t) making S stationary will satisfy these equations and as a particular consequence the quantity

$$E' = \frac{1}{4}\left(\frac{dy}{dt}\right)^2 + V(y)$$

will be a constant of the motion. Moreover, since the path y(t) is presumed to be stationary we have the usual relation $\tau = \int_a^b dy[E'-V(y)]^{-1/2}$. But $\tau$ was picked so that this relation was true for a particular number E. Therefore $E' = E$. Since E and V are real so must be $(dy/dt)^2$. This shows that even though t was allowed to wander arbitrarily in the complex t plane the path y(t) cannot stationary unless: (1) t moves parallel to the real $t$ axis for $y$ such that $E > V(y)$ and (2) $t$ moves parallel to the imaginary $t$ axis for $y$ such that $E < V(y)$.

There is little new in this description of the barrier penetration ray. Perhaps I have been a bit more explicit than previous authors in stating the variational problem.

Does being able to speak of a ray with complex time parameter allow one to think of quantum particles as having definite trajectories on their way through a barrier? If one sticks closely to the Newtonian picture of particles and trajectories the answer must be no. But if one takes the Hamilton Jacobi equations as the basis of classical particle mechanics then the relation of G of (7.6) to the solution of variational problem suggests an extension of the notion of particle. One can for example give physical meaning to the idea of the "time spent in the barrier" and to the extent such meaning can be assigned that time coincides with $\mathrm{Im}\tau$ given above (11).

There is another situation where two points are not connected by any classical path but the leading asymptotic approximation to $G$ is given by an analytic continuation of the classical action function. This occurs for caustics. Keller deals with this in terms of imaginary rays and in Sec. 15 of Ref. (4) (in particular p. 126) I show how these terms arise from analytic continuation of the action to regions where no classical paths exist. It is an exercise in complex variables and classical mechanics to relate this statement to analytic continuation in $\tau$ or $E$. The point of the exercise is that once again the leading asymptotic behavior of $G$ is given by a generalized kind of classical mechanics.

## VIII. SUMMARY

Electromagnetism - a wave phenomenon for the purposes of this paper - lends itself to a particle or ray description in a number of situations considered characteristic of the wave picture. This is the theme of Keller's geometric diffraction theory and it is a theme which the path integral context of this paper makes explicit.

Much of the paper is concerned with calculating knife edge diffraction and recovering the exact solution from what starts as a semiclassical approximation to the path integral. The full propagator then contains term identifiable as a "diffracted ray" which gives rise to traditional diffraction terms but which behaves like an ordinary ray once it is launched from the knife edge. We also discuss in Sec. VI how this diffracted term arises in the path integral from a stationary point of the action that lies outside the range of functional integration.

Sec. VII deals with two other cases where a sufficiently broad view of classical mechanics would allow one to include other characteristical wave phenomena in the category of particle behavior. The cases are barrier penetration and illumination in the caustic shadow.

ACKNOWLEDGEMENTS

I wish to thank Ady Mann and David McLaughlin for useful conversations on this work.

REFERENCES

1. J.B. Keller, A Geometrical Theory of Diffraction in "Calculus of Variations and its Applications", Proceedings of a Symposium in Applied Mathematics", ed. L.M. Graves, McGraw, New York, 1958.
2. R.G. Kouyoumjian, The Geometrical Theory of Diffraction and its Application, in "Numerical and Asymptotic Techniques in Electromagnetics", ed. R. Mittra, Springer-Verlag, Topics in Applied Physics, Vol. 3, Springer, Berlin, 1975.
3. A.J.W. Sommerfeld, "Optics", Academic Press, New York 1954.
4. L.S. Schulman, "Techniques and Applications of Path Integration", Wiley, New York 1981.
5. C. Morette (DeWitt), On the Definition and Approximation of Feynman's Path Integrals, Phys. Rev. 81, 848 (1952).
6. R.M. Lewis and J. Boersma, Uniform Asymptotic Theory of Edge Diffraction, J. Math. Phys. 10, 2291 (1969).
7. S.W. Lee, Path Integrals for Solving some Electromagnetic Edge Diffraction Problems, J. Math. Phys. 19, 1414 (1978).
8. M.C. Gutzwiller, Phase-Integral Approximation in Momentum Space and Bound States of an Atom, J. Math. Phys. 8, 1979 (1967).
9. N. Bleistein and K.A. Handelsman, "Asymptotic Expansions of Integrals", Holt, Rinehart and Winston, New York, 1975.
10. D.W. McLaughlin, Complex Time, Contour Independent Path Integrals, and Barrier Penetration, J. Math. Phys. 13, 1099 (1972).
11. M. Buttiker and R. Landauer, Traversal Time for Tunneling, IBM preprint, 1982.
12. L.S. Schulman, Exact Time-Dependent Green's Function for the Half-Plane Barrier, Phys. Lev. Lett. 49, 599 (1982).
13. See also R.S. Longhurst, "Geometrical and Physical Optics," Longmans, Green and Co. London 1957; p. 439. I thank Michael Berry for bringing this reference to my attention.

# ATOMIC COLLISION EXPERIMENTS AT THE BORDER LINE BETWEEN CLASSICAL AND QUANTUM MECHANICS

Vincenzo Aquilanti

Dipartimento di Chimica dell'Università, Via Elce di Sotto 8, 06100 Perugia, Italy

In order to understand atomic and molecular interactions, one has to learn how to live with the wave-particle duality, considering classical nuclei and quantum electrons. A better way, illustrated here by reference to experiments, is by quasiclassical (or semi-classical) mechanics, governing a world with a quasi-zero Planck's constant. One thus explains optical analogs (shadows, rainbows, glories) as interference effects in atomic collisions. Reference is also made to Wheeler's "black bird" on the inversion problem from spectroscopy and scattering to molecular structure. The paper concludes outlining a journey in the hyperspace to escape from Einstein's torus and to find interferences and resonances in three body scattering and reactions.

## 1. TO LIVE WITH THE WAVE PARTICLE DUALITY: CLASSICAL NUCLEI AND QUANTUM ELECTRONS?

"Un mobile quelconque pourrait dans certain cas se diffracter". This sentence by Louis de Broglie appears in one of his 1923 notes (1), which anticipate his famous thesis introducing wave characteristics for particles.

Although de Broglie writes "a mobile whatsoever", as the following sentences show ("Un flot d'électrons traversant une ouverture assez petite présenterait des phénomènes de diffraction. C'est de côté qu'il faudra peut-être chercher des confirmations expérimentales de nos idées."), he was definitely indicating that



*S. Diner et al. (eds.), The Wave-Particle Dualism, 273–290.*

the experimental verifications of the wave-particle behaviour had to be looked for studying electrons first.

The following years told us that he was right. As a consequence, the study of the structure of matter, at the atomic and molecular level, is nowadays an immense laboratory of applied (mainly non relativistic) quantum mechanics. This has to be regarded in the same sense that astronomy and civil engineering can be considered as a laboratory of classical mechanics. Quantum chemistry, for example, looks for the energy and distribution of electrons in molecules by solving Schrödinger's wave equation, considering nuclei as slowly moving (Born-Oppenheimer separation), with fallouts in applications ranging from heavy industry to pharmaceutics.

Chemists, among others, have thus learned how to heuristically deal with the wave-particle duality: nuclei are classical objects, quantum mechanics has to be introduced only for electrons. It is the purpose of this paper to show that a very rich world opens when we look for the wave behaviour of particles much heavier than electrons.

Interference effects in electron scattering are by now a familiar phenomenon, after the stage has been settled by the Davisson-Germer experiment, the discovering of the Ramsauer effect, the introduction of the electron microscope. Although embedded into the standard quantum mechanical theory, the actual discovery of interference effects in the behaviour of objects much heavier than electrons, such as atoms and molecules, had to wait a few decades. Only around 1960 both theoretical progress and experimental ingenuity contributed to establish a rich phenomenology, whereby atomic and molecular scattering processes exhibit effects which find their most attractive explanation by exploiting the analogy between wave optics and wave mechanics.

Reading further the 1923 de Broglie's note (1): "La nouvelle dynamique du point matériel libre est à l'ancienne dynamique (y compris celle d'Einstein) ce que l'optique ondulatoire est à l'optique géométrique". In modern research on atomic and molecular scattering, we thus hear about shadows, rainbows, glories: names for new effects borrowed from naturally occurring optical phenomena.

We will give a brief account of a few of these phenomena in the following. They confirm beautifully de Broglie's anticipation, but our interest in them lies also in other two points, which will be also discussed further.

A first point is that in dealing with particles much more massive than electrons, and therefore with much shorter wavelengths, it has become necessary to develop techniques, known as semiclassical mechanics, where Planck's constant $h$ is a small parameter (but not zero: if $h$ were exactly zero, we would fall back again into classical mechanics, and lose any opportunity of describing phase interference effects).

The second point is that the experimental study of these phenomena allows to correlate the observed behaviour to effectively probe molecular and nuclear structure: this is a very important aspect, which can hardly be overestimated. Related to it, we have a renewed interest in the inversion problem (from scattering and spectroscopy to the interaction potential), although the most significant progress so far has been achieved through trial-and-error methods, starting from the outcome of experiments of ever-increasing accuracy and sophistication.

A further point we will briefly discuss in the following is the extension of semiclassical ideas to multidimensional problems, such as those which arise in dealing with many body interactions and reactions. Although these problems are inherently very intricate we outline some recent developments, which allow, through a simple coordinate change, the formal reduction of many body interactions to that of a particle interacting with an anisotropic centre of force in a mathematical dimension higher than the physical one. In this hyperspace, semiclassical quantizations and scattering phase shifts can be introduced, thus obtaining an interpretive and computational framework for extending the concept of phase interferences and resonances in three-body elastic processes and reactions.

## 2. DE BROGLIE'S "MOBILE WHATSOEVER"

It is immediate to compute the typical de Broglie wavelength for an air molecule, say, given its mass, $m \simeq 5\text{x}10^{-23}$ g and average speed $v \simeq 5\text{x}10^{4}$ cm s$^{-1}$ at room temperature. From the formula $\lambda = h/mv$ , being $h = 6\text{x}10^{-27}$ erg s, we have that the wavelength $\lambda$ is about $3\text{x}10^{-8}$ cm, small enough compared with the dimensions of macroscopic objects such a molecule would find on its way, were we trying to discover its wave properties by setting up double slit type of experiments.

Furthermore, a single air molecule is just one of a very large number of other particles moving in the room, and so its detection is hard in the presence of so many twins, and the characterization of its velocity may be strongly affected by the fact that collisions are continuously changing it.

In other words, the observation of wave behaviour at the atomic level requires the search for suitable "diffracting devices", and poses experimental problems of vacuum, selective detection and state analysis. That both these requirements could be met was discovered around 1960 (2).

A diffracting device showing a characteristic length of the order of $10^{-8}$ cm is a free atom or molecule. In fact atoms and molecules interact through very complicated many body forces when their electron clouds overlap. However in gases, and to a certain extent in liquids and solids, it is often a very useful approximation to assume that two of them influence each other by means of a two body potential function $V(R)$ dependent only on their distance $R$. The range of typical potentials turns out to be, as anticipated, a few units in the $10^{-8}$ cm scale.

In turn, this means of course that a study of collisions showing interference effects due to the interatomic or intermolecular potentials may be a useful route to unveil properties of the latter. It will be shown in the following sections that this is indeed the case, and that such an approach is complementary to spectroscopic techniques in some instances, and unique in others.

As far as the progress in experimental techniques is concerned, vacuum technology allows to maintain apparatuses whose linear dimensions are of the order of meters under vacuum conditions, implying very large mean free paths for atoms and molecules. Therefore, single collisions can be studied.

It is also possible, although not routinely, to produce molecular beams of various stable and unstable species, to select their velocities by rotating slotted (Fizeau-like) discs and to measure them through time-of-flight techniques. Magnetic and electric fields can be used for state analysis, and intense laser fields for selective state production.

However, although modern electronics allows efficient treatments of signals and discrimination of noise, the reader of current specialized literature will find that we are still in the stage that a "universal" apparatus for the study of atomic collisions can hardly be built in practice: in other words, a given experimental arrangement for the study of collisions between given partners in a given energy range will most likely have to be dismantled, redesigned and rebuilt once different partners or different energies are to be investigated.

In conclusion, any mobile in de Broglie's sense can be made to diffract in certain cases: to do this for atoms and molecules is a job we are slowly learning to perform. What we gain is information about interactions, which will eventually be used for the characterization of chemical bonds, and for the understanding of how atoms and molecules will behave in the gas phase under high pressure, and also in the liquid and solid phases.

## 3. OBSERVABLES IN AN ATOMIC COLLISION EXPERIMENT: CLASSICAL COLLISION THEORY REVISITED

The goal of an atomic collision experiment is to measure cross sections. Since the very concept of cross section has some pitfalls in classical mechanics, it is useful to start by reviewing a few definitions.

Consider a beam of particles A (atoms or molecules), identical as far as their velocities and internal states are concerned: let $n_A$ be the number of these "monoenergetic" particles (in $cm^{-2}$ $s^{-1}$, for example), and let them be propagating along the Z axis. Consider now that around the point O along the Z axis we confined $n_B$ target particles B, likewise monoenergetic. Both A and B are sufficiently diluted, that only binary collisions are possible. Each collision will give rise to scattering of particles A and B themselves, or to scattering of new particles formed by the collision itself.

A detector, sensitive to the nature of particles and possibly to their velocity and internal states, is now placed far enough from O, relative to the typical range of interparticle forces. Suppose we are interested in monoenergetic particles of type C: the detector counts selectively their number $n_C$ (particles $s^{-1}$) scattered within a solid angle element $d\omega$. This number will be in general proportional to $d\omega$, and also to $n_A$ and $n_B$:

$$dn_C(\omega) = I(\omega)\, n_A\, n_B\, d\omega,$$

the proportionality constant $I(\omega)$($cm^2$ per steradian) is an area, usually referred to as the differential cross section for scattering of C-type particles in A+B collisions along the laboratory direction $\omega$.

Using known kinematic transformations from the laboratory to the centre-of-mass system, we can always introduce instead of $\omega$ the solid angle $\Omega$ indicating the direction of a polar axis whose origin is now moving with the centre-of-mass, thus eliminating the restriction that B is at rest. For the *differential cross section,* we have, in analogy with the above formula,

$$I(\Omega)d\Omega = dn_C(\Omega)/n_A n_B.$$

The *integral cross section* Q is given by integrating over the solid angle:

$$Q = \int_\Omega I(\Omega)\, d\Omega.$$

These quantities depend on the relative kinetic energy, E, and their measurements are frequently barred by unavoidable averaging processes due to parameters hard to control experimentally. For collisions between atoms interacting through a potential depending only on their distance (see Sec.2), scattering will be symmetric around the direction of the centre-of-mass. Therefore, of polar angles $\theta$ and $\phi$ the latter is superfluous, and $I(\Omega) \rightarrow I(\theta)$; furthermore, we now have:

$$Q = 2\pi \int_0^\pi I(\theta)\, \sin\theta\, d\theta.$$

However, when dealing with atoms having spin and orbital angular momenta, or with molecules, orientation is an important additional feature. The experimental measurement of effects associated with this feature, such as polarization and rotational excitation transfer, requires sophisticated state selection techniques, and the theoretical analysis is likewise much more complicated. The field is being actively investigated, and the interested reader is referred to a forthcoming review by the author (3).

From now on attention will therefore be restricted to the scattering by a central potential, $V(R)$, which is assumed to be, as is typical, attractive at long range and repulsive at short range: in general there will be a value of the interatomic distance $R$, for which $V(R)$ will show a minimum: the value of the energy at the minimum is the dissociation energy of the molecule formed by the two atoms, and the distance where this minimum occurs is a measure of the length of the chemical bond.

Let $m_A$ and $m_B$ be the masses of the two atoms: in the centre-of-mass system, one has to deal with scattering of a single particle, whose reduced mass is:

$$\mu = m_A m_B/(m_A + m_B),$$

by an *effective* potential

$$V_{eff} = V(R) + \mathcal{L}^2/2\mu R^2,$$

where $\mathcal{L}$ is the modulus of angular momentum. The centrifugal term gives an ever repulsive contribution to the interaction, and a scattering process will have to be interpreted by eventually including these contributions from all allowed values of $\mathcal{L}$, which is however a constant in a single collision.

From the constancy of $\mathcal{L}$, which can be written in terms of asymptotic relative velocity $v$ and impact parameter $b$,

$$\mathcal{L} = \mu v b,$$

and of the total energy,

$$\mathcal{E} = 1/2\mu v^2,$$

any textbook indicates how to derive the following expressions for the radial velocity $v_R$,

$$v_R = \frac{dR}{dt} = v\left(1 - \frac{V(R)}{\mathcal{E}} - \frac{b^2}{R^2}\right)^{1/2},$$

and for the distance of closest approach

$$R_o = b\left(1 - \frac{V(R_o)}{\mathcal{E}}\right)^{-1/2}.$$

This distance $R_o$ is the classical turning point, corresponding to $v_R = 0$; it is smaller than $b$ if the potential is attractive there ($V(R_o) < 0$), and larger than $b$ if it is repulsive ($V(R_o) > 0$). The classical particle makes a stop when it gets that close to the scattering centre, then moves away from it following a trajectory symmetric to the one previously described.

The final formula for the classical description of a scattering process is the expression for the angle of deflection $\theta$, which gives the direction of the scattered particle with respect to the initial one:

$$\theta(v,b) = \pi - 2b\int_{\infty}^{R_o}\left(1 - \frac{V(R)}{\mathcal{E}} - \frac{b^2}{R^2}\right)^{-1/2}\frac{dR}{R^2}.$$

These elementary classical formulas contain aspects which have been overlooked until around 1960. The above expression for the radial velocity will later be used to derive the formula for semiclassical phaseshift in conjunction with the idea of a de

Broglie's wave. Here we outline some consequences of the formula for the deflection angle.

It is easy to see from the formula that for a typical potential, exhibiting short range repulsion and long range attraction, the deflection will be small and negative for large impact parameters, which mostly sample a net attraction. As impact parameter decreases, attraction increases and the more negative becomes the deflection. At low energies classical particles can even be trapped by the potential, and describe several orbits about its centre (*orbiting collisions*); for collisions having energies of the order of the well depth or larger, the deflection $\theta$ can be shown to exhibit a finite minimum, which, for reasons to be explained later, is now referred to as the *rainbow angle*.

At low impact parameters a typical atom-atom collision will feel the repulsive part of the interaction: the above formula then predicts an increase in the deflection angle. In particular, this angle will be zero (no net deflection) when attraction and repulsion balance each other. There are therefore trajectories for which the particle, although passing close to the potential, will appear not deflected: associated with this classical feature, we will have a wave interference effect, to be discussed later. Because of an analogy with wave optics, this effect is known as the *glory effect*.

In an actual experiment all impact parameters are possible. The above equation, which gives the explicit classical relationship between the scattering angle $\theta$ and the impact parameter $b$, can be used to obtain an expression for the differential cross section $I(\theta)$. Particles travelling with an impact parameter in the range between $b$ and $b + db$ will be deflected by an angle between $\theta$ and $\theta + d\theta$ (for $d\theta/db > 0$) or between $\theta$ and $\theta - d\theta$ (for $d\theta/db < 0$). In the above formulas for cross sections, we now have immediately

$$2\pi I(\theta)\sin\theta\, d\theta = |2\pi b\, db|,$$

since particles incident within an impact ring of area $2\pi b\, db$ will be scattered into the element of solid angle

$$d\Omega = 2\pi \sin\theta\, d\theta,$$

provided that $\theta$ and $b$ are related by the preceeding relationship.

But, as stressed, $\theta$ is not in general a univocal function of $b$: therefore different contributions from different branches will have to be summed:

$$I(\theta) = \Sigma\, b \,/\, |\sin\theta \; d\theta/db| .$$

In a wave picture, *interference* will arise between trajectories on different branches of the deflection function leading to same scattering angle.

This formula also shows that very unphysically the classical differential cross section diverges for *glory trajectories* ($\theta = 0$) and for *rainbow trajectories* ($d\theta/db = 0$).

Finally, the classical expression for integral cross section can be written

$$Q = 2\pi \int_0^\infty b\,db ,$$

an expression which diverges for long range potentials! Of course, most of these divergencies are to be removed by quantum mechanics, a remarkable exception being the case of the integral cross section for a Coulomb potential. On the other hand, semiclassical methods, which we will approach in the following section in a heuristic way, capitalize on these divergencies, exploiting interesting analogies with wave optics.

## 4. THE PHASE SHIFT FOR A DE BROGLIE'S WAVE ASSOCIATED TO A CLASSICAL TRAJECTORY.

Since we will outline in this section how the well known semiclassical or JWKB phaseshift for potential scattering can be simply obtained by associating a de Broglie's wave to a classical trajectory, we start with the meaning which can be attached to the very concept of a trajectory in quantum mechanics.

According to the uncertainty principle, applied to the conjugate quantities position and momentum, the particle of mass $\mu$ introduced in the previous section in a central force field $V(R)$ will have a well defined velocity $v$ and a well defined distance from the centre $R$ only if

$$\mu v R \gg h/2\pi .$$

It is easy to show that for particles having atomic masses and thermal velocities, under the action of potentials with a typical range of the order of $10^{-8}$ cm, the above inequality is generally fulfilled, except of course at turning points $R_0$, where the radial velocity $v_R$ drops to zero.

Besides the failure at low velocity, the concept of a classical trajectory also fails for a possible violation of the uncertainty principle, associated to the other conjugate quantities energy and time. The introduction of a characteristic *collision time* $R/v$, implies an uncertainty in the potential

$$\Delta V \simeq hv/2\pi R .$$

This relationship *(Massey's adiabatic criterium)* semiquantitatively shows that an *adiabatic* view of a collision process will fail for high velocities and close encounters, where the very concept of a univocally determined potential fails.

It is not unusual in atomic collisions that these regions of failure of the concept of classical trajectories can be localized within restricted ranges in configuration space. The search for the characterization of these ranges, and the way to deal with phenomena occurring there, is an actively investigated area of semiclassical collision theory.

It is often possible to identify these regions as responsible for branching of trajectories, with the result that *transitions* may take place there: *interference* effects, in a sense to be explained below, will also show up as a consequence of these branchings. Further comments on how to deal within semiclassical theory with these *nonadiabatic* or *diabatic* processes is deferred to a later section.

As anticipated, a fruitful approach to understand wave effects in atomic collision is to retain as far as possible the classical picture, to spot where and why it fails, and to go on via semiclassical arguments and analogies with wave optics. For example, arguments such as those given above allow to understand how to remove the aforementioned divergence in the classical integral cross section formula: the quantum mechanical indistinguishability between small deflections and the natural wave broadening effectively limits the contribution to the cross section from very large impact parameters. Recipes can therefore be given for the experimental conditions to be met (distance of the detector from the scattering centre, its angular resolution), in order to perform a meaningful measurement of an integral cross section in the quantum mechanical sense.

Let us consider next how by associating to a classical trajectory a *local* de Broglie's wavelength $\lambda_R$ , one is most naturally lead to the concept of a *phase shift* $\delta$, which is fundamental in the quantum mechanical theory of potential scattering.

For the particle of mass $\mu$ and asymptotic velocity $v$, scattered by the potential $V(R)$ as in the previous sections, by formulas given there we have

$$\lambda_R = \frac{h}{\mu v_R} = \frac{h}{\mu v}\left(1 - \frac{V(R)}{\mathcal{E}} - \frac{b^2}{R^2}\right)^{-1/2},$$

and therefore a phase angle which varies with time as $2\pi(v_R/\lambda_R)dt$. The phase shift $\delta$ is defined as the difference, as $t \to \infty$, between the phase accumulated along the true trajectory T, that the particle follows under the action of $V(R)$, and the phase it accumulated along the (straightline) trajectory $T_o$ it would follow if the potential were absent ($V(R) = 0$)

$$\delta(v,b) = \int_T \frac{2\pi v_R}{\lambda_R}\,dt - \int_{T_o} \frac{2\pi v_R}{\lambda_R}\,dt .$$

One finally gets

$$\delta(v,b) = \frac{2\pi\mu v}{h}\left[\int_{R_o}^{\infty}\left(1 - \frac{V(R)}{\mathcal{E}} - \frac{b^2}{R^2}\right)^{1/2} dR - \int_b^{\infty}\left( - \frac{b^2}{R^2}\right)^{1/2} dR\right].$$

This expression for the phase shift happens to be identical to the one which is obtained by more sophisticated techniques, within the semiclassical method, once the equivalence between the classical impact parameter $b$ and the quantum number for orbital angular momentum $\ell$ is made explicit:

$$\mu v b = (\ell + \tfrac{1}{2})\, h/2\pi .$$

The above expression for $\delta$ is usually referred to as the JWKB phase shift.

The present approach is manifestly heuristic, but it definitely adds physical insight into our understanding of a collision process, and provides encouragements for not to abandone too soon the familiar concept of a classical trajectory.

## 5. SHADOWS, RAINBOWS, GLORIES AND OTHER INTERFERENCE EFFECTS IN ATOMIC COLLISIONS

In this section we briefly list typical wave interference effects in atomic collisions, by outlining their physical origin in the spirit of optical analogies suggested by the qualitative approach we have been following so far. Deferring to the next sections theoretical machinery needed to develop a quantitative approach, we stress the suggestive power of a language based on well known phenomena of wave optics.

Actually, we already met a divergence in the classical formula for the integral cross section, and indicated how it is removed for most potentials in a wave mechanical context: here the analogy can be shown to be with treatments of shadows cast by small objects in optics, and therefore some authors refer to the integral cross section for scattering as to the *shadow effect* : the suggestion is thus to regard cross sections as shadows cast by the scattering potential on the detector.

Particularly interesting are effects which can be associated with interferences among trajectories, to which we have to associate a wavelike behaviour by an intuitive extension of the de Broglie's concept, in the sense outlined in the previous Section. These effects produce characteristic oscillations and anomalies in the experimentally observable quantities, $I(\theta, v)$ and $Q(v)$, when they are measured as a function of the relative velocity $v$ or of the scattering angle $\theta$ .

We already mentioned the *rainbow angle* in a previous Section: its name stems from the fact that the way it arises considering trajectories deflected by the potentials can be shown to be analogous to the way used by Descartes to explain the optical rainbows by tracing sun rays refracted by water droplets. And as Airy in the mid-nineteenth century explained quantitatively by interference between rays the appearance of colours in the rainbow, so we are encouraged to do for potential scattering oscillations in the differential cross sections as a function of the deflection angle.

These oscillations are thus to be interpreted as interferences between trajectories corresponding to different impact parameters, but leading to the same scattering angle. They are known as *supernumerary rainbows* in the specialized literature.

A further divergence in classical cross sections has been introduced previously as the *glory effect*. Its physical origin lies in the fact that for an interaction typically attractive at long range and repulsive at short range classical trajectories can be individuated, for which the particle is first attracted by the potential and then repelled in such a way that there is apparently no net deflection: these particles, to which we associate a wave according to the recipe outlined in the previous section, will experience a phase shift, and therefore interference may arise with those wave particles which are likewise not deflected, for the simple reason that their impact parameter was so large that they completely missed the potential.

The glory effect can nowaday be observed in the velocity

dependence of integral cross section for atom-atom collisions: it shows as characteristic oscillations superimposed on a smooth behaviour, just like glories around a shadow!

Further interference effects are known in atomic collisions, and we will meet some of them in a following Section. In general, aside from the picturesque aspects of analogies with wave optics, their observation is invaluable for our understanding of the structure of matter by sensitively probing intermolecular forces. And, to quote from the conclusions of a recent review (4): "Intermolecular forces in simple systems have entered the age of spectroscopic precision. A few interactions are known with 1 or 2% accuracy and the first theoretical methods of comparable precision are becoming available. The extension of this precision to nonspherical systems represents the challenge of the next few years. When, in the not too distant future, the interaction potential ceases to be one of the unknowns in the problems involving energy transfer, molecular solids, and molecular liquids, these fields will undergo a rapid transformation".

## 6. SEMICLASSICAL POTENTIAL SCATTERING: WHEELER'S BLACK BIRD AND THE INVERSION PROBLEM

The outline of effects associated with potential scattering in the previous sections should be made more precise and rigorous by reviewing the details of semiclassical methods, as they have been developed in recent years. This is however outside the scope of this paper. An excellent review (5) may be referred to, giving an outlook of the field as it was about ten years ago, and collecting developments mainly carried out in the sixties.

It became clear during those years that all the results previously known had to be unified within a coherent picture. Semiclassical theory of potential bound states and scattering, as comprising venerable approximate yet strikingly accurate formulas, such as the Bohr-Sommerfeld quantization rule and the JWKB scattering phase shift, has been firmly founded in the theory of the asymptotic solution of the second order homogeneous ordinary differential equation containing a large parameter. Such an equation is of course Schrödinger equation for potential bound states and scattering under short wavelength conditions. The large parameter is then obviously $\mu/\hbar$ .

Among the successes of this approach we quote the semiclassical theory of *resonance* effects (somehow related to the *orbiting* collisions we met earlier in the classical picture), for which usable formulas have been developed, although a uniform treatment is not yet available.

Besides effectively solving this problem of computing bound state energies, scattering phase shifts and resonances, semiclassical theory has faced with success the task inherent in a short wavelength world: to have to deal with large angular momenta. As outlined in earlier Sections, to compute observable quantities one has classically to integrate over contributions coming from an extended range of impact parameters: quantum mechanically this means that one has to sum over contributions coming from several values of the orbital angular momentum.

The elegant solution to this problem consists in applying transformations of sums into integrals, and to exploit again techniques of asymptotic analysis to sort out the important contributions. The large parameter is again $\mu/\hbar$ , the techniques are the sophisticated modern versions of saddle point and steepest descent methods, and the results are a consistent picture in terms of *localized* features of the potential for most of interference effects encountered previously.

Moreover, in the latest literature, the formal analogy between these effects in potential scattering and those in wave optics is being exploited in a highly sophisticated language, including that of catastrophe theory.

The further aspect for which such an approach is extremely appealing is related to the so called *inversion problem.* In simple words, this problem asks how to find a route from observables such as bound state energies and scattering phenomena to the potential, if any, responsible for them.

The importance of these aspects should be by now clear to the reader, after we have been stressing it several times in previous Sections. Most of the work done so far for atomic and molecular scattering, has been based on more or less trial-and-error procedures.

True, the observation of interference effects, in particular, offers a powerful handle to unveil characteristic features of the potential. Much better is the situation for bound states, since semiclassical procedures from observed spectra to the potential are more familiar.

In an essay significantly entitled "Semiclassical analysis illuminates the connection between potential and bound states and scattering" (6), one of the historical masters of this field, J.A.Wheeler, reviews the subject: besides a detailed account and reference to the relevant literature, the interested reader will meet there an interesting zoological curiosity, the *black bird* of semiclassical approximation methods, opposed by Wheeler to a crimson bird, which is instead a representative of mathematical exactitude.

Although both birds "get their nourishment in the same domain of nonrelativistic scattering physics", they hardly interbreed. The black bird "feeds on experiment" and "it shows nowhere to better advantage than in molecular physics". It clearly sees and explains such phenomena as rainbows, glories and other interference effects (Section 5). In view of these successes Wheeler's thesis is therefore that we should use extensively the eyes of the black bird of semiclassical methods to bring out "the physics that is telescoped into" the standard equations of quantum mechanics.

## 7. THE NONSEPARABILITY PROBLEM IN SEMICLASSICAL MECHANICS

Appealing as is Wheeler's program, and impressive as the semiclassical theory of potential scattering is in adding to physical insight, it is however important to realize that this is only a part, and perhaps only a small one, of a long story.

We have been constantly restricting ourselves to the consideration of *separable* problems. Loosely speaking, we always invoked the existence of a "good" coordinate, specifically the interatomic distance $R$, for which we considered explicitly classical trajectories (the time dependence of $R$ under the influence of the potential) in Section 3. The association in Section 4 of a de Broglie's wavelength to these trajectories was a heuristic way to introduce the wonders of semiclassical potential scattering theory, listed as optical analogies in Section 5. As discussed in Section 6, all this can be given a respectable status, and its justification goes under the name of Born-Oppenheimer separation.

How to treat multidimensional problems, and therefore how to deal with *nonseparable* systems is *the* big challenge for modern semiclassical theory: some of the associated difficulties appear to have been borne out even in the old quantum theory, with Einstein's famous idea of quantization over invariant tori leading the way in 1917. Brillouin's contribution in the twenties, and Keller's in the fifties, marked further important steps until the exponential growth of papers on this subject in recent years. But to quote from Berry's recent review (7): "It is not known in general whether or how it is possible to specify uniquely the quantum mechanics of a system whose classical mechanics is given".

In spite of the difficulties posed by these questions, intuitive reasoning based on primitive semiclassical ideas has been actively pursued without waiting for the exact answers. Specifically, some confidence has been acquired on how to heuristically deal with the breakdowns of the concept of a trajectory, such as it was briefly outlined in Section 4. It was anticipated there that failures of the trajectory concept may be well localized in configuration space: techniques have been introduced, following the pioneer work of Landau, Zener and Stückelberg, to deal with these regions of *nonadiabaticity*.

The power and rigour of asymptotic methods as applied to these problems appears to have not been fully exploited, and approaches nearly as rude as those in Section 4 have been followed.

In one of these approaches, the concept of an average trajectory is introduced (8), with the result of transforming Schrödinger equation into a set of first order *semiclassical trajectory equations* (some analogy with methods used in optics makes this approach as one of a set of techniques known as *eikonal*).

Possible branching of trajectories in regions of nonadiabaticity leads, once the proper phases are heuristically added as in Section 4, to interesting interference effects, which are now starting to be well characterized experimentally (3).

Recent activity in this field, as summarized for example in a recent book (8), includes the idea that semiclassical mechanics can be built as a proper analytical continuation of classical mechanics, by allowing space and time to be complex. This idea, which can be traced back to Landau and Stückelberg with reference to atom-atom nonadiabatic transitions, is currently being implemented for the treatment of the intriguing but most interesting cases of rearrangement processes and reactions in atomic and molecular collisions (8).

## 8. INTERFERENCES AND RESONANCES IN REACTIONS: A PREVIEW

In concluding this brief account on the power of semiclassical methods in dealing with current problems posed by atomic and molecular scattering experiments, we give an example of possible further developments as suggested by our current work on the theory of reactions (9).

The well known problem of reactions is that an obvious candidate as a quasi-separable coordinate is not immediately available in a three-body rearrangement problem. In fact, any coordinate based on interatomic distances, as in Born-Oppenheimer separation, while describing one rearrangement channel may not describe satisfactorily the others.

To overcome this difficulty, interest has arisen in recent years in a coordinate system which is a generalization of the spherical coordinate system in three dimensions: the square of one of this coordinate, the *hyperradius*, is basically the (properly weighted by mass factors) sum over squares of physical distances.

This coordinate has the properties of describing "democratically" the possible rearrangement channels of a collision process. Its properties were known for example to Fock and to Morse and Feshbach in the fifties, and the basic formalism started to be developed in the sixties.

Applications to nuclear and atomic physics, and to chemical kinetics and spectroscopy, are slowly but steadily coming out in recent years.

A very recent paper by Ugo Fano, significantly titled "Unified treatment of collisions" (9), outlines a program to extensively apply the concept of hyperradius to look at a many-body collision process: the idea, to be implemented analytically and eventually computationally, is to look at a rearrangement process not as traditionally by following reactants evolving to products, but by extablishing their connections as both evolve from the decay of a collision complex.

The decay of the complex is to be followed in configuration hyperspace by use of this coordinate, the hyperradius. The natural question is therefore: is the hyperradius a reasonably separable coordinate in large regions of configuration space?

Our recent computational work on simulations of chemical reactions suggests that an affirmative answer can be given to this question in many cases. The obvious consequence is that it appears that we have then a coordinate, analogous to the interatomic distance of the Born-Oppenheimer separation, for which the machinery of (separable) semiclassical mechanics can be applied, including the search for bound states, resonances, scattering phases. In a sense, it is emerging the usefulness of associating de Broglie's waves to classical-like hyperspace trajectories connecting reactants and products of rearrangement processes.

The first available results (9) show that the journey in the hyperspace may indeed lead to the discovery of analogs of the interference effects and resonances we have been dealing with in previous Sections. On the other hand, there are perspectives that the experimental techniques currently available for the investigation of the dynamics of molecular rearrangement processes will soon reach a stage that these phenomena will be actually observed, and thus become a further tool for our understanding of interatomic and intermolecular forces.

## ACKNOWLEDGMENTS

The contribution by Franca Bellini Bizzarri to the preparation of the paper is gratefully acknowledged.

## REFERENCES

1. L.de Broglie's note, entitled "Quanta de lumière, diffraction et interférences", first appeared in 1923, Comptes rendus des séances de l'Académie des Sciences de Paris 177, p.548, is reprinted in the book de Broglie,L.: *Recherches d'un demi-siècle,* Editions Albin Michel, Paris 1976, pp.111-113.

2. R.B.Bernstein, Ed.: *Atom-Molecule Collision Theory, A Guide for the Experimentalist,* Plenum Press, New York 1979.

3. V.Aquilanti, in *Physics of Electronic and Atomic Collisions, Invited Papers of the* XIII *International Conference on the Physics of Electronic and Atomic Collisions,* Berlin, West Germany, 1983, to be published by North-Holland, Amsterdam.

4. G.Scoles: 1980, Ann.Rev.Phys.Chem. 31, pp. 81-96.

5. M.V.Berry and K.E.Mount: 1972, Rep.Prog.Phys. 35, pp. 315-397.

6. J.A.Wheeler, in: *Studies in Mathematical Physics, Essays in Honour of Valentine Bargmann,* Edited by E.H.Lieb, B.Simon, and A.S.Wightman, Princeton, N.J. 1976, pp. 351-422.

7. M.V.Berry: *Semiclassical Mechanics of Regular and Irregular Motion,* Lectures given at the Les Houches Summer School 1981, to be published by North-Holland.

8. M.S.Child, Ed.: *Semiclassical Methods in Molecular Scattering and Spectroscopy,* Reidel, Dordrecht, 1980.

9. V.Aquilanti *et al.:* 1982, Chem.Phys.Lett., in press.

ON THE SUPERPOSITION PRINCIPLE AND ITS PHYSICS CONTENT

Matts Roos

Department of High Energy Physics
University of Helsinki
SF-00170 Helsinki 17, Finland

Abstract: What is commonly denoted the superposition principle is shown to consist of three different physical assumptions: conservation of probability, completeness, and some phase conditions. The latter conditions form the physical assumptions of the superposition principle. These phase conditions are exemplified by the Kobayashi-Maskawa matrix. Some suggestions for testing the superposition principle are given.

The superposition principle is so well integrated into the basic framework of quantum mechanics that hardly anybody, not even authors of textbooks on quantum mechanics pause to ask the question: what exactly is the physical content of the superposition principle? Surely, it must have a physical content.

The first explicit formulation of the principle seems to have been given in 1930 by Dirac in the first edition of his book (1). In the later editions of the book, Dirac becomes less explicit in this respect.

The present article is based on a rather detailed study of the superposition principle which I published in a very little read journal (2). The essential mathematical formalism (3) also seems to be little known.

The notion of superposition of states enters the standard quantum mechanics as follows. Suppose an observation A on a physical system can have the possible outcomes $a_1, a_2, \ldots, a_n$. If the system is prepared to yield a unique result $a_i$ with certainty whenever the observation A is made, we say that the system is in a pure state $|a_i\rangle$.

The states $|a_i\rangle$ form a linear vector space in the sense that



*S. Diner et al. (eds.), The Wave-Particle Dualism, 291–295.*

- an arbitrary state $|\psi>$ is a vector

$$|\psi> = \alpha_i |a_i> \qquad (1)$$

in this space, (summation implied)
- the $\alpha_i$ are complex numbers,
- the real numbers $|\alpha_i|^2$ give the probability for the outcome $a_i$.

Departure from linear vector spaces generally imply some assumption of non-linearity. It is not our purpose here to test models of non-linear quantum mechanics which violate the superposition principle.

For simplicity, let us assume that the outcomes $a_i$ form a discrete spectrum, and that the sum in Eq. (1) contains a finite number of terms, n.

Suppose now that another possible observation, B, is made on the same system, with n possible outcomes $b_1, b_2, \ldots, b_n$. Then, in the linear vector space of the pure states $|b_i>$, the arbitrary state $|\psi>$ in Eq. (1) is the linear combination

$$|\psi> = \beta_i |b_i> \qquad i = 1, \ldots, n. \qquad (2)$$

It follows from the above that

$$\alpha_i \alpha_i^* = \beta_i \beta_i^* = 1. \qquad (3)$$

Moreover, since the physical system was assumed to be the same, regardless of whether the observation was A or B, it follows that

$$\alpha_i = U_{ij} \beta_j \qquad (4)$$

where $U = (U_{ij})$ is a unitary matrix.

The physical meaning of Eq. (4) is obviously that the probability to obtain the outcome $b_j$ in an observation of B on the system in a state $|a_i>$ is

$$p(a_i, b_j) = |U_{ij}|^2 \qquad (5)$$

How much physics has gone into this?

(i) Conservation of probability is implied by the n conditions

$$\sum_{i=1}^{n} p(a_i, b_j) = 1 \qquad j = 1, \ldots, n. \qquad (6)$$

(ii) Completeness of the set of states is implied by the further n - 1 independent conditions

$$\sum_{j=1}^{n} p(a_i, b_j) = 1 \qquad (7)$$

(iii) The superposition principle may or may not imply some

further restrictions on the remaining set of $(n-1)^2$ independent real probabilities $p(a_i, b_j)$. This we shall see below.

Recall that a unitary matrix depends on $n^2$ real independent parameters. If the moduli of its elements are fixed, one still has the freedom to choose $2n-1$ phases arbitrarily. Thus the unitary matrices U and U' have the same moduli, if

$$U' = D_\varphi U D_\psi, \tag{8}$$

where $D_\varphi$ is the diagonal matrix with elements $\delta_{jk} \exp(i\varphi_k)$, and $D_\psi$ is the diagonal matrix with elements $\delta_{jk} \exp(i\psi_k)$. The $\varphi_k$, $k = 1, \ldots, n$ and $\psi_k$, $k = 2, \ldots, n$ are $2n-1$ arbitrary real phases, and $\psi_1 = 0$. Thus, subtracting the number of arbitrary phases, there are at most $n^2 - (2n-1) = (n-1)^2$ independent moduli $|U_{ij}|$.

We have learned that the number of available independent probabilities $p(a_i, b_j)$ is the same as the number of independent moduli $|U_{ij}|$ in a unitary matrix.

Consider now an experimental setup measuring all the quantities $p(a_i, b_j)$. A Gedanken-experiment may consist of two consecutive Stern-Gerlach magnets, rotated with respect to each other. Given a molecular beam of spin j, the first magnet, A, splits the beam up into $n = 2j+1$ components, corresponding to the emergence of the pure states $|a_i\rangle$, $i = 1, \ldots, n$ behind the magnet. Subsequently the second magnet, B, splits each of these beams into n further components, corresponding to the emergence of n beams of pure states $|b_1\rangle$, n beams of $|b_2\rangle$, etc. The intensity of these latter beams will be proportional to $p(a_i, b_j)$.

Can one always form a unitary matrix with elements of moduli $|U_{ij}| = \sqrt{p(a_i, b_j)}$, given the experimental data set $p(a_i, b_j)$? If yes, the superposition principle is always valid; i.e. it has no physics content beyond Eqs. (6) and (7). If no, there are conditions which can serve as tests of the superposition principle.

Let's decompose U as follows (4,3):

$$U = D_\varphi \prod_{k=2}^{n} \prod_{j=1}^{k-1} S_{jk}(\theta_{jk}, \sigma_{jk}). \tag{9}$$

Here $D_\varphi$ is the diagonal matrix in Eq. (8) and the $S_{jk}$ are $n \times n$ unitary unimodular matrices, each a function of two parameters only: $\theta_{jk}$ and $\sigma_{jk}$. Only four elements in each matrix $S_{jk}$ are different from $\delta_{jk}$: the jj and kk elements have the value $\cos\theta_{jk}$, the jk element has the value $-\sin\theta_{jk} \exp(-i\sigma_{jk})$, and the kj element is the negative of the jk element.
In this way the matrix U is expressed very conveniently in terms of the n phases $\varphi_i$, the $\frac{1}{2}n(n-1)$ parameters $\theta_{jk}$, and the $\frac{1}{2}n(n-1)$ parameters $\sigma_{jk}$.

The reason why this parametrization is convenient is the following. The moduli of U will depend on the $\theta$ and on a well-defined subset of $\frac{1}{2}(n-1)(n-2)$ parameters $\gamma$ which are linear combinations of the $\sigma$. The remaining $n-1$ independent parameters $\sigma$ correspond to the $\psi$ in Eq. (8). Thus we can determine the $(n-1)^2$ parameters $\theta$ and $\gamma$ from the independent experimental quantities $p(a_i, b_j)$, and ignore the $2n-1$ phases $\varphi$ and $\psi$.

However, one can show (3) that there are $\frac{1}{2}(n-1)(n-2)$ inequalities of the type

$$|\cos\gamma| \leq 1 \tag{10}$$

which also must be satisfied by the data! These represent the physical content of the superposition principle.

Consider the first few cases with n small. When $n = 2$ there are no conditions of the kind of Eq. (10). Thus the superposition principle, as defined here, lacks physical content.

For $n = 3$ there is one inequality to be respected. In ref. (2) I have given this condition for the case of the double Stern-Gerlach experiment. Let me use here a more modern example: the $3 \times 3$ Kobayashi-Maskawa matrix (5) for the mixing of three quark flavors. Let's fill in only the moduli of the matrix:

$$\begin{pmatrix} A & B & \cdot \\ C & D & \cdot \\ \cdot & \cdot & \cdot \end{pmatrix} \tag{11}$$

Here the dots denote dependent elements which can be obtained from Eqs. (6) and (7). In terms of the usual parametrization of the Kobayashi-Maskawa matrix, one has (with $C_i$ for $\cos\theta_i$ and $S_i$ for $\sin\theta_i$):

$$A = C_1 ; \qquad B = S_1 C_3 ; \qquad C = -S_1 C_2$$
$$D^2 = (C_1 C_2 C_3 + S_2 C_3 \cos\delta)^2 + (S_2 C_3 \sin\delta)^2 \tag{12}$$

The quantities A, B, C, D can in principle be determined independently. The superposition principle imposes the condition

$$-1 \leq \cos\delta \leq 1, \tag{13}$$

where

$$\cos\delta = \frac{D^2(1-A^2)^2 - (ABC)^2 - B^2(1-A^2-C^2)}{-2AB^2C\sqrt{1-A^2-C^2}} . \tag{14}$$

So far the experimental information on A, B, C, D does not allow one to test the superposition principle. Rather, one uses A, B, C and the superposition principle to set limits on D.

The case $n = 4$ has been solved in ref. (3). It is too laborious to report here.

In conclusion, we note that what usually is referred to as

the superposition principle really consists of three physical assumptions: conservation of probability, completeness of the set of states, and some phase conditions in unitary matrices. These phase conditions have not to my knowledge ever been tested. A way to test them would be to measure the intensities of the $(2j+1)^2$ beams of spin j molecules passing through a double Stern-Gerlach magnet. Assuming conservation of probability and completeness, only $(2j)^2$ of the intensities are independent. Of the remaining $4j+1$ intensities, one would serve as normalization and 4j as additional constraints.

The superposition principle would be best tested in a beam of large j, since the number of phase conditions grows as $j(2j-1)$, and the number of extra constraints available grows as 4j.

In a test of the superposition principle there is no obvious need to go to high energies. Since quantum mechanics works so well, we have no reason to anticipate a violation of the superposition principle in any particular area of physics. However, we note that CP-violation 18 years after its discovery has not been satisfactorily explained, and that one obvious but unattractive explanation may still be a non-linear quantum mechanics (2,6).

## REFERENCES

(1) Dirac, P.A.M.: *The Principles of Quantum Mechanics*. 1930, Clarendon Press, Oxford.

(2) Roos, M.: 1966, *Comm. Phys.-Math. Soc. Sci. Fennicae* 33, nr. 1, pp. 1-17.

(3) Roos, M.: 1964, *J. Math. Phys.* 5, pp. 1609-1611; 1965, *J. Math. Phys.* 6, p. 1354.

(4) Murnaghan, F.D.: *The Unitary and Rotation Groups*. 1962, Spartan Books, Washington, D.C.

(5) Kobayashi, M., and Maskawa, K.: 1973, *Progr. Theor. Phys.* 49, pp. 652-657.

(6) Laurent, B., and Roos, M.: 1964, *Phys. Letters* 13, pp. 269-270; 1965, *Phys. Letters* 15, p. 104; 1965, *Nuovo Cimento* 40A, pp. 788-801; Roos, M.: 1966, *Phys. Letters* 20, pp. 59-62.

# THE PROBABILISTIC ROOTS OF THE QUANTUM MECHANICAL PARADOXES

Luigi ACCARDI

Dipartimento di Matematica
Università di Roma II
Torvergata, Roma

0. The goal of any mathematical investigation on the foundations of a physical theory is to clarify to what extent the mathematical formalism of that theory is uniquely determined by some clearly and explicitely stated physical assumptions. The achievement of that goal is particularly relevant in the case of the quantum theory where the novelty of the formalism, its being far away from any immediate intuition, the substantial failure met, for many years, by any attempt to deduce the quantum formalism from plausible physical assumptions, intersected with the never solved problems concerning the interpretation of the theory. That with quantum theory a new kind of probability theory was involved, was clear since the very beginnings of quantum mechanics (cf. [28]), even if it was not so clear which of the axioms of classical probability had to be substituted, which physically meaningful statement had to replace it, how and if a physically meaningful statement could justify the apparently strange quantum mechanical formalism. The lack of clear answers to these questions had a tremendous impact on the process of interpretation and misinterpretation of quantum theory. The attempts to answer these questions motivated the development of a new branch of probability theory -quantum probability- and led to definite mathematical answers to these questions. In the present paper we want to discuss how these mathematical results allow to solve in a rather natural way some old problems concerning the interpretation of quantum theory and its mathematical foundations.



*S. Diner et al. (eds.), The Wave-Particle Dualism, 297–330.*

## 1. Introduction

Assume that some astronomers of a civilization, which considers euclidean geometry not as one of the possible mathematical models of space but as the obvious expression of the properties of physical space and the only possible one, measure the angles among three stars which they are sure to be coplanar and find that the sum of the inner angles of the stellar triangle differs from $\pi$.

Calling "Euclid's equality" the theorem according to which the sum of the inner angles of a triangle is $\pi$, we would expect that in order to conciliate the belief that the euclidean theorems are self-evident truths with the fact that the experimental data violate "Euclid's equality", these astronomers might be led to strange statements. May be some of them would suggest that stars have only a virtual existence when we don't look at them and that only when we use our telescope to perturb this indistinct virtual being they acquire a definite position; some other might advocate the use of a "non-local astronomy" according to which we cannot describe three stars within a single classical geometrical model, but we need a different geometrical model for each pair of stars; and so on... .

Some people will probably look at the anecdote above as to a ridiculously absurd one; yet substituting physicists for astronomers, electrons and similar objects for stars, and probability for geometry, the fanciful anecdote becomes contemporary history: history of the interpretation of quantum theory.

The present paper, as well as several previous ones [ 1 ], [ 2 ], .............., [ 12 ] is devoted to the exposition of a new approach to that problem which can be synthetized in the following three statements:

i) Bayes' axiom (cf. § 3) plays for probability the role played by the parallel axiom for geometry.

ii) All the paradoxes of quantum theory arise from the implicit or explicit application of Bayes' axiom (or of the theorem of composite probabilities, which is an equivalent formulation of it, cf. Lemma (3.1)) to the statistical data of quantum theory. This application being unjustified both physically and mathematically.

iii) The whole mathematical formalism of quantum theory can be deduced, within the framework of Schwinger's algebra of measurements, from a single axiom, which is a model independent formulation of a consequence of Heisenberg indeterminacy principle.

Of course, the main conceptual points of this thesis have a long history in quantum theory. The fact that Bayes' definition of conditional probability (cf. formula (3.1) in the following) might be meaningless in a quantum mechanical context goes back at least to a 1937 paper of Strauss |44| and has been clearly stated by Suppes [45]. We have shown that this formula is mathematically incompatible with the experimental data of quantum theory (cf. [6], [8], and the following § 2). The proof given in [6] is a generalization of Wigner's proof [51] of Bell's inequality [14] ; while the more complete results in [8] use different techniques. The remark that Wigner's proof of Bell's inequality concerns in fact transition probabilities (and not, as in Bell's original version, correlations) is due to Bub [16]. The possibility of realizing a given set of transition probabilities within the complex Hilbert space framework has been studied by M. Roos (cf. [37], [38] and these proceedings) in view of a possible experimental verification of the superposition principle (cf. also [9] and § 5). Roos' inequality [38] goes in the converse direction as Bell's one: the former gives conditions for the realization of a given set of transition probabilities within the Hilbert space model of quantum theory; the latter do the same for the usual probabilistic model (cf. § [8] for a precise discussion). The first necessary and sufficient condition for the existence of a kolmogorovian model compatible with a given set of joint probabilities of pairs of two-valued observables is due to Corleo, Gutkowski, Masotto and Valdes [17], [25] ; their condition is equivalent to the one formulated in [8] in terms of transition probabilitis. Recently Suppes and Zanotti ([47][48], and these proceedings) have given an independent proof of this result using correlations as statistical data, like in the original paper by Bell. The use of transition probabilities instead of correlations has the advantage to be applicable to individual systems (and not only composite ones) and this helps in clearing the ground from confusions concerning the relevance of "locality arguments" for such questions. The suggestion to interpret Bell's inequalities as a mathematical and experimental evidence that there are in nature some statistical data which cannot be fit into the framework of the classical probabilistic model is also contained in an early paper of G. Lochak [31] and in a recently appeared paper of I. Pitowsky [33] (I am grateful to H. van den Berg for pointing out to me the latter reference).

The idea that the basic object of quantum theory are the transition (or conditional) probabilities goes back to N. Bohr (a good exposition of Bohr's thought is in [39]); while A. Landé [30] made the first attempt to deduce the Hilbert space structure from an analysis of the experimentally given transition probabilities. The remark that "... The circumstance that therefore not the probabilities themselves but their amplitudes follow the usual rules of probability calculus..." goes back to Jordan [28] who, immediately after, adds that this circumstance "...can, in a convenient way, be denoted as <u>interference</u> of probability...". Looking at the just mentioned paper by Jordan one can easily recognize that the "...usual rules of probability calculus..." he refers to are in fact the theorem of composite probabilities, which must be replaced by the principle of composite amplitudes.

This analysis is pushed forward by Feynman who further recognizes [20], [21] that the principle of composite amplitudes alone is sufficient to deduce the whole quantum mechanical formalism. Both the above mentioned authors implicitely assume the validity of Bayes' postulate and this leads to a recurrent confusion in their papers between the theorem of composite probability and the principle of additivity of probability on disjoint events: as argued in [7] (cf. also §§ 2, 3 in the following) this is not a technical point because the measurement apparatus needed to measure the two kind of probabilities are different. The fact that the act of measurement without taking into account the result re-establishes the validity of the theorem of composite probability (cf. [7] for more details) was correctly stated in probabilistic terms by W. Heisenberg (cf. [26], pg. 61) without even mentioning such notorious terminology such as "collapse", "conscience", ... . Finally we owe to J.Schwinger [ 41 ] the introduction of the useful formalism of the "algebra of measurements" in terms of which the deep connection between the algebra of quantum theory and its statistics is best understood: in fact the main result in the classification of those non-kolmogarovian models which we call "of Heisenberg type" (cf. § 9 ) is that the experimentally measured transition probabilities determine (in an essentially unique way) the usual quantum-mechanical algebra.

The monumental work of J. von Neumann gave rise to the theory of von Neumann algebras and was originated as a generalization of the quantum probabilistic formalism. Since von Neumann the developement of the theory of operator algebras has always been deeply connected with the quantum mechanical formalism. We will not mention here this development, as well as the several other approaches to the foundations of quantum theory (cf. G.W. Mackey [ 32 ], J.M. Jauch [ 27 ], C. Piron [ 34] I.E. Segal [ 42 ] , ...), because here we will not be concerned with the inner developement or generalization of the quantum probabilistic formalism but, rather, with those results, both experimental and mathematical, which prove without any possible ambiguity the necessity of introducing non-kolmogarovian models in the statistical description of nature, and the relevance of this point of view for the solution of the interpretative problems of quantum theory. Our emphasis in the present paper will be on interpretation; for the proofs of the mathematical results and for a more thorough discussion of some conceptual analyses which we shall only sketch we will refer to published papers (with a few exceptions where we refer to preprints).

In conclusion let us emphasize that the main differences between the point of view of non-kolmogorovian probability and the known proposals of abandoning the classical aristotelic logic (when we don't look at objects they do not obey the usual logic) or the usual concept of reality (when we don't look at observables they don't take their values) are the folllwing:

i) The necessity of abandoning the kolmogorovian model of probability can be proved only in terms of experimentally measurable quantities (cf. § 2); while the necessity of giving up aristotelean logic or the usual concept of realism can be motivated only postulating the validity of Bayes'axiom even in situations in which it cannot be experimentally checked.

ii) If we accept the necessity of using non-kolmogorovian models in the description of quantum theory, then we don't need to abandon neither aristotelean logic nor the classical concept of realism. While if we abandon either of these concepts, nevertheless we must abandon the kolmogarovian model because the fact that the experimental data don't fit into this model is independent on any interpretation.

iii) The traditional interpretations of quantum theory compel us to make definite statements on events which are in principle out of any experimental control (observables not looked at don't assume values or do not obey usual logic). Quantum probability doesn't need to get involved in the problem of how things behave when nobody is looking at them.

## 2. The oldest Bell's inequality: the two-slit experiment

As R. Feynman says ( [22] , pg. 146) the two-slit experiment "... is formulated so to include all the misteries of quantum mechanics...". Confirming this statement, and in order to give an intuitive support to the more general considerations which will follow, we will illustrate the main contention of quantum probability theory starting from an analysis of the 2-slit experiment.

In fact, for lack of space, we will analyze only "one-half" of the two slit experiment, i.e. the part concerning the non validity of the theorem of composite probabilities when no attempt is made to decide which of a set of alternatives is choosen.

An exhaustive discussion of the "other half" of the experiment, i.e. the fact that the formal validity of that theorem is re-established by the physical possibility of distinguishing among the alternatives, is contained in the paper [7] .

The scheme of the experiment is well known and we recall it just to fix the notations: a source S emits identically prepared particles (by this we mean that some observable of the particles - energy, frequency, ... - has a definite value) towards a screen $\Sigma_1$ with two slits, denoted 1 and 2. The particles which pass the screen $\Sigma_1$ are collected on another screen $\Sigma_2$ parallel to $\Sigma_1$ and, denoting X a small portion of the screen $\Sigma_2$ one counts the relative frequencies (probabilities) of the particles hitting the region X - i.e. the quotient $N(X)/N$ of the number N(X) of particles hitting the region X over the number N of particles hitting the screen $\Sigma_2$. This counting is carried out in three different experimental situations:

a) with both slits 1 and 2 open

b) with slit 1 closed and slit 2 open

c) with slit 2 closed and slit 1 open.

Let us denote P(X), P(X|2), P(X|1) the probability of arrival in X in case (a), (b), (c) respectively. P(X|j) (j=1,2) is the conditional probability that a particle arrives in X given that we know it passed through the slit j. The event "the particle reached the region X of the screen $\Sigma_2$" will be simply denoted X. The theorem of composite probabilities in this case simply amounts to the statement:

$$P(X) = P(1) \cdot P(X|1) + P(2) \cdot P(X|2) \tag{2.1}$$

where P(j) (j = 1,2) denotes the a priori probability that a particle passes through hole j; often one assumes that P(1) = = P(2) = 1/2, but nothing in the following analysis will depend on these values or even on our knowledge of them therefore we leave them unspecified. It is well known however that, according to the experimental data

$$P(X) \neq P(1) \cdot P(X|1) + P(2) \cdot P(X|2) \tag{2.2}$$

Since the theoretical deduction of equality (2.1) involves only elementary steps, anybody can realize that the experimental validity of inequality (2.2) mines the foundations of some fundamental principle, as it is always the case when something previously considered very elementary, turns out to be questionable or false. Let us therefore make explicit the elementary steps which lead from the experimental data (i.e. P(X), P(X|1), P(X|2)) to the theoretical equality (2.1) because we know from the experiments that at least one of these steps is founded on an empirically false prejudice. We will denote 1, 2 - the event that the particle passes through hole 1, 2 and with X the event that the particle reaches the screen $\Sigma_2$ at X.

**Step I**) - If a particle reaches the screen $\Sigma_2$ then it passed at least through one of the two slits. We will express this fact with the identity:

$$1 \vee 2 = I \quad \text{(event always realized)} \tag{2.3}$$

A particle can pass through one and only one of the slits at a time:

$$1 \wedge 2 = \emptyset \quad \text{(impossible event)} \tag{2.4}$$

**Step II**) - The rules of classical aristotelean logic allow us to deduce from (2.3) and (2.4) the identities:

$$X = (X \wedge 1) \vee (X \wedge 2) \tag{2.5}$$

$$\emptyset = (X \wedge 1) \wedge (X \wedge 2) \tag{2.6}$$

which express the fact that a particle can arrive in X only if it has passed through one and only one of the slits 1 or 2.

**Step III**) - The postulate of additivity of the probabilities of disjoint events and (2.5), (2.6) imply the identity

$$P(X) = P(X \wedge 1) + P(X \wedge 2) \tag{2.7}$$

**Step IV**) - The joint probabilities $P(X \wedge j)$ $(j = 1,2)$ are related to the conditional probabilities $P(X|j)$ through Bayes' identity

$$P(X|j) = \frac{P(X \wedge j)}{P(j)} \; ; \qquad j = 1,2 \tag{2.8}$$

**Step V**) - The identities (2.7) and (2.8) imply the theorem of composite probabilities, i.e.:

$$P(X) = P(1) \cdot P(X|1) + P(2) \cdot P(X|2) \tag{2.9}$$

The conclusion of Step V) follows, as a trivial mathematical consequence, from the preceeding four steps. Since this conclusion is experimentally wrong, at least one of these steps must be wrong. Which one? All the different interpretations of quantum theory can be classified according to which of the above four steps they consider to be the first wrong one.

The Copenhagen interpretation, the most widespread at present, confutes the first step. This point of view is clearly expressed by the words of Feynman ( [20] , pag. 536) "... we concluded on logical bases that, since

$$P(X) \neq P(1) . P(X|1) + P(2) \cdot P(X|2) \tag{2.10}$$

[here we adapt Feynman's notations to ours] it is not true that the electron passes either through hole 1 or through hole 2...". And, shortly after ([20], pg. 538), Feynman himself explains that in order to be compatible with the experimental data, this sentence should be understood in the sense that "...when no attempt is made to determine through which hole the electron passes one cannot say that it must pass through one or the other hole...". Thus the Copenhagen school from one side rightly asserts that physics should not make statements which in principle cannot be subjected to any experimental verification; from the other, to avoid contradiction with the inequality (2.10), is led to state that when no attempt is made to determine through which hole the electron passes then at least one of the events (2.3) or (2.4) must not happen - and this is a typical such statement.

Similarly the quantum logic point of view [35], even admitting Step I), confutes Step II) which uses the distributivity axiom of the boolean propositional calculus. The arguments of Feynman can be applied also to this case and show (cf. [11] for details) that the non validity of the distributive axiom should concern particles only when nobody looks at them: the experiments made to check (2.5) or (2.6) confirm their validity.

The non validity of Step III), i.e. of the additivity postulate of probability, has been considered by some authors; for example P. Jordan (cf. Postulate (C.) of [28]) and Feynman ([23], pg. 1.10) who propose to substitute for it the additivity postulate for amplitudes. However these authors tend to look at the non validity of (2.7) as a consequence of the non validity of Step I) (this is particularly explicit in Feynman [20], pg. 536). Moreover, the additivity postulate for probabilities, whenever experimentally verified holds and when no experimental verification is possible, as it is the case for the right hand side of (2.7), its validity can be mantained at a purely existencial level in the following sense: accepting Steps I) and II) and the frequency interpretation of probability, then all particles arriving in X can be subdivided into two classes, those coming from hole 1 and those coming from hole 2, the cardinalities $N(X \wedge 1)$ and $N(X \wedge 2)$ of these classes might not be experimentally measurable but, whatever they are, the relative frequencies $P(X \wedge 1)$, $P(X \wedge 2)$ must satisfy (2.7).

The basic remark of the quantum probability point of view

is that (2.7) is equivalent to (2.9) only if Bayes' identity (2.8) is postulated. Let us make more explicit what does this mean: first of all remark that, as Feynman shows, the right hand side of (2.8) is not measurable without altering the conditions of the experiment. Therefore what we are really doing when postulating the validity of (2.8) is to postulate the existence of four strictly positive numbers $P(1)$, $P(2)$, $P(X \wedge 1)$, $P(X \wedge 2)$ subjected to the relations:

$$P(1) + P(2) = 1 \tag{2.11}$$

$$P(X \wedge 1) + P(X \wedge 2) = P(X) \tag{2.12}$$

$$\frac{P(X \wedge 1)}{P(1)} = P(X|1) \tag{2.13}$$

$$\frac{P(X \wedge 2)}{P(2)} = P(X|2) \tag{2.14}$$

But this is a system of four equations in four unknowns for which the existence of a solution is not a question of postulates but something which can be uniquely decided once given the "coefficients" $P(X)$, $P(X|1)$, $P(X|2)$. It is a simple computation to show that the system above admits a strictly positive solution if and only if

$$0 < \frac{P(X) - P(X|2)}{P(X|1) - P(X|2)} < 1 \tag{2.15}$$

therefore to postulate the validity of Bayes' identity (2.8) is equivalent, in our case, to postulate the validity of the following, far from obvious, statement: the laws of nature must be such that, however choosen three events X,1,2, the probabilities $P(X)$, $P(X|1)$, $P(X|2)$ satisfy the inequalities (2.15).

But, since the probabilities $P(X)$, $P(X|1)$, $P(X|2)$ are experimentally measurable, whether or not the laws of nature satisfy (2.15) is not a question of postulates, but of experiments. Many experiments have been done (cf. in particular the papers by Gozzini, Rauch, Zeilinger, in these proceedings) and they have indisputably proved that there are natural phenomena which violate (2.15) or equivalently, that there are in nature sets of statistical data to which Bayes' formula for the conditional probabilities, and therefore the whole classical model of probability theory, cannot be applied for mathematical rather than interpretative reasons. Since the inequalities (2.15) are interpretation free, neither the Copenhagen nor the quantum logic interpretation can help preventing the use of a non-kolmogorovian probabilistic model; on the contrary once accepted (as we must in any case) the empirical groundlessness of the Bayes' postulate, one is not committed to any statement on the behaviour of things when nobody looks at them; in particular one can quietly believe, if he likes, that an observable assumes one and only one of its values even if nobody looks at it. Summing up, since the inequalities (2.15) are equivalent to the inequality (2.10), we can conclude that the whole debate on the interpretation of quantum mechanics began, many years ago, with a discussion on a Bell's type inequality.

The discussion above concerns a particular, even if important, case. We must now show that the non validity of Bayes' axiom is really universal for quantum theory and that all the interpretative problems of this theory arise from an injustified application of this axiom. This will conclude the critical part of our programme. The constructive part will consist in the explicit formulation of the postulates of the non-kolmogorovian probabilistic models and in the investigation of the structure and the applications of these models.

## 3. The hidden postulate of classical probability

Let A, B denote two events; P(A|B) will denote the conditional probability of the event A given that one knows that the event B has happened. If A, B denote states of a system one also speaks of transition probability. Physically "conditioning with the event B" means that one has prepared an ensemble of systems of the same type (say neutrons) so that on each of the systems the event B is verified (e.g. all the systems have spin +1 in the b-direction in appropriate units). The joint probability of the events A and B, i.e. the probability that both A and B happen, is denoted $P(A \cap B)$. In the following we will call conditional probability with condition B, any probability measure $P(\cdot|B)$ satisfying the identity: $P(B|B) = 1$.

Bayes' Axiom. For any two events A and B, the conditional probability $P(A|B)$ and the joint probability $P(A \cap B)$ are related by the identity

$$P(A|B) = \frac{P(A \cap B)}{P(B)} \tag{3.1}$$

In most probability textbooks the identity (3.1) is not formulated as an axiom but as a definition (even an informal definition, cf. [29]). From the physical point of view, whenever the two sides of the identity (3.1) can be experimentally measured with different physical operations, the validity of the identity (3.1) cannot be taken for granted. If, moreover, the right hand side cannot be experimentally measured but the left hand side can, then the attempt to define the former in terms of the latter is equivalent to postulate the existence of two numbers in $[0,1]$ satisfying (3.1). For a single pair A, B one can always find such numbers; but for three or more events some constraints arise and existence is no longer granted. More specifically: if $A_1, \ldots, A_n$ are events then one can trivially find $\binom{n}{2}$ probability measures $P_{jk}$ $(j,k = 1,\ldots,n)$ satisfying

$$P(A_j|A_k) = P_{jk}(A_j \cap A_k)/P_{jk}(A_k) \tag{3.2}$$

however a single probability measure $P(\cdot)$ satisfying

$$P(A_j|A_k) = P(A_j \cap A_k)/P(A_k) \tag{3.3}$$

for each pair of indices $j,k = 1,\ldots,n$ might not exist. By a "non-local" classical probabilistic model for the set of transition probabilities $\{P(A_j|A_k): j,k = 1,\ldots,n\}$ we mean a family of $\binom{n}{2}$ - or a few less - probability measures satisfying (3.2) for each j and k. The term "locality" in the above has historical reasons and has nothing to do with space-time. The use of a different probability measure for each pair $(A_j, A_k)$ is justified by saying that the probability measure should take into account the experimental setup needed to verify the events $A_j$ and $A_k$. We won't discuss in the following the advantages and disadvantages of non-local models but rather the consequences of the erroneous implicit assumption that for any set of conditional probabilities $\{P(A_j|A_k): j, k = 1,\ldots, n\}$ there is a single probability measure $P(\cdot)$ satisfying (3.3) for any pair j, k.

First of all remark that if $B_1, \ldots, B_n$ are disjoint events and $A \subset \bigsqcup_{j=1}^{n} B_j$ then the validity of Bayes' identity

$$P(A|B_j) = P(A \cap B_j)/P(B_j) \tag{3.4}$$

for each $j = 1,\dots,n$ implies the so-called theorem of composite probabilities:

$$P(A) = \sum_{j=1}^{n} P(B_j)\, P(A|B_j) \tag{3.5}$$

The converse implication also holds, more precisely one has:

Lemma 3.1 - Let $(B_j)$ be a family of disjoint events and let for each $B_j$ be given a conditional probability $P(\cdot|B_j)$ (in the sense specified at the beginning of this §). The following statements are equivalent:

i) for each $A \subseteq \bigsqcup_{j=1}^{n} B_j$, Bayes' identity (3.4) holds.

ii) for each $A \subseteq \bigsqcup_{j=1}^{n} B_j$, the theorem of composite probabilities (3.5) holds.

Proof. The implication (i) ⇒ (ii) is trivial. To prove the converse, remark that since $P(B|B) = 1$, then

$$P(A|B_j) = P(A \cap B_j|B_j) \tag{3.6}$$

hence it will be sufficient to prove (3.4) in the assumption that $A \subseteq B_k$ for some k. In this case, if (3.5) holds, using (3.6) one finds: $P(A) = P(B_k)\, P(A|B_k)$ which is the thesis.

Remark - Usually one considers the case in which $\bigsqcup_{j=1}^{n} B_j = I$ (i.e. the event always realized). In this case the inclusion $A \subseteq I$ is true for arbitrary A.

Thus the relation between Bayes' axiom and the theorem of composite probabilities is the same as the one between the parallel axiom and the theorem that the sum of the inner angles of a triangle is $\pi$: they are equivalent. This establishes in general what we have shown in the preceeding § in the case of the two-slit experiment. In particular this provides a tool to check the experimental validity of Bayes' axiom even in cases in which the right hand side of (3.4) is not experimentally measurable.

## 4. Zeno's paradox

Let us denote A a complete set of quantum mechanical observables, and $[A_t = a]$ the event that the value a is the result of a measurement, actually performed on a system, of the observable A at time t. In quantum mechanics, the event $[A_t = a]$ uniquely determines a pure state; hence the transition probabilities $P(B_u = b|A_t = a)$ where $t < u$, B is any observable and b any value of B, do not depend on the past history of the system. The mathematical formulation of this condition of lack of memory is called, in classical probability theory, a Markov property and is expressed by the validity of the equality:

$$P(B_u = b|[A_t = a] \cap [A^{(n)}_{t_n} = a_n] \cap \ldots \cap [A^{(o)}_{t_o} = a_o]) =$$

$$= P(B_u = b \mid A_t = a) \tag{4.1}$$

for any choice of the observables A, B, $A^{(o)}, \ldots, A^{(n)}$; of their values a, b, $a_o, \ldots, a_n$; and of the times $t_o < t_1 < \ldots < t_n < t < u$ (for a more detailed discussion of this point and its connection with the quantum theory of measurement, cf. [7]).

In particular, choosing in (4.1) all the observables equal to A and all the values equal to a, one finds:

$$P(A_t = a |[A_{t_n} = a] \cap [A_{t_{n-1}} = a] \cap \ldots \cap [A_{t_o} = a]) =$$

$$= P(A_t = a| A_{t_n} = a) \tag{4.2}$$

where $t_o < t_1 < \ldots < t_n < t$ are arbitrary instants of time. Now let us postulate the validity of Bayes' axiom for the left hand side of (4.2), that is:

$$P([A_t = a] \cap [A_{t_n} = a] \cap \ldots \cap [A_{t_o} = a]) =$$
$$= P(A_t = a|A_{t_n} = a) \cdot P([A_{t_n} = a] \cap \ldots \cap [A_{t_o} = a]) \tag{4.3}$$

From (4.3), by iteration, we obtain:

$$P([A_t = a] \cap [A_{t_n} = a] \cap \ldots \cap [A_{t_o} = a]) = \tag{4.4}$$

$$= P(A_t = a|A_{t_n} = a) \cdot P(A_{t_n} = a|A_{t_{n-1}} = a) \cdot \ldots \cdot P(A_{t_1} = a|A_{t_o} = a)$$
$$P(A_{t_o} = a) \qquad (4.4)$$

Conversely the validity of (4.4) for each n implies (4.3).

Assuming the translation invariance of the transition probabilities $P(A_t = a|A_s = a')$ (surely true in quantum mechanics, in absence of exterior forces) and choosing $t_o = 0$; $t_{k+1}-t_k = t/(n+1)$, (4.4) becomes

$$P([A_t = a] \cap [A_{t-\frac{t}{n+1}} = a] \cap \ldots \cap [A_o = a]) =$$
$$= P(A_{\frac{t}{n+1}} = a \mid A_o = a)^{n+1} \cdot P(A_o = a) \qquad (4.5)$$

It is a known fact that, substituting in the right hand side of (4.5) the quantum mechanical expression for the transition probability, i.e. (in obvious natations):

$$P(A_s = a|A_o = a) = |<\psi_a, e^{-isH} \psi_a>|^2 \qquad (4.6)$$

one reaches the paradoxical conclusion that

$$\lim_{n\to\infty} P([A_t = a] \cap [A_{t(\frac{n}{n+1})} = a] \cap \ldots \cap [A_{\frac{t}{n+1}} = a]|A_o = a) = 1 \qquad (4.7)$$

Here again the formula which gives rise to the paradox follows (is even equivalent to) Bayes' axiom (4.3), and it is a simple exercise to verify that the quantum transition probabilities cannot fullfil that identity. To postulate the validity of (4.3) is erroneous also for conceptual reasons: the joint probabilities in the two sides of (4.3) are evaluated under different experimental conditions. The left hand side under the condition of making n+2 measurements at times $t_o, \ldots, t_n, t$; the right hand side, under the condition of making n+1 measurements at time $t_o, \ldots, t_n$. Denoting $P_{t_o,\ldots,t_n,t}$ and $P_{t_o,\ldots,t_n}$ the corresponding probabilities, one knows that the family $(P_{t_o,\ldots,t_n})$ is not an agreeing one: for example, restricting $P_{t_o,\ldots,t_n,t}$

to $(t_o,\dots,t_n)$ means that you perform the measurement at time t and don't read the result, and since this is different from not doing the measurement, the resulting measure is not $P_{t_o,\dots,t_n}$. This means that even if we want to postulate Bayes' identity (4.3), that identityshouldconnect $P_{t_o,\dots,t_n,t}$ with its restriction to $(t_o,\dots,t_n)$ and not, as implicitely assumed in the deduction of (4.4) from (4.3), with $P_{t_o,\dots,t_n}$.

Moreover, even postulating the validity of the identity (4.1) for this restriction, there will be no reason, nor physical neither mathematical, to suppose that the transition probabilities defined by it are given by the usual quantum mechanical formula (4.6) (calculations in simple examples show that in general they are quite different).

Finally it should be remarked that the mathematical model for the quantum theory of measurement, deduced from quantum Markov theory (cf. |3|), does not lead to Zeno's paradox.

## 5. Physical Superpositions

The general situation, illustrated by the two-slit experiment, is best explained again in a paper by Feynmann |21|. Let A, B, C, ... denote some physical observables and a, b, c, ... their values. The identity:

$$P(A=a|C=c) = \sum_b P(A=a|B=b)\cdot P(B=b|C=c) \qquad (5.1)$$

(the sum being extended to all the values of B) follows from, and in fact is equivalent to, the theorem of composite probabilities plus "Dirac's jump assumption", which is nothing but the Markov Property as formulated in the preceeding § (cf. 4.1.), i.e.:

$$P(A=a|[B=b]\wedge[C=c]) = P(A=a|B=b) \qquad (5.2)$$

where, in the left-hand side it is understood that the measurement of C has been performed before that of B. All the transition probabilities in (5.1) can be checked experimentally, and the results disprove the validity of (5.1). It is known that, in order to find agreement with the experiments, one must introduce a "conditional probability amplitude", i.e. a complex number $\psi(A=a|B=b)$ related to the conditional probability by the relation:

$$P(A=a|B=b) = |\psi(A=a|B=b)|^2 \tag{5.3}$$

and satisfying the principle of composite amplitudes:

$$\psi(A=a|C=c) = \sum_b \psi(A=a|B=b)\cdot\psi(B=b|C=c) \tag{5.4}$$

Feynmann comments this situation as follows |21|: "...If (5.4) is correct, ordinarily (5.1) is incorrect [here again I have changed the numeration. NdA]. The logical error made in deducing (5.1) consisted, of course, in assuming that to get from a to c the system had to go through a condition such that B had to have some definite value b ..." and further he adds that "...Looking at probability from a frequency point of view (5.1) simply results from the statement that in each experiment giving a and c, B had some value. The only way (5.1) could be wrong is the statement "B had some value" must sometimes be meaningless. Noting that (5.4) replaces (5.1) only under the circumstance that we make no attempt to measure B, we are led to say that the statement "B had some value" may be meaningless whenever we make no attempt to measure B...".

But if B has no values when nobody looks at it, then what happens? The usually accepted answer is that the system is in a "physical superposition" of states in each of which B has a definite value. Nobody has ever seen a "physical superposition" of states. One needs such a concept only in case he has a strong emotional affection for Bayes' postulate. In fact it is only assuming that this postulate is a truth of nature, that the non validity of (5.1) becomes a physical puzzle. As remarked by many authors, first of all Einstein, the concept of "physical superposition" is not only unnecessary but harmfull. In fact, since the result of any act of measurement is always a definite value and never a superposition, the problem arises of when this physical transition: superposition $\longrightarrow$ definite value, takes place; what causes this transition; how to describe it. The attempts to answer these problems gave rise to the quantum theory of measurement. A purely probabilistic approach to that theory has been described in | 7 |. A lucid exposition of the point of view, shared nowadays by most physicists, is described by R. Schlegel in |40|. We will come back to the existence of physical superpositions in the discussion of the EPR paradox (cf § 7.); before this we need some more remarks on the respective regions of validity of the principle of composite amplitudes and the theorem of composite probabilities. The principle of composite amplitudes can be trivially deduced within the usual Hilbert space formulation of quantum theory. The converse deduction is also true, as specified by the following Proposition:

**Proposition (5.1).** Let be given a family $F = \{A, B, C, \dots\}$ of physical observables. Denote $\mathrm{sp}\,A$ the set of values of A, and assume that $\mathrm{sp}\,A$ has n (different) points for each $A \in F$. Let be given, for each pair $A, B \in F$, $a \in \mathrm{sp}\,A$, $b \in \mathrm{sp}\,B$, a complex number $\psi(A=a|B=b)$ satisfying:

$$\psi(A=a|A=a') = \delta_{a,a'} \tag{5.5}$$

The map $(i,j) \to \psi(A=a_i \mid B=b_j)$; $i,j = 1,\dots,n$ is a positive definite Kernel on $(1,\dots,n)$, (5.6)

$$\psi(A=a|C=c) = \sum_b \psi(A=a|B=b) \cdot \psi(B=b|C=c) \tag{5.7}$$

$$\sum_a |\psi(A=a|B=b)|^2 = 1 \tag{5.8}$$

Then there exists a **C**-Hilbert space $\mathcal{H}$, and for every observable A an ortho-normal basis $\{\varphi_a^A : a \in \mathrm{sp}\,A\}$ of $\mathcal{H}$ such that for any $A, B \in F$, $a \in \mathrm{sp}\,A$, $b \in \mathrm{sp}\,B$:

$$\langle \varphi_a^A, \varphi_b^B \rangle = \psi(A=a|B=b) \tag{5.9}$$

Conversely, for any **C**-Hilbert space $\mathcal{H}$ and any family $(\varphi_a^A)$ of ortho-normal bases of $\mathcal{H}$, the function $\psi$ defined by (5.9) satisfies (5.5), (5.6), (5.7), (5.8).

**Proof.** Fix an observable $A \in F$ and let $\mathcal{H}$ be an Hilbert space with an ortho-normal basis $\{\varphi_a^A : a \in \mathrm{sp}\,A\}$. Defining for any observable $B \in F$ and $b \in \mathrm{sp}\,B$:

$$\varphi_b^B = \sum_a \psi(A=a|B=b) \cdot \varphi_a^A \tag{5.10}$$

The family of o.n. bases $\{\varphi_a^A : A \in F;\ a \in \mathrm{sp}\,A\}$ has the required properties. The converse is trivial.

Remark - The Hilbert space and the family of o.n. bases $(\varphi^A_a)$ above are unique up to isomorphism, in the sense that if $\mathcal{H}'$ is another Hilbert space and $(\chi^A_a)$ another family of orthonormal bases, then there exists a unique unitary operator $V: \mathcal{H} \to \mathcal{H}'$ satisfying

$$V \varphi^A_a = \chi^A_a \ ; \qquad \forall A \in F \ ; \ \forall_a \in \mathrm{sp}\, A \tag{5.11}$$

The vector $\varphi^A_a$ (better - the associated ray) is called the state of the physical system associated to the preparation A=a. The identity

$$\varphi^A_a = \sum_b \psi(B=b \mid A=a)\ \varphi^B_b \tag{5.12}$$

is the mathematical formulation of the fact that the state $\varphi^A_a$ is a superposition of states in which B=b.

Summing up: the principle of composite amplitudes is equivalent to the superposition principle. Therefore if we can deduce the principle of composite amplitudes from some physically plausible assumptions made on experimentally measurable quantities, we will have established quantum theory on a solid conceptual ground, independently on its empirical success. In |9| we have shown how this goal can be accomplished.

## 6. Mixtures, and the physical formulation of Bayes's postulate

We will now rephrase Bayes' postulate in a language more familiar to the quantum physicists. We will deal only with a very special case, which is all we need for our discussion of the EPR paradox (of § 7), and leave the obvious generalization to readers. Consider two events, denoted 1 and 2, defining two quantum mechanical pure states, $\psi_1$, $\psi_2$. Assume that the events are disjoint and that at least one of them should happen:

$$1 \wedge 2 = \emptyset ; \qquad 1 \vee 2 = I \tag{6.1}$$

Thus $(\psi_1, \psi_2)$ is an orthonormal basis in a Hilbert space $\mathcal{H}$. As discussed at length in §§ 2, 3, Bayes' postulate in this case is reduced to the statement that for any event X one has:

$$P(X) = P(1) \cdot P(X|1) + P(2) \cdot P(X|2) \tag{6.2}$$

where P(j) denotes the probability of the event j (= 1,2). Denoting $Q_X$ the projection operator corresponding to X in the usual quantum formalism (6.2) can be written in the form:

$$P(X) = P(1)\ \mathrm{Tr}(Q_X\ P_{\psi_1}) + P(2)\ \mathrm{Tr}(Q_X\ P_{\psi_2}) \tag{6.3}$$

where $P_\psi$ denote the rank-one projection onto the state $\psi$ and Tr - the trace. The fact that (6.3) must hold for an arbitrary projection $Q_X$ means that the statistical description of the system is given by the density matrix:

$$W = P(1)\ P_{\psi_1} + P(2)\ P_{\psi_2} \tag{6.4}$$

Summing up: when translated into the quantum-mechanical formalism, Bayes' postulate reduces to the following statement:

QMB] Let 1,2 be states of a physical system corresponding to the ortho-normal basis $(\psi_1, \psi_2)$ of the quantum mechanical state space of a system. If on an ensamble of systems we know with certainty that the fraction P(1) is in the state 1 and the fraction P(2) in the state 2, then the statistical description of this ensemble is given by the mixture (6.4).

In the statement above, we purpously left unspecified how do we "know with certainty" that each system of the ensemble is either in state 1 or in state 2. In case our knowledge comes from an experiment, in the sense that all the systems have interacted with an apparatus which can in principle allow to decide which of the two alternatives is realized, then the above statement is very well verified by the experiment. In this case, using the mixture (6.4), we are not only expressing a physical property of the system, but also our knowledge that at a certain time $t_o$ a measurement was performed on it. However, until a measurement is not made, we have not the right to describe the system with the mixture (6.4) because this would be equivalent to the application of Bayes' postulate to a situation in which the experiments have shown that it is usually not applicable.

The postulate of realism has to do with the fact that each system of the ensemble is either in state 1 or in state 2, but not in both, and this is quite different from the statement that the validity of the postulate of realism is equivalent to the validity of Bayes' axiom. Summing up: the phenomenon called "collapse of the wave packet" when stated in probabilistic language sounds as follows: the act of an incomplete measurement re-establishes the validity of the theorem of composite probabilities. (cf. [7] for details).

## 7. The EPR paradox

Once accepted the idealization of "instantaneous measurement" it is clear that the notion of physical superposition implies action at distance: for example an instantaneous measurement of position would instantaneously change a particle whose position is "superposed" throughout the space into a localized particle-like object. Considering systems which are separated in space-time but connected by a conservation law, the argument of Einstein, Podolsky and Rosen |19| shows that the same conclusion can be reached without using the idealization of instantaneous measurement. This is one aspect of the EPR paradox, the other one will be discussed in the following. From the point of view of quantum probability the superposition principle, i.e. the principle of composite amplitudes means only that we are describing a set of statistical data by means of a particular non kolmogorovian model and the empirical necessity to use such a model by no means implies the necessity to postulate the existence of something like a "physical superposition". Thus, from the point of view of quantum probability the argument, according to which quantum theory should imply the existence of superluminal signals, even if at an unobservable level, stems from the fact that one has deduced a conclusion on the physical world (the existence of physical superpositions) from a demostrably wrong mathematical premise (the applicability of the theorem of composite probabilities to a given set of statistical data). A more sophisticated discussion of the EPR paradox is due to B. d'Espagnat |18| (cf. also the paper by F. Selleri and Tarozzi |43|, which we will follow in our exposition). In it one doesn't make use of the notion of physical superposition" but only needs the universally accepted difference between such mathematical objects as superpositions and mixtures. The argument of these authors starts from Bohm's version of the EPR paradox: one considers two particles, say 1 and 2, far away from each other, and the system (1.2) composed of these two particles. It is assumed that at time 0, hence at any time before the measurement, the system (1.2) is in the singlet state

$$\psi_s = \frac{1}{\sqrt{2}} (u_1^+ \otimes u_2^- - u_1^- \otimes u_2^+) \tag{7.1}$$

where $u_j^{\pm}$ (j = 1,2) are the eigenstates of the z-component of the spin of particle j. One considers an ensemble of N copies of the system (1,2) all prepared in the singlet state. With these notations the main steps of the argument are the following (|43|, pag. 10):

1. At time $t_o$ one measures the z-component of the spin of particle 1 on each of the composed systems (1,2), while particle 2 is left isolated. Quantum theory says that approximatively for 1/2 of the systems the result obtained will be + 1/2 (in appropriate units). This statement is confirmed by the experiments.

2. Denoting $\sigma_z^j(t)$ (j = 1,2) the z-component of the spin of particle j at time t, the singlet assumption and the conservation of spin implies that at each time t:

$$\sigma_z^1(t) + \sigma_z^2(t) = 0 \tag{7.2}$$

This prediction is confirmed by the experiments too. Therefore, since (7.2) is equivalent to:

$$\sigma_z^1(t) = +1/2 \Leftrightarrow \sigma_z^2(t) = -1/2 \tag{7.3}$$

whenever the result of the $\sigma_z^1(t_o)$- measurement is +1/2, we know with certainty that $\sigma_z^2(t_o) = -1/2$. Thus (since we assume that our measurement is a first kind one) we know with certainty that for any $t \geqslant t_o$ one has

$$\sigma_z^1(t) = +1/2; \quad \sigma_z^2(t) = -1/2$$

we will express this fact saying that the system (1.2)

is in the state (+, −); similarly for the state (−, +).

3. "...But, at time $t = t_o$, when particle 1 interacts with an instrument, nothing can happen to particle 2, which can be as far away as one whishes from 1. All what is true for 2 for $t \geq t_o$ must have been true before (namely for $t < t_o$)..." |43|. This means that if without acting on the z-component of the spin of particle 2 we know with certainty that $\sigma_z^2(t_o) = -1/2$ (resp. +1/2) then the same must be true before $t_o$.

4. Using the argument at step (2) we conclude that, before and after time $t_o$, the state of the system (1,2) was (+, −) (resp. (−, +)).

5. But according to quantum theory, if at time t the system (1,2) is in state (+, −) (resp. (−, +)) then, in the mathematical model, the system (1,2) is described by the density matrix $P_{u_1^+ \otimes u_2^-}$ (resp. $P_{u_1^- \otimes u_2^+}$) where, as usual, $P_\psi$ denotes the rank one projection onto the state $\psi$.

6. Because of step (1) we then know with certainty that in the ensemble of N system, approximatively 1/2 are in the state (+, −) and the other half in state (−, +).

7. From Step (6) and from the physical formulation of Bayes' postulate (cf. § 6) we deduce that at any time before or later to the statistical description of the system (1,2) is given by the density matrix:

$$W = P(1)\, P_{u_1^+ \otimes u_2^-} + P(2)\, P_{u_1^- \otimes u_2^+} \tag{7.4}$$

8. But this contradicts, in an observable manner, the initial assumption that at any time before $t_o$, the system (1,2) was described by the singlet state $W_o = P_{\psi_s}$, $\psi_s$ being the singlet state (7.1).

The conclusion of Selleri and Tarozzi is that the statement of step (3) above: "...at time $t = t_o$ noting can happen to particle 2, which is very far away from 1..." is incompatible with quantum mechanics. since the above statement is a formulation of "Einstein' locality" (cf. |43|, pag. 7), they conclude that quantum theory in incompatible with Einstein' locality.

However at step (7) above these authors have (implicitely in |43|) used the physical formulation of Bayes' postulate. Therefore a more correct restatement of their conclusion should be "...if we insist to apply Bayes' postulate to the statistical data of quantum mechanics, then quantum theory is incompatible with Einstein locality ...". But why should one insist in applying Bayes' postulate to the statistical data of quantum theory, when both mathematics and the experiments provide evidence against this application?

## 8. Inequalities and statistical invariants

Having shown that there are in nature sets of statistical data which cannot be described by the classical kolmogorovian probabilistic model, a number of questions naturally arise: given a set of statistical data, how to decide wether to describe it with the classical or with the quantum model? are there other models besides these two? etc. ... . Until now the answer to the first question has been purely empirical: if the data come, say, from CERN, one uses the quantum model; if they come from more classical environments, one uses the kolmogorovian model. The notion of statistical invariant has been introduced to find a theoretical answer to this problem. This notion is best illustrated again by the example of the two slit experiment: in this case the statistical data are the probabilities $P(X)$, $P(X|1)$, $P(X|2)$ and we have shown in § 2 that a kolmogorovian model for these data exists if and only if

$$0 < \frac{P(X) - P(X|2)}{P(X|1)-P(X|2)} < 1 \tag{8.1}$$

The inequality (8.1) is an invariant of the set of statistical data $P(X)$, $P(X|1)$, $P(X|2)$ in the sense that changing these data in a way that (8.1) is still satisfied (or still not satisfied) will not change the property of admitting (or non admitting) a kolmogorovian model. For the given set of statistical data, we can also ask the question of the existence of a complex, or real Hilbert space model, and so on... . In general, given a set of statistical data, we will call a statistical invariant relative to this set, any family of conditions which are expressed only in terms of the initial data and which are necessary and sufficient for the possibility of describing these data within the given probabilistic model. Let us formulate precisely this definition for the two best known probabilistic models:let $F=\{A,B,C, \dots\}$ be a family of n-valued observables ($n \leqslant \infty$, independent on the observable, and all the values are distinct); let, for any pair of observables A, B and of their values a, b, be given a transition probability $P(A=a|B=b)$-considered as experimental data.

Definition (8.1) - A kolmogorovian model for the family of transition probabilities $\{P(A=a|B=b): A, B \in F;\ a \in \mathrm{sp}\, A,\ b \in \mathrm{sp}\, B\}$ is defined by:

i) A probability space $(\Omega, \partial, \mu)$.

ii) For any $A \in F$ - a measurable partition $\{A_a : a \in \mathrm{sp}\, A\}$ of $\Omega$ such that

$$P(A=a|B=b) = \frac{\mu(A_a \cap B_b)}{\mu(B_b)} \tag{8.2}$$

for any $A, B \in F$, $a \in \mathrm{sp}\, A$, $b \in \mathrm{sp}\, B$.

Definition (8.2) - A **C**-(resp **R**- or **Q**-, ...) Hilbert space model for the above family of transition probabilities is defined by:

i) A **C**-(resp. **R**-, **Q**-, ...) Hilbert space $\mathcal{H}$.

ii) For any $A \in F$ - an orthonormal basis $\{\varphi_a^A : a \in \mathrm{sp}\, A\}$ such that

$$P(A=a|B=b) = |\langle \varphi_a^A, \varphi_b^B \rangle|^2 \tag{8.3}$$

for any $A, B \in F$, $a \in \mathrm{sp}\, A$, $b \in \mathrm{sp}\, B$.

The quantum mechanical transition probabilities must therefore satisfy the symmetry condition:

$$P(A=a|B=b) = P(B=b|A=a) \tag{8.4}$$

Under this condition for two observables a kolmogorovian model always exists if $n < +\infty$, but a **C**-Hilbert space might not exist if $n \geqslant 3$ (cf. |6|). For three two-valued observables the statistical invariants have been explicitely computed for the kolmogorovian, **C**-, **R**-, and quaternion models. They are fine enough to distinguish among the first three cases. For lack of space we refer to |8|, |9| for a precise formulation and comments; however, what should be clear from the above few remarks, is that all the Bell's type inequalities |14|, |17|, |18|, |24|, |25|, |9|,[50][51] have a simple probabilistic interpretation: they are necessary conditions on a set of statistical data for the existence of a kolmogorovian model.

## 9. Outline of the deduction of the quantum formalism

Let A be an observable relative to a given system, $a \in \text{sp}\, A$ - a value of A. To the pair (A,a) we associate the elementary filter $A_a$ corresponding to the ideal selective measurement that from an ensemble of independent, identically prepared systems accepts those for which the observable A takes the value a and rejects all the other ones (cf. |41|). Elementary filters can be applied in series, and the symbol $A_a \cdot B_b$ means that to the given ensemble one applies first the filter $A_a$ and then $B_b$. In order to avoid possible confusions it is important to keep in mind that the apparatus $A_a \cdot B_b$ should be considered as a whole, in the sense that one counts the particles coming out from the filter $B_b$ <u>but not</u> those who pass the filter $A_a$ and enter $B_b$. The analogous convenction will be stipulated for products of more than two filters. The above shows that a notion of time is implicit in the multiplication of filters, however in first approximation we will not make it explicit, assuming that the effect due to time variation are negligible. In other words, what we are assuming is that, for very small $\varepsilon > 0$, the results of the application of the filters $A_a(t) \cdot B_b(t+\varepsilon)$ will be independent on $\varepsilon$ and, by time stationariety, on t. This is what one means with the heuristic statement "the filter $B_b$ is applied "immediately after" the filter $A_a$".

Two elementary filters $A_a$, $B_b$ are called <u>compatible</u> if the following conditions are satisfied:

i) it is possible to realize an apparatus, denoted $A_a + B_b$, such that for each single system emerging from it one knows with certainty that either A=a or B=b.

ii) $A_a \cdot B_b = B_b \cdot A_a$

As shown by J. Schwinger |41|, the operations + and · enjoy the usual properties of addition and multiplication. The multiplication of an elementary filter $A_a$ for a scalar p, between 0 and 1, can be defined as that apparatus - denoted $pA_a$ - which from an ensemble of systems accepts at random only a fraction p of those for which A=a, and rejects all the other ones. These operations allow to embed the family of elementary filters in a larger mathematical structure, i.e. an algebra $\mathcal{A}$ with unity, over the reals - the so-called Schwinger's algebra of mesurements.

With the goal of formulating in this context Heisenberg's indeterminacy principle, let us distinguish in it two aspects: a purely existential statement, namely

HP1) There exist in nature pairs of incompatible observables,

which is independent on the pair of observables involved; and a more specific statement concerning "how much" the two observables are incompatible - this one necessarily depends on the pair of observables.

The statement (HP1) says only that the algebra of measurements cannot be commutative, but this is too little to give an idea of its structure. A more quantitative consequence of Heisenberg' principle is the so-called Dirac's jump assumption which, as seen in § 4. can be formulated as a purely statistical property: the Markov property (cf. 4.1). This implies, in particular, that about any system emerging from the filter $A_a \cdot B_b \cdot A_a$, we can only say with certainty that for that system A=a and not, as in the classical case, that A=a and B=b. Moreover, in general, not all the systems which passed the first filter $A_a$ will pass also the filter $B_b$. Thus, if we consider two identically prepared ensembles and apply to one of them the filter $A_a$ and to the other one the filter $A_a \cdot B_b \cdot A_a$, then only a fraction of the systems which passed the former will pass the latter. It is an experimentally verifiable fact that this fraction is just the transition probability $P(A=a|B=b)$. In symbols:

$$A_a \cdot B_b \cdot A_a = P(A=a|B=b)\, A_a \qquad \text{(HP2)}$$

This statement, which establishes a definite connection between the structure of the algebra of measurements and the empirically given statistical data, will be assumed as the basic postulate of the quantum mechanical algebra of measurements. More precisely:

Definition (8.1) - Let T be any set and let be given, for any $x \in T$, an observable $A(x)$ with values $a_1(x), \dots, a_n(x)$ ($n < +\infty$, independent on x). Let be given, for any $x,y \in T$, a transition probability matrix

$$P(x,y) = \{ p_{\alpha\beta}(x,y) : \alpha, \beta = 1, \dots, n \}$$

$$p_{\alpha\beta}(x,y) = P(A(x) = a_\alpha(x) | A(y) = a_\beta(y))$$

(by definition $P(x,x)$ = the identity matrix).
We will say that the family of transition probability matrices $\{ P(x,y) : x,y \in T \}$ admits an Heisenberg model if there exist:

i) a real associative algebra with unity, $\mathcal{A}$.

ii) for each $x \in T$, a family $\{ A_\alpha(x) : \alpha = 1, \dots, n \}$ of elements of $\mathcal{A}$ satisfying:

$$A_\alpha(x) \cdot A_{\alpha'}(x) = \delta_{\alpha\alpha'} A_\alpha(x) ; \ \forall \alpha, \alpha' \tag{9.1}$$

$$\Sigma_\alpha A_\alpha(x) = 1 \tag{9.2}$$

$$A_\alpha(x) \cdot A_\beta(y) \cdot A_\alpha(x) = p_{\alpha\beta}(x,y) A_\alpha(x) \tag{9.3}$$

The Heisenberg model above will be called standard if, denoting $\kappa$ the center of the algebra $\mathcal{A}$, one has:

iii) For any $A \in \mathcal{A}$ and $\lambda \in \kappa$
$\lambda \cdot A = 0 \Leftrightarrow \lambda = 0$ or $A = 0$

iv) For any pair $x,y \in T$, $x \neq y$, the products $\{ A_\alpha(x) \cdot A_\beta(y) : \alpha, \beta = 1, \dots, n \}$ are a linear basis of $\mathcal{A}$ over $\kappa$.

It will be called minimal if it is standard and $\kappa$ has the smallest possible dimension over **R**, i.e.

v) If $\{\mathcal{A}', \kappa'\}$ is another standard Heisenberg model for the family $\{P(x,y)\}$ then $\kappa$ can be isomorphically embedded into $\kappa'$.

The classification problem for Heisenberg models consists in describing which families of transition probabilities admit an Heisenberg model and which Schwinger algebras arise as Heisenberg models of some set of transition probabilities. Our main result is the solution of the classification problem for standard Heisenberg models. To state the result we need one more definition:

Definition (8.2) - Let $P = (p_{\alpha\beta})$ be an n x n transition probability matrix, and let $\kappa$ be a commutative, associative algebra over the reals with unity. A $\kappa$-valued matrix $U = (u_{\alpha\beta})$ satisfying:

$$\Sigma_\alpha (u_{\alpha\beta})^* u_{\alpha\beta'} = \delta_{\beta\beta'} \tag{9.4}$$

$$\Sigma_\beta u_{\alpha\beta} \cdot (u_{\alpha'\beta})^* = \delta_{\alpha\alpha'} \tag{9.5}$$

where:

$$(u_{\alpha\beta})^* = p_{\alpha\beta}/u_{\alpha\beta} \tag{9.6}$$

will be called a $\kappa$-valued transition amplitude matrix for P (we assume $u_{\alpha\beta} \neq 0$).

Theorem (8.3) - Let T be a set; $\{P(x,y): x, y \in T\}$ be a family of transition probability matrices. The following are equivalent:

i) The family $\{P(x,y)\}$ admits a standard Heisenberg model with center $\kappa$.

ii) For each pair, $x, y \in T$, one has

$$P_{\alpha\beta}(x,y) = p_{\beta\alpha}(y,x) \tag{9.7}$$

and there exists a $\kappa$-valued transition amplitude $U(x,y)$ for $P(x,y)$ such that the family $\{U(x,y)\}$ satisfies the conditions:

$$U(x,y) \cdot U(y,z) = U(x,z); \ \forall x,y,z \in T \tag{9.8}$$

$$U(x,x) = \text{id}$$

iii) There exists a $\kappa$-module $\mathcal{H}$ and, for each $x \in T$, a $\kappa$-basis $(a_\alpha(x))$ of $\mathcal{H}$ such that, denoting $A_\alpha(x)$ the linear rank-one projection onto $\kappa \cdot a_\alpha(x)$, defined with respect to the $(a_\alpha(x))$-basis, one has:

$$A_\alpha(x) \cdot A_\beta(y) \cdot A_\alpha(x) = p_{\alpha\beta}(x,y)\ A_\alpha(x) \tag{9.9}$$

for any $x, y \in T$; $\alpha, \beta = 1, \ldots, n$.

Writing equation (9.8) in terms of the coefficients $(u_{\alpha\beta}(x,y))$ of $U(x,y)$, one immediately recognizes that (9.8) is a generalized form of the principle of composite amplitudes. Moreover, using the techniques mentioned in § 8 it is possible, at least in principle, to determine the minimal dimension of $\kappa$ over $\mathbf{R}$ as a function only of the transition probabilities $\{p_{\alpha\beta}(x,y)\}$. Finally, a simple generalization of Proposition (5.1) allows to characterize those Heisenberg models in which $\mathcal{H}$ can be choosen to be an Hilbert space.

Summing up: we have shown that, within the framework of Schwinger's algebra of measurements, and assuming as a postulate only a simple and experimentally verifiable consequence of the Heinsenberg indeterminacy principle, the whole mathematical apparatus of quantum theory can be deduced only from a detailed mathematical analysis of the transition probabilities. Our analysis yields moreover, some examples of non-kolmogorovian models of a new type (i.e. those minimal Heisenberg models which are not of Hilbert space type) whose structure seems to be interesting to investigate, at least from a mathematical point of view.

Some more refined non-Kolmogorovian models can be constructed and lead to a generalization of gauge theories (cf. the author's paper in : Proceeding of the workshop : "Quantum probability and applications to the quantum theory of irreversible processes" Rome, Sept. 1982 (to appear in Springer Lecture Notes in Mathematics).

## References

|1| Accardi, L. Non-Commutative Markov chains. in: Proceedings School of Math. Phys. Università di Camerino, Sept.-Oct. 1974.

|2| Accardi, L. Non relativistic quantum mechanics as a non-commutative Markov process. Advances in Math. 20 (1976), 329-366.

|3| Accardi, L. Noncommutative Markov chains Associated to a preassigned evolution: an application to the quantum theory of measurement. Advances in Math. 29 (1978), 226-243.

|4| Accardi, L. On the quantum Feynman-Kac formula. Rendiconti del Seminario Matematico e Fisico di Milano, 48 (1980), 135-179.

|5| Accardi, L. Quantum Markov Processes. in: Dynamical Systems and Microphysics; A. Blaquiere, F. Fer, A. Marzollo eds., Springer 1980.

|6| Accardi, L. Topics in quantum probability. Physics Reports 77 (1981), 169-192.

|7| Accardi, L. Probabilità e teoria quantistica. Physis 23 (1981), 485-524.

|8| Accardi, L., Fedullo A. On the statistical meaning of complex numbers in quantum theory. Lett. Nuovo Cimento 34 (1982), 161-172.

|9| Accardi, L. Foundations of Quantum Probability. Rend. Sem. Mat. Univers. Politecn. Torino (1982), 249-270.

|10| Accardi, L., Cecchini C. Conditional expectations in von Neumann algebras and a theorem of Takesaki. Journ. of Funct. anal. 45 (1982), 245-273.

|11| Accardi, L. Dio gioca ai dadi? (Ciò che è vivo e ciò che è morto nel dibattito sull'interpretazione della teoria quantistica). Il Saggiatore (to appear).

|12| Accardi, L. History of the theorem of composite probabilities from 1925 to 1948. Communication to the Congress "La storia della Matematica in Italia", Cagliari, Ottobre 1982.

|13| Bell, J.S. On the Problem of Hidden variables in quantum Mechanics. Rev. of Mod. Phys. 38 (1966), 447-452.

|14| Bell, J.S. Physics 1 (1965), 195.

|15| Beltrametti E., Cassinelli C. The logic of Quantum Mechanics. Encyclopedia of Mathematics, Addison-Wesley (1982).

|16| Bub J. The statistics of non-boolean event structure. Foundations of Phylosophy of Statistical Theories in the physical sciences. vol. III; W.L. Harper, C.A. Hooker (eds.) Reidel (1973).

|17| Corleo G., Gutkowski D., Masotto G., Valdes M.V. Are Bell's type inequalities sufficient conditions for local hidden variable theories? Nuovo Cimento 25 B (1975), 413-424.

|18| D'Espagnat, B. Conceptions de la physique contemporaine. Paris 1965.

|19| Einstein A., Podolsky B., Rosen N. Phys. Rev. 47 (1935), 777.

|20| Feynman R.P. The concept of probability in quantum mechanics. Proceedings II-d Berkeley Symp. on math. Stat. and Prob. University of California Press, Berkeley, Los Angeles (1951), 533-541.

|21| Feynman R.P. Space-time approach to non-relativistic quantum mechanics. rev. of Mod. Phys. 20 (1948), 367-385.

|22| Feynman R.P. The character of physical law (1965). Ed. Boringhieri (1971).

|23| Feynman R.P., Leighton R.B., Sands M.; Lectures on physics vol. III. Addison-Wesley (1966).

|24| Garuccio A., Selleri F., Systematic derivation of all the Inequalities of Einstein Locality. foundations of Physics 10 (1980), 209-216.

|25| Gutkowski D., Masotto G. An inequality stronger than Bell's Inequality. Nuovo Cimento B 22 (1974), 121.

|26| Heisenberg W. The physical principles of the quantum theory. Dover (1949).

|27| Jauch J.M. foundations of Quantum Mechanics. Addison-Wesley (1968).

|28| Jordan P. Uber eine neue Begründigung der Quantenmechanik. Zeitschr. für Physik 40 (1927), 809.

|29| Kolmogorov A.N. Foundations of the theory of Probability. Chelsea, New York (1950).

|30| Landé A. New foundations of Quantum Mechanics. Cambridge University Press (1965).

|31| Lochak G. Has Bell's inequality a general meaning for Hidden-variable theories? Foundations of Physics 6 (1976), 173-184.

|32| Mackey G.W. Mathematical foundations of quantum mechanics. W.A. Benjamin Inc. (1963.

|33| Pitowsky I. Resolution of the einstein-Podolsky-Rosen paradox. Phys. Rev. Lett. 48 (1982), 1299-1302.

|34| Piron C. Foundations of quantum physics. W.A. Benjamin Inc. (1976).

|35| Putnam H. Is logic empirical? in: The Logico-algebraic approach to quantum mechanics. C.A. Hooker (ed.), vol. II, Reidel.

|36| Putnam H. Three valued logic. in: The Logico-algebraic approach to quantum mechanics. C.A. Hooker (ed.), vol. I, Reidel.

|37| Roos M. On the construction of a unitary matrix with elements of given moduli. Journ. of Math. Phys. 5 (1964), 1609-1611.

|38| Roos M. Studies of the principle of superposition in quantum mechanics. Commentationes Physico-Matematical 33 (1966) N. 1.

|39| Scheibe, E. The logical analysis of quantum mechanics. Pergamon Press.

|40| Schlegel, R. Superposition and interaction. The University of Chicago Press (1980).

|41| Schwinger, J. Quantum kinematics and Dynamics Academic Press (1970).

|42| Segal, I.E. Mathematical problems of relativistic physics. Lectures in Appl. Math. Amer. Math. Soc. (1963).

|43| Selleri F., Tarozzi G. Quantum mechanics reality and separability. La Rivista del Nuovo Cimento (1981), N.2.

|44| Strauss, M. Mathematics as Logical Syntax - A method to formalize the language of a physical theory (1937). in: The Logico-Algebraic approach to quantum mechanics. C.A. Hooker (ed.), Reidel (1975).

|45| Suppes, P. The probabilistic argument for a non-classical logic of quantum mechanics (1966). cf. ref. |44|.

|46| Suppes P., Zanotti M. Stochastic incompleteness of quantum mechanics. in: Logic and Probability in quantum mechanics. P. Suppes (ed.), Reidel

|47| Suppes P., Zanotti M. Bell's inequalities are necessary and sufficient conditions for extension of probability on the partial boolean algebra of four observables to the

full boolean algebra. (preprint, March 1982).

|48| Suppes P., Zanotti M. Bell's inequality is implied by existence of a joint probability distribution of the four observables. (submitted to Nuovo Cimento).

|49| Santos, E. Local hidden variable theories violating Bell inequalities (preprint).

|50| Svetlichny, G. A Bell-type inequality distinguishes Three-Body from two-body non separability. (preprint).

|51| Wigner, E.P. On hidden variables in quantum mechanics. Amer. Journn. of Phys. 38 (1970), 1005-1009.

# CAUSALITY AND SYMMETRY

Patrick Suppes and Mario Zanotti

Institute for Mathematical Studies in the Social Sciences
Stanford University

This paper is concerned with inferences from phenomenological variables to hidden causes or hidden variables. A number of theorems of a general sort are stated. The paper concludes with a treatment of Bell's inequalities and their generalization to more than four observables.

1. *Introduction.* In this paper we are concerned to present a number of theorems about inferences from phenomenological correlations to hidden causes. In other terms, the theorems are mainly theorems about hidden variables. Most of the proofs will not be given but references will be cited where they may be found.

To emphasize conceptual matters and to keep technical simplicity in the forefront, we consider only two-valued random variables taking the values $\pm 1$. We shall also assume symmetry for these random variables in that their expectations will be zero and thus they will each have a positive variance of one. For emphasis we state:

GENERAL ASSUMPTION. *The phenomenological random variables* $X_1,\ldots,X_N$ *have possible values* $\pm 1$, *with means* $E(X_i) = 0$, $1 \le i \le N$.

We also use the notation $X$, $Y$, and $Z$ for phenomenological random variables. We use the notation $E(XY)$ for covariance which for these symmetric random variables is also the same as their correlation $\rho(X,Y)$.

The basic meaning of causality that we shall assume is that when two random variables, say $X$ and $Y$ are given, then in order



*S. Diner et al. (eds.), The Wave-Particle Dualism, 331–340.*

for a hidden variable $\lambda$ to be labeled a cause, it must render the random variables conditionally independent, that is,

$$E(XY|\lambda) = E(X|\lambda)E(Y|\lambda) \quad . \tag{1}$$

It is worth noting that although the context for discussion of most of what we have to say is quantum mechanics, the reasoning from effects to causes is classical in science and has been an important explicit methodology since the appearance of Laplace's famous memoir of 1774 (1).

2. *Two deterministic theorems.* We begin with a theorem asserting a deterministic result. It says if two random variables have a strict negative correlation then any cause in the sense of (1) must be deterministic, that is, the conditional variances of the two random variables, given the hidden variable $\lambda$ must be zero. We use the notation $\sigma(X|\lambda)$ for the conditional standard deviation of $X$ given $\lambda$, and its square is, of course, the conditional variance.

THEOREM 1 (Suppes and Zanotti (2)). *If*

*(i)* $E(XY|\lambda) = E(X|\lambda)E(Y|\lambda)$

*(ii)* $\rho(X,Y) = -1$

*then*

$$\sigma(X|\lambda) = \sigma(Y|\lambda) = 0 \; .$$

The second theorem asserts that the only thing required to have a hidden variable for $N$ random variables is that they have a joint probability distribution. This theorem is conceptually important in relation to the long history of hidden-variable theorems in quantum mechanics. For example, in the original proof of Bell's inequalities, Bell assumed a causal hidden variable in the sense of (1) and derived from this assumption his inequalities. What Theorem 2 shows is that the assumption of a hidden variable is not necessary in such discussions--it is sufficient to remain at the phenomenological level. Once we know that there exists a joint probability distribution then there must be a causal hidden variable and in fact this hidden variable may be constructed so as to be deterministic. This theorem shows how fundamental the question of the existence of joint probability distributions is in quantum mechanics. Once again another foundational question, namely, in this case the existence of a hidden variable, in fact reduces to the existence of a joint probability distribution.

On the other hand, it is important to emphasize that the

hidden variable constructed in the proof of this theorem is physically very unsatisfactory. This is of course natural when one is concerned to show that no hidden-variable theory of even the weakest sort can be consistent with quantum mechanics, but when one wants a positive causal theory it is important to impose additional constraints on the hidden variable. Some of the theorems we consider later do this.

THEOREM 2 (Suppes and Zanotti (3)). *Given phenomenological random variables $X_1,\ldots,X_N$, then there exists a hidden variable $\lambda$ such that*

$$E(X_1,\ldots,X_N|\lambda) = E(X_1|\lambda)\cdots E(X_N|\lambda)$$

*if and only if there exists a joint probability distribution of $X_1,\ldots,X_N$. Moreover, $\lambda$ may be constructed as a deterministic cause, i.e., for $1 \leq i \leq N$*

$$\sigma(X_i|\lambda) = 0 .$$

3. *Exchangeability.* We now turn to imposing some natural symmetry conditions both at a phenomenological and at a theoretical level. The main principle of symmetry we shall use is that of exchangeability. Two random variables $X$ and $Y$ of the class we are studying are said to be exchangeable if the following probabilistic equality is satisfied.

$$P(X = 1, Y = -1) = P(X = -1, Y = 1) . \qquad (2)$$

The first theorem we state shows that if two random variables are exchangeable at the phenomenological level then there exists a hidden causal variable satisfying the additional restriction that they have the same conditional expectation if and only if their correlation is nonnegative.

THEOREM 3 (Suppes and Zanotti (4)). *If $X$ and $Y$ are exchangeable, then there exists a hidden variable $\lambda$ such that*

$$(i) \quad E(XY|\lambda) = E(X|\lambda)E(Y|\lambda)$$

$$(ii) \quad E(X|\lambda) = E(Y|\lambda)$$

*if and only if*

$$\rho(X,Y) \geq 0 .$$

There are several useful remarks we can make about this theorem. First, the phenomenological principle of symmetry, namely, the principle of exchangeability, has not been used in physics as explicitly as one might expect. In the context of the kinds of experiments ordinarily used to test hidden-variable theories the requirement of phenomenological exchangeability is uncontroversial. On the other hand, the theoretical requirement of identity of conditional distributions does not have the same status. We emphasize that we refer here to the expected causal effect of $\lambda$. Obviously the actual causal effects will in general be quite different. We certainly would concede that in many physical situations this principle may be too strong. The point of our theorems about it is to show that once such a strong theoretical principle of symmetry is required then exchangeable and negatively correlated random variables cannot satisfy this theoretical principle of symmetry.

We now turn to the only theorem whose proof we give in this paper. This is the strengthening of Theorem 3 to show that when the correlations are strictly between zero and one then the causal variable cannot be deterministic. What is important from our conceptual standpoint is to show that the principles of symmetry used in Theorem 3 force us to stochastic hidden variables for correlations that are not deterministic or strictly zero.

THEOREM 4. *Given the conditions of Theorem 3, if* $0 < \rho(X,Y) < 1$, *then* $\lambda$ *cannot be deterministic, i.e.,* $\sigma(X|\lambda), \sigma(Y|\lambda) \neq 0$.

PROOF. We first observe that under the assumptions we have made:

$$Min\{P(X = 1, Y = -1), P(X = 1, Y = 1), P(X = -1, Y = -1)\} > 0 .$$

Now, let $\Omega$ be the probability space on which all random variables are defined. Let $\mathcal{A} = \{A_i\}$, $1 \leq i \leq N$ and $\mathcal{H} = \{H_j\}$, $1 \leq j \leq M$ be two partitions of $\Omega$. We say that $\mathcal{H}$ is a *refinement of* $\mathcal{A}$ *in probability* if and only if for all $i$'s and $j$'s we have:

$$\text{If } P(A_i \cap H_j) > 0 \text{ then } P(A_i \cap H_j) = P(H_j) .$$

Now let $\lambda$ be a causal random variable for $X$ and $Y$ in the sense of Theorem 3, and let $\lambda$ have induced partition $\mathcal{H} = \{H_j\}$, which without loss of generality may be assumed finite. Then $\lambda$ is deterministic if and only if $\mathcal{H}$ is a refinement in probability of the partition $\mathcal{A} = \{A_i\}$ generated by $X$ and $Y$, for assume, by way of contradiction that this is not the case. Then there must exist $i$ and $j$ such that $P(A_i \cap H_j) > 0$ and

$$P(A_i \cap H_j) < P(H_j) ,$$

but then $0 < P(A_i|H_j) < 1$.

We next show that if $\lambda$ is deterministic then $E(X|\lambda) \neq E(Y|\lambda)$, which will complete the proof.

Let, as before, $\mathcal{H} = \{H_j\}$ be the partition generated by $\lambda$. Since we know that

$$\sum_j P(X = 1, Y = -1, H_j) = P(X = 1, Y = -1) > 0$$

there must be an $H_j$ such that

$$P(X = 1, Y = -1, H_j) > 0 ,$$

but since $\lambda$ is deterministic, $\mathcal{H}$ must be a refinement of $\mathcal{A}$ and thus as already proved

$$P(X = 1, Y = -1 | H_j) = 1 ,$$

whence

$$P(X = 1, Y = 1 | H_j) = 0$$

$$P(X = -1, Y = 1 | H_j) = 0$$

$$P(X = -1, Y = -1 | H_j) = 0 ,$$

and consequently we have

$$\begin{aligned} P(X = 1 | H_j) = P(Y = -1 | H_j) = 1 \\ P(X = -1 | H_j) = P(Y = 1 | H_j) = 0 . \end{aligned} \tag{3}$$

Remembering that $E(X|\lambda)$ is a function of $\lambda$ and thus of the partition $\mathcal{H}$, we have from (3) at once that

$$E(X|\lambda) \neq E(Y|\lambda) .$$ *Q.E.D.*

4. *Joint distribution.* Given the covariances (or correlations) of $N$ random variables of the sort we are considering, it is natural to ask when a compatible joint distribution exists. For $N = 2$, the answer is "always" whatever the correlation, but already for $N = 3$ restrictions are required. For example, if three random variables have identical pairwise correlations of $-1/2$, no compatible joint distribution exists, as is easily checked. A condition for $N = 3$ is this:

THEOREM 5 (Suppes and Zanotti (3)). *A necessary and sufficient condition for the existence of a joint probability distribution compatible with the given covariances of three phenomenological random variables X, Y and Z is that the following two inequalities be satisfied:*

$$-1 \leq E(XY) + E(YZ) + E(XZ) \leq 1 + 2\mathrm{Min}\{E(XY), E(YZ), E(XZ)\} .$$

in view of the following theorem, the condition for three is necessary and sufficient for four.

THEOREM 6 (Suppes and Zanotti (5)). *Let* $X_1, \ldots, X_N$ *phenomenological random variables be given and let* $N$ *be even. Then a necessary and sufficient condition that there exist a joint probability distribution compatible with the given covariances of all pairs of the phenomenological random variables is that there exist such a compatible distribution for each subset of* $N - 1$ *variables.*

Finally, we state the only really difficult theorem in this paper. The set of inequalities we give for each $N$ is, we think, about as simple as can be expected purely in terms of the covariances. The inequalities for $N = 3$ given in the theorem are easily shown to be equivalent to the condition of Theorem 5.

THEOREM 7 (Suppes and Zanotti (5)). *A necessary and sufficient condition that there exist a joint probability distribution compatible with the given covariances of all pairs of* $N$ *phenomenological random variables is that*

$$\sum_{i<j} a_i a_j E(X_i X_j) \geq (1-n)/2 \tag{4}$$

*for all subsets of odd cardinality* $n \leq N$ *and with* $a_i, a_j = \pm 1$.

(The subscript notation stands for summation over $1 \leq i < j \leq n$.)

It is easy to show that in the symmetric case of all the covariances being equal Theorem 7 can be simplified as follows:

THEOREM 8. *If the given covariances of all pairs of phenomenological random variables are equal, then there exists a joint probability distribution compatible with the given covariances if and only if for all* $1 \leq i < j \leq N$

$$E(X_i X_j) \geq \begin{cases} -\dfrac{1}{N-1} & \text{if } N \text{ is even,} \\ -\dfrac{1}{N} & \text{if } N \text{ is odd.} \end{cases}$$

5. *Bell covariances.* First, we recall Bell's inequalities are specifically formulated for measurements of spin of pairs of particles originally in the singlet state. Let $A$ and $A'$ be two possible orientations of apparatus I, and let $B$ and $B'$ be two possible orientations of apparatus II. Let the measurement of spin by either apparatus be 1 or -1, corresponding to spin 1/2 or -1/2, respectively. By $E(AB)$, for example, we mean the expectation of the product of the two measurements of spin, with

apparatus I having orientation $A$ and II having orientation $B$. By axial symmetry, we have $E(A) = E(A') = E(B) = E(B') = 0$, i.e., the expected spin for either apparatus is 0. It is, on the other hand, a well-known result of quantum mechanics that the covariance term $E(AB)$ is:

$$E(AB) = -\cos\theta_{AB},$$

where $\theta$ is the difference in angles of orientation $A$ and $B$. Again, by axial symmetry only the difference in the two orientations matters, not the actual values $A$ and $B$. (To follow the literature, we begin with the notation $A$, $B$, $A'$ and $B'$ for phenomenological random variables, rather than $X_1,\ldots,X_N$, which we go back to later.)

On the assumption that there is a hidden variable that renders the spin results conditionally independent, i.e., that there is a causal hidden variable $\lambda$ in the sense of equation (1) in the first section, Bell (6) derives the following inequalities:

$$\begin{aligned}
-2 &\leq -E(AB) + E(AB') + E(A'B) + E(A'B') \leq 2,\\
-2 &\leq E(AB) - E(AB') + E(A'B) + E(A'B') \leq 2,\\
-2 &\leq E(AB) + E(AB') - E(A'B) + E(A'B') \leq 2,\\
-2 &\leq E(AB) + E(AB') + E(A'B) - E(A'B') \leq 2.
\end{aligned} \tag{5}$$

This form of the inequalities is due to Clauser, Horne, Shimony and Holt (8).

We first prove a theorem that uses the condition of Theorem 5 to derive Bell's inequalities. The theorem is essentially equivalent to one stated by Arthur Fine (7). The proof we give is different from his but not radically so. We include the proof because it is simple and it shows how an elementary, purely phenomenological approach to Bell's inequalities is possible--a fact perhaps not as widely known as it should be.

THEOREM 9. *Bell's inequalities are a necessary consequence of the existence of joint probability distributions for any three of the four phenomenological random variables $A$, $A'$, $B$, and $B'$.*

PROOF. We apply directly Theorem 5. For each subset of three, $\{A,A',B\}$, $\{A,A',B'\}$, $\{A,B,B'\}$, and $\{A',B,B'\}$ inequalities of Theorem 5 hold. Adding the four sets of inequalities, and dividing by 2, we have:

$$-2 \leq E(AA') + E(AB) + E(AB') + E(A'B) + E(A'B') + E(BB') \tag{6}$$

$$\leq 2 + Min\{E(AA'),E(AB),E(A'B)\} + Min\{E(AA'),E(AB'),E(A'B')\}$$

$$+ Min\{E(AB),E(AB'),E(BB')\} + Min\{E(A'B),E(A'B'),E(BB')\}$$

To obtain Bell's inequalities from (6), we merely select from each of the four sets that covariance we desire. To exhibit details, we derive the second inequality of (5). Since the right inequality of (6) is satisfied in each case by the minimum of each set, necessarily it is satisfied by any one of the three. Thus we get:

$$E(AA') + E(AB) + E(AB') + E(A'B) + E(A'B') + E(BB') \tag{7}$$

$$\leq 2 + E(AA') + E(AB') + E(AB') + E(BB') .$$

Simplifying (7) we obtain the right-hand side of the second Bell's inequality (4).

$$E(AB) - E(AB') + E(A'B) + E(A'B') \leq 2 . \tag{8}$$

To derive the left-hand side of this Bell inequality, we use two of the inequalities that follow from Theorem 5. First, for the subset $\{A,A',B\}$, we have

$$-1 \leq E(AA') + E(AB) + E(A'B) , \tag{9}$$

and for the subset $\{A,A',B'\}$

$$E(AA') + E(AB') + E(A'B') \leq 1 + 2\, Min\{E(AA'),E(AB'),E(A'B')\} \tag{10}$$

$$\leq 1 + 2E(A'B') .$$

From (10), we obtain at once by elementary operations:

$$-1 \leq - E(AA') - E(AB') + E(A'B') . \tag{11}$$

Adding now (9) and (11), we get the desired result:

$$-2 \leq E(AB) - E(AB') + E(A'B) + E(A'B') . \qquad Q.E.D.$$

The natural converse of Theorem 9 also holds--the proof is due to Arthur Fine.

THEOREM 10 (Fine (7)). *Bell's inequalities are sufficient for the existence of a joint probability distribution compatible with the six given covariances of the phenomenological random variables A, A', B, and B'.*

It is obvious that Theorem 10, not Theorem 9, is physically interesting and rather surprising. Unfortunately the situation becomes more complicated as we go beyond $N = 4$. To examine the general case we need to make explicit the idea of what covariances are given phenomenologically. Garg and Mermin (9) have given a counterexample to Bell's inequalities being sufficient for eight random variables when what we term *Bell covariances* are given. By Bell covariances we mean covariances $E(X_iX_j)$ for $1 \leq i \leq i_0 < j \leq N$, for some integer $i_0$. Garg and Mermin's counterexample is for $N = 8$ and $i_0 = 4$. Let $E(X_1X_5) = E(X_2X_6) = E(X_3X_7) = 1$ and $E(X_iX_j) = -1/3$ for $1 \leq i \leq i_0 < j \leq N$. Then it is easy to show that for the quintuple $(\overline{X}_1, \overline{X}_3, X_4, \overline{X}_6, X_8)$ all covariances must be $-1/3$. But it follows at once from Theorem 7 that with $N = 5$, $a_i = a_j = 1$, for existence of a compatible joint distribution $\Sigma E(X_iX_j) \geq -2$, and so there can be no joint distribution compatible with the given covariances all equal to $-1/3$.

We conclude with a probabilistic analysis of Bell covariances. Given such a restricted set of covariances, which arise naturally in quantum mechanics, it is natural to ask under what conditions there exists a compatible joint distribution. We have as an immediate consequence of Theorem 7 the following result.

THEOREM 11 (Suppes and Zanotti (6)). *Let Bell covariances* $E(X_iX_j)$, $1 \leq i \leq i_0 < j \leq N$, *be given. Then a necessary and sufficient condition that there exist a joint distribution of the N variables compatible with the given covariances is that there exist a solution of the inequalities of Theorem 7, with the non-Bell covariances as unknowns.*

## REFERENCES


(1) Laplace, P. S.: 1774, *Mémoire sur la probabilité des causes par les événements. Mémoires de l'Académie royale des Sciences de Paris (Savants étrangers)*, Tome VI, p. 621.

(2) Suppes, P., and Zanotti, M.: 1976, *On the determinism of hidden variable theories with strict correlation and conditional statistical independence of observables.* In P. Suppes (Ed.), *Logic and probability in quantum mechanics.* Dordrecht: Reidel, pp. 445-455.

(3) Suppes, P., and Zanotti, M.: 1981, *When are probabilistic explanations possible?* Synthese 48, pp. 191-199.

(4) Suppes, P., and Zanotti, M.: 1980, *A new proof of the impossibility of hidden variables using the principles of exchangeability and identity of conditional distribution.* In P. Suppes (Ed.), *Studies in the foundations of quantum mechanics.* East Lansing, Mich.: Philosophy of Science Association, pp. 173-191.

(5) Suppes, P., and Zanotti, M.: To appear, *Existence of joint distribution with given covariances and Bell's inequalities in quantum mechanics.*

(6) Bell, J. S.: 1964, *On the Einstein Podolsky Rosen paradox*, Physics 1, pp. 195-200.

(7) Fine, A.: 1982, *Hidden variables, joint probability, and the Bell inequalities*, Phys. Rev. Lett. 48, pp. 291-295.

(8) Clauser, J. F., Horne, M. A., Shimony, A., and Holt, R. A.: 1969, *Proposed experiment to test local hidden-variable theories*, Phys. Rev. Lett. 23, pp. 880-884.

(9) Garg, A., and Mermin, N. D.: 1982, *Correlation inequalities and hidden variables*, Phys. Rev. Lett. 49, pp. 1220-1223.

# A CRITICAL ANALYSIS OF THE QUANTUM THEORY OF MEASUREMENT

Francis FER

Ecole des Mines de Paris

"You can't learn too soon that the most useful thing about a principle is that it can always be sacrificed to expendiency".

Somerset Maugham, The circle.

*First of all I must advise the reader that hereafter we keep strictly in the positivist and probabilistic, hence hilbertian frame of Quantum Mechanics. We shall pay no attention at all to questions of interpretation, as well realistic as any other else. The only question discussed here is to ascertain whether or not Quantum Mechanics, starting from its axioms, reaches the aim of any physical theory, that is, comparison with experiment. The answer is: no, as long as it keeps close to the existing axiomatics, and also to accurate mathematics.*

## I.- THE AXIOMATICS OF MEASUREMENT

To refer easily to them, I reproduce briefly the classical axioms.

Consider an observable and let $T$, defined in the suitable Hilbert space $\mathcal{H}$, be the corresponding self-adjoint operator (as usual we shall call also $T$ the observable itself). We assume that the spectral decomposition of $T$ has the form



*S. Diner et al. (eds.), The Wave-Particle Dualism, 341–366.*

(1) $$T = \Sigma_j \lambda_j E_j$$

where all the projectors $E_j$ have infinite multiplicities (for the reason of these assumtions, see § II and appendix I). We have the three axioms:

V (for value): the result of a measurement of $T$ is one of the eigenvalues $\lambda_j$;

P (for probability): if the physical system is described by a state-vector $\psi_t \in \mathcal{H}$, function of the time $t$, the probability of observing the value $\lambda_j$ at $t$ is

(2) $$Pr(\lambda_j ; t) = (E_j \psi_t, \psi_t)$$

I do not insist upon the extension to the representation by a statistical operator,which is simple;

R (for reduction of state): immediately after the observation of a value $\lambda_j$, the state-vector is the normalized projection, on the manifold $E_j \mathcal{H}$, of the state-vector before the measurement, that is

(3) $$\psi_{t+0} = E_j \psi_{t-0} / \| E_j \psi_{t-0} \|$$

As a particular case, if we know nothing at all about $\psi_{t-0}$,the state-vector after the measurement is any vector of the manifold $E_j \mathcal{H}$, that is, is submitted to the only conditions

(3') $$E_j \psi_{t+0} = \psi_{t+0} \qquad \| \psi_{t+0} \| = 1$$

As is well known, this axiomatics generates a synthetic and beautiful formalism. But this is not sufficient: *any theoretical prediction relating to a so-called "observable" must be actually observed.* As soon as this question is raised, the axiomatics appears to be lacking of two fundamental precisions (in addition to some others of less importance).

*Question M.*- The axiomatics uses the word "measurement" without having defined it previously. This is not a small gap. Quantum Mechanics tells us that there exist two sorts of evolution of the state-vector: the continuous one, ruled by Schrödinger's equation or similar, and the discontinuous process of measurement ruled by the axiom R, which we must *decide* to apply at such or such place. Assuredly Von Neumann said ([1], p. 420) that the boundary between the two can be displaced "to a large extent"; but this extent must stop somewhere, and a time comes when the experimenter must give (at least he thinks so) a definite datum and not a catalogue, and accordingly the theoretician must determine the suitable corresponding operator. Therefore we should need *an axiomatic universal criterion in order to say: here exactly we have measured such physical quantity. But the existing axiomatics does not provide this criterion.*

*Question T*.- For any measurement we should know the value of $t$ to be introduced in the right member of (2), more especially because the probability can depend strongly on $t$. Then two cases occur.

Either we are able to measure the time $t$ at which a measurement is made (taking as origin the instant of emission of the particle). This is actually the case with the various methods of time of flight. Here the comparison between the theoretical prediction and experiment is plain.

Or we cannot measure $t$, as for instance in the case of a particle striking a photographic plate. On the other hand the axiomatics does not provide us *any indication at all* about $t$; neither is it correct to take, as is sometimes done, an estimate of $t$ by dividing the length of a (supposed) trajectory by the mean velocity: this a reasoning of classical Mechanics, and we have no right to mix two separate theories at our convenience. The axiomatics being dumb and the experiment helpless, what we should need is, not the instantaneous probability (2), but the probability that the particle be in such or such region of the plate at any time from $t = 0$ up to $+\infty$. But there is no means to derive from an instantaneous probability a density of probability with respect to time. Thus the ignorance of $t$ raises a real problem we shall have to overcome in some way or another.

## II.- MEASUREMENT IN EXPERIMENTAL REALITY

We have first to answer to the question M, that is, to examine what a measurement is, not in the world of abstractions, but in the laboratory.

As far as I know, the only clear answer was given, for a *determinative* measurement, by L. de Broglie ([2], p. 83): "toute observation ou toute mesure relative à un corpuscule microphysique s'opère toujours en observant un phénomène macroscopique très localisé déclenché par l'action de ce corpuscule". I think no experimenter will reject this statement, and even at first sight it may appear to him as trivial; but in fact it goes very far, as we shall see now.

An immediate consequence is that, the measurements of time set apart, the *only direct observations* - let us not speak yet of measurements - that we can make are *observations of position*.

Now, when we draw measurements from the observation, two cases occur.

Either all we can derive, with the use of a scale, from the observation, is the knowledge of the three coordinates of the observed region (which is the case of most of the detectors); we have then a measurement of position.

Or, as for a track in a cloud chamber, we may hesitate between two interpretations of the observation. We can consider the track as a set of successive *measurable positions* (each at an unknown time, but this point is irrelevant at the present stage); or we may as well consider the track as giving us *the direction (not the magnitude) of the momentum* of the particle. This ambiguous case deserves a special comment.

Adopt first the position point of view and consider one droplet of the track; it fills a region $D$ of the space (fig. 1). Let $\mathfrak{I}_D(x)$ be the characteristic function of $D$ in the $x$-space, and $Q_D$ the corresponding projector of the position operator in the space $L^2_x$ ($Q_D$ is the multiplication by $\mathfrak{I}_D$). Applying the axiom R of reduction of state, we see that, if $\psi(x)$ is the wave-function of the particle immediately after the observation, we must have $Q_D\psi = \psi$ (whether we use (3) or (3')).

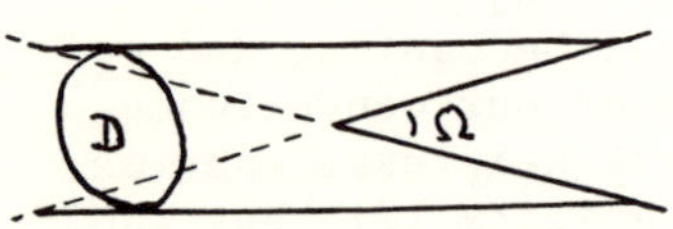


fig. 1

Adopt now the momentum point of view; let $\Omega$ be the solid angle enclosing all the directions of the momentum which are compatible with the visible track, and $\mathfrak{I}_\Omega(p)$ be the characteristic function of $\Omega$ in the $p$-space. If $\Phi$ is the Fourier transformation mapping $\psi(x)$ on $\hat{\psi}(p) = \Phi\psi(x)$, we must have $\mathfrak{I}_\Omega(p)\hat{\psi}(p) = \hat{\psi}(p)$; hence, if $P_\Omega = \Phi^{-1}\,\mathfrak{I}_\Omega\,\Phi$ is the projector of momentum in the space $L^2_x$, we must have $P_\Omega\psi(x) = \psi(x)$.

Now, if the region $D$ is bounded, as is the case here, the two conditions $Q_D\psi = \psi$ and $P_\Omega\psi = \psi$ are *totally incompatible*, that is, there exists *no* function $\psi$ satisfying both conditions. The reason is very simple ([3]): if $Q_D\psi = \psi$, that is, if $\psi$ cancels outside $D$, the Fourier transform $\hat{\psi}(p)$ is an analytic and entire function of $p$; hence its zeros are isolated and we cannot have

$\mathfrak{I}_\Omega(p)\ \hat{\psi}(p) = \hat{\psi}(p)$, that is, $\hat{\psi}(p) = 0$ outside $\Omega$ (a domain of measure $> 0$).

One must notice that *this conclusion has nothing to do with the uncertainty relations*: it is quite independent of the magnitude of $D$ and $\Omega$, and as well of $h$.

We meet an analogous situation in the preparatory measurement. Consider a source $S$ emitting through a collimator $C$. At the time a particle is emitted by the source, we make in fact no observation at all, hence no more measurement (fig. 2). But, according to our general views on particles and solid matter, we can say that, at the time of its emission, the particle is *surely* in the domain $D$ of the space exterior to the walls of the collimator and of the rest of the apparatus. If we made a measurement of the position of the particle, the probability of finding it in $D$ would be 1, so that the axiom P entails $(Q_D\psi,\psi) = 1$, hence $Q_D\psi = \psi$.

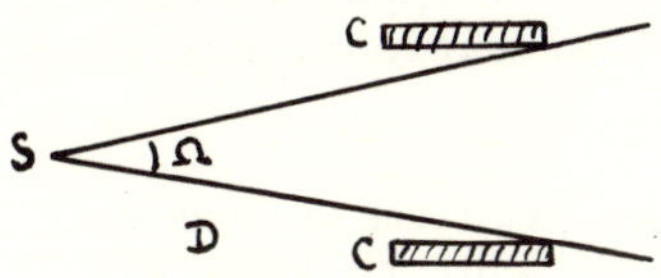

fig. 2

But we could also say - it is may be disputable, however it does not violate any axiom - that the initial direction of the momentum is *surely* inside the solid angle $\Omega$, from which we should deduce $P_\Omega\psi = \psi$ .

As the domain $D$ is unbounded here the above reasoning does not hold any more; but it can be shown (I omit the proof for the sake of brevity, it is similar to the preceding one) that, as the complement of $D$ is bounded, there exists at least a submanifold of infinite dimension of $P_\Omega\mathcal{H}$ which has no vector in common with $Q_D\mathcal{H}$; moreover the vectors of $P_\Omega\mathcal{H} \cap Q_D\mathcal{H}$, if they exist, represent particles whose momentum has a non-zero probability to be unbounded in magnitude. We find again an incompatibility between the conditions $Q_D\psi = \psi$ and $P_\Omega\psi = \psi$ which, if not rigorously total, is nevertheless a *practical* one.

To summarize we see that, in all these ambiguous cases, *we can and we must choose* between a measurement of position and a measurement of direction of momentum, the two choices being *separately* compatible with the axiomatics. How to decide between them is a question we can leave open for the moment.

But an essential conclusion is to be drawn from the above discussion: besides the measurements of time, *the only measure-*

*ments we can really make in practice are measurements of position or of direction of momentum.* Hence two consequences.

The first one, if not fundamental for this paper, is so for other points of the theory of measurement ([4]), and deserves to be emphasized casually. Both operators $x$. and $-i\hbar\ grad$ have pure continuous spectra; when approximated by the procedure of Von Neumann (app. I), the projectors of the decomposition have all infinite dimensions, as was assumed in (1). Accordingly the operators with discrete simple spectrum which are often used in the litterature must be considered - at least in measurement theory - as reflecting purely mathematical situations, not physical ones.

The second consequence leads us to the problem we had in view since the beginning. In microphysics we want to know the value of "observables" such as the magnitude of a velocity, an orbital momentum or a spin. But we have no "velocity-meter" nor "spin-meter"; we can only measure positions or directions of momentum. Therefore we have to imagine devices which enable us to *deduce theoretically* the former quantities from the latter ones. Of course the problem is nearly the same in macrophysics; but Quantum Mechanics, owing to its particular probabilistic background, needs a theory of its own.

## III.- THE SUBSTITUTED MEASUREMENT

Making an indirect measurement is obtained by a well-known procedure, whose importance has been particularly emphasized by L. de Broglie ([2], p. 84).

Let $T$, explicited by (1), the *theoretical* operator we want to measure at a time taken as origin, and that we cannot measure. We submit the system to the action of a device which we call an analyzer, to give it a name. Afterwards, at a time $t \geqslant 0$, we measure *really* an operator $R$, which we write

(4) $$R = \Sigma_k \mu_k F_k$$

$R$ is in fact the operator position or momentum, but we keep the general symbol $R$; we assume yet that every projector $F_k$ has an infinite dimension.

Let $H$ be the hamiltonian expressing the action of the analyzer on the system; we suppose *$H$ fixed and independent of the time*. The evolution of the system is ruled by an unitary operator which we write, adopting Schrödinger's equation in order to handle explicit formulas

$$U_t = exp(-i \frac{t}{\hbar} H) \tag{5}$$

At a time $t \geqslant 0$ which, may be, we do not know, but which we can always suppose to be fixed at an arbitrary value, the experimenter gives us the observed values $\mu_k$ and, on a statistical ensemble of measurements, their frequencies $q_k$ which, according to Bernoulli's theorem, we equal to the theoretical probabilities.

The question is: knowing by the experiment the quantities $\mu_k$ and $q_k$ *measured at the time* $t$, and on the other hand the evolution operator $U_t$, can we *deduce without any indeterminacy* the values and probabilities that the measurement of $T$ would have given to us *at the time* $t = 0$ in the same set of experiments if this measurement had been possible? *It is clear that this inference is only theoretical and not liable to any direct verification.*

If this determination is possible, we shall say that we have made a *substituted measurement* of $T$ by $R$.

This problem is complicated by the diversity of accompanying circumstances, which we must review.

a.- The system may be described initially by a state-vector or a statistical operator. This distinction is not very important, because a careful study shows that the results obtained with a state-vector are easily extended without any change to a statistical operator. Therefore I shall treat essentially the former, which is easier to handle.

b.- A much more important question is to distinguish between two sorts of measurements.

The most frequent in the practice is the statistical ensemble of measurements, which provides the probabilities $q_k$ of the eigenvalues $\mu_k$ of $R$. The possible values of $T$ are known a priori: they are the eigenvalues $\lambda_j$ of $T$, and all that we require from the substituted measurement is the determination of the probabilities $p_j$ of the $\lambda_j$'s. We must obviously suppose that the initial state-vector $\psi$ is always the same over all the statistical ensemble. We expect then that the $p_j$'s have the form

$$p_j = f_j (q_1, \ldots q_k \ldots ; \psi) \tag{6}$$

the functions $f_j$ being well-determined, depending eventually on $\Psi$.

But we must consider also another sort of measurement, which we call *"isolated"*. Given *one* measurement of $R$, whose *result is an eigenvalue of $R$, say $\mu_l$*, we can ask whether it is possible to tell *with certainty* that in the same experiment the measurement of $T$ would have given a well-determined value of $T$, say $\lambda_{l'}$. In order to be able to make such a statement for all the possible values of $R$ and $T$, we must suppose that an isolated measurement is actually an element of a large ensemble of successive experiments which yield all the values of $R$ and $T$. Moreover, for the same reason (i.e. not to repeat constantly the same experiment), we must allow the initial state-vector $\psi$ to vary arbitrarily from a measurement to another.

In this case the probabilities are irrevelant; all we want to know is *a determined mapping of the $\mu_k$'s on the $\lambda_j$'s*; several $\mu_k$'s may correspond to the same $\lambda_j$, but the converse is obviously to be rejected.

c.- Let us come back to the equations (6). In order that they give us uambiguously the $p_j$'s, their right members must be independent of $\psi$ as long as $\psi$ varies according to the information we could have on it previously to the measurement.

In many cases a particle, when emitted, undergoes a preparatory measurement such as the one we examined in § II. This gives us the information in question, which is expressed by a condition of the form $Q\psi = \psi$, where $Q$ is a projector of position or of momentum in the space $\mathcal{H} = L^2$ (or a tensorial product containing this projector for a particle with spin). Then *the right members of (6) must be independent of $\psi$ for any $\psi$ obeying $Q\psi = \psi$.*

But, for a same measurement apparatus, the conditions of preparation may vary to a very large extent. On the other hand it is difficult to accept that the substituted measurement needs in any case an anterior information, since we should hardly accept this necessity for a direct measurement itself. Then we are led to require that *the right members of (6) be independent of $\psi$ whatever be $\psi \in \mathcal{H}$*. This is the case we shall consider as the essential one; it will be solved in § IV and V. Nevertheless, in order not to miss any possibility, we shall also examine briefly the case when we take into account an information anterior to the measurement (§ VI).

The same remarks can be made for an isolated measurement: we can require that the mapping $\{\mu_k\} \to \{\lambda_j\}$ is independent of $\psi$, either for any $\psi \in \mathcal{H}$, or for any $\psi$ satisfying to $Q\psi = \psi$.

d.- Finally we must examine the influence of the time $t$ of the measurement of $R$ on the deduced measurement of $T$ at $t = 0$ (of course, if $t$ is measurable, we have only to see whether the conditions of the substituted measurement are fulfilled at such or such time).

If $t$ cannot be measured we know, as already pointed out in § I, question T, that the axiomatics does not provide us any more any indication about its value; nor is it possible to derive this value from the subsequent formalism. Thus we are faced with the following alternative:

- either we add to the existing axiomatics an extra-axiom which would determine $t$; but really one wonders how this could be done;

- or we require that the deduced measurement of $T$ is not altered by a variation of $t$ (at least in some intervel of time) in the measurement of $R$.

## IV.- CONDITIONS FOR THE SUBSTITUTED MEASUREMENT WHEN THERE IS NO PREVIOUS INFORMATION

### A.- *Statistical ensemble of measurements*

In this case, as already seen (§ III, b), all that we want is that the values (6) of the $p_j$'s are independent of the initial state-vector of the system.

Let $\psi$ be the initial state-vector; after a time $t$ it becomes $\psi_t = U_t\psi$, where $U_t$ is given by (5); for convenience we drop until further notice the suffix $t$ of $U_t$. The probability of getting the value $\mu_k$ of $R$ at the time $t$ is

$$q_k = (F_k\, U\psi,\, U\psi)$$

From this formula, and due to the fact that the $q_k$'s are the only knowledge that the measurement of $R$ gives us, we deduce

$$U\psi = \Sigma_k \sqrt{q_k}\, e^{i\alpha_k}\, \chi_k \tag{7}$$

where the $\alpha_k$'s are arbitrary phases, and $\chi_k$ is a normalized vector which is arbitrary in the manifold $F_k\mathcal{H}$ . Now $p_j$ is equal to $(E_j\psi\ ,\psi)$, hence, by setting

$$\tilde{E}_j = UE_jU^{-1} \tag{8}$$

$p_j$ is given by

$$(9) \qquad p_j = (\tilde{E}_j U\psi , U\psi)$$

and thus finally by

$$(10) \qquad p_j = \Sigma_{kl} \sqrt{q_k q_l}\, e^{i(\alpha_k - \alpha_l)} (\tilde{E}_j \chi_k , \chi_l)$$

(it is easily seen that the permutation of the sums in $k$ and $l$ is sound).

It is clear on this equation that *the knowledge of the* $q_k$*'s learns us practically nothing about the values of the* $p_j$*'s* because of the arbitrairness of $\alpha_k$ and $\chi_k$ (this point can be quantitatively precised on some simple examples). Therefore we have to look for the conditions which annihilate the effect of this arbitrariness. The result is as follows.

*The n.a.s.c. in order that the probabilities* $p_j$ *of the values* $\lambda_j$ *of* $T$ *be determined unambiguously as functions of the probabilities* $q_k$ *of the values* $\mu_k$ *of* $R$ *whatever be* $\psi \in \mathcal{H}$ *is that*

*α) there exists a mapping of the set of the projectors* $F_k$ *of* $R$ *onto the set of the projectors* $E_j$ *of* $T$ *such that each* $F_k$ *has an image in* $\{E_j\}$ *and that to each* $E_j$ *there corresponds a non-empty sequence* $F_{j_1}, F_{j_2}, \ldots;$

*β) if we set* $F'_j = F_{j_1} + F_{j_2} + \ldots$, *we have*

$$(11) \qquad UE_j U^{-1} = F'_j \qquad \forall j$$

*If these conditions are fulfilled, one has*

$$p_j = q_{j_1} + q_{j_2} + \ldots \qquad \forall j$$

The proof of this theorem is sketched in appendix II.

Let us make a remark in order to avoid a misunderstanding. The mapping $F_{j_1}, F_{j_2}, .. \to E_j$, which is in fact a pure mapping of the suffixes $k$ of the $\mu$'s over the suffixes $j$ of the $\lambda$'s, induces a mapping of the eigenvalues $\mu_k$'s onto the $\lambda_j$'s. In the present case of a statistical ensemble of measurements, this latter mapping has only a mathematical content, not a physical one, because the observation of a $\mu_k$ in one measurement does not entail anything for the value of $T$ in the same experiment. It will be different for the isolated measurement.

B.- *Isolated measurement*

We have seen (§ II, b) that this type of substituted measurement requires that there exists a mapping of the $\mu_k$'s onto the $\lambda_j$'s which makes correspond to each $\lambda_j$ a non-empty sequence $\mu_{j_1}, \mu_{j_2}, \ldots$ We want that the observation of $\mu_{j_1}$ or $\mu_{j_2}$ or ... on $R$ entails with certainty the value $\lambda_j$ for $T$. The conditions for this implication are given by the following theorem.

*The n.a.s.c. in order that we can conclude with certainty to the value $\lambda_j$ of $T$ if we have observed the values $\mu_{j_1}$ or $\mu_{j_2}$ or ... for $R$, and this for every $j$, is that*

*- the conditions $\alpha$ and $\beta$ of the preceding theorem be still fulfilled;*

*- moreover, at each isolated measurement, the initial state-vector $\psi$ belong to one and only one of the manifolds $E_j\mathcal{H}$ (in other words $\psi$ must be an eigenvector of $T$).*

For the proof see appendix III.

If the system is initially represented by a statistical operator, the theorem remains valid, provided that the last condition is applied to each of the vectors constituting the statistical operator.

C.- *Another form of the conditions (11)*

The ensemble of the conditions (11) for any $j$ *is equivalent to the following statement: there exists an operatorial function $f(R)$ of $R$ such that we have the equality*

$$UTU^{-1} = f(R) \qquad (12)$$

*The scalar function $\lambda = f(\mu)$, defined over the set of the $\mu_k$'s, is nothing else than the mapping $\mu_{j_1}, \mu_{j_2}, \ldots \rightarrow \lambda_j$ that we have displayed at the end of the paragraph A or assumed as existing in § B.*

The proof of the direct proposition, that is, (11) implies (12), is elementary. Let us multiply (11) by $\lambda_j$ and sum over $j$. The left member yields $UTU^{-1}$. In the right member we have a sum of the form $\Sigma_j \lambda_j F'_j$, which can be written $f(\mu_{j_1})F_{j_1} + f(\mu_{j_2})F_{j_2} + \ldots$, since $f(\mu_{j_1}) = f(\mu_{j_2}) = \ldots = \lambda_j$; this sum is therefore

$\Sigma_k f(\mu_k)\ F_k$, which is $f(R)$ by definition.

The proof of the converse statement, (12) implies (11), is slightly complicated by the fact that the function $f(\mu)$ is not necessarily monotonous; but a re-ordering of the suffixes of the right member yields a monotonous growth, and the equations (11) follow then from the unicity of the spectral decomposition.

D.- *Influence of the time t of the measurement of R*

Let us dismiss momentarily the case when $t$ is measurable, to which we come back in the next paragraph (V, b), and consider the case when $t$ is not measurable, or not measured.

From the discussion of § III, d, we know that there must exist at least an interval $[t_1, t_2]$ whose all points are acceptable as the instant of the measurement of $R$. Therefore the conditions (11) or (12), in which we restore the suffix $t$ of $U$, must remain verified whatever be $t$ in $[t_1, t_2]$. Hence the following consequence.

*The n.a.s.c. in order that the substituted measurement of T by R be valid whatever be t even in a small interval is that*

*α) T commute with the hamiltonian H of the analyzer;*

*β) one have T =f(R), f being the function defined above*¶.

The proof of the necessary condition is short. If (11) is satisfied for two instants $t'$ and $t' + t''$ belonging to $[t_1, t_2]$ we have, using $U_{t'+t''} = U_{t'} U_{t''}$

$$U_{t'}\ E_j = F'_j\ U_{t'}\quad ,\quad U_{t'} U_{t''}\ E_j = F'_j\ U_{t'} U_{t''}$$

which yields, by replacing in the second equation $F'_j\ U_{t'}$ by $U_{t'}\ E_j$, the condition $U_{t'} U_{t''}\ E_j = U_{t'}\ E_j U_{t''}$, hence $U_{t''} E_j = E_j U_{t''}$ : $U_{t''}$ commutes with $E_j$ for any $j$. As $t''$ is variable in some interval, we see, using the group property (for example $U_{2t''} = U_{t''} U_{t''}$) that $U_t$ commutes with any $E_j$ for any value of $t$. It results then from Stone's theorem ([5], p. 380) that the resolution of the identity of $U_t$ commutes with any $E_j$, and likewise the resolution of the identity of $H$; hence $H$ and $T$ commute. The

¶ which entails that $T$ and $R$ commute; but this necessary condition is not sufficient to find again $T = f(R)$.

equation $T = f(R)$ follows then directly from (12).

The proof of the sufficiency of the conditions $\alpha$ and $\beta$ does not present much difficulty.

## V.- CONCLUSION FOR THE PRECEDING TYPE OF MEASUREMENT

It results from this study *that the substituted measurement of T by R, when we impose the condition that it should be valid whatever be the initial state-vector* $\psi$, *is radically impossible, in any case.*

a.- *The case when t is not measurable*

The impossibility derives here from two reasons. The first one is that $T$ and $H$ must commute, which implies ([1], p. 173) that $T$ and $H$ are operatorial functions of a third suitable operator, which means in turn that the physical quantity represented by $T$ and the energy of the particle in the field of the analyzer are themselves scalar functions of a third physical quantity. It is clear that in general this cannot happen. For example if we want to measure the magnitude $|p|$ of a momentum by the means of a deflecting magnetic field, the two operators $T = p^2/2m$ and $H = (1/2m)(p - eA)^2$ do not commute, for the physical reason just given.

However it may happen exceptionnally that $T$ and $H$ commute. Consider a particle emitted by a point-source in the field-free space, and suppose we want to measure the coordinates of its initial momentum by a measurement of position on a target. Here $T = -i\hbar\, \partial/\partial x$ (for example) and $H = -(\hbar^2/2m)\Delta$ : $T$ and $H$ commute. But $R = x.$ or $y.$ or $z.$, or some combination of these operators, and it is easily seen that there exists no square-integrable function $\psi$ such that $-i\hbar\, \partial\psi/\partial x = f(xyz)\psi$ (whatever be $f$), so that the equality $T = f(R)$ is impossible. The fundamental reason of this impossibility is the same as above: $T = f(R)$ means that the physical quantity represented by $T$ is a scalar function of the physical quantity represented by $R$, which is impossible if these quantities have different physical nature.

b.- *t is measurable*

We have here only to satisfy conditions (11) or (12). It is more difficult to draw as general conclusions as above, but the examination of important particular cases for which (11) or (12) can be tested directly shows that the situation is not improved.

Let us consider anew the measurement of the magnitude of a momentum (operator $T$) by the means of a deflecting magnet of hamiltonian $H = (1/2m)(p - eA)^2$ and a subsequent measurement of position (operator $R$). Set $H = H_o + \varepsilon H$ , where $H_o = -(\hbar^2/2m)\Delta$ and $\varepsilon H_1$ is the magnetic term before which we have inserted the parameter $\varepsilon$. For $\varepsilon = 0$ the evolution operator is $U_t = exp(-i\frac{t}{\hbar} H_o)$, which commutes with $T$, so that the equation (12) reduces to $T = f(R)$, an equality we have just seen to be impossible (§a). Casually remark that this shows that *the momentum of a free particle can never be measured through a measurement of position, whether t is measured or not.*

Let us now introduce again $\varepsilon H_1$ in the hamiltonian $H$. As $H_o$ has a pure continuous spectrum, it follows from a theorem of Rellich ([6], p. 55) that the resolution of the identity of $H$ is a continuous function of $\varepsilon$ near $\varepsilon = 0$ and consequently that $U_t = exp\left[-i\frac{t}{\hbar}(H_o+\varepsilon H_1)\right]$ is also a continuous function of $\varepsilon$ for every fixed $t$. Therefore, as the conditions (11), $U_t E_j U_t^{-1}$, are not satisfied for $\varepsilon = 0$, there exists a neighbourhood of this value for which they cannot be any more verified. Physically this means that for sufficiently small magnetic fields the substitute of $T$ by $R$ is impossible, whatever be the time $t$ which could be observed for the measurement of $R$. For larger fields the proof fails; but the study of this case would need a rather hard mathematical research, which does not seem very useful since the question of principle has been settled.

A second particular case is the measurement of the spin by the Stern-Gerlach method, for which we shall find in § VII,B the same sort of impossibility.

c.- *Some remarks about the isolated measurement*

As shown previously (§ IV,B), the isolated measurement requires first the same conditions (11) or (12) as above, so that the preceding conclusions remain fully valid. Moreover the last condition of theorem IV,B entails that the substituted isolated measurement is superfluous or impossible: either we know that the initial state-vector $\psi$ belongs to a determined manifold $E_j\mathcal{H}$ of the operator $T$, and it is useless to measure $R$, which can only confirm this knowledge without bringing anything new as for the measurement; or we know nothing about $\psi$, and an isolated measurement of $R$ does not allow us to bear a retroactive judgment on $T$. This conclusion may seem strange because we have a "physical" tendency to deduce causes from effects, but the probabilistic ground of Quantum Mechanics does not permit such deductions.

## VI.- CONDITIONS FOR THE SUBSTITUTED MEASUREMENT WHEN A PREVIOUS INFORMATION IS TAKEN INTO ACCOUNT

Let us consider now the case, foreseen in § III, c, when we take into account the fact that, at the time $t = 0$, the state-vector $\psi$ belongs to a *known* manifold $Q\mathcal{H}$.

a.- *Statistical ensemble of measurements*

Now the $p_j$'s of the equation (10) must be independent of the $\alpha_k$'s and the $\chi_k$'s whatever be $\psi$ such that $Q\psi = \psi$. I state only the result, the proof being similar to that of appendix II. Let us introduce the convenient notation

$$\tilde{F}_k = U^{-1}F_kU \tag{13}$$

($U$ for $U_t$ as usual). Then

*The n.a.s.c. in order that the probabilities $p_j$ be determined unambiguously as functions of the $q_k$'s whatever be $\psi \in Q\mathcal{H}$ is that*

- *one have*

$$Q\,(E_j\tilde{F}_k - \tilde{F}_kE_j\tilde{F}_k)\,Q = 0 \qquad \forall\, j \text{ and } k \tag{14}$$

- *there exist constants $K_{jk}$ real, belonging to $[0,1]$ and satisfying $\Sigma_j\, K_{jk} = 1\ \ \forall k$, and such that*

$$Q\,(E_j - K_{jk})\tilde{F}_k\,Q = 0 \qquad \forall\, j \text{ and } k \tag{15}$$

The probabilities $p_j$ are then given by $p_j = \Sigma_k\, K_{jk}q_k$.

It is easily seen that, by setting $Q = I$ (identity in $\mathcal{H}$), one finds again the results of theorem IV A (with $K_{jk} = 0$ or $1$).

These conditions are more complicated than those of § IV A, and in particular, regarding the influence of the time $t$, they do not yield as simple results as in § IV D. However, in spite of this complication, they lead to significant conclusions.

Generally speaking, they mean that *the four operators $T$, $R$, $U$ and $Q$ must be fitted one to another*. Now $T$ is given by definition; in the most frquent cases $R$ and $U$ are determined in the first place by the experimental possibilities in constructing an analyzer. On the other hand $Q$ is determined by the experimental conditions of preparation, which have nothing in common with those of the analyzer. Therefore, in general, it cannot be expected that the equations (14) and (15) be fulfilled.

As a physically important example, consider anew the case when $T$ is the momentum operator, $R$ the position operator, and $U$ the evolution operator in the field-free space. Suppose that $Q$ is a projector of *momentum* corresponding to a *bounded* domain of the $p$-space. Then, if we take for the $E_j$'s and the $F_k$'s projectors which correspond equally to *bounded* domains (respectively of the $p$-space and of the $x$-space), it can be shown that the whole of equations (14) and (15) cannot be satisfied, and this whatever be the (fixed) time $t$ of the measurement of $R$.

The proof is somewhat intricate and long but, owing to the interest of the subject, I sketch it briefly in appendix IV. This proof fails if $Q$ corresponds to an unbounded domain of the $p$-space (as for the domains corresponding to the $E_j$'s and $F_k$'s, they can be considered as bounded in any reasonable measurement); but bounded does not signify small, and the hypothesis of boundedness rejects only momenta of practically infinite magnitude.

This results shows that, as in § V b, *it is impossible to measure the initial momentum of a free particle by a subsequent measurement of position, even if we have already some information about the range of initial momenta. Moreover this impossibility can be extended, by the same reasoning as in §V b, to the measurement of the magnitude of the initial momentum by a deflecting magnet*, provided that the magnetic field is sufficiently small.

It is not difficult to find more general examples of operators $T$, $R$, $U$, $Q$ for which the conditions (14) and (15) cannot be fulfilled; but it is not necessary to insist upon them, since the preceding example is sufficiently eloquent.

We see finally that the conclusion is nearly the same as in § V: *the substituted measurement of $T$ by $R$ is yet, at least in general, impossible.*

*It may happen, however, that the adaptation of $Q$ to the triplet $T,R,U$ be satisfied*, and it is the case for the measurement of the spin by the Stern-Gerlach method, as will be shown in § VII B; but it is at the price of choosing for $R$ and $Q$ other operators than those which would have been expected from experimental conditions.

b.- *Isolated measurement*

This case does not deserve a long discussion, because its mathematical treatment is analogous to the one of § IV B, and the conclusions are the same as in § V c. In short the mathematical result is as follows: it is necessary and sufficient that:

α) the initial state-vector $\psi$ belong to an eigenmanifold $E_j\mathcal{H}$ of the theoretical operator $\tilde{T}$, hence $\psi \in Q\mathcal{H} \cap E_j\mathcal{H}$ since $Q\psi = \psi$;

β) one have $Q\mathcal{H} \cap E_j\mathcal{H} \subseteq F'_j\mathcal{H}$, $\tilde{F}'_j = U^{-1}F'_j\, U$ being defined as in § IV B.

*The negative conclusion follows immediately:* the substituted measurement of $T$ by $R$, either is superfluous if we know already the manifold $E_j\mathcal{H}$ to which $\psi$ belongs, or does not provide any certitude about the value of $T$ if we know only that $\psi$ belongs to $Q\mathcal{H}$.

## VII.- HOW QUANTUM MECHANICS DODGES THE DIFFICULTY

After the conclusions we drew in the last two sections, one can wonder how Quantum Mechanics manages to do measurements which seem acceptable, or are accepted, by most of the physicists. We shall clear up this question on two examples.

A.- *Measurement of the direction of a momentum by a measurement of position for a free particle.*

Let us take again the example of § V of a source emitting in the field-free space; we have seen that the substituted measurement is impossible if we do not know anything on the initial $\psi$, or even if we know its initial range of momenta.

But suppose we know, no matter how for the moment, that the initial $\psi$ is a gaussian wave-function $\psi(x) = (\alpha\sqrt{2\pi})^{-3/2} \times \exp(-x^2/4\alpha^2)\, \exp(ip_0.x/\hbar)$, for which the average position is zero, the origin O of the coordinates. It is well-known (see e.g. [7], p. 127) that, due to the orders of magnitude of the quantities involved, after an infinitely short while (say $10^{-20}$ to $10^{-30}$ s) the wave-packet spreads as if it were a bunch of classical particles. On this basis an elementary calculation leads to the following result.

*Let $\tilde{D}$ be any domain of the p-space, and $D_t = \frac{t}{m}\tilde{D}$ its homothetic in the x-space,* O *being the centre of homothety; m is the mass of the particle. Then the probability of observing the particle in $D_t$ is equal to the probability that the extremity of its momentum at $t = 0$ is in $\tilde{D}$.*

Let then $\Delta$ be some domain of the $x$-space. If we are able to measure the time of flight of a particle between its emission and its arrival in $\Delta$, we can deduce, from the probability of position in $\Delta$ at the time $t$, the probability for the initial momentum of being in the homothetic domain $(t/m)^{-1} \times \Delta$ of the $p$-space, and therefore the probability distribution of this momentum.

But, for many detectors or beams, we cannot measure $t$, so that, given $\Delta$, the region $(t/m)^{-1} \times \Delta$ in the $p$-space cannot in general be determined. But there is a case when this determination is possible: if we take for $\Delta$ a solid angle $\Omega$ *having the mean point* 0 *of* $\psi$ *for its apex,* $(t/m)^{-1} \times \Delta$ is the same solid angle in the $p$-space, and this *whatever be* $t$. If then we admit, as does explicitly Blokhintsev ([7],p. 451), that the probability of observing the particle in a detector transverse to $\Omega$ at an *unknown* time is equal to the probability of the particle to be in $\Omega$ at *any fixed* time, we know the distribution of probability of the direction of the momentum at the time zero.

It is practically in this - implicit - manner that Quantum Mechanics explains "successfully" many experiments, particularly experiments of collisions.

But it is clear that all this reasoning is completely outside Quantum Mechanics. Firstly we have given up the axiom P, by removing the time $t$ which figures in equation (2) and by replacing the region of real observation by a cone in which we do not observe anything. Secundly we have assumed that the particle has a globular shape which can be reasonably approximated by a gaussian wave. And, above all, if we want to place the cone $\Omega$ in the $x$-space, we are obliged to suppose *that its apex -that is, the mean point of the initial* $\psi$ *- is in the source*. But we have no logical reason - I mean, derived from the axiomatics - to set the mean point of $\psi$ in the boundary of the source rather than in any other point of the free space. In other words we have used classical trajectories and, moreover, placed their origin where we were be pleased to do. All that is not at all coherent with our fundamental axiomatics.

B.- *Measurement of the spin in the Stern-Gerlach experiment*

It is customary, when dealing with measurements of spin, to reason on Pauli matrices alone and so to obtain easily satisfying formulas. Unfortunately the Stern-Gerlach apparatus does not observe any spin at all, only spots on a screen; that is, it measures positions or, if admitted as in § II, directions of momentum. Therefore a coherent theory of the measurement needs compulsorily the association of spin and space variables.

The suitable Hilbert space is then the tensorial product of the 2-dimensional spin-space $s$ by the $L^2$-space of wave-functions of $xyz$: $\mathcal{H} = s \otimes L^2$. Let $I$ be the identity in $L^2$, $J$ the identity in $s$. Suppose we want to measure the $z$-component of the spin;

let us introduce the eigenvectors $\varepsilon_+$ and $\varepsilon_-$ of the matrix $\sigma_z$. A normalized vector $\omega$ of $\mathcal{H}$ has the form

$$\omega = \varepsilon_+ \otimes \zeta_+ + \varepsilon_- \otimes \zeta_- \qquad \|\zeta_+\|^2 + \|\zeta_-\|^2 = 1 \tag{16}$$

The theoretical operator $T$ of the general theory is here $T = \sigma_z \otimes I$. Let $\Pi_+$ and $\Pi_-$ be the projectors on $\varepsilon_+$ and $\varepsilon_-$ in the space $s$; the projectors $E_j$ of the general theory reduce to

$$E_+ = \Pi_+ \otimes I \quad ; \quad E_- = \Pi_- \otimes I \tag{17}$$

The exact hamiltonian in the magnetic field $B$ is $H = - J \otimes (\hbar^2/2m)\, \Delta + \mu\Sigma_{xyz}\sigma_x \otimes B_x$ . As this value would lead to unworkable calculations, it is customary (see e.g. [8], p. 593) - and reasonable - to approximate it by $H = \mu\sigma_z \otimes B_z$. Then the calculation of $U_t$ is easy and yields $U_t = \Pi_+ \otimes V_+ + \Pi_- \otimes V_-$ , where

$$V_+ = exp(-i \frac{t}{\hbar} \mu B_z) \quad , \quad V_- = exp(i \frac{t}{\hbar} \mu B_z) \tag{18}$$

From (8) and (17) we get therefore

$$\tilde{E}_+ = U_t\, E_+\, U_t^{-1} = \Pi_+ \otimes I \quad , \quad \tilde{E}_- = \Pi_- \otimes I \tag{19}$$

that is, $\tilde{E}_+ = E_+$ , $\tilde{E}_- = E_-$, *independent of* $t$.

Now the real measurement bears on position or momentum. Let us the choice open for the moment. Call $X$ the operator of measurement in $L^2$, $M_k$ its projectors. The operator $R$ of the general theory is $R = J \otimes X$ and its projectors $F_k = J \otimes M_k$.

Let us apply then the results of § IV. Whatever be the manner of gathering the projectors $F_k$ in two sums $F'_+$ and $F'_-$ (the $F'_j$ of the general theory), the conditions (11) become, owing to (19)

$$\Pi_+ \otimes I = J \otimes M_+ \; , \quad \Pi_- \otimes I = J \otimes M_-$$

which is impossible, and this *whatever be the time* $t$ of the measurement of $R = J \otimes X$ (and also whatever be $X$). *Consequently the substituted measurement of* $\sigma_z \otimes I$ *by the means of the Stern-Gerlach apparatus is always impossible if we know nothing about the initial state-vector* $\omega$.

Let us try then to take advantage of a previous information if available. Suppose that the preparatory measurement provides us the indication $(J \otimes Q)\omega = \omega$, where $Q$ is a projector in $L^2$ (of position or of momentum, we do not choose yet). Owing to (16) this condition is equivalent to

(20) $$Q\zeta_+ = \zeta_+ \quad , \quad Q\zeta_- = \zeta_-$$

whose physical meaning is obvious. Now we could attempt to test the equations (14) and (15), but it is shorter here to reason directly.

By the action of the evolution operator $\omega$ becomes, at the time $t$, $\omega' = U_t\omega$, which, if we set $\phi_+ = V_+\zeta_+, \phi_- = V_-\zeta_-$ , yields

(21) $$\omega' = \varepsilon_+ \otimes \phi_+ + \varepsilon_- \otimes \phi_- \qquad \|\phi_+\|^2 + \|\phi_-\|^2 = 1$$

Moreover, by setting

(22) $$Q_+ = V_+QV_- \quad , \quad Q_- = V_-QV_+$$

the conditions (20) are transformed into

(23) $$Q_+ \phi_+ = \phi_+ \, , \quad Q_- \phi_- = \phi_-$$

Now the probabilities $p_+$ and $p_-$ of the eigenvalues of the spin are derived from the general formula (9), which becomes here $p_+ = (E_+\omega' , \omega')$, $p_- = (E_-\omega' , \omega')$, that is, finally and owing to (19) and (21)

(24) $$p_+ = \|\phi_+\|^2, \quad p_- = \|\phi_-\|^2$$

On the other hand let us calculate the probabilities $q_k = (F_k\omega',\omega')$ for $R$; one finds easily, from $F_k = J \otimes M_k$ and (21)

(25) $$q_k = (M_k\phi_+ \, , \, \phi_+) + (M_k\phi_- \, , \, \phi_-)$$

We are ready now to examine the two cases: $Q$ is a projector of position, or of momentum.

Suppose first that $Q$ is a projector of position. The multiplication operators $V_+$ and $V_-$ of (18) commute obviously with $Q$, so that (22) yields $Q_+ = Q_- = Q$. Refering to (20) and (23) we see therefore that the wave-functions $\phi_+$ and $\phi_-$ of the analyzed beam belong to the same manifold $QL^2$ as the functions $\zeta_+$ and $\zeta_-$ of the initial beam, which means: the theory of the analyzer cannot explain the separation of the beams, in disagreement with the experiment.

But we can go further. Suppose that, in spite of this failure, we want to take into account the fact that we observe two spots on the screen, defining two disjoint domains $D_1$ and $D_2$ of the space. Accordingly we take for $X$ the position operator; to $D_1$ and $D_2$ there correspond two well-determined projectors $M_1$ and $M_2$ of $X$ and, applying (25), we obtain

$$q_1 = (M_1\phi_+,\phi_+) + (M_1\phi_-,\phi_-)\ ,\ q_2 = (M_2\phi_+,\phi_+) + (M_2\phi_-,\phi_-)$$

It appears immediately that the knowledge of $q_1$, $q_2$ and $M_1$, $M_2$ does not determine $\|\phi_+\|^2$ and $\|\phi_-\|^2$, that is, $p_+$ and $p_-$. Therefore the real measurement of position does not learn us anything about the probabilities of the spin.

But suppose now that $Q$ is a momentum projector determined by the collimation. $V_+$ and $V_-$ do not commute any more with $Q$, hence $Q_+$ and $Q_-$ defined by (22) are different; moreover an approximate calculation whose principle is in Bohm ([8], p. 595) shows that $Q_+$ and $Q_-$ are yet practically momentum projectors corresponding to two disjoint solid angles, so that $Q_+Q_- = 0$. Hence, taking for $X$ the momentum operator and identifying $M_1$ to $Q_+$, $M_2$ to $Q_-$, we get, from (23) and (25)

$$q_1 = \|\phi_+\|^2 \quad , \quad q_2 = \|\phi_-\|^2$$

and therefore, by virtue of (24), $p_+ = q_1$ , $p_- = q_2$ : the probabilities of the final momenta of being in the two solid angles determined by the two spots on the screen are equal to the probabilities of the initial spin.

We have thus finally succeeded. But, as regards our initial axiomatics, the cost is very high. We must first consider the preparatory measurement as a measurement of momentum instead of position; and likewise we must secundly convert the measurement of position on the screen into data of directions of momentum, exactly as in § A, and this by placing, somewhat arbitrarily, the apexes of the solid angles on the axis of the magnet; and finally we have disregarded the time $t$ which figures in the equation (2).

## VIII.- GENERAL CONCLUSIONS

Quantum Mechanics considers, in its basic postulates, a lot of self-adjoint operators, which are supposed to represent "observables". But in fact two observables only can be really measured: position or, eventually, direction of momentum.

Any other quantity requires to be measured indirectly, by the means of a theoretical link with one of the preceding measurements. Unfortunately, when starting from the classical axiomatics, this link fails, in general.

It is only in some particular cases that it is possible to obtain indirect measurements, but under a triple condition: firstly by converting experimental data of position into data of direction of momentum; secundly by attributing to the particles strongly condensed wave-functions and placing their mean point in a conveniently chosen source; and thirdly by disregarding the time $t$ which figures in the axiom P. In fact Quantum Mechanics can make measurements only by calling upon the help of classical Mechanics during a short - but essential - stage.

Thus Quantum Mechanics offers the queer example of a theory which abandons its axiomatics at the very moment it needs it, and which does not measure really what it claims it measures.

## APPENDIX I

As stated in § II, the only operators that we can measure *directly* are operators of position or of direction of momentum. Both of them have *purely continuous spectra*. Let us consider then an operator $A = \int \lambda \, d\mathcal{E}_\lambda$, where the projector $\mathcal{E}(\lambda)$ is a continuous function of $\lambda$. Let us divide the interval of variation of $\lambda$ in sub-intervals $[\lambda_j, \lambda_{j+1})$ ($j = -\infty$ or $M$ to $+\infty$ or $N$), define the step-function $e(\lambda)$ by $e(\lambda) = \lambda_j$ when $\lambda_j \leqslant \lambda < \lambda_{j+1}$, and consider the operator $B = e(A) = \int e(\lambda) \, d\mathcal{E}_\lambda$. If we define the projector $E_j$ as $E_j = \mathcal{E}(\lambda_{j+1}) - \mathcal{E}(\lambda_j)$, $B$ has also the form $B = \Sigma_j \lambda_j E_j$, that is, the discrete form assumed in the equation (1) of the text. This is Von Neumann's procedure ([1], p. 220).

For physical uses we can say, after Von Neumann, that "the approximate measurement of $A$ is equivalent to the exact measurement of $B$". From a mathematical point of view it is readily seen that $B$ can be made as close (strongly) to $A$ as desired by taking the intervals $[\lambda_j, \lambda_{j+1})$ sufficiently small. Thus the replacement of $A$ by $B$ must be considered as a technical device which provides simpler notations, and which we can abandon at will by making $B$ converge to $A$.

Now it must be noted, according to a general property of the limit-points of a spectrum ([5], p. 361), that the multiplicity of any projector $E_j = \mathcal{E}(\lambda_{j+1}) - \mathcal{E}(\lambda_j)$ is *always infinite*. This finishes justifying the assumed form of the operator $T$, *§ I*.

## APPENDIX II

Let us start from the formula (7). It can be written $U\psi = \sqrt{q_1}\, e^{i\alpha_1} \chi_1 + \phi$, where $\phi$ is independent of $\alpha_1$. It is easily seen that the probability $p_j$ has the value

$$(1) \qquad p_j = q_1 (\tilde{E}_j \chi_1, \chi_1) + 2\, Re\left[\sqrt{q_1}\; e^{i\alpha_1} (\tilde{E}_j \chi_1, \phi)\right] + (\tilde{E}_j \phi, \phi)$$

The right member is independent of $\alpha_1$ if and only if $q_1 = 0$ or $(\tilde{E}_j \chi_1, \phi) = 0$. Dismiss momentarily the condition $q_1 = 0$ and consider $(\tilde{E}_j \chi_1, \phi) = 0$. It must be satisfied whatever be $\chi_1 \in F_1 \mathcal{H}$ and $\phi \in (I - F_1)\mathcal{H}$ ; then it is equivalent to the equation $(\tilde{E}_j F_1 u, (I-F_1)v) = 0$ whatever be $u$ and $v \in \mathcal{H}$ . Hence the equa- $(I-F_1)\, \tilde{E}_j\, F_1 = 0$; as this is a relation between the projectors of $T$ and $R$, it remains valid even if it could happen that $q_1 = 0$ for a particular $\psi$, so that we can forsake this particular case. Finally, as in the above equation we can replace the suffix 1 by any suffix $k$, we obtain the general condition

$$F_k\, \tilde{E}_j\, F_k = \tilde{E}_j\, F_k$$

Now the operator of the left member is obviously self-adjoint, hence also the right member. Therefore $\tilde{E}_j$ and $F_k$ commute. It follows that $(\tilde{E}_j \chi_k, \chi_l) = (\tilde{E}_j F_k \chi_k, F_l \chi_l) = (\tilde{E}_j F_l F_k \chi_k, \chi_l) = 0$ if $k \neq l$, so that (1) becomes

$$p_j = \Sigma_k\, q_k\, (\tilde{E}_j \chi_k, \chi_k) \qquad \forall j$$

where $p_j$ is now independent of the $\alpha_k$'s, as was expected.

We must now require that, for $j$ and $k$ fixed, $(\tilde{E}_j \chi_k, \chi_k)$ does not depend on the (normalized) vector $\chi_k$ as long as $F_k \chi_k = \chi_k$; hence the equation $(\tilde{E}_j u, u) = K\,(u,u)$, $K$ constant, for any $u \in F_k \mathcal{H}$. This can be written $((\tilde{E}_j - K)u, u) = 0$ $\forall u \in F_k \mathcal{H}$, from which it is easily shown that $((\tilde{E}_j - K)v, w) = 0$ whatever be $v$ and $w$ belonging to $F_k \mathcal{H}$. Hence $(\tilde{E}_j - K)v = 0$ for any $v \in F_k \mathcal{H}$. This means that any vector of the manifold $F_k \mathcal{H}$ is an eigenvector of $\tilde{E}_j$; hence $K = 0$ or $1$. We see then finally that, given any $j$, and whatever be $v \in F_k \mathcal{H}$, we must have either $\tilde{E}_j v = 0$ or $\tilde{E}_j v = v$. Therefore

$$\tilde{E}_j F_k = 0 \quad \text{or} \quad \tilde{E}_j F_k = F_k \qquad (j \text{ fixed, for all } k)$$

Let $F_{j_1}$, $F_{j_2}$, ... be the projectors of $R$ which satisfy the second relation. By setting $F'_j = F_{j_1} + F_{j_2} + \dots$ one obtains $\tilde{E}_j F'_j = F'_j \ \forall j$, or

$$F'_j < \tilde{E}_j \qquad \forall j$$

Add now these relations for all $j$; as $\Sigma_j F'_j = \Sigma_j \tilde{E}_j = I$, it follows that only the equality is possible; hence $F'_j = \tilde{E}_j$, that is, the equation (11) of the text.

The sufficiency of the conditions $\alpha$ and $\beta$ does not present any difficulty.

## APPENDIX III

Suppose that in one measurement of $R$ we observe one value of the subset $\mu_{j_1}$, $\mu_{j_2}$, ... ; in order to be sure that the corresponding value of $T$ is $\lambda_j$, we must have $(E_j\psi,\psi) = 1$, that is, $E_j\psi = \psi$; as this must be true whatever be $j$ the last condition of the theorem is proved. Now consider some vector $\psi$ of the manifold $E_j\mathcal{H}$. I assert that we cannot have $(F_{l_1}U\psi,U\psi) \neq 0$, $(F_{l_2}U\psi,U\psi) \neq 0$, ...for any $l \neq j$; otherwise we should observe $\mu_{l_1}$ or $\mu_{l_2}$ or ... for some of the successive measurements, so that we could say that $\mu_{l_1}$ or $\mu_{l_2}$ or ... implies $\lambda_j$, which is impossible by definition of the subsets $\mu_{j_1}$, $\mu_{j_2}$, ... Therefore we have $(F'_l\, U\psi,\, U\psi) = 0$ for any $l \neq j$. Hence $(F'_j\, U\psi,\, U\psi) = 1$, hence $F'_j\, U\psi = U\psi$. Therefore $E_j\psi = \psi$ implies $U^{-1}F'_jU\psi = \psi$; hence $E_j \leqslant U^{-1}F'_j\, U$ and, by summation over $j$ as in appendix II, $E_j = U^{-1}F'_j\, U$ or (11).

The proof of the sufficiency of the conditions is plain.

## APPENDIX IV

We begin by a short lemma. Let $Q$ be a projector of momentum corresponding to a domain $\Delta$ of the $p$-space, and $F$ a projector of position corresponding to a domain $D$ of the $x$-space. Then, if $\Delta$ and $D$ are both bounded, the equality

(1) $$QFQ = 0$$

is impossible.

The proof rests on the same property of analyticity as in § II (refer to it for the notations). $Q$ has the form $\Phi^{-1}\mathfrak{I}_{\Delta}(p)\Phi$, and $F$ is the multiplication by $\mathfrak{I}_{D}(x)$. Hence (1) is equivalent to

$$\mathfrak{I}_{\Delta}(p)\ \Phi\ \mathfrak{I}_{D}(x)\ \Phi^{-1}\ \mathfrak{I}_{\Delta}(p) = 0 \tag{2}$$

Now, whatever be the function $\hat{u}(p) \in L^2_p$, $\Phi^{-1}\ \mathfrak{I}_{\Delta}(p)\ \hat{u}(p)$ is an entire function of $x$, which cannot cancel in the whole domain $D$ (if this latter is of positive measure, of course); hence $\Phi\ \mathfrak{I}_{D}\ \Phi^{-1}\ \mathfrak{I}_{\Delta}\ \hat{u}$ is a non-zero function of $p$ which is entire, hence cannot cancel in the whole $\Delta$; therefore (2) is impossible, and likewise (1). It results from this impossibility that there exists at least a non-zero function $f(x) \in L^2$ such that $g(x) = QFQ\ f(x) \neq 0$.

Let us now come back to equations (15). Owing to (13) and to the commutativity of $E_j$, $Q$ and $U$ (and for this latter, whatever be $t$), (15) may be written, as is easily seen

$$E_j\ Q\ F_k\ Q = K_{jk}\ Q\ F_k\ Q \tag{3}$$

where we find again the product $QF_kQ$ (for any $k$) instead of the $QFQ$ above.

Let us multiply the two members of (3) by a function $f(x)$ such that $g(x) = QF_kQ\ f(x) \neq 0$; we obtain $E_jg = K_{jk}g$, which means that $K_{jk}$ is an eigenvalue of $E_j$, hence is 0 or 1. But from the theorem VI we have $K_{jk} \geqslant 0$, $\Sigma_j K_{jk} = 1\ \forall k$. Then, for any fixed $k$, there exists one and only one suffix $j_k$ such that $K_{j_k k} = 1$, and, for any $j \neq j_k$, we have $K_{jk} = 0$. Accordingly the equations (3) separate in two groups

$$E_{j_k}\ Q\ F_k\ Q = Q\ F_k\ Q \tag{4}$$

$$E_j\ Q\ F_k\ Q = 0 \qquad \forall j \neq j_k \tag{4'}$$

Let us now use the equations (14). Owing to (13) and to the commutativity of $Q$, $E_j$, $U$ they become, as is easily seen, $Q\ F_k\ E_j\ F_k\ Q - E_j\ Q\ F_k\ Q = 0$, so that, for a fixed $k$ and $j \neq j_k$, (4') yields $Q\ F_k\ E_j\ F_k\ Q = 0\ \ \forall j \neq j_k$, or, by introducing the characteristic functions $\mathfrak{I}_j(p)$ and $\mathfrak{I}_k(x)$ corresponding to $E_j$ and

$F_k$ and the Fourier operator $\Phi$

$$\mathfrak{I}_\Delta(p)\ \Phi\ \mathfrak{I}_k(x)\ \Phi^{-1}\ \mathfrak{I}_j(p)\ \Phi\ \mathfrak{I}_k(x)\ \Phi^{-1}\ \mathfrak{I}_\Delta(p) = 0 \quad \forall\, j \neq j_k$$

The same property of analyticity as above shows that this equality is impossible, since it would require that the characteristic functions $\mathfrak{I}_k(x)$ or $\mathfrak{I}_j(p)$ be identically zero, that is, that the corresponding domains of the $x$-space or of the $p$-space do not exist.

It must be remarked that this result is independent of the time $t$ which is inserted in the evolution operator $U_t$.

*References*

1.- Von Neumann - Mathematical foundations of Quantum Mechanics. Transl. by R. Beyer. Princeton Univ. press, 1955.

2.- L. de Broglie - La théorie de la mesure en Mécanique ondulatoire. Gauthier-Villars, Paris, 1957.

3.- D. Fargue, private communication.

4.- F. Fer - Annales de la Fondation Louis de Broglie. 6, 87 and 217, 1981.

5.- F. Riesz et B. Sz-Nagy - Leçons d'analyse fonctionnelle. Akademiai Kiadó, Budapest, 1953.

6.- V. P. Maslov - Théorie des perturbations et méthodes asymptotiques. Trad. Lascoux et Seneor. Dunod - Gauthier-Villars, Paris, 1972.

7.- D. I. Blokhintsev - Mécanique quantique. Masson et Cie, Paris 1967.

8.- D. Bohm - Quantum theory. Constable and Cy, London, 1954.

# QUANTUM INTERFERENCE OF PROBABILITIES AND HIDDEN VARIABLE THEORIES

M.D. Srinivas

Department of Theoretical Physics
University of Madras
Guindy Campus
Madras-600025

One of the fundamental contributions of Louis de Broglie, which does not get cited often, has been his analysis of the basic difference between the calculus of the probabilities as predicted by quantum theory and the usual calculus of probabilities -the one employed by most mathematicians, in its standard axiomatised version due to Kolmogorov. In a seminal paper[1] published in 1948, de Broglie showed that the basic difference between the 'Quantum' probability theory and the 'classical' probability theory manifests itself in a certain 'interference of probabilities', exhibited by the quantum-theoretic probabilities, and which is totally absent in classical probability theory. From this, de Broglie was led to the very important conclusion that the 'usual mathematical statistics is based on postulates which cease to be exact in wave mechanics in such a way that, to make the formalism of wave mechanics enter into the general picture of the calculus of probabilities, it is necessary to construct a mathematical statistical theory more comprehensive than that of which use is generally made -by abandoning certain too restricted postulates'[2]. The realisation of this programme of developing a 'quantum probability theory', is still only in its initial stages[3-5].

This talk is basically devoted to a discussion of the 'quantum interference of probabilities', discovered by de Broglie. In particular, we shall show that it is this feature of the quantum-theoretic probabilities which leads to some serious constraints[6] on the possible 'hidden-variable formulations' of quantum mechanics, including the celebrated theorem of Bell[7].

## I. THE QUANTUM INTERFERENCE OF PROBABILITIES



S. Diner et al. (eds.), The Wave-Particle Dualism, 367–375.

As is well known, in conventional non-relativistic quantum mechanics, the states are represented by density operators on a separable Hilbert space $\mathcal{H}$, and observables by self-adjoint operators on $\mathcal{H}$. The probability that the outcome of an experiment to measure the observable A, lies in the Borel set $\Delta \subset R$, is given by

$$Pr^{\rho}_{A}(\Delta) = Tr(\rho\ P^{A}(\Delta)) \quad , \tag{1}$$

where $\Delta \rightarrow P^{A}$ is the spectral measure associated with A and where $\rho$ is the state of the system. Conventional quantum theory also predicts the joint-probabilities for observing various outcomes in a sequence of measurements, via the so-called 'collapse postulate' which fixes the post-measurement state of the system, conditional on a particular outcome being realised. This collapse postulate, in its usual form given by von Neumann[8] and Lüders[9], is available only for observables with purely discrete spectra[10]. Let A,B be two such observables, with the corresponding spectral resolutions

$$A = \sum_{\alpha} \alpha\ P^{A}(\alpha) \tag{2a}$$

$$B = \sum_{\beta} \beta\ P^{B}(\beta) \tag{2b}$$

where $\alpha,\beta$ are the eigenvalues and $P^{A}(\alpha)$ and $P^{B}(\beta)$ are the projectors onto the associated eigen-subspaces. From the von Neumann -Lüders collapse postulate, it follows that the joint-probability $Pr^{\rho}_{A,B}(\alpha,\beta)$ that the outcomes $\alpha,\beta$ are found in a measurement of A followed by that of B, is given by[4,11]

$$Pr^{\rho}_{A,B}(\alpha,\beta) = Tr(\rho\ P^{A}(\alpha)\ P^{B}(\beta)\ P^{A}(\alpha)). \tag{3}$$

In the above, we have assumed that the observables and states are in the Heisenberg picture where all the time-evolution, in between measurements, is carried by the observables.

The joint-probability $Pr^{\rho}_{A,B}(\alpha,\beta)$, as given by eq.(3), has the property that

$$\sum_{\beta} Pr^{\rho}_{A,B}(\alpha,\beta) = Tr(\rho\ P^{A}(\alpha)) = Pr^{\rho}_{A}(\alpha). \tag{4}$$

However, when the same joint-probability is summed over the various outcomes of the earlier A-measurement, we do not in general have a relation similar to eq.(4). In fact, if the observables A,B are incompatible (i.e., $AB-BA \neq 0$), there always exists a state $\rho$ such that

$$\sum_{\alpha} Pr^{\rho}_{A,B}(\alpha,\beta) \neq Pr^{\rho}_{B}(\beta) = Tr(\rho\ P^{B}(\beta)). \tag{5}$$

The above relation (5) expresses the basic non-classical feature of the quantum-theoretic joint-probabilities, that one of the standard 'marginal probability conditions' of classical probability theory is not valid any more. In relation (5), $Pr^{\rho}_{B}(\beta)$ represents the probability that an ensemble of systems prepared in state $\rho$ yield the outcome $\beta$ in a B-measurement, with no other measurements being made on the ensemble earlier. On the other hand, the sum $\sum_{\alpha} Pr^{\rho}_{A,B}(\alpha,\beta)$ is the probability that an ensemble of systems prepared in state $\rho$ and subjected first to an A-measurement, yield the outcome $\beta$ in a later $\beta$-measurement. That these two probabilities, $Pr^{\rho}_{B}(\beta)$ and $\sum_{\alpha} Pr^{\rho}_{A,B}(\alpha,\beta)$ are in general different, is the essential feature of the quantum-theoretic probabilities which was termed the 'quantum interference of probabilities' by de Broglie. In essence, in quantum theory the statistics of the outcomes of a later B-measurement depends not only on the state $\rho$ in which the ensemble was prepared to start with, but also on whether or not this ensemble was first subjected to an A-measurement. Obviously, this 'interference of probabilities' is manifest also in the conditional and other joint-probabilities predicted by quantum theory[12].

It can be easily checked that the interference of probabilities as expressed in relation (5) will not be there (i.e., the inequality in (5) will become equality) if the observables A,B are compatible (i.e. AB-BA=0). Also, even when A,B are incompatible, the interference will be absent if the initial state of the system $\rho$ is such that

$$\sum_{\alpha} P^{A}(\alpha)\ \rho\ P^{A}(\alpha) = \rho \tag{6}$$

Clearly, eq.(6) is satisfied only by those density operators which commute with all the eigen-projectors $\{P^{A}(\alpha)\}$ of A and hence with A itself. Therefore we can see that the 'quantum interference of probabilities' arises from two basic features of quantum theory :

(i) Existence of incompatible observables.

(ii) The transformation in the state of an ensemble due to measurement, in the manner prescribed by the 'collapse postulate'.

We now consider whether it is possible to do away with the interference of probabilities, by say postulating joint-probabilities which are different from those given by eq.(3). In fact it is often emphasised that eq.(3) after all is based on the controversial 'collapse postulate'. We shall now cite the following result which is essentially due to Wigner[13] and Davies[14] :

Theorem 1 (Wigner-Davies Theorem) : Let A,B be two self-adjoint operators which have a purely discrete spectrum[15]. Suppose, there is associated with each density operator a joint probability $P^{\rho}_{A,B}(\alpha,\beta)$ such that

$$\text{(i)} \sum_{\beta} P^{\rho}_{A,B}(\alpha,\beta) = \mathrm{Tr}(\rho\, P^{A}(\alpha)) = \mathrm{Pr}^{\rho}_{A}(\alpha) \tag{7}$$

$$\text{(ii)} \sum_{\alpha} P^{\rho}_{A,B}(\alpha,\beta) = \mathrm{Tr}(\rho\, P^{B}(\beta)) = \mathrm{Pr}^{\rho}_{B}(\beta) \tag{8}$$

$$\text{(iii)}\ P^{\nu\rho_1+(1-\nu)\rho_2}_{A,B}(\alpha,\beta) = \nu P^{\rho_1}_{A,B}(\alpha,\beta) + (1-\nu)\, P^{\rho}_{A,B}(\alpha,\beta) \tag{9}$$

for all density operators $\rho_1$, $\rho_2$ and $0 < \nu < 1$.
Then A,B are compatible and

$$P^{\rho}_{A,B}(\alpha,\beta) = \mathrm{Tr}(\rho\, P^{A}(\alpha)\, P^{B}(\beta)). \tag{10}$$

As can clearly be seen, the above theorem is a statement of non-existence of joint-probabilities for incompatible observables, whenever it is demanded that the joint-probabilities also satisfy the conditions (i)-(iii). Now, the usual joint-probability $\mathrm{Pr}^{\rho}_{A,B}(\alpha,\beta)$ given by eq.(3), escapes the 'negative' conclusion of the Wigner-Davies theorem precisely because it exhibits the 'interference of probabilities', which is nothing but the violation of condition (ii) (or, condition (i) if it is the B-measurement which is performed earlier). The question then is, can we retain the conditions (i) & (ii) (so that there will be no 'interference of probabilities') but drop condition (iii) so that the very existence of joint-probabilities will not get ruled out.

However, what does dropping of condition (iii) mean ? This condition represents the requirement that the probabilities have a linear, or more appropriately affine, dependence on the density operator. This requirement in fact is a consequence of the usual description for the density operator associated with a mixed-ensemble (to be referred to as the 'Mixture Postulate'), which is the following :

Mixture Postulate : Suppose we have two ensembles $\Sigma_1$, $\Sigma_2$ that are characterized by the density operators $\rho_1$ and $\rho_2$ respectively. Let the ensemble $\Sigma$ be formed by selecting (randomly) a fraction $(0 \leqslant \nu \leqslant 1)$ of systems from $\Sigma_1$ and a fraction $(1-\nu)$ of systems from $\Sigma_2$. Then the ensemble $\Sigma$ will be characterized by the density operator[16] $\nu\rho_1 + (1-\nu)\rho_2$.

It is because of the above prescription for the state of a mixed-ensemble, that the probabilities in quantum theory are required to be affine functions of the density operator -i.e., satisfy the condition (iii). Once this is accepted, the Wigner-Davies theorem inevitably leads to the conclusion that, for incompatible observables to have a joint-probability, it is necessary to violate at least one of the conditions (i), (ii). In other words, we cannot do away with the 'quantum interference of probabilities' unless we also gives up the usual 'Mixture Postulate' and, this would indeed call for a rather fundamental modification in the conventional formulation of quantum theory.

## II. IMPLICATIONS FOR HIDDEN VARIABLE THEORIES[17]

In a hidden variable formulation of quantum mechanics, one usually demands the following :

There exists a measure space $(\Gamma, B)$ such that the following are satisfied :

(I) With each density operator $\rho$ is associated a probability measure $\mu^\rho$ on $(\Gamma, B)$ ; with each self-adjoint operator A is associated a measurable real-valued function (random variable) $X_A : \Gamma \to R$.

(II) For each state $\rho$ and observable A,

$$Pr^{\rho}_{A,B}(\Delta) = Tr(\rho\, P^A(\Delta)) = \mu^{\rho}(X_A^{-1}(\Delta)) \tag{11}$$

for all Borel sets $\Delta \subset R$.

In the above framework, $\Gamma$ refers to the set of all 'completely-specified states' (or the so-called 'dispersion-free states') of the system. If an ensemble of systems is prepared in state $\rho$, then each individual system in the ensemble is supposed to have a completely specified state given by some point say $\lambda \in \Gamma$, and the various systems in the ensemble are distributed over the points in $\Gamma$ as per the probability measure $\mu^\rho$. Also, if an individual system is in the completely specified state $\lambda \in \Gamma$, then the value of the observable A (that would result in an A-measurement performed on the system) is the number $X_A(\lambda)$. The requirement II now ensures that given any ensemble of systems prepared in state $\rho$, the individual systems in the ensemble would be distributed over the points in $\Gamma$ precisely in such a way that the statistical distribution of the values of any observable A would be the same as predicted by quantum theory.

Suppose we now consider two ensembles of systems prepared in states $\rho_1$ and $\rho_2$ respectively, and mix them in the ratio $\nu : 1-\nu$ to form a new ensemble. From what we said above, the individual systems of the new ensemble are distributed over $\Gamma$ as per the probability measure $\nu\mu^{\rho_1} + (1-\nu)\mu^{\rho_2}$. Also, from the 'Mixture Postulate' stated in the last section, it follows that this ensemble is characterised by the density operator $\nu\rho_1 + (1-\nu)\rho_2$. Hence it follows that the 'Mixture Postulate' or the conventional prescription for the density operator associated with a mixed-ensemble is equivalent to the condition

$$\mu^{\nu\rho_1 + (1-\nu)\rho_2} = \nu\, \mu^{\rho_1} + (1-\nu)\mu^{\rho_2} \quad , \tag{12}$$

for all density operators $\rho_1$, $\rho_2$ and $0 \leqslant \nu \leqslant 1$.

Now, it is well known that there are several hidden variable models which satisfy the requirements (I) and (II). However, what the following theorem shows is that <u>all</u> these models are necessarily incompatible with eq.(12), or equivalently, with the standard prescription for the density operator associated with a mixed-ensemble in quantum theory.

Theorem 2 : Any hidden variable theory satisfying (I) and (II), is incompatible with the condition that :

$$\mu^{\nu\rho_1+(1-\nu)\rho_2} = \nu\mu^{\rho_1} + (1-\nu)\mu^{\rho_2} \tag{12}$$

for all density operators $\rho_1$, $\rho_2$ and $0 \leqslant \nu \leqslant 1$.

Proof : Let us assume that condition (12) is also satisfied in addition to the requirements (I) and (II). Let A, B be two incompatible observables with a purely discrete spectrum and let $P^{\rho}_{A,B}(\alpha,\beta)$ be given by

$$P^{\rho}_{A,B}(\alpha,\beta) = \mu^{\rho}(X_A^{-1}(\{\alpha\}) \cap X_B^{-1}(\{\beta\})) \tag{13}$$

Now, from (I), (II) and eq.(12) it is easy to check that $P^{\rho}_{A,B}(\alpha,\beta)$ given by eq.(13) is a joint-probability and that it satisfies the conditions (i) - (ii) of the Wigner-Davies theorem. It then follows that A, B are compatible, thereby establishing a contradiction. Hence the condition (12) or equivalently the 'Mixture Postulate' is incompatible with the basic requirements (I), (II) characterizing a hidden variable theory.

What we saw in the last section was that we can not do away with the 'quantum interference of probabilities' unless we give up the 'Mixture Postulate'. What the above theorem shows is that all hidden variable theories satisfying (I) and (II) are ruled out for the same reason. However, it is possible to modify the usual framework of the hidden variable theories so that the above problem does not arise. All that we need to do, for this purpose, is to replace (I) by the following requirement (I') :

(I') with each pure state $|\psi\rangle$ is associated a probability measure $\mu^{|\psi\rangle}$ on $(\Gamma,B)$ ; ... (rest as in (I)).

It is indeed possible to have a hidden variable theory which satisfies (I'), (II) and is compatible with quantum theory, including the 'Mixture Postulate' which only serves now the purpose of fixing the probability measures on $(\Gamma,B)$ associated with mixed ensembles. What is however interesting is the following :

Consider a density operator $\rho$ which has the following two decompositions :

$$\begin{aligned}\rho &= \nu|\psi_1\rangle\langle\psi_1| + (1-\nu)|\psi_2\rangle\langle\psi_2| \\ &= \nu'|\psi_1'\rangle\langle\psi_1'| + (1-\nu')|\psi_2'\rangle\langle\psi_2'| \quad ,\end{aligned} \tag{14}$$

where $|\psi_1\rangle$, $|\psi_2\rangle$, $|\psi_1'\rangle$ and $|\psi_2'\rangle$ are all different. We can prepare two ensembles of systems $\Sigma$, $\Sigma'$ both characterized by the same density operator $\rho$ ; $\Sigma$ is prepared by mixing appropriately two ensembles in states $|\psi_1\rangle$ and $|\psi_2\rangle$, and $\Sigma'$ is prepared by mixing appropriately two ensembles in states $|\psi_1'\rangle$ and $|\psi_2'\rangle$. As far as quantum theory is concerned, $\Sigma$ and $\Sigma'$ cannot be distinguished by (future) measurements performed on them. However any hidden variable theory satisfying (I') and (II) will have to necessarily distinguish between the two ensembles $\Sigma$ and $\Sigma'$. For $\Sigma$ it would associate (because of the 'Mixture Postulate') the probability measure $\mu$, given by

$$\mu = \nu\mu^{|\psi_1>} + (1 - \nu)\mu^{|\psi_2>} \quad . \tag{15}$$

For the ensemble $\Sigma'$, the associated probability measure $\mu'$ would be given by

$$\mu' = \nu'\mu^{|\psi_1'>} + (1 - \nu')\mu^{|\psi_2'>} \tag{16}$$

That the probability measures $\mu$ and $\mu'$ are in general different follows as a direct corollary of the theorem proved above. In essence, <u>in a hidden variable theory, it is not just that the density operator provides an incomplete characterization of the state of individual systems ; it is also the case that the density operator provides only an incomplete characterization of even ensembles of systems if they happen to be mixed ensembles</u>.

## III. RELATION WITH BELL'S THEOREM

We shall now show that the celebrated theorem of Bell (that every hidden variable theory, which reproduces the statistical predictions of quantum theory, has to violate 'local causality') can also be viewed as arising mainly from the 'quantum interference of probabilities'. For this purpose, let us consider the usual EPR set up[18,19], where a source emits back-to-back two photons such that the composite two-photon system is in S-state. These photons are detected by polarisation sensitive detectors at space-like separated space-time regions $O_1$, $O_2$. Now, if in region $O_2$ the detector measures whether the photons which arrive there are linearly polarised (say in the vertical plane), then the ensemble of photons which arrive at $O_1$ will be a mixture, with equal weightage 1/2, of two ensembles of photons in states $|V>$ and $|H>$, where $|V>$ and $|H>$ refer to states with linear polarisation in the vertical and horizontal planes respectively. On the other hand if the detector in $O_2$ is set to measure circular polarisation, then the ensemble of photons which arrive in $O_1$ will be a mixture again with equal weightage 1/2, of two ensembles of photons in states $|R>$ and $|L>$, where $|R>$ and $|L>$ refer to states of right and left circular polarisation respectively. As far as quantum theory is concerned these two ensembles (which arrive at $O_1$ depending on which experiment is performed at $O_2$) are <u>indistinguishable</u> as they are characterised by the <u>same</u> density operator

$$\frac{1}{2}|V><V| + \frac{1}{2}|H><H| = \frac{1}{2}|R><R| + \frac{1}{2}|L><L| = \frac{1}{2}\, I, \tag{17}$$

where I is the identity operator in the polarisation space. Hence, using quantum mechanics alone, it is not possible to find out which experiment was carried out at $0_2$ by examining the ensemble of photons which arrive in $0_1$.

The crucial point is, however, that a hidden variable theory (as was shown in the last section) has to necessarily distinguish between these two ensembles -for, though both these ensembles are characterised by the same density operator, they have been prepared very differently by mixing different sets of ensembles in different pure states[20]. Hence any hidden variable theory (compatible with quantum mechanics) distinguishes these two ensembles and therefore also provides us a way of finding out what experiment is carried out at $0_2$ by examining the ensemble which arrives at $0_1$. This is precisely the break down of local causality that Bell's theorem also shows as a necessary feature of all hidden variable theories which are compatible with quantum mechanics. An interesting feature of the above demonstration of Bell's theorem is that it is based only on requirements (I') & (II) with no assumptions being made regarding joint distributions etc., of two observables.

REFERENCES AND FOOT NOTES

1. L. de Broglie, La Revue Scientifique, N°.3292 fasc.5, 87, 259 (1948)
2. L. de Broglie, Physics and Microphysics (Hutchinson, London, 1955)
3. E.B. Davies and J.T. Lewis, Commun.Math.Phys. 15, 305 (1969)
4. M.D. Srinivas, Jour. Math. Phys. 16, 1672 (1975)
5. See the lectures of L. Accardi in this volume, p. 297.
6. Some implications of the quantum interference of probabilities for the hidden variable theories were discussed in G. Lochak, Found. Phys. 6, 173 (1976)
7. J.S. Bell, Physics 1, 195 (1964) ; for recent reviews see : J.F. Clauser and A. Shimony, Rep. Prog. Phys. 41, 1881 (1978) ; F. Selleri and G. Tarozzi, Rev. del Nuovo Cim. 4, I(1981)
8. J. Von Neumann, Mathematical Foundations of Quantum Mechanics (Princeton University Press 1955)
9. G. Lüders, Ann. Phys. (Leipzig) 8, 322 (1951)
10. See for example, M.D. Srinivas, Commun. Math. Phys. 71, 131 (1980)
11. See for example, E.P. Wigner, in Foundations of Quantum Mechanics, Ed.B. d'Espagnat (Academic Press, New York, 1971)
12. See for example, M.D. Srinivas, Jour.Math.Phys. 19, 1705(1978)
13. E.P. Wigner, Phys.Rev. 40, 749 (1932) ; see also : E.P. Wigner, in Perspectives in Quantum Theory, Eds. W. Yourgrau and T. Vander Merwe (MIT Press, Cambridge, 1971) ; M.D. Srinivas and E. Wolf, Phys.Rev. D11, 1477 (1975) ; M. Mugur-Schuachter, in Quantum Mechanics a Half Century Later, Eds. J. Leite-Lopes and M. Paty (Reidel, Dordrecht, 1977).
14. E.B. Davies, Quantum Theory of open systems (Academic Press,

New York, 1976) Ch.III, Theorem 2.1.

15. The theorem is actually valid for arbitrary self-adjoint operators.
16. We may also note that according to conventional quantum theory, the density operator provides a complete characterization of an ensemble of systems.
17. The contents of this section are based on : M.D.Srinivas, Pramana 19, 159 (1982)
18. The following discussion mainly resulted from an examination of a recent paper of Herbert (reference 19) which contains a proposal for a superluminal communication device. Of course the theoretical and experimental possibility of such a device, are being widely contested.
19. N.Herbert, 'FLASH - A superluminal communicator based upon a new kind of quantum measurement', Boulder Creek Preprint (Revised) Dec.1981. The author is grateful to Professor F. Selleri for bringing to his attention the preprint of Herbert.
20. Infact Herbert introduces the terms 'plane-unpolarised light' and 'circularly-unpolarised light' to distinguish between these two ensembles, while in the quantum mechanical jargon, both the ensembles will be referred to as merely 'unpolarised light'.

EXPERIMENTAL TESTS OF BELL'S INEQUALITIES

Alain Aspect

Institut d'Optique Théorique et Appliquée
Bâtiment 503 - Centre Universitaire d'Orsay - B.P. 43
F 91406 ORSAY Cédex France

My aim is to speak of the experiments related to Bell's inequalities. But since I am the first speaker who invokes Bell's inequalities, I must first introduce Bell's theorem (at least the way in which I understand it). Then, I will present a brief review of the experiments carried out between 1970 and 1976. In the third part, I will give some details on experiments that are in progress in the Institut d'Optique d'Orsay.

## 1. BELL'S THEOREM

### 1.1. Einstein-Podolsky-Rosen Correlations

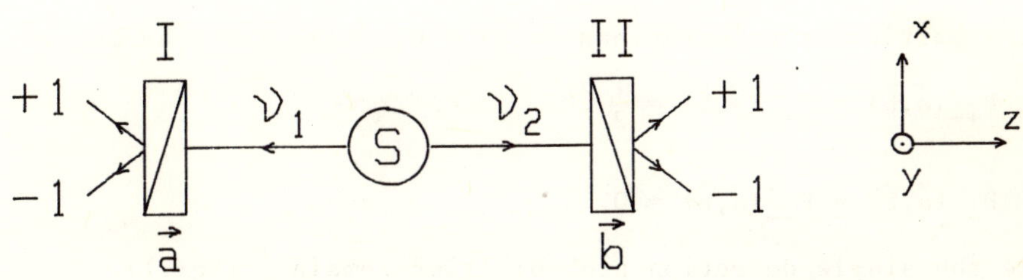


*Figure 1. Optical version of the Einstein-Podolsky-Rosen-Bohm thought experiment. The two photons $\nu_1$ and $\nu_2$, emitted in state (1) are analyzed by linear polarizers in orientations $\vec{a}$ and $\vec{b}$. One can measure the probabilities of single or joint detections after the polarizers.*



*S. Diner et al. (eds.), The Wave-Particle Dualism, 377–390.*

Let us consider the optical version of the Einstein-Podolsky-Rosen[1] Gedankenexperiment as modified by Bohm[2]. Two photons with different energies, $\nu_1$ and $\nu_2$, are emitted by the source S in a state

$$|\psi(\nu_1,\nu_2)\rangle = \frac{1}{\sqrt{2}} \{|x,x\rangle + |y,y\rangle\} \tag{1}$$

where $|x\rangle$ and $|y\rangle$ are linear polarization states. Photons $\nu_1$ is propagating along $+\vec{Oz}$, while $\nu_2$ is propagating towards $-\vec{Oz}$.

The polarization analyzer I, in orientation $\vec{a}$, yields two possible results, + 1 or − 1, corresponding to a polarization found parallel or perpendicular to $\vec{a}$. Similarly, the analyzer II measures (along $\vec{b}$) the linear polarization of photon $\nu_2$.*

Standard Quantum Mechanical calculations yield, as probabilities of single or joint measurements :

$$\begin{cases} P_+(\vec{a}) = P_-(\vec{a}) = \frac{1}{2} \\ P_+(\vec{b}) = P_-(\vec{b}) = \frac{1}{2} \end{cases} \tag{2}$$

and

$$\begin{cases} P_{++}(\vec{a},\vec{b}) = P_{--}(\vec{a},\vec{b}) = \frac{1}{2}\cos^2(\vec{a},\vec{b}) \\ P_{+-}(\vec{a},\vec{b}) = P_{-+}(\vec{a},\vec{b}) = \frac{1}{2}\sin^2(\vec{a},\vec{b}) \end{cases} \tag{3}$$

($P_{+-}(\vec{a},\vec{b})$ being the probability of finding + 1 for $\nu_1$ and − 1 for $\nu_2$ with polarizers I and II in orientations $\vec{a}$ and $\vec{b}$).

In the particular orientations $(\vec{a},\vec{b}) = 0$, these formulas become

$$\begin{cases} P_{++}(\vec{a},\vec{b}) = P_{--}(\vec{a},\vec{b}) = \frac{1}{2} \\ P_{+-}(\vec{a},\vec{b}) = P_{-+}(\vec{a},\vec{b}) = 0 \end{cases}$$

while the single detection probabilities remain $\frac{1}{2}$ (Eq(2)).

---

* These measurements are obviously similar to measurements of spin components by Stern-Gerlach filters acting on spin 1/2 particles.

So, there is 50 % chances of getting $\nu_1$ in channel + 1, but if it is so, then we are sure to find $\nu_2$ in channel + 1. There is a complete correlation between the results of measurements at I and II, although each result looks random.

A convenient way of exhibiting these correlations is the coefficient of correlation of polarization

$$E(\vec{a},\vec{b}) = P_{++}(\vec{a},\vec{b}) + P_{--}(\vec{a},\vec{b}) - P_{+-}(\vec{a},\vec{b}) - P_{-+}(\vec{a},\vec{b}) \quad (4)$$

Its predicted value (according to Quantum Mechanics) is

$$E_{MQ}(\vec{a},\vec{b}) = \cos 2(\vec{a},\vec{b}) \quad (5)$$

It reaches ± 1 (complete correlation) for $(\vec{a},\vec{b}) = 0$ and $(\vec{a},\vec{b}) = \pi/2$.

There is a classical picture for understanding correlations between remote measurements on two particles that have interacted. The two particles of a given pair bring some common property (related to their common past) that determines the results of the measurements.

For explaining the randomness of individual results, we have just to admit that this property is different for the various pairs. By averaging over all the pairs, we can hope to recover the Quantum Mechanical predictions. On the other hand, in Quantum Mechanics, all the pairs are described by the same state vector (1). So, our classical looking explanation for the EPR correlations leads to complete Quantum Mechanics with some supplementary properties (supplementary parameters, or hidden variables).

## 1.2 Bell's inequalities

Bell tried to express in mathematical terms this classical-looking picture. Each emitted pair is characterized by a supplementary parameter $\lambda$, which determines the results of the measurements by the analyzers. These results are :

$$\begin{cases} A(\lambda,\vec{a}) = \pm 1 & \text{at analyzer I} \\ B(\lambda,\vec{b}) = \pm 1 & \text{at analyzer II} \end{cases} \quad (6)$$

The pairs are emitted with differents $\lambda$. The distribution of the supplementary parameters is specified by a probability distribution $\rho(\lambda)$, such that

$$\rho(\lambda) \geq 0 \qquad \text{and} \qquad \int d\lambda\ \rho(\lambda) = 1 \quad (6')$$

Any particular supplementary parameters theory will explicity specify the functions $\rho(\lambda)$, $A(\lambda,\vec{a})$ and $B(\lambda,\vec{b})$. It will then be easy to compute, in the frame of this theory, the probabilities of the various possible results of measurements. The function of correlation of polarization can for instance be written

$$E(\vec{a},\vec{b}) = \int d\lambda\ \rho(\lambda)\ A(\lambda,\vec{a})\ B(\lambda,\vec{b}) \tag{7}$$

The formulas (6), (6') and (7) are sufficient for derivating Bell's inequalities[3,4]. For any theory embeded in the formalism (6) and (6'), there are restrictions on the function of correlation. A useful form for these rectrictions is the so-called BCHSH inequalities (Bell-Clauser-Horne-Shimony-Holt inequalities[4]) :

$$-2 \leqq S \leqq 2$$

with (8)

$$S = E(\vec{a},\vec{b}) - E(\vec{a},\vec{b}') + E(\vec{a}',\vec{b}) + E(\vec{a}',\vec{b}')$$

The quantity S involves four measurements of the function of correlation of polarization, in four different orientations of the polarizers.

The demonstration is very straight forward[5]. Let us consider four numbers x, x', y and y' with an absolute value equal to 1. It is then obvious to show that

$$s = xy - xy' + x'y + x'y' = \pm 2 \tag{9}$$

We can apply this result to the quantities :

$$\begin{cases} x = A(\lambda,\vec{a}) \\ y = B(\lambda,\vec{b}) \end{cases} \qquad \begin{cases} x' = A(\lambda,\vec{a}') \\ y' = B(\lambda,\vec{b}') \end{cases}$$

The quantity s can be averaged over $\lambda$ by multiplication by $\rho(\lambda)$ and integration, and this leads to the inequality (8).

### 1.3 Conflict with quantum mechanics

As we have already said, there is an a priori hope that a supplementary parameter theory yields, on the average, the same predictions as Quantum Mechanics. Thanks to the Bell's inequalities, we can show that it is impossible. Let us consider the particular set of orientations displayed of Fig. 2.a.

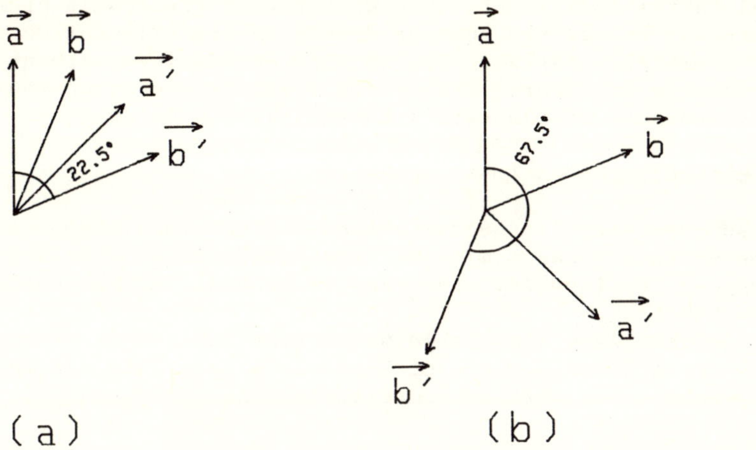


*Figure 2. Orientations giving the maximum violation of Bell's inequalities.*

The quantity S computed for these orientations, with the Quantum Mechanical results (5), takes the value

$$S_{QM} = 2\sqrt{2}$$

This value clearly violates the BCHSH inequalities (8).

1.4. Discussion

We are then allowed to conclude that no supplementary parameter theory following the formalism (6) and (6') can yield results in agreement with all the Quantum Mechanical predictions. Since the formalism (6) and (6') considered up to now might appear restrictive, it is important to know that generalizations of this result exist[5/9], that deal with larger classes of theories, for instance stochastic (non-deterministic) supplementary parameters theories. Thanks to these discussions, we can point out a crucial hypothesis for derivating Bell's inequalities (and thus obtaining a conflict with Quantum Mechanics) : it is the locality assumption already stressed by Bell in his first paper. The locality condition claims that the results of measurements at I, $A(\lambda,\vec{a})$, do not depend on the orientation $\vec{b}$ at II, and vice-versa, nor does $\rho(\lambda)$ (i.e. the way in which pairs are emitted) depend on $\vec{a}$ and $\vec{b}$. The formalism (6) (6') obviously obeys this condition.

The locality condition can be taken as an hypothesis, as we have done here. But, it would be better to derive it as a consequence of a basic physical law. In his first paper, Bell remarked that this is possible, if we consider experiments in which the orientations of the polarizers are changed quickly and at random. If, according to Einstein's causality, we admit that no influence can travel faster than light, then we see that the result of a measurement on one side cannot depend on the orientation of the other (remote) polarizer. So, in such an ideal experiment with time-varying analyzers, the Bell's locality condition would be the consequence of Einstein's causality.

### 1.5 Conclusion

As a summary, Bell's theorem states that no local supplementary parameter theory can mimick all the predictions of Quantum Mechanics. The locality condition, essential for the conflict, can be taken

- either as a reasonable assumption ;
- or as a consequence of Einstein's causality, in an experiment with random time-varying analyzers.

## 2. FROM GEDANKENEXPERIMENT TO ACTUAL EXPERIMENTS

Quantum Mechanics is such a successful theory that one might believe that Bell's theorem states the impossibility of local supplementary parameters theories. But situations in which there is a conflict (we will call them "sensitive situations") are very rare ; in 1965, none had been realized. One could then think that Bell's theorem points out a limit of the validity of Quantum Mechanics. By yielding a quantitative criterium (Bell's inequalities), Bell's theorem allowed to give an experimental answer to the problem.

As a matter of fact, sensitive situations are rare, even when we consider ideal experiments. Moreover, when considering an actually workable experiment, the possibility of a sensitive experiment looks even more seldom since all the know defects, lead to a decrease of the correlations predicted by Quantum Mechanics. With experimental imperfections, the conflict thus decreases, or even disappears.

A first series of experiments[9] made use of pairs of $\gamma$ photons emitted by annihilation of positronium in its fundamental (singlet) state. As shown by Bohm and Aharonov[10], such pairs of photons are in an EPR type state, similar to state (1). Unfortunately, there are no true analyzers of polarization for $\gamma$ photons at such an energy (0.5 MeV). So the polarization measurements were replaced by the study of Compton scattering of the photons (which depends on the polarization). Then, by an indirect reasoning (using a Quantum Mechanical calculation), it is possible to test the experimental results versus Bell's inequalities.

Although the first results were conflicting, most of these experiments finally gave results in agreement with Quantum

Mechanics[11], and it was claimed that Bell's inequalities were violated. Concerning this last claim, we must remember that the reasoning is very indirect since we have not true measurements of polarization, and this conclusion has been challenged[9,12].

An experiment subjected to the same kind of criticisms has been performed onto pairs of protons scattered in a singlet state[13]. The results are in good agreement with Quantum Mechanics.

Stimulated by the famous CHSH paper[4], another series of experiments was performed, using visible photons emitted in atomic radiative cascades. For visible photons, there are true polarizers, and the situation looks better. However, for practical reasons, the four experiments carried out between 1972 and 1977 made use of single channel polarizer. Such polarizers transmit light polarized along $\vec{a}$ (cf. Figure 1) but they block light polarized perpendicular to $\vec{a}$. The measurements are thus incomplete and one must use an indirect reasoning (and admit a supplementary assumption)[4] to be able to test the experimental results versus Bell's inequalities.

Although the two first results were conflicting, these experiments[9,14,15] clearly favour Quantum Mechanics, and show a violation of Bell's inequalities (modified for the one-channel-polarizers case). Even if they do not exactly reproduce the ideal experimental scheme of Fig. 1, these experiments are quite convincing, and closer to the thought experiment than the $\gamma$ photons experiments. The discrepancies between the first results can be easily understood as due to the very low signal obtained. But in the last of these experiments[15] the signal was larger, and the experiment seems very conclusive ; the results are in excellent agreement with Quantum Mechanics, and they violate Bell's inequalities by 4 standard deviations.

## 3. ORSAY EXPERIMENTS

In designing a new experiment, our main purpose was to realize more sophisticated experimental schemes, closer to the ideal thought experiment. But we had first to build a high efficiency source of pairs of photons correlated in polarization.

### 3.1. Source of pairs of photons

As shown by CHSH[4], a $J = 0 \rightarrow J = 1 \rightarrow J = 0$ atomic cascade yields pairs of photons in a state (1), i.e. good candidates for a sensitive experiment. When one takes into account the finite solid angles of detection of the fluorescence light, one finds as correlation coefficient

$$E_{MQ}(\vec{a},\vec{b}) = F(u)\ \cos 2(\vec{a},\vec{b}) \tag{10}$$

instead of formula (5). With an half-angle of detection $u = 32°$ (our experiments), the function $F(u)$ remains close to 1 (0.984) and the correlation remains large enough for a sensitive experiment.

We have chosen the $4p^2\ {}^1S_0 - 4s4p\ {}^1P_1 - 4s^2\ {}^1S_0$ cascade of Calcium, already used by Clauser and Freedman[14]. The atoms in the ground state (in an atomic beam) are excited to the upper state by a two-photon absorption, involving a Krypton ion laser at 406 nm and a tunable dye laser at 580 nm (Figure 3).

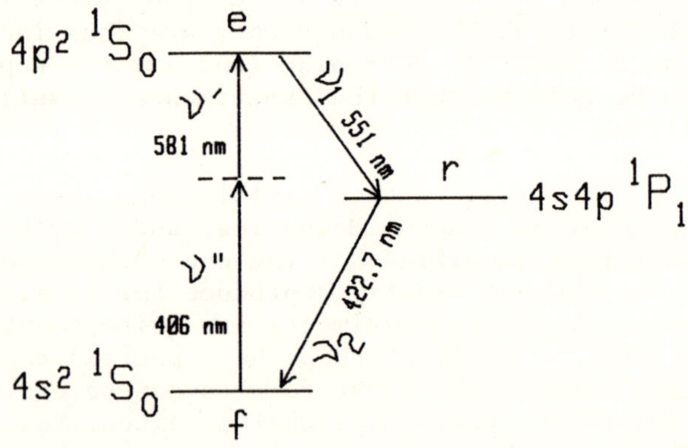


*Figure 3. Two photons excitation of the e → r → f radiative cascade. The excitation process is efficient when $\nu' + \nu'' = \nu_1 + \nu_2$*

The two lasers are focussed onto the calcium atomic beam, and the resulting source has a small size (0,5 mm x 0,05 mm x 0,05 mm). This feature is convenient for the optics of the detection channels.

A feedback loop controls the wavelength $\lambda'$ of the tunable laser for ensuring the resonance conditon ($\nu' + \nu'' = \nu_1 + \nu_2$). A second loop controls the power of one laser, for a constant rate of emission of the pairs of photons. The remaining fluctuations, and drifts are less than 1 % for several hours.

With a few tens of milliwatts from each laser, we easily obtain a cascade rate N of several $10^7\ s^{-1}$. An increase of this rate would not significantly improve the signal-to-noise ratio for coïncidence counting, since the accidental coïncidence rate varies as $N^2$, while the true coïncidence rate varies as N.

With such a signal, we obtain a coïncidence rate over 100 $s^{-1}$. In only 100 s of data accumulation, we reach a statistical accuracy of 1 % (in the first experiments, a similar accuracy needed several tens of hours of experiment). This allows us to perform various test and auxiliary checks.

3.2 Experiment with one channel polarizers[16]

Similarly to previous experiments, our first experiment used one-channel polarizers, that were pile-of-plates polarizers (ten glass plates at Brewster angle). The optical and mechanical quality of these polarizers was good enough to avoid any displacement of the light beams when rotating or withdrawing the polarizers, avoiding any spurious effect related to the heterogeneity of the photocathodes.

A test of the Bell's inequalities suited for one channel polarisers[4] exhibited a violation by 9 standard deviations. Figure 4 displays a comparison of our measured values with the Quantum Mechanical predictions taking into account the solid angles and the measured actual efficiencies of our polarizers.

The agreement is obviously excellent.

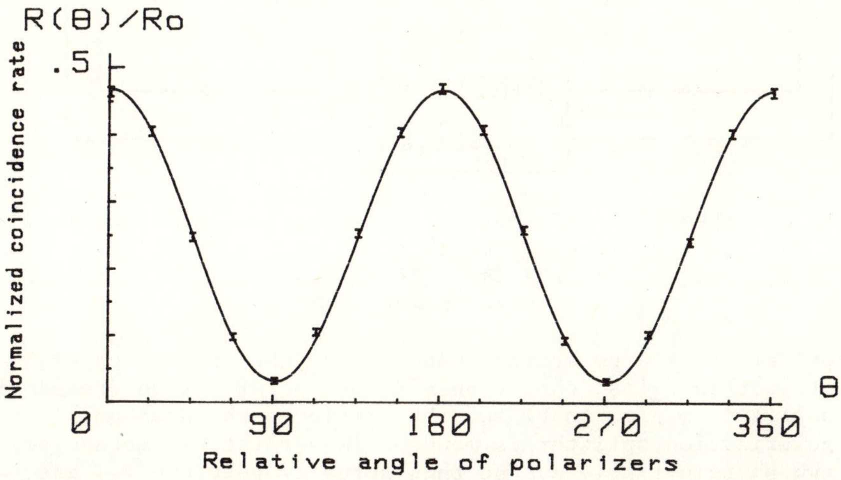


*Figure 4. Experiment with one channel polarizers. Indicated errors are ±1 standard deviation. The solid curve is not a fit to the data but the predictions by Quantum Mechanics.*

Additionally, we could perform several auxiliary tests. The most interesting consisted in removing both polarizers at 6.5 meters from the source. This distance is larger than the coherence length associated to the photon $\nu_2$ ($c\tau_r$ = 1,5 m), and it had been suggested that the correlations could decrease at such distances (this point had been subject to controversy in experiments with γ photons)[17]. Actually, we observed no modification of the experimental results.

### 3.3 Experiment with two channel polarizers*

We have already seen that measurements of polarization with single channel polarizers are incomplete, since one does not detect the -1 result.

By using two-channel polarizers, we have performed an experiment following much more closely the ideal scheme of Figure 1.[18]

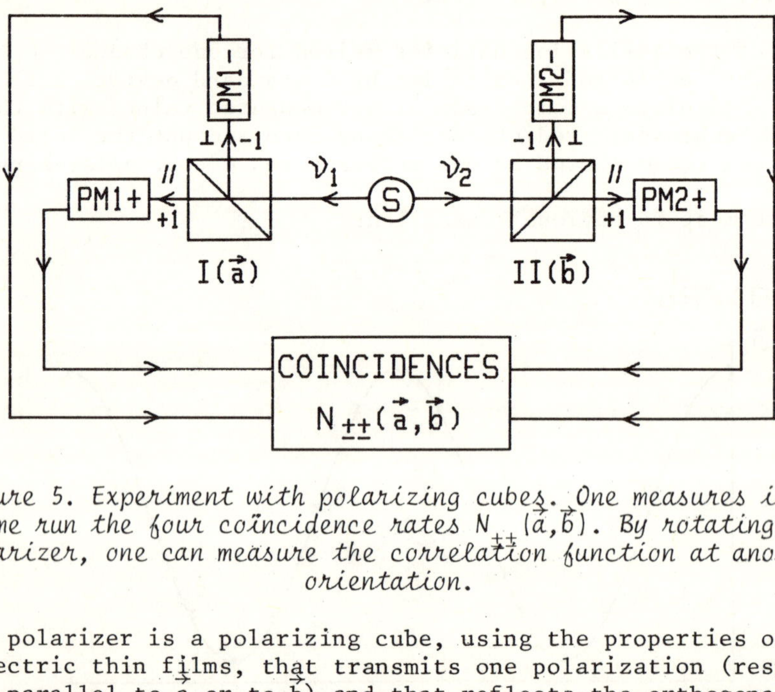


*Figure 5. Experiment with polarizing cubes. One measures in the same run the four coïncidence rates $N_{\pm\pm}(\vec{a},\vec{b})$. By rotating any polarizer, one can measure the correlation function at another orientation.*

Each polarizer is a polarizing cube, using the properties of dielectric thin films, that transmits one polarization (respectively parallel to $\vec{a}$ or to $\vec{b}$) and that reflects the orthogonal one. This polarization splitter is mounted in a rotatable mechanism, with two photomultipliers. One thus actually gets the + 1 and - 1 results for a measurement of linear polarization. With a fourfold coïncidence system, we are able to measure the four coïncidence

* A similar experiment, using calcite two-channel polarizers, is in progress at the university of Catania (cf. the paper by Falciglia et al. in the same volume).

rates $N_{++}(\vec{a},\vec{b})$ (the polarizers I and II being orientated along $\vec{a}$ and $\vec{b}$). This yields directly the polarization correlation coefficient.

$$E(\vec{a},\vec{b}) = \frac{N_{++}(\vec{a},\vec{b}) + N_{--}(\vec{a},\vec{b}) - N_{+-}(\vec{a},\vec{b}) - N_{-+}(\vec{a},\vec{b})}{N_{++}(\vec{a},\vec{b}) + N_{--}(\vec{a},\vec{b}) + N_{+-}(\vec{a},\vec{b}) + N_{-+}(\vec{a},\vec{b})} \qquad (11)$$

Repeating this measurement for the four sets of orientations of Figure 2, we can directly test the B.C.H.S.H. inequalities (8).

The results of this measurement are

$$S_{exp} = 2.697 \pm 0.015 \qquad (12)$$

The inequalities (8) are violated by about 40 standard deviations, and the result is in excellent agreement with the predictions of Quantum Mechanics taking into account the polarizers efficiencies and the solid angles of detection :

$$S_{QM} = 2.70 \pm 0.05$$

Here, we must remark that we can use the formula (11) for testing inequalities (8) only if we admit that the ensemble of actually detected pairs is a faithfull sample of all emitted pairs (only a weak fraction of which is detected, since the detection efficiencies are low). This is a reasonable assumption, but it cannot be proved to be true. However, we have performed some tests that are in agreement with this assumption (and that could have disproved it). The most convincing consisted in checking that the sum of the four coïncidence rates $\sum_{\pm\pm} N_{\pm\pm}(\vec{a},\vec{b})$ is constant, i.e. that the size of the selected sample is constant when the orientations are changed. (it is important for this test that the source is very stable).

In addition to this test of Bell's inequalities, we have also measured the polarization correlation coefficient in various orientations, for comparison with the Quantum Mechanical calculations. Figure 6 displays the results : the agreement is obviously excellent.

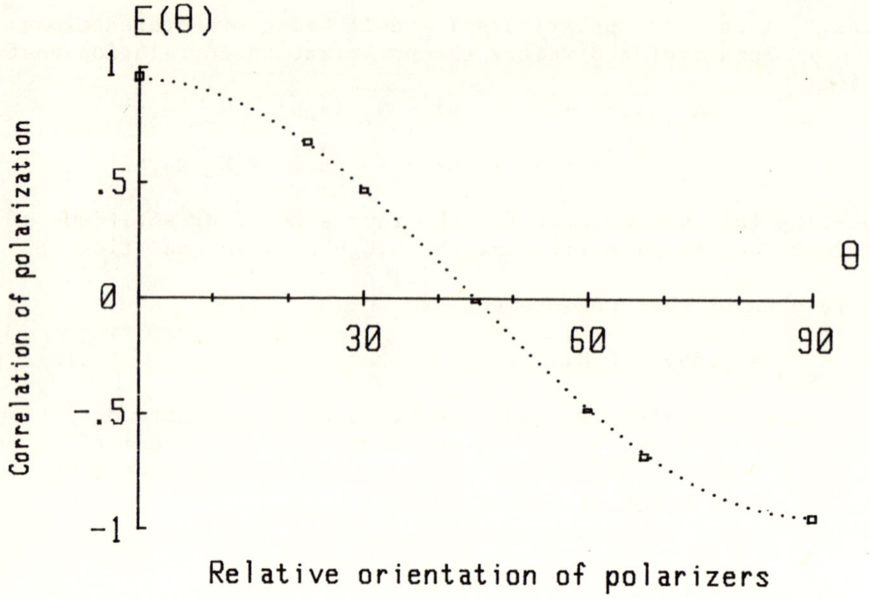


*Figure 6. Experiment with two-channel polarizers. The indicated errors are ±2 standard deviations. The dashed curve is the prediction of Quantum Mechanics for the actual experiment.*

### 3.4 Experiment with time-varying polarizers

As emphasized in § 1, it would be interesting to carry out an experiment in which the orientations of the polarizers are changed quickly at random. A step towards such an experiment can be done by using a modified experimental scheme[19] (Figure 7).

Each polarizer is replaced by an optical switch which is able to rapidly redirect the incident light to one of two polarizers. Each set up is obviously equivalent to a variable polarizer switched between two orientations.

In the experiment now in progress*, the switches use the acousto-optical interaction of the light with an ultrasonic standing wave at 25 MHz. A switching occurs about each 10 ns, which is smaller than L/c (40 ns). Unfortunately, due to this principle, the switching is not effected at random but rather in a periodic way (however, the two switches function in an uncorrelated way).

* Note added in proof. The experiment has been completed an yields a result in good agreement with Quantum Mechanics, violating Bell's inequalities by 5 standard deviations (Alain Aspect, Jean Dalibard, and Gérard Roger, Phys. Rev. Lett 49, 1804 (20/12/82)).

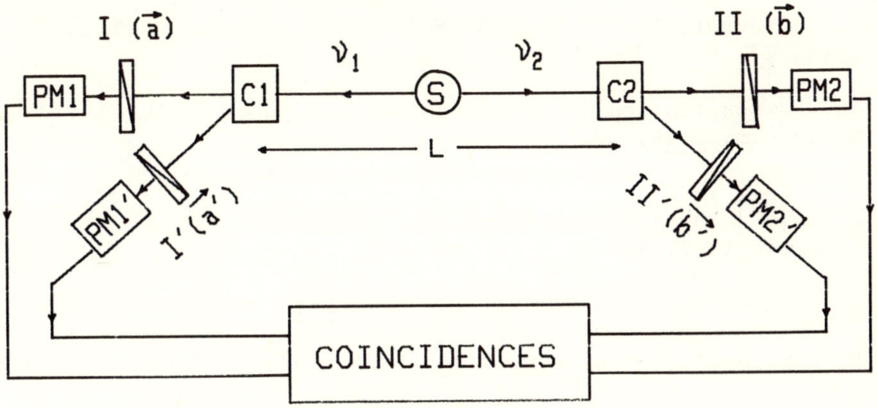


*Figure 7. Experiment with optical switches.*

In spite of its defects this experiment is, as far as I know, the first one that includes time-varying polarizers

4. CONCLUSION

We have reviewed a series of experiments of increasing accuracy and/or leading closer to the ideal thought-experiment. Actually, all these experiments have some imperfections that leave open some loopholes for the advocates of local supplementary parameters theories. Indeed, one can imagine theories that would exploit these defects for restoring results in agreement with the Quantum Mechanical predictions. But, such theories appear somewhat unnatural since the results should change with improved apparatus.

So, it is more natural to conclude that local supplementary parameters theories are not a possible picture for understanding the E.P.R. correlations that appear as a typically quantum phenomenon. A little more generally, we can say that these experiments show that we must renounce interpreting E.P.R. correlations with a picture in the spirit of Einstein's ideas of separability and causality. Quantum Mechanics demand pictures more alien from our common sense.

REFERENCES

1 A. Einstein, B. Podolsky et N. Rosen, Phys. Rev. 47, 777 (1935)

2 D. Bohm, Quantum Theory (prentice Hall, Englewood Cliffs, N.J., 1951).

3 J.S. Bell, Physics 1, 195 (1964).

4 J.F. Clauser, M.A. Horne, A. Shimony et R.A. Holt, Phys. Rev. Lett. 23, 880 (1969).

5 This demonstration is inspired from the paper by P. Eberhard, Nuovo Cim. B 46, 392 (1978).

6 J.S. Bell in Foundations of Quantum Mechanics, proceedings of the 49th course Enrico Fermi, B. d'Espagnat edit., Academic (N.Y. 1971).

7 J. Clauser and M.A. Horne, Phys Rev. D 10, 526 (1974).

8 B. d'Espagnat, Phys. Rev. D 11, 1454 (1975) ; D 18, 349 (1978).

9 J.F. Clauser et A. Shimony, Rep. Progr. Phys. 41, 1881 (1978).
F.M. Pipkin, Advances in Atomic and Molecular Physics, D.R. Bates and B. Bederson ed., Academic (1978).
F. Selleri and G. Tarozzi, Riv. Nuovo Cim 4, 1 (1981).

10 D. Bohm et Y. Aharonov, Phys. Rev. 108, 1070 (1957).

11 In addition to the results reported in the reviews 9, see also K. Meisenheimer, Diplomarbeit (Freiburg 1979).

12 M. Froissart, Nuovo Cim. B 64, 241 (1981)

13 M. Lamehi-Rachti and W. Mittig, Phys. Rev. D 14, 2543 (1976).

14 S.J. Freedman and J.F. Clauser, Phys. Rev. Lett 28, 938 (1972).
J.F. Clauser, Phys. Rev. Lett. 36, 1223 (1976).
For the Holt Pipkin experiment, see F. Pipkin, réf. 9.

15 E.S. Fry et R.C. Thompson, Phys. Rev. Lett. 37, 465 (1976).

16 A. Aspect, P. Grangier et G. Roger, Phys. Rev. Lett. 47, 460 (1981).

17 D. Bohm et B.J. Hiley, Nuovo Cim. B 35, 137 (1976).

18 A. Aspect, P. Grangier et G. Roger, Phys. Rev. Lett. 49, 91 (1982).

19 A. Aspect, Phys. Lett. 54 A, 117 (1975)
A. Aspect, Phys. Rev. D 14, 1944 (1976).

# TEST OF THE NON SEPARABILITY OF THE $K^0\bar{K}^0$ SYSTEM*[1]

SIX J.

Laboratoire de l'Accélérateur Linéaire, Orsay, France

At present there is no experimental verification that the $K^0\bar{K}^0$ system remains a non-separate system when the particles fly apart. For example, Furry's hypothesis may be considered. An experimental test of this non-separability requires producing $K^0$ and $\bar{K}^0$ interactions. This can be done for the annihilation $p\bar{p} \to K^0\bar{K}^0$ with the CERN LEAR facility.

The non separability of a system of particles in quantum mechanics, linked to the theory of measurement, causes a problem which was recognized at the origin of quantum mechanics[1]. This property, conceivable for interacting particles, remains true in quantum mechanics even if the particles are separated in space. The impossibility of dividing such a system into subsamples of independent systems led Einstein, Podolsky and Rosen to formulate the so-called EPR paradox[2]. This paradox refers to the hypothesis of "elements of reality" of subsample systems separated by space-time distances.

The best illustration of the paradox is given by a spin J=0 system decaying into two spin 1/2 particles. To have a singlet state J=0, the final-state vector must have the antisymmetric form :

$$|\psi\rangle=(1/\sqrt{2})(|u_+\rangle|v_-\rangle-|u_-\rangle|v_+\rangle), \quad (1)$$

where $|u_+\rangle$ is the state for one particle U with z component equal to $+\frac{1}{2}\hbar$ and so on. If one measures the spin component of one particle, quantum mechanics says that the state vector is reduced by the measurement to $|u_+\rangle|v_-\rangle$ or $|u_-\rangle|v_+\rangle$. The spin component of the other distant particle is then instantaneously fixed. The



*S. Diner et al. (eds.), The Wave-Particle Dualism, 391–396.*

paradox lies in this wave packet reduction without an "action at a distance". To avoid this paradox many studies have been made in the direction of hidden variables theories[3]. One of them is the so-called Furry hypothesis[4]. The state vector would have the non-separate form (1) in the interaction but it would split after that, independently of measurement, in an equal mixture of two independent states $|\psi_1\rangle = |u_+\rangle|v_-\rangle$ and $|\psi_2\rangle = |u_-\rangle|v_+\rangle$.

The very important contribution of Bell[5] was to propose inequalities which must be verified by any deterministic local theory (as for example by Furry's hypothesis) and which may be violated by quantum mechanics. Experiments have been made, in particular on the correlation between two photons emitted in a radioactive cascade of an atom $J=0\to 1\to 0$. The existing results[6]*2 strongly favour quantum mechanics to the detriment of any local deterministic theory. Thus it seems that in a system of particles at least the spin component of a particle has a non-separate character.

In particle physics, the system $K^0\bar{K}^0$ presents some similarities to the case discussed above. For a $J^{PC} = 1^{--}$ state decaying into $K^0\bar{K}^0$ —this is the case for the $\phi$ meson or for $\bar{p}$ annihilation in the s state— the state function must have the form :

$$|\psi\rangle=(1/\sqrt{2})(|K^0\rangle_a|\bar{K}^0\rangle_b-|\bar{K}^0\rangle_a|K^0\rangle_b)$$

$$= (1/\sqrt{2})[|K_S\rangle_a|K_L)_b - |K_L)_a|K_S\rangle_b] \qquad (2)$$

where a and b indicate the directions of particles and $K_S = (1/\sqrt{2})(K^0 + \bar{K}^0)$ and $K_L = (1/\sqrt{2})(K^0 - \bar{K}^0)$ are, respectively, the short-and long-lived states of the $K^0$ or $\bar{K}^0$ decay (we can neglect the small CP non-conservation in $K^0$ decay).

The form (2) is imposed by CP conservation and the decay must contain only the $K_SK_L$ mode. The non-separate form (2) has the peculiarities of the EPR paradox. Before its decay, the particle a or b does not know if it has to decay in $K_S$ or $K_L$. When one particle decays for example in $K_S$ the other is instantaneously forced to decay in the other mode.

Due to strangeness non-conserving weak interactions a $K^0(\bar{K}^0)$, after its production, oscillates between $K^0$ and $\bar{K}^0$ states. These oscillations give interference effects in $K^0$ interactions[8], in CP violation[9] and in $K_S$ regeneration[10]. However all this experimental evidence[8-10] do not concern the interference between two K, as may be considered in a $K^0\bar{K}^0$ system represented by the non-separate expression (2). To avoid the problem of the non separability, one can then tentatively envisage Furry's hypothesis applied to the $K^0\bar{K}^0$ system. In $K^0\bar{K}^0$ production, one would have the non-separated form (2) but, after that, a rapid splitting would occur with an equal mixture of two states, for example, $|\psi_1\rangle=|K_S\rangle_a|K_L\rangle_b$ and $|\psi_2\rangle=|K_L\rangle_a|K_S\rangle_b$.

This "Furry's" hypothesis as well as quantum mechanics predict the decay in $K_S K_L$ only. To distinguish the two hypotheses it is thus necessary to consider $K^0$ and $\bar{K}^0$ interactions and to produce interference terms between the two parts of the expression (2). Taking into account that $K_S = (1/\sqrt{2})(K^0+\bar{K}^0)$ and $K_L = (1/\sqrt{2})(K^0-\bar{K}^0)$, these interference terms can occur for example in $K^0K^0$, $K^0\bar{K}^0$, $\bar{K}^0\bar{K}^0$ interaction events or in $K_S K_S$ or $K_L K_L$ regeneration events but only if there is an absorber on the a and b paths.

The clearest example and probably the easiest one for experimental purposes is given by interactions of $\bar{K}^0$ giving a $\Lambda^0$ i.e a $\bar{K}^0$ signature. The predictions of quantum mechanics and "Furry's" hypothesis may be easily calculated.

The evolution of the state vector (2) in vacuum with time (or space) is given by

$$|\psi(x_a,x_b)\rangle = (1/\sqrt{2})\,|K_S\rangle_a|K_L\rangle_b \exp(\alpha_S x_a + \alpha_L x_b)$$

$$- |K_L\rangle_a|K_S\rangle_b \exp(\alpha_L x_a + \alpha_S x_b)],$$

with the following conventions :

$x_a$, $x_b$ are the distances of particles from the origin,

$$\alpha_S = -\,(im_{S(L)}\ \tau_{S(L)}/\hbar + \tfrac{1}{2})\,\frac{1}{\Lambda_{S(L)}}$$

where $\Lambda_S = \beta\gamma c\tau_{S(L)}$ and $m_{S(L)}$ are, respectively, the lifetime and the mass of $K_{S(L)}$.

If we put a thin absorber in $x_a$ and $x_b$ positions, we may calculate the intensity of $\bar{K}^0\bar{K}^0$ wave by the expression $|\langle\bar{K}^0\bar{K}^0|\psi\rangle|^2$, taking into account the $\bar{K}^0$ content of $K_S$ and $K_L$. We then have :

$$I(\bar{K}^0\bar{K}^0) = \frac{1}{8}\left|\exp(\alpha_S x_a + \alpha_L x_b) - \exp(\alpha_L x_a + \alpha_S x_b)\right|^2,$$

or

$$I(\bar{K}^0\bar{K}^0) = \frac{1}{8}\left[\exp -\left(\frac{x_a}{\Lambda_S} + \frac{x_b}{\Lambda_L}\right) + \exp -\left(\frac{x_a}{\Lambda_L} + \frac{x_b}{\Lambda_S}\right)\right.$$

$$\left. - 2\exp -\left(\frac{x_a + x_b}{2\Lambda}\right)\cos\frac{\delta}{\Lambda_S}(x_a - x_b)\right],$$

where $1/\Lambda = 1/\Lambda_S + 1/\Lambda_L$ and $\delta = (m_L - m_S)\tau_S/\hbar$ is the usual $K_L - K_S$ mass difference.

With "Furry's" hypothesis, we obtain :

$$I(\bar{K}^0\bar{K}^0) = \frac{1}{8}(|\exp(\alpha_S x_a + \alpha_L x_b)|^2 + |\exp(\alpha_L x_a + \alpha_S x_b)|^2),$$

with no interference term.

Calling $x_1$ the smaller of the distance $x_a$ and $x_b$, $x_2$ the greater one and $x = x_2 - x_1$ and neglecting $1/\Lambda_L \ll 1/\Lambda_S$ we find for quantum mechanics :

$$I_{QM} = \frac{1}{8}\exp(-x_1/\Lambda_S)y_{QM}$$

with

$$y_{QM} = 1 + \exp(-x/\Lambda_S) - 2\exp(-x/2\Lambda_S)\cos(\delta x/\Lambda_S)$$

and for "Furry's" hypothesis

$$I_F = \frac{1}{8}\exp(-x_1/\Lambda_S)y_F$$

with

$$y_F = 1 + \exp(-x/\Lambda_S).$$

These functions $y_{QM}$ and $y_F$ are represented in fig. 1.

With a double absorber, one might also attempt to apply some of Bell's inequalities in order to test quantum mechanics against any local deterministic theory. Unfortunately due to K decay, it is impossible to have a valid test[11].

At this point, we can summarize the motivation of this paper by the following remarks :

1) There is at present no experimental verification that the $K^0\bar{K}^0$ system has the non-separate form given by eq.(2). A "Furry's" hypothesis may be envisaged. Even if non separability is a property of the spin component of particles, it could be interesting to verify it in the case of CP conservation.

2) The verification of the non-separate form given by quantum mechanics requires an absorber for the $K^0$ and $\bar{K}^0$ as described above. This sort of experiment is difficult but it may be envisaged with the new LEAR facility at CERN.

The annihilation of antiprotons at rest in liquid hydrogen takes place mainly in the state for $K^0\bar{K}^0$ production and one has $p\bar{p} \to K_S K_L$ at $\sim$ 99 %[12]. The $K_S$ of this reaction has $\Lambda_S \sim 4$ cm

(6 mm in the case of $e^+e^- \to \phi \to K_S K_L$) and thus it is possible to detect interactions of $K_S$ before its decay. The CERN LEAR facility designed to produce $10^6$ $\bar{p}$/s with very low momenta may be used to envisage such an experiment.

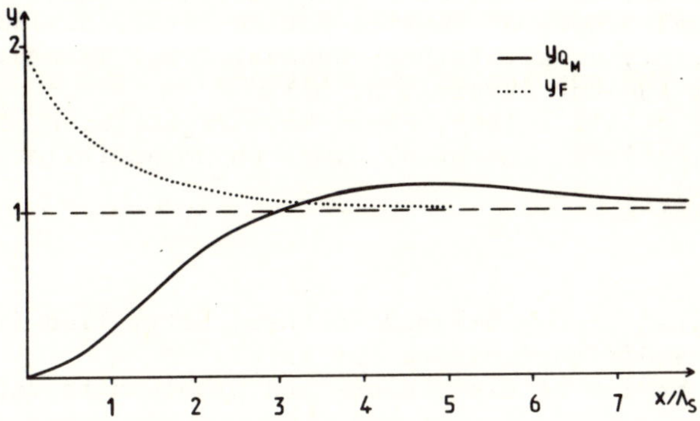


Fig. 1. The curves $y_{QM}$ and $y_F$ predicted, respectively, by quantum mechanics and by "Furry's" hypothesis versus $x/\Lambda_S$. See the text for the definition of x and y.

It would be easy to rule out "Furry's" hypothesis (region of small x in fig. 1) and it would be possible to verify the maximum of the curve in fig. 1, predicted by the interference term of quantum mechanics.

The author is grateful to Professor B. D'Espagnat for interesting and fruitful discussions.

REFERENCES

1. Schrödinger, E. : 1935, Naturwiss 23, p. 807.
2. Einstein, A., Podolsky, B., and Rosen, N. : 1935, Phys. Rev. 47, p. 777.
3. For a summary see : Selleri, F., and Tarozzi, G. : 1981, Riv. Nuovo Cimento, vol. 4, n°2.
4. Furry, W.H. : 1936, Phys. Rev. 49, p. 393.
   Bohm, D., and Aharonov, V. : 1957, Phys. Rev. 103, p. 1070.
5. Bell, J.S. : 1965, Physics 1, p. 135.
6. Aspect, A., Grangier, P., and Roger, G. : 1981, Phys. Rev. Lett. 47, p. 460.
7. Clauser J.F., and Schimony A. : 1978, Rep. Progr. Phys. 41, p. 1881.
8. See for example Chang., C.V., et al. : 1966, Phys. Lett. 23, p. 702.

9. See for example Geweniger, C., et al. : 1974, Phys. Lett. 48B, p. 487.
10. See for example Good, R.M., et al. : 1961, Phys. Rev. 124, p. 1223.
11. d'Espagnat, B., ed. : 1970, Foundations of quantum mechanics, International School of Physics Enrico Fermi, p. 180.
12. Armenteros, R., et al. : 1962, Geneva, Proc. Intern. Conf. on High Energy Nuclear Physics, p. 351.
Baltay, C., et al. : 1965, Phys. Rev. Lett. 15, p. 532.
Bizzari, R. : 1972, Chexbres, Symp. on $N\bar{N}$ annihilation, (CERN 72-10) p. 161.

*[1] This paper has been published in Phys. Lett. 114B (1982) 200 Only minor modifications are put here.

*[2] See also a review of experiments and problems in ref. [7].

A NEW APPROACH TO TESTING THE SEPARABILITY IN MICROPHYSICS: RAPISARDA'S EXPERIMENT (*)

F. Falciglia (Δ), L. Fornari (§), A. Garuccio (§), G. Iaci (∇), L. Pappalardo (∇)

(Δ) Ist. Fisica, Fac. Ingegneria, Università di Catania
(§) Dipartimento di Fisica, Università di Bari
(∇) Ist. Fisica, Facoltà Scienze, Università di Catania
Sezioni I.N.F.N. di Bari e Catania

ABSTRACT: The two experimental configurations up to now proposed for testing the separability in microphysics via Bell's type inequalities are described. The theoretical assumptions of CHSH and FOCA frames are analysed; in particular in the last configuration it is derived a new method to treat discriminating inequalities. A method is discussed which permits to measure the speed of the eventual superluminal correlation propagation. The Rapisarda's experiment, based on FOCA configuration, is described.

## 1. INTRODUCTION

Bell's paper (1964) (1) opened a route to an experimental check of the problems raised by the EPR paradox (2). Unfortunately the inequality proved in that work is based on two hypothesis hard to realize experimentally, i.e.:

---

(*) This experiment (FOCA2) was proposed and directed, until his untimely death occurred on february 21st, 1982, by Prof. V. A. RAPISARDA.

The experiment is supported by I.N.F.N., M.P.I. and C.N.R.



*S. Diner et al. (eds.), The Wave-Particle Dualism, 397–412.*

1) In any given experiment measuring the correlation between physical quantities of different microscopical systems, an experimental configuration yielding total correlation exists.
2) It is possible to perform measurements on each physical system and such measurements always yield well defined results.

The aim of the researchers since 1964 was directed on one hand toward the definition of new experimentally suitable inequalities, on the other hand toward new experimental designs. The papers by CHSH, H, CH, Fry (3) represent a main step toward that goal. Schematically an experiment on the EPR paradox based on polarization correlation measurements of optical photons emitted in a atomic decay is made as follows:

a) A source S of properly excited atoms, emitting polarization correlated photon-couples.
b) Two optical systems which collect part of the emitted photons and send them to the measuring apparata.
c) Two measuring apparata, each made by a polarizer (analyser) followed by one or two photomultipliers (detectors). The orientation of the polarizer can be changed at will.
d) An electronic apparatus which analyses in coincidence the detector's signals.

The theoretical discussion of the separabilty and locality problem in Quantum Mechanics (QM) has been formulated in terms of a mathematical formalism known as "Hidden Variable Theories" (HVT). In this paper we will not use the HVT formalism, but the one of the "Probabilistic Hidden Variable Theories" (PHVT) defined as follows [1]:

I) $\Lambda$ is the set of values of the hidden variable $\lambda$.

II) To each couple $(\gamma_A, \gamma_B)$ of particles emitted in S the value $\lambda \in \Lambda$ is associated.

III) $\rho(\lambda)$ is the normalized density function of $\lambda$ in $\Lambda$. We have: $\int \rho(\lambda)\, d\lambda = 1$.

IV) Let $\phi_A$ be the angle between the analyser's axis and a reference plane.
The probability that the event k is given by the measuring apparatus set into the configuration labeled by the parameter $\phi_A$ is denoted by $P[k|\phi_A,\lambda] = P^k(\phi_A,\lambda)$.

V) The probability that the value k is obtained as result of a measurement on the first particle and a value j on the second one is:

$$P[k,j|\phi_A,\phi_B,\lambda] = P[k|\phi_A,\lambda]\cdot P[j|\phi_B,\lambda] = P^k(\phi_A,\lambda)\cdot P^j(\phi_B,\lambda)$$

Assumptions IV and V are respectively the definitions of PHVT and of microscopic factorizability (4).

## 2. CHSH CONFIGURATION

Let us analyse the explicite and implicite assumptions of the experimental scheme proposed by H, CHSH and later completed by CH (CHSH scheme), as outlined in fig. 1.

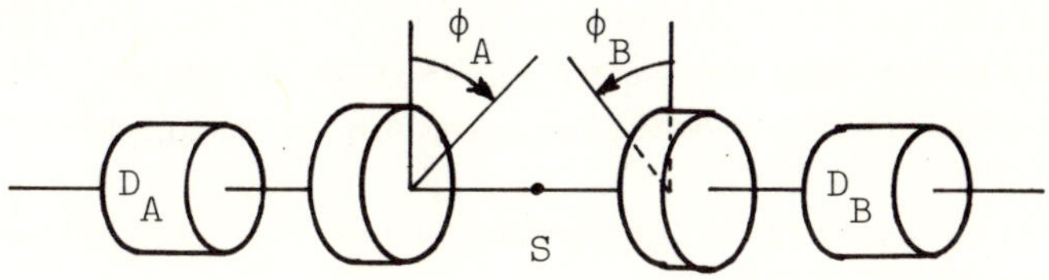


Figure 1. CHSH scheme

A0) The dicothomic observable used is: "the photon has / has not gone through the polarizer".

A1) The probabilty of joint detection of a photon-couple emerging out of the polarizers is independent of $\phi_A$ and $\phi_B$.

A2) The overall detection probability of a photon-couple A and B, i.e. the correlation function $P(\phi_A,\phi_B)$, is only function of the difference between the parameters $\phi_A$ , $\phi_B$:

$$P(\phi_A,\phi_B) = P(\phi_A+\phi_B)$$

A3) The flux of photons impinging on the polarizers is independent on $\phi_A$ , $\phi_B$.

A4) The density distribution $\rho$ and in general the flux of the analyzed photon-couples is the same whether the polarizers are in place or not.

A5) The detection probability of a photon of given $\lambda$ when the polarizer is in place is always less or equal to the one when the polarizer is removed.

Assumption A0 is the definition of a one-coincidence experiment.

Assumption A1 can be equivalently decomposed in two assumptions:

A1.1) The detection probability of a photon which has passed through a polarizer set at the angle $\phi_A$ is independent of $\phi_A$, and it is equal to the one when the polarizer is removed.

A1.2) The photon detection probability is independent of value of the hidden variable associated to the photon.

Assumption A2 imposes symmetry conditions on the physical phenomenon studied.

Assumptions A3 and A4 imply too the validity of conditions III of PHVT.
From A5 the following inequality can be defined:

$$- P(\infty,\infty) \leq P(\phi_A,\phi_B) - P(\phi_A,\phi'_B) + P(\phi'_A,\phi_B) + P(\phi'_A,\phi'_B) + - P(\phi'_A, \infty) - P(\infty, \phi'_B) \leq 0 \tag{2.1}$$

The symbol $\infty$ means that the polarizer is missing.
The quantities in (2.1) cannot be directely measured since is not possible to known the number of particles impinging on the analyzers. Therefore the following supplementary assumption is necessary:

A6) The source is stable during the experiment and it emits a constant rate.

Then: $$\frac{P(i,j)}{P(\infty,\infty)} \triangleq \frac{N(i,j)}{N(\infty,\infty)} \equiv \omega(i,j) \tag{2.2}$$

where: $i = \phi_A, \phi'_A, \infty$ ; $j = \phi_B, \phi'_B, \infty$ and $N(i,j)$ is the measured coincidence number in the configuration $(i,j)$.
Now it is possible to write the following experimental verifiable inequality:

$$- 1 \leq \omega(\phi_A,\phi_B) - \omega(\phi_A,\phi'_B) + \omega(\phi'_A,\phi_B) + \omega(\phi'_A,\phi_B) + - \omega(\phi'_A, \infty) - \omega(\infty ,\phi'_B) \leq 0 \tag{2.3}$$

QM violates the previous inequality in some experimental configurations.
We note that the only experimentally verifiable condition is A2; condition A0 is the definition of the experiment; all the others are not experimentally verifiable.

## 3. FOCA CONFIGURATION

We refer now to the experimental configuration in fig. 2.
The analyser decomposes the impinging light into an ordinary and an extraordinary ray. The parameter $\phi_A$ is the relative angle between the polarization plane of the ordinary ray and a reference plane. Along each ray is placed a photomultiplier. By convention we will say that the event "-1" has taken place if a photon is detected by the photomultiplier along the extraordinary channel; the event "+1" has taken place if a photon is detected by

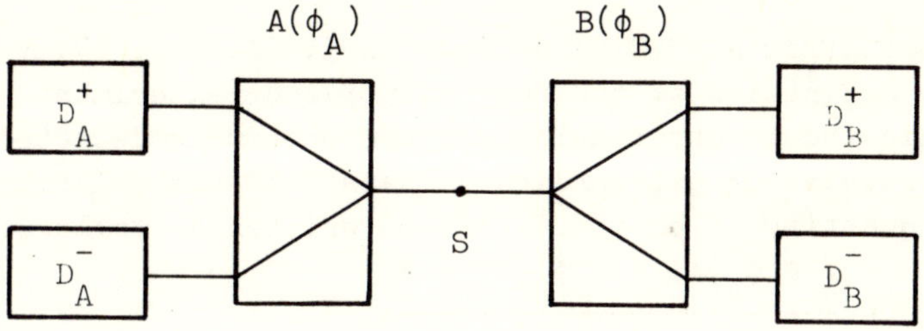


Figure 2. FOCA scheme

the photomultiplier along the ordinary channel. The events detected by the apparatus labeled by A are analyzed in coincidence with the ones detected by the other apparatus labeled from now on by B.

We have therefore a four-coincidence system that allows us to perform at the same time the four measurements:

$$N_{++} \qquad N_{+-} \qquad N_{-+} \qquad N_{--} \tag{3.1}$$

where $N_{ij}$ is the number of coincidences between the event i in the apparatus A and the event j in the apparatus B.

If

$$N = N_{++} + N_{+-} + N_{-+} + N_{--} \tag{3.2}$$

the measured frequencies of the event (i,j) are given by:

$$f_{ij} = N_{ij} \; / \; N \tag{3.3}$$

and the correlation function will be:

$$E\,(\phi_A,\phi_B) = \sum_{ij} i\; j\; f_{ij} = \frac{N_{++} + N_{--} - N_{+-} - N_{-+}}{N_{++} + N_{--} + N_{+-} + N_{-+}} \tag{3.4}$$

Eq. (3.4) allows a direct and straightforward check of the QM prediction. A comparison of the experimental values against the Bell's inequality to test the EPR paradox problematics is not so simple.

## 4. PROBABILISTIC HIDDEN VARIABLE THEORY IN THE FOCA CONFIGURATION

Let us start from a PHVT as defined in I, II, III, IV, V and the following definition of the four-coincidence experiment:

BO) The dicothomic observable is: "the photon is detected along the ordinary or extraordinary channel of the polarizer".

Then the quantity in eq. (3.4) can be written as follows:

$$E(\phi_A,\phi_B) = \frac{\int F(\phi_A,\lambda) \cdot F(\phi_B,\lambda)\, \rho\, d\lambda}{\int S(\phi_A,\lambda) \cdot S(\phi_B,\lambda)\, \rho\, d\lambda} \qquad (4.1)$$

where:

$$S(\phi_A,\lambda) = P^+(\phi_A,\lambda) + P^-(\phi_A,\lambda) \qquad (4.2)$$

$$F(\phi_A,\lambda) = P^+(\phi_A,\lambda) - P^-(\phi_A,\lambda) \qquad (4.3)$$

The (4.1) denominator in general depends on the parameters $\phi_A$, $\phi_B$ and this makes impossible to prove, without further assumptions, a Bell's inequality with eq. (4.1).

Let $C_A^+(\phi_A,\lambda)$ be the probability for a photon to go through the ordinary channel and $D_A^+(\phi_A,\lambda)$ the probability to be detected once passed. Then:

$$P_A^+(\phi_A,\lambda) = C_A^+(\phi_A,\lambda) \cdot D_A^+(\phi_A,\lambda) \qquad (4.4)$$

Now, with the assumptions:

B1.1) identical with A1.1

B1.2) identical with A1.2

eq. (4.4) becames:

$$P_A^+(\phi_A,\lambda) = D_A^+ \cdot C_A^+(\phi_A,\lambda) \qquad (4.5)$$

We note that in the FOCA configuration the assumption B1.2 is still not verifiable, while B1.1 is derivable from the experimental configuration if:

a) The extinction ratio of the polarizer is small.

b) The photomultiplier sees the photons always in the same area of the photocathode and with the same polarization, whatever is $\phi_A$ setting.

This is obtained e.g. if the photomultiplier rotates with the polarizer.

If we define, for K = A , B :

$$G_K(\phi_K,\lambda) = C_K^+(\phi_K,\lambda) + C_K^-(\phi_K,\lambda) \tag{4.6}$$

$$g_K(\phi_K,\lambda) = C_K^+(\phi_K,\lambda) - C_K^-(\phi_K,\lambda) \tag{4.7}$$

$$\alpha_K = \frac{D_K^+ - D_K^-}{D_K^+ + D_K^-} \tag{4.8}$$

$$x(\phi_A,\phi_B) = \frac{\int g_A(\phi_A,\lambda)\; G_B(\phi_B,\lambda)\; \rho(\lambda)\; d\lambda}{\int G_A(\phi_A,\lambda)\; G_B(\phi_B,\lambda)\; \rho(\lambda)\; d\lambda} \tag{4.9}$$

$$y(\phi_A,\phi_B) = \frac{\int G_A(\phi_A,\lambda)\; g_B(\phi_B,\lambda)\; \rho(\lambda)\; d\lambda}{\int G_A(\phi_A,\lambda)\; G_B(\phi_B,\lambda)\; \rho(\lambda)\; d\lambda} \tag{4.10}$$

$$z(\phi_A,\phi_B) = \frac{\int g_A(\phi_A,\lambda)\; g_B(\phi_B,\lambda)\; \rho(\lambda)\; d\lambda}{\int G_A(\phi_A,\lambda)\; G_B(\phi_B,\lambda)\; \rho(\lambda)\; d\lambda} \tag{4.11}$$

then we have:

$$E(\phi_A,\phi_B) = \frac{\alpha_A\alpha_B + \alpha_A\; y + \alpha_B\; x + z}{1 + \alpha_B\; y + \alpha_A\; x + \alpha_A\alpha_B\; z} \equiv E(A,B) \tag{4.12}$$

Moreover on the set of the coincidence detected photon-couples it is possible to define the average value of the polarization of each photon as:

$$E(A) = \frac{(N_{++} + N_{+-}) - (N_{-+} + N_{--})}{N} \tag{4.13}$$

$$E(B) = \frac{(N_{++} + N_{-+}) - (N_{+-} + N_{--})}{N} \tag{4.14}$$

The (4.13) and (4.14) can be written, as function of the hidden variables, as:

$$E(A) = \frac{\alpha_A + \alpha_A\alpha_B\, y + x + \alpha_B\, z}{1 + \alpha_B\, y + \alpha_A\, x + \alpha_A\alpha_B\, z} \qquad (4.15)$$

$$E(B) = \frac{\alpha_B + \alpha_A\alpha_B\, x + y + \alpha_A\, z}{1 + \alpha_B\, y + \alpha_A\, x + \alpha_A\alpha_B\, z} \qquad (4.16)$$

If we solve the system of eq. (4.12), (4.15), (4.16) for z we get:

$$z(\phi_A,\phi_B) = \frac{E(A,B) - \alpha_B\, E(A) + \alpha_A\, E(B) + \alpha_A\alpha_B}{1 - \alpha_A\, E(A) - \alpha_B\, E(B) + \alpha_A\alpha_B\, E(A,B)} \qquad (4.17)$$

Finally with the help of the following assumption:
B2) The integral $\int G_A(\phi_A,\lambda)\, G_B(\phi_B,\lambda)\, \rho\, d\lambda$ is independent of $\phi_A$, $\phi_B$.
it is possible to derive the following inequality:

$$Q(z) = |z(\phi_A,\phi_B)-z(\phi_A,\phi'_B)| + |z(\phi'_A,\phi_B)+z(\phi'_A,\phi'_B)| \leq 2 \qquad (4.18)$$

This inequality is violated by QM.
It is interesting to observe that in the r.h.s. of (4.17) are present the parameters $\alpha_K$, while the l.h.s. is, by definition, $\alpha_K$-independent. Being the r.h.s. directly measurable it is possible to verify this independence. So (4.17) permits to check the validity of B1.1 and B1.2 assumptions.

## 5. QUANTUM MECHANICS PREDICTION

To simplify the reasoning let us suppose that:
a) The polarizers have a small extinction ratio.
b) We can neglect the depolarization due to the finite solid angle of photon collecting optics.

If $T^i_K$ is the K calcite transmittance along the i channel, putting

$$\beta_K = \frac{T^+_K - T^-_K}{T^+_K + T^-_K} \qquad (5.1)$$

The QM correlation function became:

$$E(A,B)_{QM} = \frac{(\alpha_A+\beta_A)(\alpha_B+\beta_B)+(1+\alpha_A\beta_A)(1+\alpha_B\beta_B)\ \cos 2(\phi_A+\phi_B)}{(1+\alpha_A\beta_A)(1+\alpha_B\beta_B)+(\alpha_A+\beta_A)(\alpha_B+\beta_B)\ \cos 2(\phi_A+\phi_B)} \tag{5.2}$$

and

$$z_{QM} = \frac{\beta_A\beta_B + \cos 2\ (\phi_A + \phi_B)}{1 + \beta_A\beta_B\ \cos 2\ (\phi_A + \phi_B)} \tag{5.3}$$

It is possible to verify (6) that, for an appropriate choice of the angles $\phi_A$ $\phi_A'$, $\phi_B$, $\phi_B'$, the value of $z_{QM}$ doesn't satisfy inequality (4.18).
The problem of showing the existence of a Bell's type inequality in conditions more general than the one we assumed is still wide open.

## 6. SUPERLUMINAL SIGNALS

The results of the experiments performed up to now (7) seem to confirm the non-separabilty in microphysics.
The possibility that these results are due to a superluminal transmission of real physical correlations - as suggested by several Authors (8) - immediately raises the crucial question: "How it is possible to define - and eventually to measure the (superluminal) speed of something unknown?" To solving this question is fruitful to remark that, in the above hypothesis, only if the communication between the two photons of the couple has taken place, can QM correlations be observed (and Bell's inequality violated); whereas, if this communication is not established before the detection of the two photons, locality must be mantained (i.e. Bell's inequality respected) and QM correlations dropped.
To apply this idea we suppose that the length of the measuring apparata are small with respect to their distance and that the correlation propagation travels only foreward in the time (9).
So, with reference to fig. 3, where:

$L_A$ is the distance from the source S of the A-apparatus

$\varepsilon$ is the delay between the emission of the first and the second photon

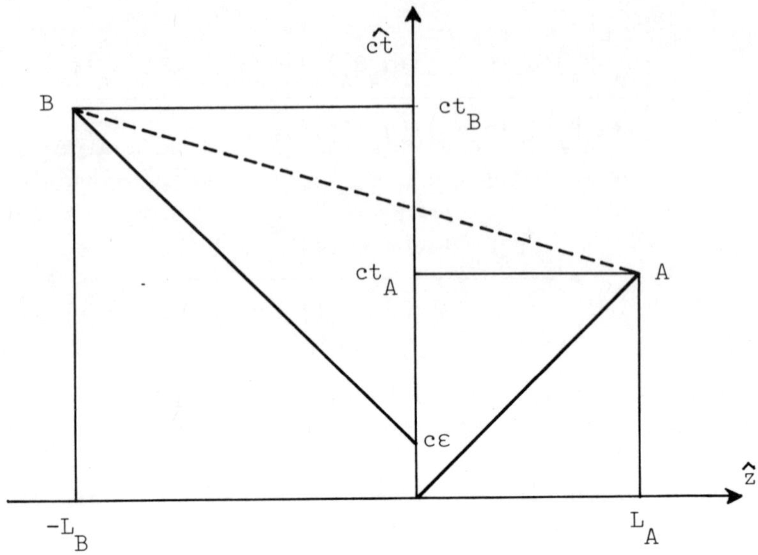


Figure 3.

$\eta$ is the speed of this "signal"

$L_B$ is the distance at which the signal starting from $L_A$, reaches the world-line of photon $\gamma_B$

if the analyzer B is placed at $L'_B < L_B$ Bell's inequality must hold, whereas it is violated for $L'_B > L_B$.
From the figure it is obvious that:

$$\eta = \frac{L_A + L_B}{L_B - L_A + c\,\varepsilon}\, c \tag{6.1}$$

Then varying the space-time configuration it is possible to determine the speed of this correlation propagation. It can be verified that the accuracy with which this speed can be measured is related to the space-time resolution of the events A and B. For example, if one wish to know $\eta$ with an error less than 10%, must be $|L_A - L_B| \geq 10\,\tau c$, where $\tau$ is the resolution time of the electronics.

## 7. THE FOCA2 EXPERIMENT

An experiment based on FOCA configuration is under accomplishment at the Physics Institute of the Engineering Faculty of the State University of Catania (Italy).

The photon-couples whose polarization is to be analyzed, are derived from the atomic cascade $4s^2\ {}^1S_0 \rightarrow 4s4p\ {}^1P_1 \rightarrow 4s^2\ {}^1S_0$ of the neutral Calcium (Ca I) in which are emitted two electric dipole photons $\gamma_A$ and $\gamma_B$ of wavelength $\lambda_A = 551.3$ nm and $\lambda_B = 422.7$ nm.

The QM state vector for the polarization of a couple $(\gamma_A,\gamma_B)$ of photons travelling in opposite directions is:

$$|\psi_{\gamma_A\gamma_B}> = \frac{1}{\sqrt{2}} \left[|R_A>|R_B> + |L_A>|L_B>\right] \qquad (7.1)$$

where $|R_K>$ ($|L_K>$) is the right (left) -handed circular polarization state.

For this state the QM correlation function is:

$$E\ (\phi_A,\ \phi_B) = \cos 2\ (\phi_A + \phi_B) \qquad (7.2)$$

where $\phi_A$, $\phi_B$ are the orientations of the two analysing apparata. It is easy to prove that (7.2) violates Bell's inequality.

## 8. EXPERIMENTAL SET-UP

The source (see fig. 4) consists of an oven where is thermically generated a Ca atomic beam successively excited, via a two-photon process, by the radiation of a dye-laser.

The photons emitted in the above mentioned cascade are analysed by two systems each made by a calcite, two interference filters and two photomultipliers (PM), whose signals are analysed by suitable electronic chains.

### 8.1 Source

The oven is made of a Tantalum "bottle" brought up to the required temperature (≃ 700 °C) by an heating element electronically

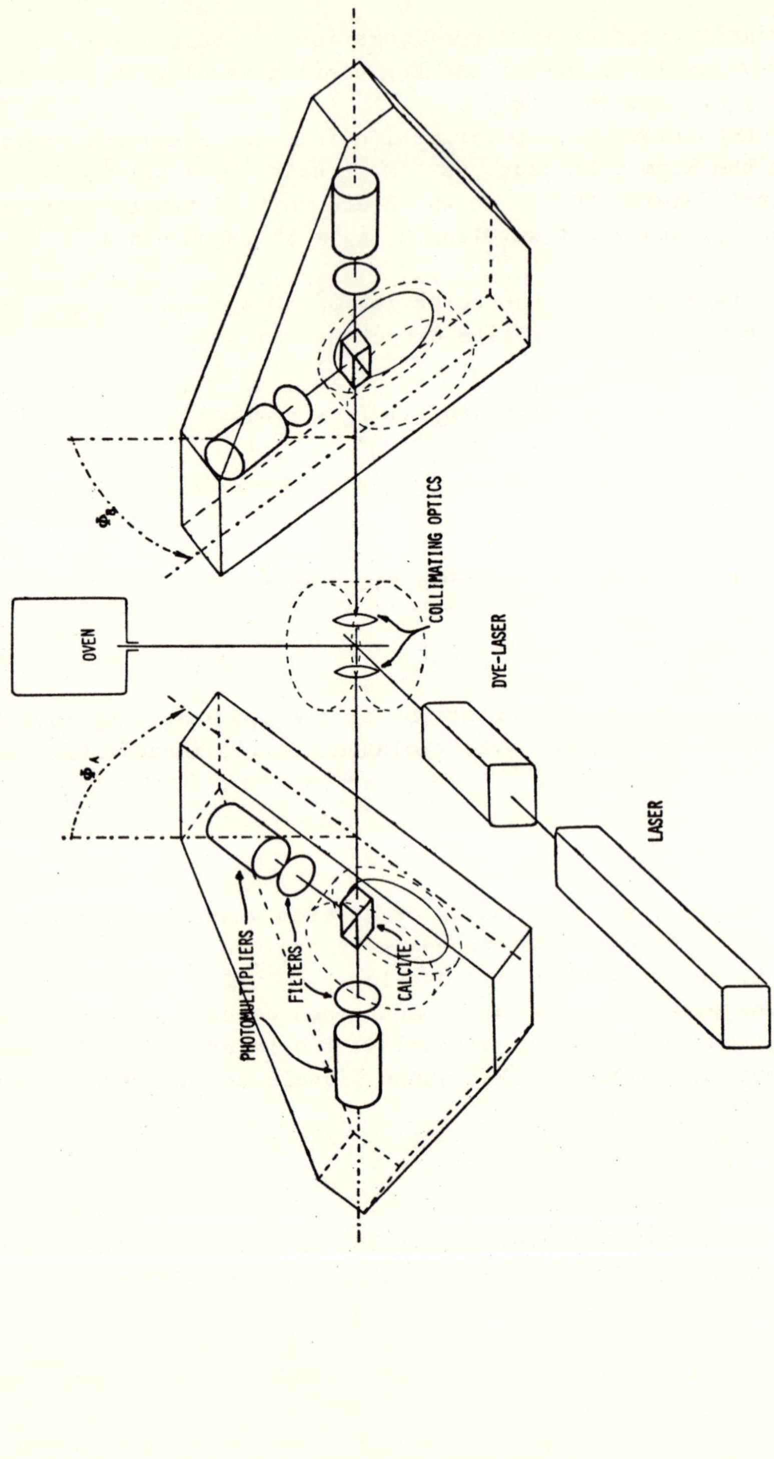


Figure 4. FOCA experimental set-up

feed-backed so as to attain a temperature stability of ± 0.5 °C. The oven actual volume, about 300 $cm^3$, ensures long term operation at a beam density of ≃ $10^{12}$ atoms/$cm^3$.
The system geometry allows to adjust by optical methods the beam alignement and collimation before the oven loading, that can be done without removing the crucible.
The upper excited level of Ca I is reached by absorption of two equal wavelength photons ($\lambda$ = 478.4 nm) from the radiation emitted by a Coherent 599-21 tunable laser, using as dye the Coumarine 480, pumped by a Coherent CR 3000K Krypton-ion laser.
With an output power of 100 mW and the beforesaid atomic density a rate of about $10^6$ cascades/s has been estimated. A better yield could be obtained exciting by two photons of wavelength closer to $\lambda_A$ and $\lambda_B$ but we have preferred that solution for experimental plainness and, mainly, for more easily filtering of diffused radiation.

## 8.2 Analysing apparatus

There are two identical optical systems, opposite each other, each consisting of two aspherical lenses for minimizing polarization distortion (10), with a collection solid angle of ∿ 1.12 sr corresponding to a geometrical efficiency of ∿ 9%.
They output parallel meams 1.1 nm in diameter.
The polarization analysers are two Foster-type cut calcites with an extinction ratio of ∿ 5 $10^{-5}$, followed by interference filters with suitable peak wavelengths and bandwith of ∿ 1.1 nm.
The four beams so obtained are detected by two pairs of low-noise PMs chosen on the basis of their spectral response and placed in cooled housings. The PMs, rigidly connected with calcites and filters, rotate jointly with them, so as ensure that the photons hit always the same region of the photocathode with a polarization independent of the angle. The small beam diameter reduces noticeably the time-spread. The relative distance between the two analysing systems can be varied from ∿ 120 cm to ∿ 13 m in order to measure the speed of the eventual superluminal signals.
In fig. 5 is shown the logical scheme of the electronics.
The PM signals are amplified and sent to fast amplitude discriminators whose output signals are fed into four double-input coincidence circuits with variable resolving time in the range from 2 ns to 100 ns.
The coincidence events $N_{++}$, $N_{+-}$, $N_{-+}$ and $N_{--}$ are totalized by

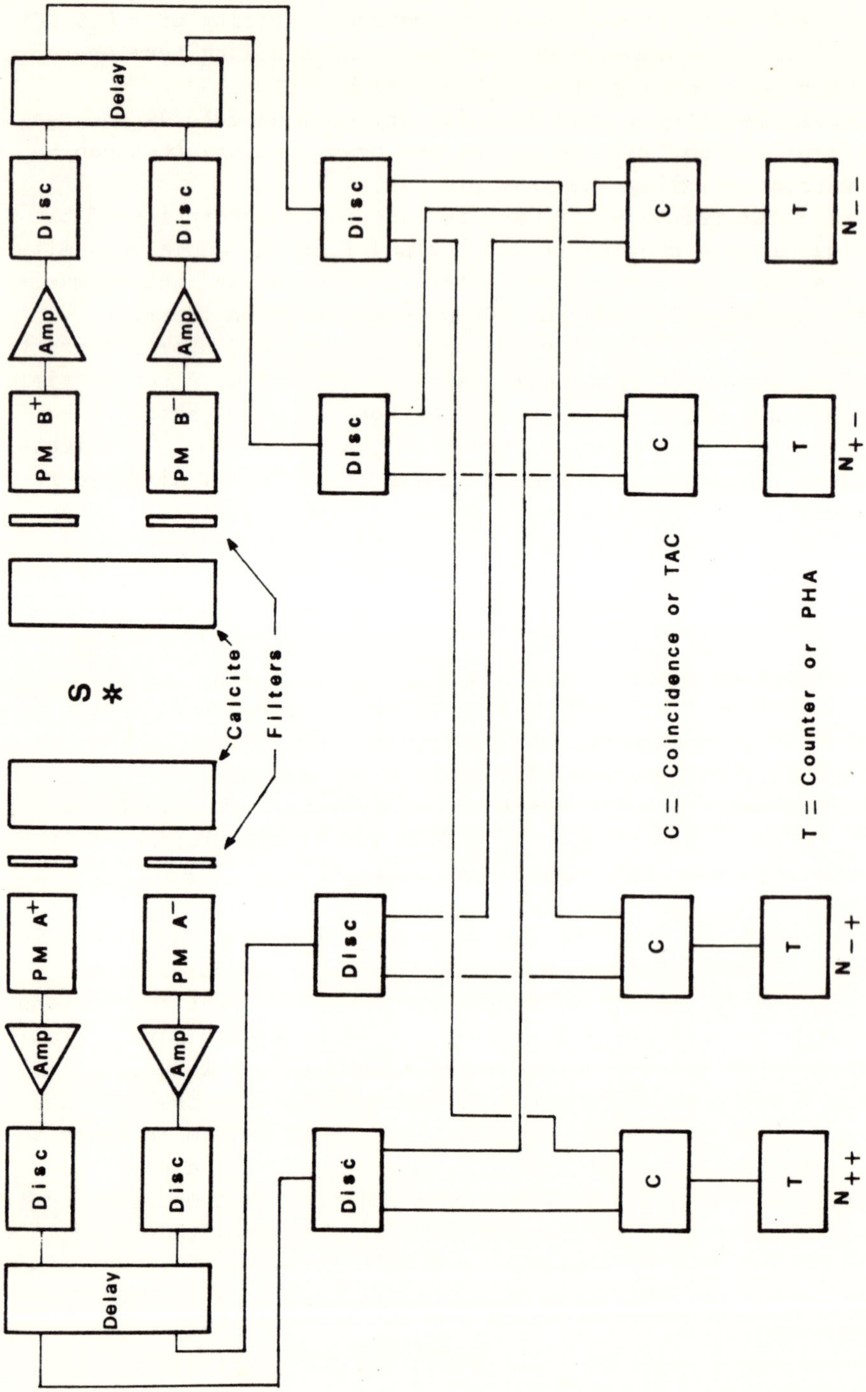


Figure 5. Scheme of electronics

four counters.
We intend to perform also a more detailed time-distribution analysis using four time-to-amplitude converters.
The preliminary measurements on the parameters of interest (e.g. PM efficiency, calcite transmittances) are to be performed in conditions similar to the working ones and therefore with single--photon techniques.
This can be done using a pulsed light source and enabling a the right moment the counting circuits. In so doing it is possible to reduce nearly to zero the background contribution.

## 9. NOTE

([1]) It has been demonstrated the formal equivalence between the two formulations. See e.g. (5)

## REFEREENCES

1) J. S. Bell: Physics, 1, 195 (1965)
2) A. Einstein, B. Podolsky, N. Rosen: Phys. Rev., 47, 777 (1935)
3) a) J. F. Clauser, M. A. Horn, A. Shimony, R. A. Holt: Phys. Rev. Lett., 23, 880 (1969)
   b) M. A. Horne: Experimental consequences of local hidden variable theories (Ph. D. Thesis, Boston University, 1970)
   c) E. S. Fry: Phys. Rev. A: 8, 1219 (1973)
4) J. F. Clauser, M. A. Horne: Phys. Rev. D, 10, 526 (1974)
5) a) A. Garuccio, F. Selleri: Lett. Nuovo Cimento, 23, 555 (1978)
   b) F. Selleri, G. Tarozzi: Lett. Nuovo Cimento, 29, 533 (1980)
   c) A. Garuccio, V. A. Rapisarda: Lett. Nuovo Cimento, 30, 291 (1980)
   d) A. Fine: Phys. Rev. Lett., 48, 291 (1982)
6) A. Garuccio, V. A. Rapisarda: Nuovo Cimento A, 65, 269 (1981)
7) a) S. J. Freedman, J. F. Clauser: Phys. Rev. Lett., 28, 935 (1972)
   b) R. A. Holt, F. M. Pipkin: Quantum Mechanics vs. hidden variables: polarization correlation measurement on an atomic mercury cascade (Preprint, Harvard University, 1974)
   c) J. F. Clauser: Phys. Rev. Lett., 36, 1223 (1976)
   d) E. S. Fry, R. C. Thompson: Phys. Rev. Lett., 37, 465 (1976)
   e) A. Aspect, P. Grangier, R. Roger: Phys. Rev. Lett., 47, 460 (1981)

8) a) D. Bohm, B. J. Hiley: Found. Phys., 5, 453 (1966)
b) D. Bohm, B. J. Hiley: Quantum Mechanics a Half Century Lateral (edited by S. Leite-Lopez and M. Paty, 1970)
c) J. P. Vigier: Lett. Nuovo Cimento, 24, 258 (1979)
d) N. Cufaro-Petroni, J. P. Vigier: Lett. Nuovo Cimento, 26, 149 (1979)
e) F. Falciglia, G. Iaci, V. A. Rapisarda: Lett. Nuovo Cimento, 26, 327 (1979)
f) A. Shimony: Report Geneva Conference (1979)
g) B. D'Espagnat: Found. Phys., 11, 205 (1981)
h) K. Nambrai: J. Phys. A, 14, 1307 (1981)

9) A. Garuccio, V. A. Rapisarda, J. P. Vigier: Lett. Nuovo Cimento, 32, 451 (1981)

10) M. Baldo, F. Falciglia, M. E. Oliveri, V. A. Rapisarda: Nuovo Cimento B, 63, 679 (1981)

# ON THE POSSIBILITY OF EXTENDING THE TESTS OF QUANTUM MECHANICAL CORRELATIONS

Silvio Bergia

Istituto di Fisica dell'Università, Bologna.
Istituto Nazionale di Fisica Nucleare, Sez. di Bologna

ABSTRACT

Experimental tests of quantum mechanical correlations in connection with Bell's inequality have generally considered decays in sub-systems characterized by two-valued observables. We analyse here the possibility of extending these tests to a much wider class of cases.

In his 1964 paper (1), Bell had already suggested that his theorem, concerning the incompatibility between local hidden variable theories (LHVT) and quantum mechanics (QM), could be extended to systems different from those initially examined by Bohm (2) and characterized by a decay in sub-systems with two-valued observables.

Clearly any attempt for an extension in this direction implies necessarily three points:

i) in the first place one has to grasp the essential aspects of the structure of the quantum mechanical correlations in the Bohm system, and analyse to what systems, and with what features, correlations of these nature can be extended;

ii) secondly, one should check whether, on the basis of hypotheses similar to Bell's, his inequality could be extended to these more general situations;

iii) thirdly, one should enquire whether there exist systems, i.e. states $|\psi>$ of the overall system, and dynamical variables $D^{I}$ and $D^{II}$ of the sub-systems, such that their quantum mechanical description leads to expectation values $<\psi|D^{I} \otimes D^{II}|\psi>$ violating the extended inequality.

If this programme succeeds, one would have achieved the result of extending, at least in principle (i.e. apart from experimental difficulties), the possibility of a comparison



S. Diner et al. (eds.), The Wave-Particle Dualism, 413–417.

between LMV, QM and experiment in cases different from those already examined.

In this note, we argue that the three points of the programme can indeed be achieved. However, we shall not go into a detailed analysis of points (ii) and (iii), and will limit ourselves to quote the results concerning point (ii) and to sketch the conceptual and practical problems concerning the possibility of actually measuring the observables mentioned in point (iii) (a more detailed discussion of these points is to be found in ref. (3)). In our opinion, these results have, on the other hand, mostly a formal value, since it seems that, in the cases analysed till now, the experiments have already discarted alternatives to quantum mechanics (4).

In view of this fact, it is point (i) that deserves the greatest attention, since it concerns a purely quantum mechanical analysis. As far as we know, little, if any, emphasis was given in the literature to the fact that a theorem established by von Neumann in connection with the theory of measurement contains in principle a complete answer to the problem posed in point (i). It seems therefore appropriate to devote a contribution to a discussion session on the fundamentals of quantum mechanics to an analysis of von Neumann's theorem as adapted to this problem.

The essential nature of the correlation which (in obvious notations) the quantum mechanical singlet state

$$|\psi\rangle = \frac{|+\rangle|-\rangle - |-\rangle|+\rangle}{\sqrt{2}} \tag{1}$$

introduces among the sub-systems is twofold. On the one hand it establishes that there exist observables belonging to the sub-systems (in this case the components of the spins along given directions) such that the measurement of an observable belonging to one sub-system determines uniquely the value of the corresponding observable of the other sub-system (the overall second kind state (1) collapses to the direct product of the corresponding eigenvectors); on the other hand it establishes a definite phase relation between the direct products in terms of which the overall state is built.

Von Neummann's theorem mentioned above shows that these features can be extended to extremely general situations. It can be stated as follows: for any system made up of two sub-systems, one can always find observables such that:
1) the sets of the corresponding eigenvectors (at least in the case of non-degenerate eigenvalues), $\{|\phi_\mu\rangle\}$ and $\{|\xi_\nu\rangle\}$, can be put in a one-to-one correspondence:$^\mu$

$$|\phi_{\mu_k}\rangle \longleftrightarrow |\xi_{\nu_k}\rangle$$

2) the state vector of the overall system can be written in the form

$$|\psi > = \sum_k w_k^{\frac{1}{2}} \; |\phi_{\mu_k} > |\xi_{\nu_k} > \qquad (2)$$

that is as a second kind state built as a linear combination of the states that the choice of observables puts in a one-to-one correspondence. In Eq. (2), the $w_k^{\frac{1}{2}}$ are the - not necessarily arithmetic - square roots of the positive numbers $w_k$; signs and absolute values of the $w_k^{\frac{1}{2}}$ shall be determined for any specific problem.

Obviously Eq. (2) represents a far more cogent result than the expansion

$$|\psi > = \sum_{m,n} f_{mn} \; |\phi_m > |\xi_n > \qquad (3)$$

which is always valid provided the sets $\{|\phi_m >\}$, $\{|\xi_n >\}$ span bases in the subspaces of the sub-systems.

The theorem was established by von Neumann in connection with an analysis of the measuring process (the sub-systems must then be identified with the system under observation and the measuring apparatus), and we refer to his treatise for a discussion of its relevance for the theory of measurement (5).

If, on the other hand, the sub-systems are identified with the decay products of a composite system, it is apparent that the theorem provides an extreme generalization of the quantum mechanical correlations arising in the Bohm case. This generalization is however achieved at the cost of a certain abstractness. Von Neumman's result becomes transparent again — and almost trivial — in the case for which:
- the only observables of interest are the angular momentum variables;
- the overall system is in an eigenstate of J and M.

It is fairly obvious that in this case the expansion of the state in the form (2) shall have to coincide with the Clebsch-Gordan expansion:

$$| j_1 j_2 \; JM > = \sum_m < j_1 j_2 \; M-m \; m \,|\, j_1 j_2 \; JM> <j_1 \; M-m | j_2 m > \qquad (4)$$

This result is indeed found by adapting von Neumann's general procedure to the case at hand (which is not quite trivial; see ref. (6)). No matter how expected is the conclusion once the problem is specified as above, the procedure exhibits some points that are worth stressing:
- going from form (3) to form (2) corresponds to going from the choice of observables $\vec{J}_1 \cdot \hat{a}$ and $\vec{J}_2 \cdot \hat{b}$ ($\hat{a}$ and $\hat{b}$ arbitrary directions), and of the corresponding basic sets $\{|\phi_m >\}$ and $\{|\xi_n >\}$, for the syb-systems, to the observables $J_{1z}$ and $J_{2z}$, of eigenstates $|\phi_\mu >$ and $|\xi_\nu >$;
- the coefficients $w_k^{\frac{1}{2}}$ in Eq. (2) coincide, in this case, with

the Clebsch-Gordan coefficients (other quantities appearing in von Neumman's procedure also receive an interpretation (6)).

In von Neumann's original version the $w_k^{\frac{1}{2}}$ were to be intended as arithmetic square roots. This appears to be at first sight incompatible with what we have said concerning their identification, in the case just examined, with the Clebsch- Gordan coefficients. The difficulty is however only apparent, since compatibility can be re-established via an — inessential — redefinition of phases in the sets $\{|\phi_{\mu_k}>$ and/or $\{|\xi_{\nu_k}>\}$(6). The apparent discrepancy may have been responsible for the fact that the immediate physical interpretation of the theorem for the case discussed above has been, as far as we know, fairly generally ignored.

- In the case above, the one-to-one correspondence between the states of the sub-systems is established on the basis of the conservation of the angular momentum component in a given direction. However, the validity of von Neumann's result does not depend in general on the validity of conservation laws; nor does it depend, on the other hand, on the fact that the state of the overall system belongs to an irreducible representation of a Lie group. Therefore, the interpretation of von Neumann's theorem given in ref. (6) refers to a sub-set of the set of systems to which the theorem applies. Nevertheless, the most interesting physical cases, as far as an analysis of the correlations between sub-systems is concerned, will be of the type discussed above, and the expansion (2) will coincide, in these cases, with the Clebsch-Gordan expansion valid for a given non-abelian Lie group.

Let us now turn to a brief discussion of points (i) and (ii). An extension of Bell's inequality to cases in which the observables to be measured on the sub-systems are not necessarily two-valued was obtained in ref. (3).

It should be noted that, as soon as the sub-systems have spins higher than ½, they will possess other dynamical variables besides angular momenta (the deuteron, for instance, has, besided its magnetic moment, an electric quadrupole moment). Therefore, the extension we are considering here does not concern only the state of the overall system, but also the nature of the observable $D^{I}\otimes D^{II}$.

In ref. (3) it was shown that $D^{I}\otimes D^{II}$ can be chosen in such a way as to produce a violation of the inequality in the case of the decay of a J = 2, M = 1 state into two massive spin 1 sub-systems.

Concerning the choice of $D^{I}\otimes D^{II}$, it should be noted that $D^{I}$ and $D^{II}$ will be described in general in terms of linear combinations of irreducible tensors $\Omega_k^{I,II}$:

$$D^{I,II} = \sum_k \omega_k^{I,II}\, \Omega_k^{I,II} \tag{5}$$

In order to test QM against the extended Bell's inequality, one must evaluate the expectation values

$$< D^{I} \otimes D^{II} > = \sum_{k,k'} \omega^{I}_{k} \, \omega^{II}_{k'} < \psi | \Omega^{I}_{k} \otimes \Omega^{II}_{k'} | \psi > \qquad (6)$$

in the overall state $|\psi>$, and check whether they lead to a violation of the inequality.

The experimental determination of the average value (6) thus requires the simultaneous measurement of two irreducible tensors belonging to the sub-systems for each run of the experiment. In the Bohm example, two Stern-Gerlach devices guarantee the feasability in principle of the experiment. Swift and Wright (7) have shown that generalized Stern-Gerlach devices can be designed for more general cases. However, these experiments might be exceedingly difficult to carry out in the laboratory.

Due to these difficulties, one would be generally forced to make "non selective" measurements, that is experiments which can only be performed on an ensemble of sub-systems and do not provide an eigenvalue but the average value of the observable involved (as in the case of polarization measurements via a rescattering). But, of course, determining the average values of $\Omega^{I}_{k}$ and $\Omega^{II}_{k}$ in the sub-systems is not the same as measuring the average value of $\Omega^{I}_{k} \otimes \Omega^{II}_{k'}$ needed in Eq. (6).

In conclusion, I wish to thank F. Cannata for useful discussion during the preparation of this note.

REFERENCES

1) Bell, J.S.: 1965, Physics 1, 195.
2) Bohm, D.: 1951, Quantum Theory (Prentis Hall, Englewood, Cliffs, N.J.), pp. 614-622.
3) Bergia, S., Cannata, F.: 1982, "Higher order tensors and tests of quantum mechanics", to appear in Foundations of Physics 12, No. 9.
4) Clauser, J.F., Shimony, A.: 1978, Rep. Prog. Phys. 41, 1881; see also: Aspect, A., this volume, p. 377.
5) von Neumann, J.: 1955, Mathematical Foundations of Quantum Mechanics (Princeton University Press, New Jersey); see also: Zurek, W.H., this volume, p. 515.
6) Bergia, S., Cannata, F., Ruffo, S., Savoia, M.: 1979, Am. J. Phys. 47, 548.
7) Swift, A.R., Wright, R.: 1980, J. Math. Phys. 21, 77.

# HOW DO WE HAVE TO CHANGE QUANTUM MECHANICS IN ORDER TO DESCRIBE SEPARATED SYSTEMS?

Dirk AERTS

Theoretische Natuurkunde, Vrije Universiteit Brussel.

Since we were able to show recently that quantum mechanics can not describe separated physical systems we analyse again the reasoning of Einstein-Podolsky-Rosen, and find that the most straith forward conclusion of this paradox is not correct. We indicate the missing elements of reality in the quantum mechanical description of separated physical systems. We show that Bell inequalities are satisfied iff the two physical systems are separated, whether the systems are quantum systems or classical systems is of no matter. We give an example of a classical macroscopical situation where Bell inequalities are violated.

## Introduction.

We were intrigued to analyse again the reasoning of Einstein Podolsky and Rosen [ 1] because we think that we recently were able to proof a result that trows another light on the problem in question. What we could proof is that the quantum mechanical description of two separated systems is wrong. And even more, that it is not possible in quantum mechanics to describe two separated systems.

The way in which we were able to derive this result is not by an impossibility proof. No, we can give a description of two separated systems in a more general theory as quantum mechanics. We say more general theory, because it is possible to define a set of conditions in this theory such that when these conditions are satisfied the theory reduces to quantum mechanics (eventually with superselection rules). We did this analysis in quantum logic, and more precisely in Pirons approach to quantum logic [ 2]. In this approach a physical system is described by the collection of



*S. Diner et al. (eds.), The Wave-Particle Dualism, 419–431.*

its properties. The way in which Piron introduces this concept of property is very similar to the idea of Einstein on the concept of element of reality. So that with every property of the system corresponds an element of reality and with every element of reality corresponds a property. This is important to notice since it shows that the theory is complete from the start. The set of all properties of a physical system has the structure of a complete lattice. Piron then defines some axioms such that when these axioms are satisfied the lattice gets the structure of an atomic, orthocomplemented weakly modular lattice that satisfies the covering law and by means of a representation theorem shows that in this case its theory reduces to quantum mechanics in Hilbert space with eventually superselection rules. What we found is that two of these axioms, the weak modularity and the covering law make it impossible for the theory to describe separated systems. Let us state this in the form of a theorem.

Theorem :
Suppose that S is a physical system composed of two separated physical systems $S_1$ and $S_2$. If the lattice of properties of S is weakly modular <u>or</u> if the lattice of properties satisfies the covering law, then $S_1$ or $S_2$ is a classical system of which the lattice of properties is a Boolean lattice.

The lattice of properties of a system described by quantum mechanics satisfies both axioms (weak modularity and the covering law) but is never a Boolean lattice. Hence from this theorem follows that S can never be described by quantum mechanics if $S_1$ and $S_2$ are described by quantum mechanics. We want to draw the attention to the fact that this is not a question of interpretation. No, it is really the mathematical structure of quantum mechanics that makes it impossible to describe separated systems. The covering law is the axiom that induces the vectorspace structure of the set of states. Hence the set of states of two separated systems will not have a vectorspace structure. As a consequence the superposition principle will not be valid. It is possible to see that the superselection rules introduced are non abelian. The weak modularity more or less corresponds to the possibility of representing observables by operators. Indeed for example von Neumann algebras give a weak modular structure for the set of projections. Hence also in the algebraic approach separated systems will not be describable. Piron's approach can be made sufficiently general to allow the description of separated systems. Indeed we just have to drop these two wrong axioms and to formulate the other axioms in a different way. We will not make this construction neither proof the theorem because our aim is to analyse the Einstein Podolsky Rosen paradox knowing this result. The reformulation of the good axioms in this more general theory which is not weakly modular and the proof of the theorem can be found in [3] and [4]. I will just mention also the following results : i) The states of the system S consisting of two separated systems are product states.

ii) Two such product states are separated by a superselection rule even when they are not orthogonal.

## 1. THE EINSTEIN PODOLSKY ROSEN ARGUMENT.

In discussing the Einstein Podolsky Rosen paradox I should like to make a distinction between two aspects of this paradox. The first aspect is the reasoning of E.P.R. [1] that leads to their conclusion that quantum mechanics is not a complete theory. The second aspect is the strange effect consisting of the fact that if we have a system described by quantum mechanics consisting of two systems the system can be put in such a state such that a measurement performed on one system, also determines the possible states of the other system. Schrödinger, more than E.P.R. studied this effect [5]. Therefore we should like to call this effect the Schrödinger effect.

We will now analyse again the E.P.R. reasoning.

E.P.R. consider the following two sentences

(1) quantum mechanics is not complete

(2) physical quantities that are not compatible cannot have simultaneous reality.

Obviously these two sentences cannot both be wrong. Indeed if two non compatible quantities can have simultaneous reality, then quantum mechanics is not complete.

So we have one of the three cases

| | | |
|---|---|---|
| A | (1) false | (2) true |
| B | (1) true | (2) false |
| C | (1) true | (2) true |

Once Einstein Podolsky and Rosen come to this conclusion they consider the situation of two separated systems $S_1$ and $S_2$. By applying quantum mechanics to describe these two separated systems, and using the Schrödinger effect they can show that it is possible to attach simultaneously elements of reality to non compatible quantities. Hence they show that (2) is false. Hence (1) has to be true. So what E.P.R. show is the following.

Quantum mechanics describes correctly and in a complete way separated systems.

$\Rightarrow$ quantities that are not compatible can have simultaneous reality

$\Rightarrow$ quantum mechanics is not complete

From this they can conclude that quantum mechanics is not correct or not complete. Einstein Podolsky and Rosen mention in the beginning of their paper that they suppose quantum mechanics to be

correct. Then they can indeed conclude that quantum mechanics is not complete. If we suppose quantum mechanics to be correct, then this reasoning of E.P.R. also indicates which are missing elements of reality in quantum mechanics. This made a lot of people think that it should be possible to solve the problem by introducing hidden variables that take into account these missing elements of reality, and in this way complete quantum mechanics. If we take into account that quantum mechanics is perhaps wrong in describing separated systems, a thing we certainly want to do since this is exactly what we show in [3] and [4], then although the reasoning of E.P.R. is still valid, it does not show anything since the premise is false. So then this reasoning does not show us a way to solve the problem and it does not indicate missing elements of reality. Let us repeat. The statement "quantum mechanics is not complete or wrong because quantities that are not compatible can have simultaneous reality" is a wrong statement. I want to put the attention to this fact because often people are not interested in the conclusion of E.P.R. namely that quantum mechanics is incomplete if it is correct, but in the conclusion that quantities that are not compatible can have simultaneous reality if quantum mechanics is correct. And it is this conclusion that leads directly to the thought that to avoid the E.P.R. problem we have to build a theory that takes into account the fact that non compatible physical quantities can have simultaneous reality. We can ask ourselves now which of these three cases A, B or C is the correct one, because since we showed in [3] and [4] that the quantum mechanical description of separated systems is wrong we cannot draw any conclusion from the E.P.R. reasoning concerning sentences (1) and (2). We will in the following however indeed see that quantum mechanics is also incomplete because we will construct the elements of reality that are missing in the quantum mechanical description of two separated physical systems. The incompleteness is however of totaly different nature as the one indicated wrongly by the E.P.R. reasoning and occurs possibly (but this is perhaps not true) only when one wants to describe separated systems. Whether to sentence (2), we show in [3] that it is true. Hence C is the right choice.
Let us sum up this all and see that there is no paradox present anymore. There are two possible situations for two systems.

First situation :
The two systems are separated. Then quantum mechanics gives not a correct description of this situation. Completing quantum mechanics does not happen by adding states to the subsystems but by taking states away of the compound system.

Second situation :
The two systems are not separated. This means that a measurement performed on one of the two systems changes the state of the other system. In this case it is clear that we cannot make a E.P.R. reasoning. Indeed it is not possible to give an element of reality

to the one system by performing a measurement on the other system. Because for a system to have an element of reality means that we must be able to predict the outcome of the experiment without performing the experiment on the system in question.

## 2. WHAT GOES WRONG IN THE DESCRIPTION OF QUANTUM MECHANICS OF TWO SEPARATED SYSTEMS.

Till now we did not explain what we mean by separated physical systems. The intuitive idea of separated systems that we shall formalize is the following. Two systems $S_1$ and $S_2$ are separated if a measurement performed on one system does not change the state of the other system and vice versa. This means that it is possible to give a description of one system without taking into account the other system. In this sense we could say that one physical system is separated of the rest of the universe. Remark that this does not mean that there can be no interaction between two separated systems. We shall even introduce a weaker notion of separation of systems, this on the hand of separated yes-no observables (or questions). Let us explain what we mean by separated yes-no observables.

### 2.1. Separated yes-no observables.

If $\alpha$ and $\beta$ are two yes-no observables on a system then in general it makes no sense to perform both yes-no observables together. Sometimes however it makes sence to do this.
What we do then is construct an experiment $E(\alpha,\beta)$ having four possible outcomes that we shall label by {yes, yes}, {yes, no}, {no, yes} and {no, no} and such that
$\alpha$ gives with certainty the answer {yes} if $E(\alpha,\beta)$ gives with certainty one of the outcomes {yes, yes} or {yes, no}.
$\alpha$ gives with certainty the answer {no} iff $E(\alpha,\beta)$ gives with certainty one of the outcomes {no, yes} or {no, no}.
$\beta$ gives with certainty the answer {yes} iff $E(\alpha,\beta)$ gives with certainty one of the outcomes {yes, yes} or {no, yes}.
$\beta$ gives with certainty the answer {no} iff $E(\alpha,\beta)$ gives with certainty one of the outcomes {yes, no} or {no, no}.
These requirements just indicates that $E(\alpha,\beta)$ is the experiment that consist in performing $\alpha$ and $\beta$ together. We can construct new yes-no observables out of $E(\alpha,\beta)$ (as we can out of every observable).

| | $\beta$ | |
|---|---|---|
| $\alpha$ | yes yes | no yes |
| | yes no | no no |

Let us define the following yes-no observables.

$\alpha \Delta \beta$ which consist of performing the experiment E and attributing the answer "yes" if we get the outcome {yes, yes}. If we get one of the outcomes {yes, no}, {no, yes} or {no, no} we attribute the answer "no".

We also define the yes-no observable

$\alpha \nabla \beta$ which consist of performing the experiment E and attributing the answer "yes" if we get one of the outcomes {yes, yes}, {yes, no} or {no, yes}. If we get {no, no} we attribute the answer "no".

What is now the meaning of these new yes-no observables ? We can proof the following theorem :

Theorem :
If $\alpha$ and $\beta$ are two yes-no observables that can be performed together then

$\alpha \Delta \beta$ gives with certainty the answer yes
$\Leftrightarrow$ $\alpha$ and $\beta$ give with certainty the answer yes.

So $\alpha \Delta \beta$ is the yes-no observable that tests $\alpha$ and $\beta$. Intuitively we would think perhaps that $\alpha \nabla \beta$ is the yes-no observable that tests $\alpha$ or $\beta$. Let us see that this is in general not the case. Suppose that the system S can be put in a state such that one of the outcomes, for example the outcome {no, no} never occurs for $E(\alpha,\beta)$. This is what happens for example if we have a system S consisting of two spin 1/2 particles in the singlet state, and we perform measurements $\alpha$ of the spin of one of the particles in a certain direction in one region of space, and a measurement $\beta$ of the spin of the other particle in the same direction in an opposite region of space. What we refer to is the well known experiment that has been carried out several times to test Bell inequalities. $E(\alpha,\beta)$ consists of performing $\alpha$ and $\beta$ together and clearly for E we always have one of the outcomes {yes, no} or {no, yes} if the system is in this state. Hence $\alpha \nabla \beta$ gives with certainty the answer yes, even when $\alpha$ neither $\beta$ give with certainty the answer yes. To be able to have such a situation, one of the experiments has to influence the outcome of the other experiment. We shall now introduce the concept of separated observables. This will be observables where this influence is not present.

Definition :
We will say that two yes-no observables $\alpha$ and $\beta$ are separated iff, they can be performed together and such that if for an arbitrary state of the system there is a certain chance to obtain one answer for $\alpha$ and another one for $\beta$, then for this state of the system there is a certain chance to obtain this combination for $E(\alpha,\beta)$.

So for example if the system is in a state such that yes is a possible answer for $\alpha$ and no is a possible answer for $\beta$, then {yes, no} is a possible answer for $E(\alpha,\beta)$. We can prove now the following theorem.

Theorem :
If $\alpha$ and $\beta$ are two yes-no observables that are separated then $\alpha \nabla \beta$ gives with certainty the answer yes iff
$\alpha$ gives with certainty the answer yes <u>or</u>
$\beta$ gives with certainty the answer yes.
Hence $\alpha \nabla \beta$ is the yes-no experiment that tests $\alpha$ or $\beta$.

Proof :
If we can predict yes for $\alpha$ and we can predict yes for $\beta$ then we can predict yes for $\alpha \nabla \beta$.
Suppose we can not predict yes for $\alpha$ and we can not predict yes for $\beta$. Then there is a certain chance to have the answer "no" for $\alpha$ and there is a certain chance to have the outcome "no" for $\beta$, hence there is a certain chance to have the outcome "no" for $E(\alpha,\beta)$. This shows that $\alpha \nabla \beta$ is not true.

■

2.2. Separated yes-no observables cannot be represented by quantum mechanics or yes-no observables described by quantum mechanics are never separated.

Let us suppose that we have a physical system S described by quantum mechanics in a Hilbert space $\mathcal{H}$. Suppose that we have two separated yes-no observables $\alpha$ and $\beta$. We shall now show that the yes-no observable $\alpha \nabla \beta$ can not be described by quantum mechanics.

Theorem :
i) If $\alpha$ and $\beta$ are two separated yes-no observables of a physical system S, then the quantum mechanical description of S in a complex Hilbert space cannot represent the yes-no observable $\alpha \nabla \beta$, and as a consequence this description cannot represent the element of reality corresponding to $\alpha \nabla \beta$.
ii) If we have a system S described by quantum mechanics and $\alpha$ and $\beta$ are arbitrary yes-no observables of this system, that can be performed together, then quantum mechanics forces the existence of a state of the system such that in this state the two yes-no observables are completely correlated (for example we can only find the answer {yes, no} or {no, yes} if we perform both yes-no observables at once). This shows that the yes-no observables are not separated.

Proof :
i) If $\alpha \nabla \beta$ is describable by quantum mechanics there has to be a projection operator $P_{\alpha\nabla\beta}$ such that $\alpha \nabla \beta$ gives with certainty the answer yes, iff the system is in an eigenstate of $P_{\alpha\nabla\beta}$.
$\alpha \nabla \beta$ is a yes-no observable that has the following property : if

there is a yes-no observable $\gamma$ such that

- if we can predict yes for $\alpha$ then we can predict yes for $\gamma$
- if we can predict yes for $\beta$ then we can predict yes for $\gamma$

then if we can predict yes for $\alpha \nabla \beta$ then we can predict yes for $\gamma$. If we represent this yes-no observable $\gamma$ by the projection operator Q we must have the following property.
If Q is an arbitrary projection operator such that $QP_\alpha = P_\alpha$ and $QP_\beta = P_\beta$, then $QP_{\alpha\nabla\beta} = P_{\alpha\nabla\beta}$.
This determines uniquely $P_{\alpha\nabla\beta}$, since from this follows that $P_{\alpha\nabla\beta}$ is the projection on the closed subspace generated by $P_\alpha(\mathcal{H})$ and $P_\beta(\mathcal{H})$. Since $P_\alpha$ and $P_\beta$ commute we have

$$P_{\alpha\nabla\beta} = P_\alpha + P_\beta - P_\alpha . P_\beta$$

Take now a vector $x \in P_\alpha(\mathcal{H})$ such that $x \notin P_\beta(\mathcal{H})$ and $x \perp P_\alpha(\mathcal{H}) \cap P_\beta(\mathcal{H})$. Take also a vector $y \in P_\beta(\mathcal{H})$ such that $y \notin P_\alpha(\mathcal{H})$ and $y \perp P_\alpha(\mathcal{H}) \cap P_\beta(\mathcal{H})$. Suppose now that the system is in a state $z = x + y$. Then :

$$\begin{aligned} P_{\alpha\nabla\beta}(z) &= (P_\alpha + P_\beta - P_\alpha P_\beta)(z) \\ &= x + P_\beta(x) - P_\beta(x) + y + P_\alpha(y) - P_\alpha(y) \\ &= x + y = z \end{aligned}$$

This means that for the system in the state z we can predict yes for $\alpha \nabla \beta$. Hence we can predict yes for $\alpha$ or we can predict yes for $\beta$. Let us see that this is not the case.

$$\begin{aligned} P_\alpha(z) &= x + P_\alpha(y) \neq z \\ P_\beta(z) &= P_\beta(x) + y \neq z \end{aligned}$$

This shows that $\alpha \nabla \beta$ is not described by $P_{\alpha\nabla\beta}$.

ii) Suppose we have two yes-no observables $\alpha$ and $\beta$ that can be performed together. They are then represented by projection operators $P_\alpha$ and $P_\beta$ such that $[P_\alpha, P_\beta] = 0$. The experiment $E(\alpha,\beta)$ has four possible outcomes {yes, yes}, {yes, no}, {no, yes} and {no, no}, let us label them by 3, 2, 1, 0. The experiment E must then be represented by the selfadjoint operator

$$\begin{aligned} H &= 3\, P_\alpha . P_\beta + 2\, P_\alpha\, (\mathbb{1} - P_\beta) + 1\, (\mathbb{1} - P_\alpha)\, P_\beta \\ &= 3\, P_\alpha . P_\beta + 2\, P_\alpha - 2\, P_\alpha . P_\beta + P_\beta - P_\alpha . P_\beta \\ &= 2\, P_\alpha + P_\beta \end{aligned}$$

We shall now show that it is possible to construct a state such that when the system is in this state $\alpha$ and $\beta$ are strictly correlated. Suppose that $P_\alpha(\mathcal{H}) \cap P_\beta(\mathcal{H})^\perp \neq 0$ and $P_\beta(\mathcal{H}) \cap P_\alpha(\mathcal{H})^\perp \neq 0$ and take $x \in P_\alpha(\mathcal{H}) \cap P_\beta(\mathcal{H})^\perp$ and $y \in P_\beta(\mathcal{H}) \cap P_\alpha(\mathcal{H})^\perp$.
Put $z = x + y$.
Then

$$\begin{aligned} P_\alpha(z) &= x \\ (\mathbb{1} - P_\alpha)(z) &= y \\ P_\beta(z) &= y \\ (\mathbb{1} - P_\beta)(z) &= x \end{aligned}$$

Hence z is not an eigenvector of $P_\alpha$, $P_\beta$, $\mathbb{1} - P_\alpha$, $\mathbb{1} - P_\beta$ such that for $\alpha$ and $\beta$ yes and no are possible outcomes. We can even choose very easily z in such a way that we have for $\alpha$ and for $\beta$ probability 1/2 to have the answer yes and propability 1/2 to have the answer no.

$$P_\alpha . P_\beta\ (z) = P_\alpha(y) = 0$$
$$(\mathbb{1} - P_\alpha).(\mathbb{1} - P_\beta)(z) = (\mathbb{1} - P_\alpha)\ (x) = 0$$

Hence if we perform the experiment $E(\alpha,\beta)$, then {yes, yes} and {no, no} are not possible outcomes. We always have as outcome {yes, no} or {no, yes}. Indeed

$$P_\alpha\ (\mathbb{1} - P_\beta)\ (z) = x$$
$$\text{and}\quad (\mathbb{1} - P_\alpha)\ P_\beta\ (z) = y$$

Hence the probability to have {yes, no} for $E(\alpha,\beta)$ is the same as the probability to have yes for $\alpha$ and the probability to have {no, yes} is the same as the probability to have yes for $\beta$.
Suppose now that $P_\alpha(\mathcal{H}) \cap P_\beta(\mathcal{H})^\perp = 0$.
Then $P_\alpha(\mathbb{1} - P_\beta) = 0$. Hence $P_\alpha = P_\alpha P_\beta$. Since $P_\alpha \neq 0$ we have $P_\alpha P_\beta \neq 0$. As a consequence $P_\alpha(\mathcal{H}) \cap P_\beta(\mathcal{H}) \neq 0$.
We then also have

$$(\mathbb{1} - P_\beta)(\mathbb{1} - P_\alpha) = \mathbb{1} - P_\alpha - P_\beta - P_\beta P_\alpha$$
$$= \mathbb{1} - P_\beta \neq 0$$

As a consequence $P_\alpha(\mathcal{H})^\perp \cap P_\beta(\mathcal{H})^\perp \neq 0$.
If we take now $x \in P_\alpha(\mathcal{H}) \cap P_\beta(\mathcal{H})$ and $y \in P_\alpha^\perp(\mathcal{H}) \cap P_\beta(\mathcal{H})^\perp$ and $z = x + y$ then z is in a state of the system such that $E(\alpha,\beta)$ gives always the answer {yes, yes} or {no, no} while yes and no are possible outcomes of $\alpha$ and $\beta$. If $P_\alpha(\mathcal{H})^\perp \cap P_\beta(\mathcal{H}) = 0$ we construct the same vector z.

■

## 3. NON SEPARATED SYSTEMS VIOLATE BELL INEQUALITIES WHILE SEPARATED SYSTEMS SATISFY BELL INEQUALITIES.

We consider a physical system described by a hidden variable formalism. This means that for every yes-no observable $\alpha$ and for every state p of the system $\alpha$ gives with certainty the answer yes or $\alpha$ gives with certainty the answer no.
We can make correspond with every observable $\alpha$ two random variables

$$X_\alpha : \Sigma \to \{-1, +1\}$$

such that $X_\alpha(p) = +1$ if $\alpha$ gives the answer yes and $X_\alpha(p) = -1$ if $\alpha$ gives the answer no.

$$Y_\alpha : \Sigma \to \{0, 1\}$$

such that $Y_\alpha(p) = 1$ if $\alpha$ gives the answer yes and $Y_\alpha(p) = 0$ if $\alpha$ gives the answer no.
If we have two observables $\alpha$ and $\beta$ that can be performed together then we define for the experiment $E(\alpha,\beta)$ the random variables

$$\begin{cases} X_{\alpha\beta}(p) = +1 \text{ if E gives the answer \{yes, yes\} or \{no, no\}} \\ X_{\alpha\beta}(p) = -1 \text{ if E gives the answer \{yes, no\} or \{no, yes\}} \end{cases}$$

$$\begin{cases} Y_{\alpha\beta}(p) = +1 \text{ if E gives the answer \{yes, yes\}, \{yes, no\} or \{no, yes\}} \\ Y_{\alpha\beta}(p) = 0 \text{ if E gives the answer \{no, no\}} \end{cases}$$

We suppose now that we have four yes-no observables $\alpha$, $\beta$, $\gamma$, $\delta$ such that $E(\alpha,\beta)$, $E(\alpha,\gamma)$, $E(\delta,\beta)$, $E(\delta,\gamma)$ exist.
Then the Bell inequality is the following inequality :

$$|X_{\alpha\beta}(p) - X_{\alpha\gamma}(p)| + |X_{\delta\beta}(p) + X_{\delta\gamma}(p)| \leqslant 2$$

for every state p of the system.

Clanser and Horn inequality is the following inequality :

$$\frac{Y_{\alpha\gamma}(p) - Y_{\alpha\beta}(p) + Y_{\delta\beta}(p) + Y_{\delta\gamma}(p)}{Y_{\delta}(p) + Y_{\gamma}(p)} \leqslant 1$$

We should like to reformulate these inequalities by using only two yes-no observables $\alpha$ and $\beta$ and their inverses which we shall denote $\alpha^{\sim}$ and $\beta^{\sim}$. ($\alpha^{\sim}$ is the experiment that consist of performing the same experiment as the one proposed by $\alpha$ but changing the role of yes and no).

We can find immediately the corresponding inequalities by considering for $\delta$ and $\gamma$ experiments that always give us the answer yes. Then

$$X_{\alpha\gamma}(p) = X_{\alpha}(p) \qquad Y_{\delta\beta}(p) = Y_{\beta}(p)$$
$$X_{\delta\gamma}(p) = 1 \qquad Y_{\alpha\gamma}(p) = Y_{\alpha}(p)$$
$$X_{\delta\beta}(p) = X_{\beta}(p) \qquad Y_{\delta\gamma}(p) = 1$$

Bell inequalities become

$$|X_{\alpha\beta}(p) - X_{\alpha}(p)| + |X_{\beta}(p) + 1| \leqslant 2$$

Clauser Horn inequalities become

$$Y_{\alpha}(p) + Y_{\beta}(p) - Y_{\alpha\beta}(p) \leqslant 1$$

Theorem :
If $\alpha$ and $\beta$ are two yes-no observables on a physical system S described by an hidden variable formalism, then $\alpha$ and $\beta$ are separated iff

$\alpha$, $\beta$ and $\alpha$, $\beta^{\sim}$ and $\alpha^{\sim}$, $\beta$ and $\alpha^{\sim}$, $\beta^{\sim}$

satisfy the simplified inequalities of Clauser and Horn for every state of the entity S.

Proof :
Suppose $\alpha$ and $\beta$ are separated then it is clear that $Y_{\alpha\beta}(p) = Y_{\alpha}(p).Y_{\beta}(p)$ and in this case Clauser Horn inequality is satisfied. Suppose that $\alpha$ and $\beta$ are not separated. Then there exists a state p such that a certain outcome for the coincidence experiment, let us say for example the outcome {yes, no} never occurs while $\alpha$ is true and $\beta^{\sim}$ is true. Hence the outcome for the coincidence experiment has to be {yes, yes}, {no, yes} or {no, no}. In each of these cases we have

$$Y_{\alpha}(p) = 1 \ , \ Y_{\beta^{\sim}}(p) = 1 \quad \text{and} \quad Y_{\alpha\beta^{\sim}}(p) = 0$$

Hence $Y_{\alpha}(p) + Y_{\beta^{\sim}}(p) - Y_{\alpha\beta^{\sim}}(p) = 2.$ ∎

4. EXAMPLE OF NON SEPARATED SYSTEMS.

Consider a system S consisting of two spin 1/2 particles in the singlet spin state. Then the two spin 1/2 particles are not separated. Indeed if we perform measurements $\alpha$ of the spin of one of the particles in a certain direction in one region of space and a measurement $\beta$ of the spin of the other particle in the same direction in an opposite region of space, then $\alpha$ and $\beta$ are non separated observables. We already mentioned this example in 2 a).
As we concluded in 1 for this system S it is indeed not possible to make an E.P.R. reasoning. What we mean is that it is not possible to give an element of reality to one of the particles spin direction in relation to a measurement of the opposite spin direction performed on the other particle.
We want to give another example of two non separated systems that shows a certain analogy with the two spin 1/2 particles in the singlet state.
We consider a physical system that is a vessel that contains water.
We consider the following experiment :
<u>experiment $\alpha$:</u>
consists of testing whether the volume of the water contained in the vessel is less than 10 l. We perform this experiment by emptying the vessel by means of a siphon and collection the water in a reference vessel of 10 l. We give the answer "yes" if the water stops flowing before it depasses 10 l. and we give the answer "no" if the water depasses 10 l.
We consider now a physical system consisting of two vessels of water. The two vessels of water are connected by a tube.

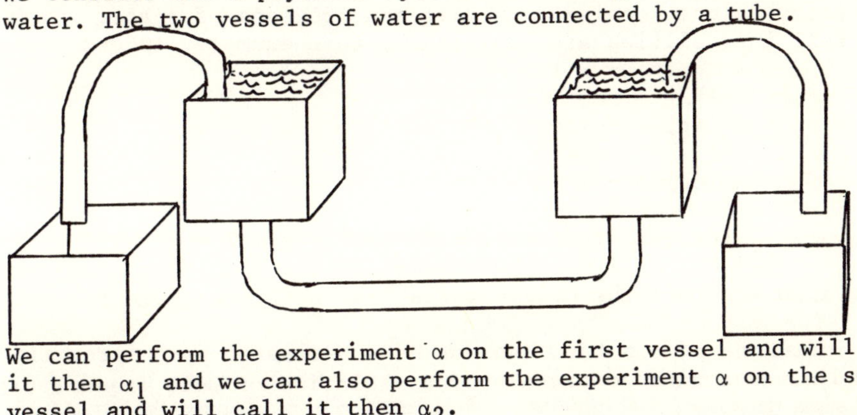

We can perform the experiment $\alpha$ on the first vessel and will call it then $\alpha_1$ and we can also perform the experiment $\alpha$ on the second vessel and will call it then $\alpha_2$.
We can also perform the coincidence experiment $E(\alpha_1, \alpha_2)$ which consists of performing $\alpha_1$ and $\alpha_2$ together. Suppose now that the system of the two vessels is in such a state, that there is 10 l. of water in each vessel and of course also some water in the tube. If we perform then the experiment $\alpha_1$, we will always find the answer no. Also for the experiment $\alpha_2$ we will find the answer no. But for the experiment $E(\alpha_1, \alpha_2)$ we will always find one of the answers {yes, no} or {no, yes}. Hence when $\alpha_1$ and $\alpha_2$ are performed together we find a strict correlation between the outcomes, just as in the case of the spin 1/2 particles.

As we go back to the construction made in 2.2 we see that $\alpha_1 \nabla \alpha_2$ is true while neither $\alpha_1$ nor $\alpha_2$ is true.
So also in this example we see that $\alpha_1 \nabla \alpha_2$ does not test the property "has the first vessel or the second vessel less than 10 l. of water". Clearly $\alpha_1$ and $\alpha_2$ are not separated yes-no observables and as a consequence the two vessels are not separated systems.
We can now easily see that it is not possible to make the E.P.R. reasoning on this example. Indeed although we can predict after having done one experiment $\alpha_1$ on one vessel the outcome of the other experiment $\alpha_2$ on the other vessel, because the outcomes are strictly correlated, we can not predict this outcome without changing the state of the second vessel. Hence this procedure does not define an element of reality for the second vessel.
Let us make some remarks on this example that can because of the analogy also be made on the system of two spin 1/2 particles in the singlet state.
- The result of the experiments does not depend on the distance, however it is possible to have two vessels without a tube connecting them.
- The signal that accounts for the strict correlation propagates with the speed of sound in water. If we consider the water as incompressible, which is an idealisation that does not change the results of the experiments, then such a theory would predict a propagation with infinite speed.
- Some properties of the water are separated and others not.
- The water is not localized in different regions of space.
- The Bell inequalities are violated by this example. Indeed

$$X_{\alpha_1\alpha_2}(p) = -1$$

$$X_{\alpha_1}(p) = -1$$

$$X_{\alpha_2}(p) = -1$$

$$|X_{\alpha_1\alpha_2}(p) + X_{\alpha_1}(p)| + |1 - X_{\alpha_2}(p)| = 2 + 2 = 4$$

and we also see that the hidden variables in this example are not local.
- The influence of the tube can not be substituted by a force field.
- Ref [6] gives a similar example of a violation of Bell inequalities.

Conclusion :

Quantum mechanics is not complete and not correct for the description of separated systems.
Bell inequalities are valid if and only if the two systems are separated. Whether the systems are classical macroscopical systems or quantum systems makes no difference.
The incompleteness of quantum mechanics is of a totally different nature than an incompleteness that could be solved by introducing hidden variables.

References :

(1) Einstein, Podolsky and Rosen : 1935, Physical Review 47, pp. 777.
(2) C. Piron : 1976, Foundations of quantum physics, W.A. Bejamin Inc.
(3) D. Aerts : 1981, The one and the many. Towards a unification of the classical and quantum description of one and many physical entities. Doctoral Thesis, TENA, Vrije Universiteit Brussel.
(4) D. Aerts : Description of many separated entities without the paradoxes encountered in quantum mechanics. Accepted for publication in Foundations of Physics.
(5) E. Schrödinger : 1935, Proc. Cambridge Phylosophycal soc., 31, pp. 555.
(6) D. Aerts : 1982, Example of a macroscopical classical situation that violates Bell inequalities, Lett. al Nuovo Cimento, Vol. 34, n° 4, pp. 107.

References


(1) [illegible] and Rosen (1935), Physical Review [illegible]
(2) [illegible] (1970), [illegible]
(3) [illegible] (1961), [illegible]
(4) [illegible]
(5) [illegible]

# ON THE FOUR-DIMENSIONAL CHARACTER OF MICRO-PHYSICAL PHENOMENA

C.W.Rietdijk

Pinellaan 7,Santpoort-Zuid,The Netherlands

Abstract:It is proved that retroactive effects exist in Nature.This emphasizes the fact that micro-processes constitute integrated wholes so much that it is no longer far-fetched to posit the hypothesis that events,that is,action, rather than objects,constitute the proper stuff of the (four-dimensional) Universe.Mind here,too,that retroactivity implies that the future and future parts of events "exist already".Then,distances between (e.g.,alternative) events A and B have to be measured by the quantity of "occurring" or action that is needed in order to transform event A into event B.The action metric so introduced appears to be in a position to solve the nonlocality paradoxes of quantum mechanics such as wave-particle "duality" and the EPR paradox.In this connection,the Minkowski metric corresponds to a macro scheme which cannot be "interpolated" to within a micro-process,i.e.,to within action quanta,without producing serious metrical distortions.Generally,metric is considered to be a property of events,it having no existence independent of them as a "pre-existing scheme".Planck's elementary quantities of action h are seen as real entities in the four-dimensional world,i.e.,as the "atoms of occurring".By intersecting (dilated) series of them with a now-hyperplane we in an imaginable way get the wave patterns satisfying the relevant wave equation.

## 1. AN INFLUENCE OF THE FUTURE ON THE PRESENT

Bohr's idea that micro-processes constitute coherent wholes may be true in an even more far-reaching and concrete sense



*S. Diner et al. (eds.), The Wave-Particle Dualism, 433–456.*

than most of us suspect now.

Actually, there are familiar experiments which, after some variation and/or consequent analysis, do nothing less than demonstrate that, e. g., a measurement decision of an observer can in certain cases influence stages of the relevant process that are in the absolute past of the measurement-event in question, at least if momentum or angular momentum are conserved in micro-processes, too.

A first example of such variation is the one proposed for other purposes by Wootters and Zurek (WZ) (1).

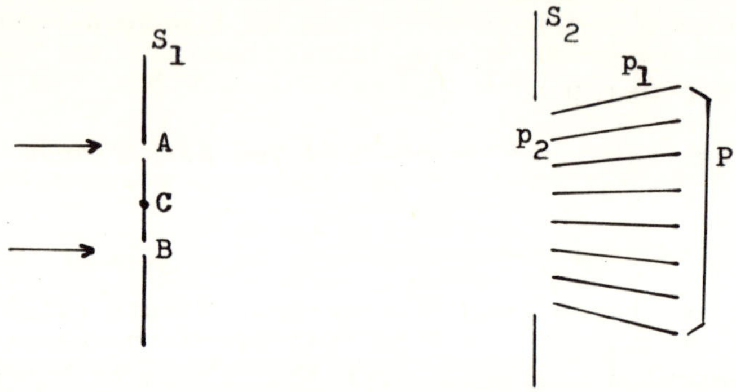


Figure 1

Consider a Young double-slit experiment (with photons or particles) in which a part of the second screen (or the photographic plate) $S_2$ has been substituted by a system P of plates $p_1, p_2, p_3, \ldots$, the produced parts of all of which pass through C between the slits A and B. (See figure 1. ) Then a point on the upper side of each plate $p_k$ can only "see" A, whereas points on $p_k$'s lower side see only B. As WZ observe, this will blur any interference pattern "on" $S_2$ in the P region. For if most "lower" impacts on a $p_k$ stem from B particles and most "upper" impacts from A momentum carriers, interference diminishes accordingly. (We say "most" impacts because diffraction, e. g., causes some A momentum carriers to hit lower sides. ) The consequence of this course of matters is far-reaching: it proves the existence of a retroactive effect. For suppose that $S_1S_2$ is so long, or the momentum carriers are so slow, that there is an hour between the interactions of those carriers with $S_1$ and $S_2$, respectively. If, then, observer O at $S_2$ makes a last-minute choice between setting up the system P or keeping $S_2$ "closed", it is clear that at the momentum exchange between the carriers and $S_1$ an hour before O's decision, the latter had already a physical influence. For if the carriers move through vacuum between $S_1$ and $S_2$ (or P), their momentum does not change any more after their interaction with $S_1$ whereas, nevertheless, such momentum determines where the carriers make impacts on $S_2$ or P. At the same time, the impact locations vary widely according to whether P has been set up or not: In the latter case they produce the interference fringes, and in the former they produce "grey". The conclusion is that O's decision - or rather, his measurement action - influenced how much momentum the momentum carriers absorbed from $S_1$ an hour before (2).

The existence of retroactivity can also be derived, e. g., from the behaviour of two mutually orthogonally polarized photons P and Q constituting an Einstein-Podolsky-Rosen system and being (quasi-)simultaneously emitted from E at point-event $E_1$. (See figures 2 a and 2 b, the latter of which is a space-time picture. ) A and B are two Nichol prisms hit by P and Q, respectively, at the events $A_1$ and $B_1$, at which z-polarized photons pass and y-polarized ones

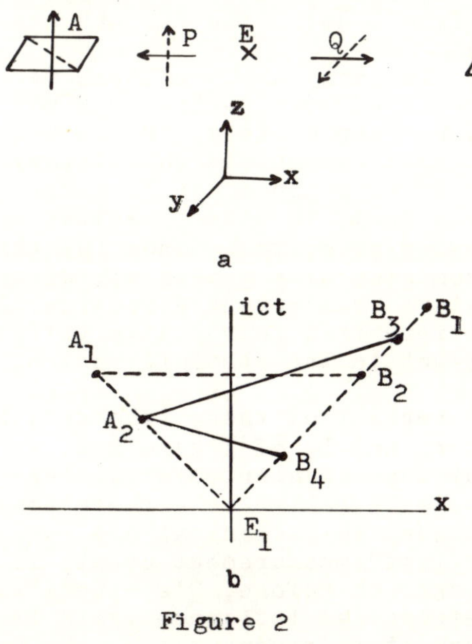


Figure 2

are stopped. Take EA < EB. Then it is well-known that if P passes A, Q will not pass B. The latter means that we know with certainty that Q is polarized in the y direction as soon as we know that P passed A. That is, Q is certainly y-polarized between $B_2$ and $B_1$ of its world-line. However, relativity requires that natural laws are equally valid in all inertial systems. That is, if, of the two correlated photons P and Q, Q is polarized with certainty in the y direction at point-event $B_3$, P has to be polarized in the z direction at $A_2$ of its world-line because an inertial system exists in which $A_2$ and $B_3$ are simultaneous. In another inertial system $B_4$ and $A_2$ are simultaneous, from which it follows now that at $B_4$ Q has to be polarized in the y direction. And so on. The foregoing implies that P's z-polarization and Q's y-polarization cannot but have existed from their common emission-event $E_1$. This circumstance must have its origin in a retroactive effect because if the Nichol prisms A and B would have been rotated on AB through a common angle of, say, $\frac{1}{4}\pi$, we can argue in the same way as above that P and Q from their origin at $E_1$ must have had other polarization directions than z and y, these directions therefore appearing to depend on event $A_1$ and possibly event $B_1$ already at earlier point-events of their (P and Q's) existence. (Figure 2 b, the above argument, and conservation of linear momentum make it also clear that P "hiddenly" passes point-event $A_2$ instead of occupying a "fundamentally uncertain" position in a spherical wave packet emanative from E until the "first" measurement at $A_1$ takes place. )

We can argue similarly about a Stern-Gerlach experiment with two correlated spin particles P and Q showing spins $\frac{1}{2}\hbar$ and $-\frac{1}{2}\hbar$, respectively, in the z direction. Then we can prove that such components had already the same values from the common emission-event because conservation of angular momentum holds in other inertial systems than the observer's rest system, too. Since the particles cannot have an angular momentum of $\pm\frac{1}{2}\hbar$ in all directions at the same time, the future measurement direction, i. e., the z one, exerted a preferential retroactive influence already at the emission-event in the absolute past of the measurement (3).

An important variant of these (thought) experiments originates if, at $A_1$ and $B_1$ of figure 2 b, we do not measure the linear but the circular polarizations of P and Q. Then there are two alternative interpretations:

a) The usual quantum-mechanical one, according to which it is the "first" measurement-event, i. e., $A_1$, which determines P's – and, therefore, Q's – helicity. Then, our accepting of a retroactive influence again is inevitable because if measurement $A_1$ produces $+\hbar$, the helicity at point-event $B_3$ will be $+\hbar$, too (both in "forward" direction) and then special relativity and conservation of angular momentum by the system (P, Q) imply $+\hbar$ in $A_2$, in $B_4$ etc., up to immediately after emission-event $E_1$.

b) An "ordinary" local hidden variable exists, viz. the helicities $+\hbar$ of P and Q, these not being produced by the measurements $A_1$ and $B_1$ but by nonretroactive, that is, causal factors. In this interpretation, $A_1$ and $B_1$ do nothing but "register" the values of the hidden helicities.

An analogue of this variant is constituted by de Broglie's paradox, which starts from two halves A and B of a formerly united box AB of which we do not know in which half, A or B, a particle P is contained. (See figure 3. The halves were separated by a partition with a hole when they still constituted a unity; after the hole had been closed, A and B were separated completely. ) A and B are brought

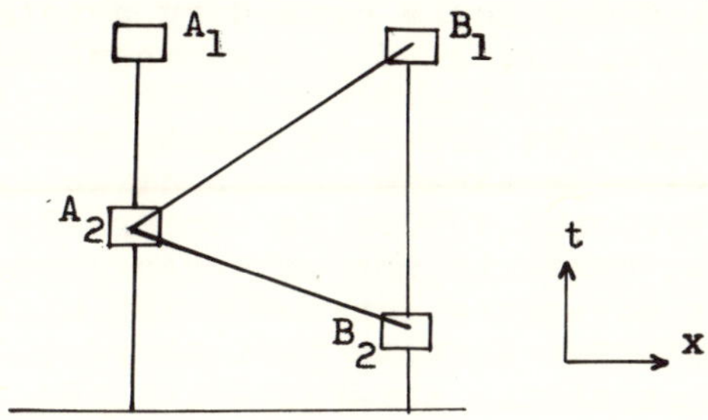


Figure 3

to Tokyo and Paris, respectively, and we still do not know in which one P is. Now we look into B. (This constitutes event $B_1$.) If we find P, does this mean that the Tokyo situation suddenly changes, "some degree of presence" being removed from A?

Considering this rather paradoxical situation, we can again only choose one out of the above alternatives a) and b): either there is a retroactive effect of our looking into B, or an "ordinary" hidden variable did exist, P having been in B all the time (in which case the status of the wave function half in A, which collapses at our B look, becomes rather unclear at situations like $A_2$). In both cases there is a hidden variable; in the a) case, it is the retroactive influence of our looking into B.

## 2. RELATIVISTIC LENGTH CONTRACTION IMPLIES AN ALREADY-EXISTING FUTURE

Certainly, present-day thinking is not very open to the ideas of retroactivity and determinism. Mind here that retroactivity implies determinism as an inevitable consequence. For if the future – e. g., a decision by an observer or one by, say, a computer who may decide either to set up P or to "close" $S_2$ in the experiment of figure 1 – appears to actually influence present events (such as the momentum transfer from $S_1$ to momentum carriers), it cannot be but already existing as well as determined, though existing outside our (present physical) horizon. (It can be proved that an observer at $S_1$ of figure 1 can in no way measure the momentum transfer from $S_1$ to the carriers accurately enough to predict observer $O_1$'s decision as regards the set-up of P. We can easily see that a paradox would arise if he could.)

There is a rather easy way to see clearly that the future has to pre-exist indeed. Actually, it may be called a "blind spot" of physics – caused by a rather general aversion against the idea of a predetermined future which, among other things, calls into question "free will" – that it did not see earlier that the relativistic length contraction of, say, a moving arrow is directly connected with the fact that the future is comparable with those parts of a film that have not yet been projected. At the same time it is necessary for our truly accepting the above result on retroactive effects that we realize and understand that future events are something real indeed and, therefore, are in principle in a position of influencing other events, notwithstanding the fact that they are within the future part of our light cone.

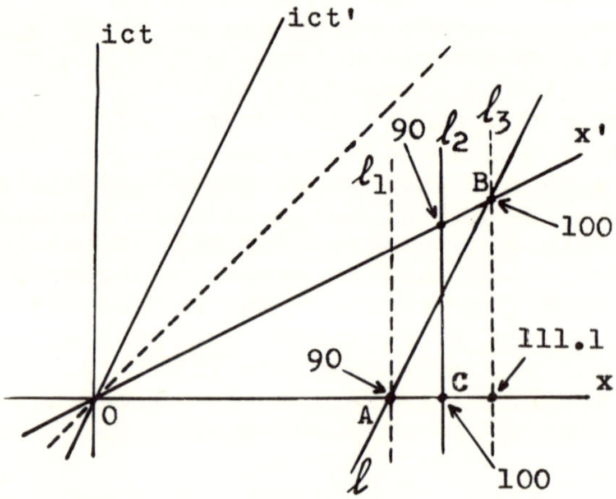


Figure 4

For a proof of this see figure 4, in which the rest systems S(x, 0, t) and S'(x', 0, t') of the two observers W and $W_1$ are pictured. Consider an arrow P with a length OB = 100 cm and which is at rest in S'. Measuring rod M has a length of 120 cm in its rest system S. At t = 0, its markings 0 and 100 are at O and C, respectively. W sees P, that is, OB, contracted to 90 cm; that is, he sees it as OA. $\ell_1$, $\ell_2$ and $\ell_3$ are the world-lines of the markings 90, 100 and $10/9 \times 100 \approx 111.1$ on W's measuring rod M, respectively. Then we realize that, P's left end O being for both W and

$W_1$ at marking 0 of M, its right end R is at A and B for W and $W_1$, respectively, if they describe and measure P "now" from their common position at 0. Now it is essential to realize, too, that the point-events A and B represent two successive stages in R's existence, viz. its being at the markings 90 and 111.1 of the same measuring rod M, respectively. In other words: the phenomenon by which $W_1$ sees P as being longer (100 cm) than W sees it, is the circumstance that in $W_1$'s "now" the point R of P has advanced 111.1 - 90 = 21.1 marking lines further on the same M than in W's "now". Because, moreover, rod M is shortened by a factor 90/100 for $W_1$ as compared to W, $W_1$ actually sees P only $\frac{90}{100} \times \frac{111.1}{90} = \frac{10}{9}$ times longer than W sees it (both as compared with their own standard measuring rod at rest), instead of the 111.1/90 times there would be without M's length contraction for $W_1$. If a third observer at 0 swiftly alternates his velocity, making it by turns equal to W's and $W_1$'s velocities, stages of R's existence in the environment of A and B (in each other's absolute past and future), respectively, would be "now" for him, this proving the "simultaneous" reality of the point-events A and B (4).

One might object that such argument is only "metrical", the "nows" A and B at a distance from 0 being not "real physical nows" for an observer at 0 but only nows in a metrical, "theoretical" sense.

This objection can be refuted as follows.

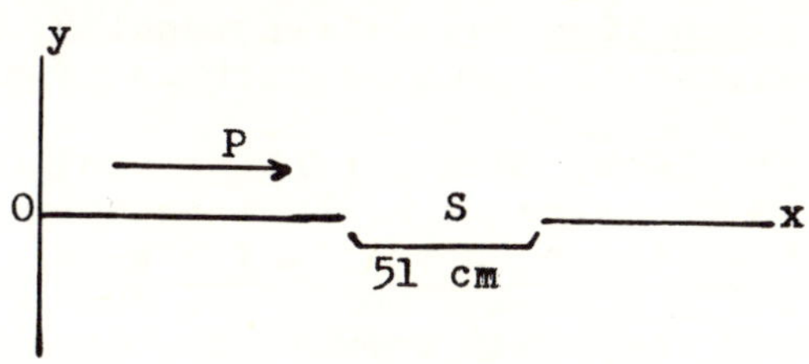


Figure 5

In the first place, relativistic length contractions are physically so real that if an arrow P with rest length 100 cm moves in the x-direction at a velocity v such that $\sqrt{1 - \frac{v^2}{c^2}} = \frac{1}{2}$, whereas it has an (arbitrarily small) additional velocity in the -y direction, it can indeed pass the hole S of 51 cm, P being contracted to 50 cm. (See figure 5. ) That is, Nature actually works with "metrical" contraction, which is indeed "only" based upon another ordering of point-events, but another ordering that can indeed make future events "nows" really operative in processes.

In the second place, the above observers W and $W_1$, with their heads and brains simultaneously being, say, at O of figure 4, both can hold the left end of arrow P with their left hands and its right end R with their right hands. Then the consequence of the standpoint considering "now" at one meter distance from my brains in my rest system as "only metrical", not really physically existing, would be that my own right hand now (at marking 111.1 of M if I am $W_1$) would be only "metrically" now for me, would not be physically present in such rest system if it holds the end of an arrow at rest with respect to me! Only one point-event (suppose, in my brains at O) of my body would be "real" here and now! The problem where my consciousness is exactly, in my brain, would become actual again. And, if my right hand is "only metrically now" at one meter from my brain in my rest system, in which system and in what sense would it exist really, then? (We supposed that my arm is one meter.)

## 3. CONSEQUENCES OF THE EFFECTIVELY FOUR-DIMENSIONAL CHARACTER OF NATURAL PROCESSES, PARTICULARLY IN CONNECTION WITH THE WAVE-PARTICLE PROBLEM AND NONLOCALITY

Our main result up to now, viz. the existence of retroactive effects showing that Nature not only is four-dimensional but can actually function four-dimensionally, too, has some radical consequences for our conception of micro-processes.

a) In the first place, Bohr's earlier mentioned idea that a micro-process constitutes a whole becomes so much an actual reality that there is not only EPR-Bell nonlocality in space – mutually distant "parts" of a process being so narrowly connected that they seem to be physically contiguous –, but that it can also occur that past, present and future stages of the process are mutually connected in two opposite directions: causally in the + t and retroactively in the - t direction. Such "feedback" means a real four-dimensional integration of it, completing the "wholes" idea.

b) Such space-time integration, making otherwise mere four-dimensional "configurations of objects traveling in time" truly four-dimensional processes, events, suggests that it may no longer be a far-fetched hypothesis to posit that it is four-dimensional events, processes, rather than three-dimensional objects, "things", which constitute real Nature.

Then energy times time, "occurring", action, emerges

as the real stuff or constituent of the four-dimensional world, and it becomes obvious now to ask for elementary events rather than for elementary objects (particles). So Planck's constant h appears as the "atom of events" and the physically most relevant elementary part of the Universe, objects being, moreover, mere three-dimensional and possibly (sometimes) "distorted" sections of four-dimensional processes, events. The "distortion" of three-dimensional Euclidean metric as compared with four-dimensional relativistic metric could even appear to be merely an aspect, or special case, of such distorting effect of only considering three-dimensional hyperplane sections of the four-dimensional structure or lattice of events (that is, of the action quanta which form that lattice) constituting reality. (Compare below. )

The indivisibility of the quantum of action, moreover, gets something equally natural and imaginable as an indivisibility of "elementary" particles.

A crucial role of action in Nature is further strongly suggested by the circumstance that both the mechanical equations of motion and Maxwell's equations can be derived from principles of least action (5).

c) Most important of all are the consequences of substituting configurations of objects by configurations of events (by structures of which action and its quanta are the stuff, constituents) for metric. For as it is natural in configurations of objects to measure distances by means of objects, i. e., measuring rods, so for the four-dimensional configurations of elementary events (action quanta) that processes - and also the four-dimensional Universe as a whole - are it is equally obvious to measure distances by means of events, that is, by establishing how much "occurring" or action it implies to get to event B from event A, or to transform event A into event B. Actually, this is the central point of this dissertation.

Mind here that we regard amounts of action, such as Planck's quantum h, as "ordinary things" in the four-dimensional world of events, and the most relevant things at that, with which we can measure the physically most relevant distances (i. e., those between four-dimensional events), the quanta, moreover, being organized in four-dimensional lattices (that are more complicated events) in a similar way as elementary objects (particles) form our familiar objects. The superlattice including all such lattices is the real four-dimensional Universe of all events in the world.

In order to illustrate the above concretely we consider a few pictures. In the first place, see figure 6. This is the most simple (and simplified) way to depict action quanta four-dimensionally, viz. as "slices", the validity of

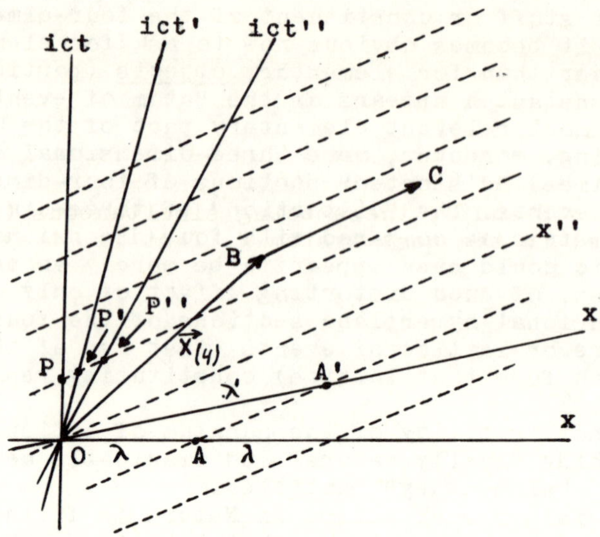


Figure 6

the essential formulas $\lambda = \frac{h}{p}$ and $E = h\nu$ being integrated in the picture for all inertial systems such as (x, O, t) and (x', O, t'). E. g., $OA = \lambda$, $OA' = \lambda'$, $OP = ic\,\Delta t = ic\frac{1}{\nu} = ic\frac{h}{E}$ and $OP' = ic\,\Delta t' = ic\frac{1}{\nu'} = ic\frac{h}{E'}$ in the two systems, respectively. This gives the picture a high degree of four-dimensional reality. By varying the slope of the monochromatic wave train (that is, of the series of slices) that has been drawn within the margin determined by the Heisenberg uncertainty $\Delta_H p_x$ (etc. ) we can four-dimensionally picture a realistic wave packet going with

$$\psi(\vec{x}, t) = \int F(\vec{p})\, e^{i(Et - \vec{p}.\vec{x})} d\vec{p}.$$

Figure 6 depicts a part of the history of a freely moving particle M in terms of the successive action quanta such history or world-line consists of. Mind here that the existence in time of M, like all other processes in Nature, is a series of events, the elementary events being quanta of a duration derivable from $h\nu = mc^2$, that is, $\nu = \frac{mc^2}{h}$, so that the duration of one quantal event is $\Delta t'' = \frac{1}{\nu''} = \frac{h}{mc^2}$ in the rest system W(x'', O, t'') of M. In figure 6, $\Delta t''$ is equal to $\frac{OP''}{ic}$. [$\frac{OP}{ic}$ is the duration of a one-quantum period of M's existence in system (x, O, t). Mind also the difference between events and point-events, the former having an extension in space-time and being associated with energy-momentum.]

Now let us consider the above-mentioned action metric (that is, the metric meant in c above, which uses action, or "amounts of occurring", as a measure to define distances between events) by separately depicting a part of figure 6, viz. a part of one of the quantum slices with the point-events B and C, these being simultaneous (and of equal phase) in the rest system W. That is, they are simultaneous in the rest system of M's Fourier component corresponding to $\vec{p}'' = 0$. (See figure 7. ) Then the physically relevant distance, the action distance, between B and C

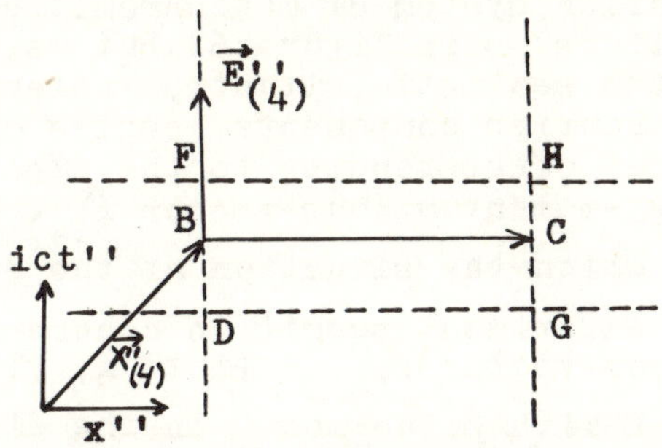


Figure 7

<u>is zero</u>, the action $S = E''t'' - \vec{p}''.\vec{x}''$ of the relevant process (i. e., the moving particle M) being exactly equal in B and in C; that is, only a zero event, "occurring", action, is necessary for transforming one of the undermentioned alternative events a) and b) into the other:

a) The proper quantal process (as a moment of M's existence) acts at B; that is, M's world-line as it appears to be situated after an impact or measurement of M (and that is parallel to OP'' of figure 6, which means: parallel to M's energy-momentum four-vector) passes through B, and

b) The same, equal-phasal proper quantal process acts at C.

Of course, if the action at B and C is the same in W, it is the same in other inertial systems, too, because action is relativistically invariant. (We may start from an action zero at a certain stage of the process P that the freely moving M constitutes, say, from O and, therewith, from the now-"hyperplane" Ox''. )

In our conception, the very action-metrical contiguity or even equivalence, for Nature, of the above situations a) and b) – compare the fundamental hypothesis below – causes the "dilation" of the "proper" quanta – the world-line or world-tube segments of duration $\Delta t'' = \frac{h}{mc^2}$ –, so that they seem to be slices for us. E. g., if a proper quantum acts between D and F of figure 7, such process can make physical contact in region GH, too; it may interact in the whole slice-shaped region of which DGHF is a part. Such very interactions retroactively may contribute to determining M's "proper" world-line as it becomes known to us after, say, an impact. It may even be justified to formulate it thus: because the physically relevant action distance between the alternative (point-)events B and C is zero for the process, the particle is, or acts, both at B and at C.

The figures 6 and 7 and the argument above dealt with one particular Fourier system of monochromatic waves, that is, of parallel slices as in figure 6. But we can generalize the argument for the realistic situation referred to above and in which many Fourier components, series of slices, are there, viz. the ones corresponding to the (Heisenberg) margins for the energy-momentum four-vector $\vec{E}_{(4)} \equiv (i\,\frac{E}{c}, p_x, p_y, p_z)$ which the situation at the emitter and the set-up of the experiment permit,in coherence with the margins for the four-vector $\vec{x}_{(4)} \equiv (ict, x, y, z)$. If we compare situation B with situation C in the figures 6 and 7 the action $S = Et - \vec{p}.\vec{x} = - \vec{E}_{(4)}.\vec{x}_{(4)}$ at both is the same because we get the action at C from the one at B by only adding the translation four-vector $\vec{BC}$ perpendicular to $\vec{E}_{(4)}$ to the location four-vector $\vec{x}_{(4)}$ of B. (We omitted the prime marks with $\vec{x}$ etc. ) If we generalize to other changes in the proper quantal situation – the proper quanta being segments of M's world-line corresponding to one slice – which leave the action invariant, we have to consider, e. g., the

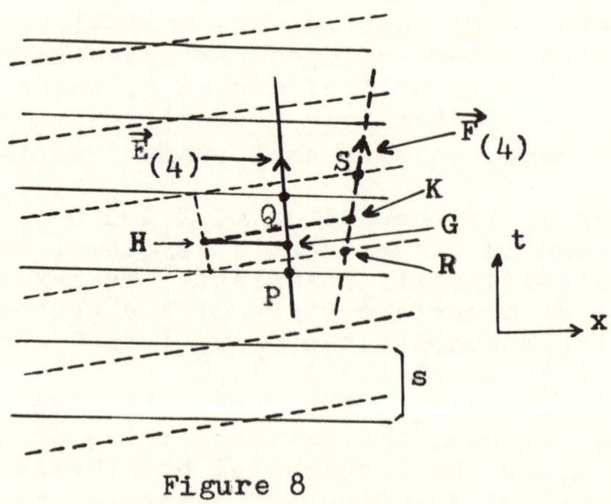


Figure 8

point-events G, H and K of figure 8, all representing, say, the stage of action $2\frac{2}{5}$ h of the relevant process. G does so in the drawn Fourier series beginning with the slice s, and K does in the alternative dotted series (which quantal series is allowed by the experimental conditions, too), whereas H represents "phase" $2\frac{2}{5}$ h in both series at a time. Then, e. g., the "action distance" between the situations a) and b) below is zero:

a) M's world-line passes through G, its momentum corresponding to the slices beginning with s, the process's phase being $2\frac{2}{5}$ (action quanta passed) and

b) M's world-line passes through K, its momentum corresponding to the dotted slices, whereas the phase of the process is again $2\frac{2}{5}$.

In figure 8 we illustrated these two alternative, action-metrically contiguous (or even equivalent) elements of the process constituted by a freely moving momentum carrier – that is, the elements corresponding to the point-events G and K, respectively – by drawing the energy-momentum four-vectors $\overrightarrow{E}_{(4)}$ and $\overrightarrow{F}_{(4)}$ and world-line segments corresponding to the elements in question. The world-line segments PQ and RS represent the "proper" quanta going with the situations a) and b), respectively.

In conformity with point c) in the first part of this section it is our fundamental hypothesis now, intended to solve the nonlocality paradoxes in micro-physics – that are actually at the origin of most quantum paradoxes at all –, <u>that the action metric in which differences, not as regards "space" but as regards action, are taken to be "distances", is the physically most relevant one within a coherent process, e. g., within action quanta</u>. This means that, among other things, the above situations a) and b) are <u>physically contiguous</u>, an infinitesimal physical distance being between them (6). We refer to Ref.6 for an explanation how the ordinary relativistic metric (the Minkowski scheme) can be constructed from the four-dimensional structure A of events, that is, action quanta, as a <u>macro metric</u> that is not operative within quantal processes but constitutes <u>a classical approximation for macro situations</u> in which small numbers of action quanta and their mutual topology, connection, within A can be neglected. In Ref. 6 our macro metric is shown to be based upon symmetries, regularities, of A. It is further argued there that only entities and concepts that can be constructed from action, quanta thereof and the lattice A these are organized in, make physical sense, all other ones being mere schematic extrapolations, or metaphysics. Extrapolation – or "interpolation" – of the macro Minkowski metric to the "difference" of the elements B and C of the quantum of the figures 6 and 7 is an example of such metaphysical interpolation, and senseless.

Actually, the various Fourier components of a wave packet are different representations in or sections with Minkowski space of the same series of action quanta constituting a process, each component corresponding to a different momentum within the permitted "uncertainty" margins. All components correspond phase by phase (think of G and K of figure 8, respectively) to an equal action and, therefore, are physically contiguous representations of the same process. In a similar way as H and K of figure 8 or B and C of figures 6 and 7 correspond to a space-time (or space) "dilation" of (a part of) the process, the set of Fourier components represents a "momentum dilation" of it in our current ordering scheme of the world, i. e. the Minkowski one that wrongly used to be interpolated too far into the sphere of elementary processes, then producing the "dilations" or discrepances in question. Mind in this connection that the four-vectors $\vec{x}_{(4)} \equiv (ict, x, y, z)$ and $\vec{E}_{(4)} \equiv (i\frac{E}{c}, p_x, p_y, p_z)$ which produce the action $S = -\vec{E}_{(4)} \cdot \vec{x}_{(4)}$ <u>are symmetric as regards the "dilation", too</u>: both a "shift" of $\vec{x}_{(4)}$ and one of $\vec{E}_{(4)}$ may leave S invariant.

What our hypothesis aims at is giving a quantization of metric and space, metric being only derived from real events, i. e., action and the action-quantal structure A. At the same time, we abandon the traditional absolute Minkowski space as a kind of "theoretical ether" which up to now was thought to be permeating the Universe of events instead of only corresponding to a frame in which it is practical to coordinate macro events in a coherent way, but which has no existence independent of the lattice of quanta A. On the contrary, we see distances in the physically relevant action metric as "only" a property of events, not as something being prior to events and the action they consist of.

What we also do is taking the words very seriously with which Ramond begins his <u>Field Theory</u> (7): "It is a most beautiful and awe-inspiring fact that all the fundamental laws of Classical Physics can be understood in terms of one mathematical construct called the Action. ... In addition, as Dirac and Feynman have shown, the Action acquires its full importance in Quantum Physics".

Well, starting from – or encouraged by – a four-dimensional reality of events, why not venture the hypothesis that it is the central, <u>physically real</u> quantity or entity in Nature as well, from which the other ones can be derived and that defines, e. g., metric, too?

## 4. FURTHER ELABORATION AND CONSEQUENCES OF THE PRIMACY OF EVENTS (ACTION) AND ACTION METRIC

a) We integrated space-time, $\vec{x}_{(4)} \equiv (ict, x, y, z)$, and energy-momentum, $\vec{E}_{(4)} \equiv (i\frac{E}{c}, p_x, p_y, p_z)$, into action $S = -\vec{x}_{(4)} \cdot \vec{E}_{(4)} = Et - \vec{p}.\vec{x}$, positing that the latter (that is, events) is primary, also as regards the physically most relevant metric, whereas space-time and energy-momentum are derived, secondary entities, originating from sectioning events by or projecting them upon, our relativistic space-time scheme, this implying radical distortions as regards metric. E. g., the physical, that is, action distance zero between the "elements" of events B and C in the figures 6 and 7 is stretched out unto the finite distance BC in our macro ordering scheme we call relativistic metric.

The above integration-with-metrical-distortion is nothing else but one similar step further than the integration-with-metrical-distortion special relativity performed with respect to time and space (and at the same time with energy and momentum), forming therewith space-time with its own metric. In this metric, s = OA has length zero, whereas its "projection" on our Euclidean now-hyperplane is $OX_1 \neq 0$, and $ict_1$ is unequal to zero, too. (See figure 9.)

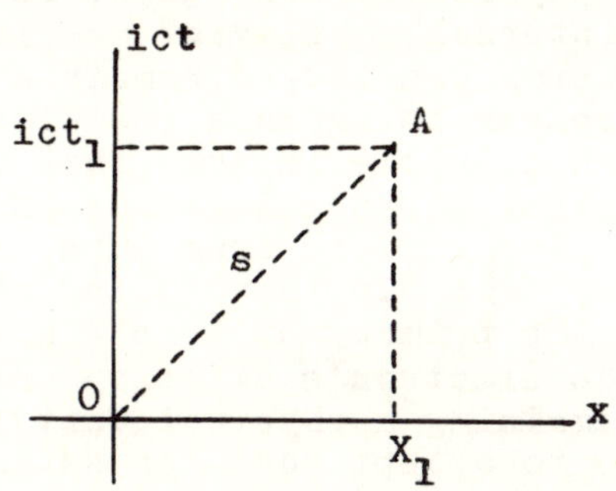


Figure 9

By considering the "real" relativistic metric, paradoxes in the Euclidean scheme, such as the constancy of the velocity of light, could be solved.

In a similar way our new integration of $\vec{x}_{(4)}$ and $\vec{E}_{(4)}$ solves the "nonlocality" paradoxes of quantum mechanics: If we see a particle interact with a whole grating, it seeming to be stretched out to become "waves" (see also b below), we only witness a similar dilation in our (relativistic and/or Euclidean) ordering scheme as OA = 0 of figure 9 shows in becoming $OX_1$ for us. If we coordinate, order, objects or events without reckoning with the physical relations (i. e., the integrations mentioned), we distort the physically relevant metric, creating paradoxes. The integration of space-time and energy-momentum in particular contains that it is incorrect to posit metric, distances, as existing independently of events. Actually, distances in the macro Minkowski scheme become mere line segments in an ordering scheme, mere constructs, if we "interpolate" such scheme into the quanta of a micro-process. Real physically based distances between two (e. g., alternative) situations measure how much of the most relevant physical entity, e. g., objects (measuring sticks) or events (action), fits in-between them. Here the measurement of distances by means of objects is a classical, or a kind of "Euclidean", approximation that is practical for macro situations with many quanta. The rest is metaphysics.

b) Actually, the essential paradox of an electron reflected by a whole grating is that it is stretched out and, at the same time, continues to be an integrated unity. Well, the message of this state of matters to us is that, for the electron (or,more precisely, for the action quanta its existence in time is a series of), otherwise than for us, the distances between many mutually distant parts of its "waves" (quanta) are by no means large. On the contrary, those parts are physically contiguous in the electron's (or: its quanta's) internally relevant metric, so that the "stretched-out" electron can indeed remain a coherent whole, the wave pattern moreover being in a position to "collapse instantaneously" at a possible impact. That is, the only solution of the relevant paradox is to assume that the electron's "internal metric" differs from ours, which assumption follows naturally from our introducing action metric as a consequence of the primacy of events and action for Nature. Actually, the electron's dilation unto the size of a grating and its remaining a physical unity at the same time almost force us to accept that for Nature, e. g., the physical situations corresponding to the point-events B and C of the figures 6 and 7 have an infinitesimal distance so far as the process in question is concerned. (Mind here that in our conception distances do not exist independent of events.) That is, it takes only a zero event (an action zero) to transform the "B" process into the "C" one. In fact, one can hardly imagine another alternative for the corresponding metrical distortion we discussed as an explanation than material stretching of the particle.

In figure 10, where the particle waves of length $\lambda$

approach a grating G, the spatial "slice" s is a section of a four-dimensional slice as drawn in figure 6 with the now hyperplane of the observer. The mutually alternative equal-phase stage-events B and $C_1$ of the process consisting of the moving particle (momentum carrier) of figure 10 are so situated that $BC_1$ would be perpendicular to the plane of figure 6, B being the same in both figures.

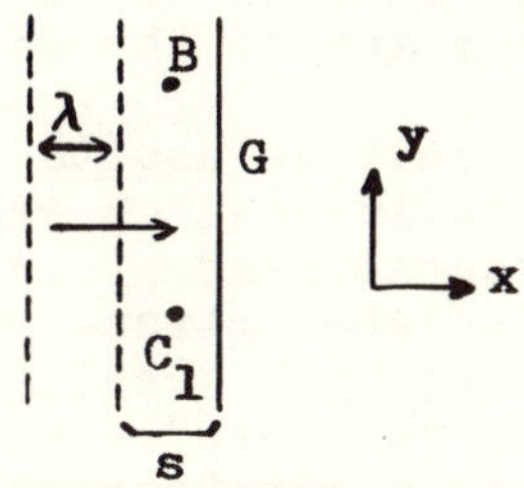


Figure 10

Also Young's double-slit experiment, nonlocal EPR correlations and nonlocality and wave-particle "duality" in general find a natural explanation by the assumption of an action micro metric. E. g., the two correlated systems in figure 11 are action-metrically contiguous along the dotted line AB which is an equi-action line; that is, a line along which the action corresponding to the emission process is constant and the action distance is zero. This very fact of being each other's direct physical neighbours enables the situations A and B (say, measurement-events) to correlate.

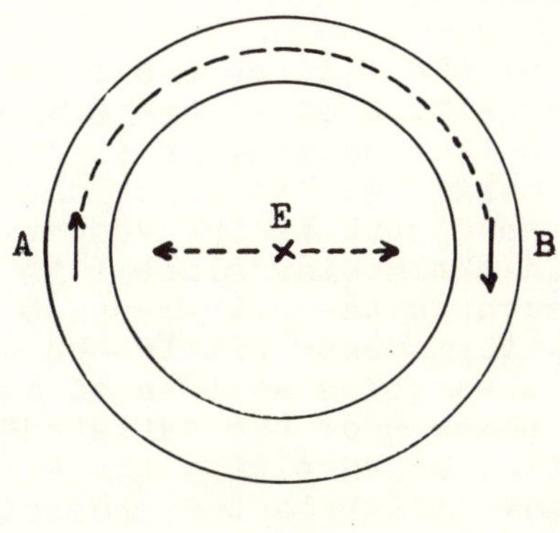


Figure 11

Further, in particular figure 6 illustrates how the wave phenomenon is a consequence of the dilation or nonlocality the action metric implies.

The latter also opens possibilities for explaining

"action at a distance", the "distance" between two charged particles or masses being physically the action distance zero that exists within the virtual photons or gravitons transmitting the relevant forces. (Compare the distance zero between B and C in the figures 6 and 7. ) Again, the metric of our macro-scheme is not the one physically operative in the micro-process in question, i. e., in the quanta constituting the "virtual" photons etc. (8).

c) In Ref. 6 it is enunciated how retroactivity, too, can be explained by means of the action metric. (See also below. )

It is somewhat obvious now to assume that the direct, "nonlocal", mutual influences on each other of different parts of a micro-process, and of which the retroactive influence of, e. g., an absorption-event on the corresponding emission-event is a special case, precisely constitute the - nonlocal - "hidden variable". In the experiment of figure 1, the alternative absorption-events on $S_2$ and P, respectively, had indeed different influences on the momentum transfers from $S_1$, <u>within the momentum uncertainty margins</u>, which can be expected of a hidden variable. Generally, we may say that hidden variables are constituted by the influences that patterns as a whole of - physically integrated - micro-processes exert on the parts, on the sub-events into which we will separate them (emission-event, reflections, absorption-event,... ).

## 5. THE SPACE-TIME REPRESENTATION OF QUANTA IN GENERAL; THE SCHRÖDINGER EQUATION AND THE FUNCTION $\psi(\vec{x}, t)$; REMAINING PROBLEMS

In the discussion of the figures 6-8 it became clear that the "monochromatic" slices of figure 6 are too simple a representation of action quanta, as it does not reckon with the "uncertainty" margin within which the momentum can vary.

Already more realistic is the "dovetail" picture of figure 12, in which the median slice 1 is accompanied by secondary slices such as the extremes 2-5 that are just still compatible with momenta within the Heisenberg margins; the picture also gives an idea of how, e. g., in point-event A the phase $\frac{1}{4}$ of the "dilated" quantum pictured can make physical contact with phase $\frac{1}{4}$ of its predecessor - another dovetail with the property that its median slice borders on slice 1 - as it is elaborated in Ref. 6. The same holds for the other phases and this makes retroactive influences possible, viz. in a zigzag way from an action quantum to its predecessor. For point-event A is physically contiguous to the phases (stages) $\frac{1}{4}$ of both successive quanta (the drawn one and its predecessor),

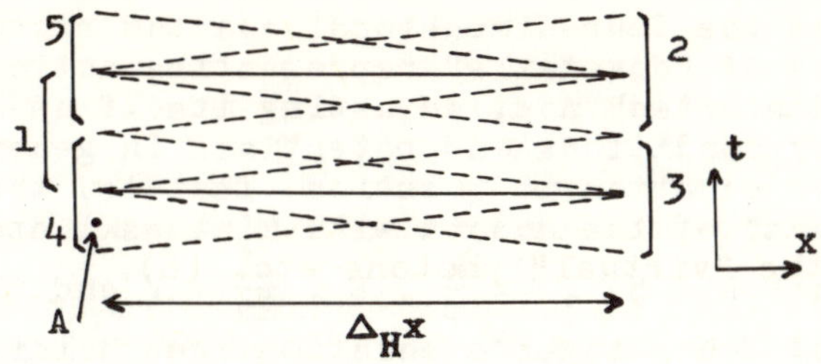


Figure 12

which in principle makes exchanges of information possible (e. g., via "hinges" in the A regions). We can also say: In the action metric both successive ("proper") quanta (quantal processes) reach unto the A region, where they can make direct contact.

Of course, also the principle of least action becomes better understandable from the point of view that both causal and retroactive influences cooperate in "vertically integrating" series of action quanta, as considered above.

In figure 12 all slices combined, that is, the waves they represent, interfere mutually extinguishingly at the left and right ends (if we also consider the other quanta of the process), the pictured quantum constituting a stage-event in the history of a particle "at rest", with the "uncertainties" $\Delta_H x$ and $\Delta_H p_x$ as to space and momentum, respectively, which uncertainties are actually dilations of the proper quantal process in our macro Minkowskian ordering scheme that are implied by discrepances between the latter scheme's metric and action metric if the scheme is "interpolated" into the micro domain. (In the figure, the slices 2-5 correspond to "extreme" positions of the energy-momentum four-vector that differ $\Delta_H p_x$ as regards their x-components. )

The relation of this picture to the wave function $\psi(\vec{x}, t)$ and the Schrödinger equation can be seen as follows. Feynman proved that for matter waves Huygens' principle can be formulated so as to be equivalent to the Schrödinger equation (9). Because the slices pictures such as the ones of figure 6 and – already a stage more general – figure 12 are, in turn, equivalent to the Huygens picture (the slices are envelopes of Huygens "wavicles"), we can say that the four-dimensional wave pattern (partly) pictured e. g. by figures 6 and 12 is the same as the one determined by the Schrödinger equation and, therefore, by the wave function (that corresponds to a superposition of waves) for the free-particle case in question. In the general case of Schrödinger's equation with potentials, the only difference as regards the four-dimensional quantum pictures is that the series of "dovetails" representing action quanta can be more "distorted" now, adjusting itself as a complex to the boundary conditions and potentials in general of the Schrödinger equation in question. That is, the slice-shaped "sections" of the quanta with Minkowski space are deformed so that $E = h\nu$, $\lambda = \frac{h}{p}$, $E = \frac{p^2}{2m} + V$ and the other determinants of Schrödinger's equation keep holding. (Because figures 6-8 and 12 are relativistic, they actually correspond to the Klein-Gordon equation. )

For the rest, Nature demonstrates the fact that an emission process reckons only with the internal action metric of the quanta in question by emitting all Fourier components (representing the "momentum dilation" referred to in the last part of Section 3) "at the same time", that is, by emitting waves and parts thereof which correspond to all E, t, $\vec{p}$ and $\vec{x}$ values permitted by the conditions of the emitter and the experiment in general: all such "versions" of it are equivalent for the emission process so far as their point-events relate to the same action. It is only we who "unravel" them, projecting those versions on our macrometrical scheme according to the corresponding E, t, $\vec{p}$ and $\vec{x}$ values. (Mind that there is indeed a difference between, e. g., the "versions" of an emission process as regards the recoil. The latter may be produced retroactively by the corresponding absorption-event. )

At a certain time t, $\psi(\vec{x}, t)$ describes a section of a now hyperplane with the action-quantal lattice or structure of quanta corresponding to a process, precisely so as in figure 6 OA is the section of the hyperplane t = 0 with one particular quantum slice. In short: waves are nothing but spatial sections of metrically distorted, "stretched", action quanta.

We see from the above that our hypothesis of the primacy of "occurring", events, above objects and the consequent action metric suffice for explaining the "wave character" of particles and for constructing an imaginable model of micro-processes, the waves and the wave function "only" corresponding to space(-time) sections of the series of action quanta particles' existences in time consist of.

In essence, this solves the quantum paradoxes relating to nonlocality and the wave-particle "duality" in general, without, e. g., our appealing to any "new way of thinking" which abandons determinism, imaginable models and microrealism.

At the same time, the wave function contains all information about a system because $\psi(\vec{x}, t)$ represents the whole four-dimensional series (or lattice) of the action quanta the process in question consists of whereas, according to our theory, there is nothing but action quanta in the physical world, so that describing the ("stretched") quanta is describing the process completely, apart from where the "proper quanta", that is, the world-line sections, in the quantal slices are, or, where the particle is in the wave packet. This hidden variable is determined by the integrated process as a whole, in probably much the same way as classical random decisions come about, though now causal and retroactive factors cooperate.

Because the Schrödinger equation appears to be equivalent to the Huygens principle (see above) it will be important to explain what the Huygens "wavicles" are. It is suggested by our theory of action metric that they are spatial representations or sections of a certain stage of the proper-quantal process which for us is dilated, "multiplied", in all directions in which the action difference or distance is zero. I. e., if the action difference between each two corresponding parts of the wavicles $W_1$ and $W_2$ is zero, $W_1$ and $W_2$ represent a same stage or part of the relevant process (e. g., what happens at a certain small segment of a particle's world-line). That is, all wavicles having their centres in an equi-action region are "copies" of a same stage of an action-quantal process. This is not the whole truth because $W_1$ and $W_2$ may experience different fates, e. g., reflections. Of course, the unsolved problem what quanta precisely are is closely connected with the "wavicles" problem. Also a part of the problem what quanta are is what the physical meaning is of the imaginary component of the wave function $\psi(\vec{x}, t)$.

## 6. SOME PHILOSOPHICAL ELABORATIONS

As we already observed, the above theory dispenses with any "new way of thinking" in the sense of introducing fundamental uncertainty, a-causality, "fundamentally probabilistic laws", an impossibility of constructing imaginable models etc. Instead, one of its consequences is some shift in the direction of "super determinism", for it does not only deny that there is such a thing as pure chance in the local sphere, but also suggests that pure chance may not even exist on a nonlocal scale in the somewhat loose sense of common parlance, i. a. because retroactivity – in fact, finalism or teleological effects – may see to it that at the "crossroads" of mutually unconnected causal chains "no unforeseen accidents occur" that would not be in harmony with the master laws or symmetries of greater wholes.

Einstein would appear to be more right with his "God

does not play dice" than he himself and his "local" thinking intended to be, and the conviction that science and understanding mean getting the Aha-Erlebnis about imaginable models would predominate.

In the twentieth century we saw a reaction against "the rationalism of the nineteenth century" in philosophy, in arts, in many ideologies and in publicly popular ideas, a movement into the direction of indeterminism, of scaling down the competence and status of reason, imaginability and "mechanistic" models, and of stressing the role of chance or even the absurd, of accepting the incoherent and/or incomprehensible. This philosophy not seldom appealed to "the new way of thinking" in quantum physics. In many spheres of life and thinking the idea gained influence that rationality, reason, coherence and, therefore, science found their limits in fundamental indeterminism, chance, unimaginability, unanalysability, unclearness, paradoxes and an inaccessibility to the Aha-Erlebnis in general (think of the role of pure formalism, too). Science had to concentrate on description instead of explaining, understanding, even "rationalists" like Wittgenstein said.

The above microrealistic theory might be a factor in the relevant philosophical controversy, too. It may even suggest, in its possibly "superdeterministic" consequences, that the rationalists of the nineteenth century were not wrong in the sense of being too rationalistic, too deterministic and too model-minded, but by being too little so. It might be that they were wrong by leaving too little room for coherence and rational functioning of the Universe because they, though considering the law of cause and effect as determining all events on a local scale, still left greater and more complicated processes such as those relevant to man's fate, such as evolution and history and such as the Universe as a whole to chaos: unconnected causal chains would meet and give "random" effects, no nonlocal and retroactive (teleological) coherences and natural laws being existent. God still would play dice with fates, with evolutions and with the Universe as a whole. Coherence and science would be very restricted in scope, and there could never be any scientific way of verifying the truth of Einstein's words: "The most surprising of all is that the world almost certainly has a meaning".

With these problems retroactive effects, action metric and nonlocal coherences of the EPR-Bell type may be connected, too. Actually, they may suggest that the indeterminism seemingly existing on the local scale precisely delineates such margins within which local factors do not determine the course of events but within which greater wholes, master coherences, nonlocal effects such as retroactive ones can exert their coordinating influence.

Remark: In Ref. 10 some experiments are suggested by means of which we may verify the appearance of action quanta as real "elements of occurring".

One cannot completely shake off the impression that quite a few physicists are primarily kept from accepting the above theory by an emotional difficulty in accommodating to the four-dimensional reality of events, i. e., the pre- and post-existence of the future and the past, respectively. Such an extra-physical motive may also be the cause of the somewhat strange circumstance that the proofs of such four-dimensional reality which earlier had been given in References 4, 3 and 2, though they never have been refuted, were neglected notwithstanding their radical physical (and philosophical) consequences.

A similar "unphysical" reaction manifested itself when the above was enunciated at the Conference in Perugia: very little explicit opposition, hardly any critical argument, and at the same time an unspoken "resistance" from a part of the audience.

References:

1. Wootters, W.K., and Zurek, W.H.: "Complementarity in the double-slit experiment; Quantum nonseparability and a quantitative statement of Bohr's principle", 1979, Phys. Rev. D 19, pp. 473-84.

2. See also Rietdijk, C.W.: "Another proof that the future can influence the present", 1981, Found. Phys. 11, pp. 783-90.

3. See Rietdijk, C.W.: "Proof of a retroactive influence", 1978, Found. Phys. 8, pp. 615-28.

4. As regards the pre-existence of the future, see also Rietdijk, C.W.: "A rigorous proof of determinism derived from the special theory of relativity", 1966, Philos. of Sci. 33, pp. 403-6 and Rietdijk, C.W.: "Special relativity and determinism", 1976, Philos. of Sci. 43, pp. 598-616.

5. As regards Maxwell's equations see, e. g., Feynman, R.P., and Hibbs, A.R.: Quantum Mechanics and Path Integrals, N. York, 1965, p. 241.

6. For a more detailed elaboration of the action metric and related problems of wave-particle "duality" (e. g., the

interference problem) see Rietdijk, C.W.: "A microrealistic explanation of fundamental quantum phenomena", 1980, Found. Phys. 10, pp. 403-57.

7. Ramond, P.: Field Theory; a Modern Primer, Reading, 1981.

8. See Rietdijk, C.W.: "How do 'virtual' photons and mesons transmit forces between charged particles and nucleons?", 1977, Found. Phys. 7, pp. 351-74.

9. Feynman, R.P.: "Space-time approach to non-relativistic quantum mechanics", 1948, Rev. of Mod. Phys. 20, pp. 367-87; see in particular pp. 374 and 377.

10. Rietdijk, C.W.: "Suggestions for experiments on action quanta", 1981, Il Nuovo Cimento 63B, pp. 541-64.

# DYNAMICS OF THE REDUCTION OF THE STATEVECTOR

Philip Pearle

Mathematical Institute, Oxford University, England
Permanent Address:
Hamilton College, Clinton, New York 13323, USA

A theory describing the dynamics of the reduction of the statevector is presented. A nonlinear term is added to the Schrödinger equation which causes an interaction between the superposed vectors making up the statevector. The indeterminism in quantum theory is ascribed to randomly fluctuating coefficients governing the nonlinear term. Three further investigations into the antecedents and consequences of this theory are reported. First, it is argued that there exists a relation between the nonlinear term and an antilinear operator analogous to the relation between the usual linear term and the linear Hamiltonian operator. Second, a possible mechanism for producing the randomly fluctuating coefficients is presented: it is ascribed to the complicated fluctuations of the statevector describing an apparatus. Third, the predictions of this theory are discussed for rapidly repeated experiments: remarkably, it is shown that they are identical to those of quantum theory, even when the time between experiments is comparable to the reduction time. Some experiments which can test this theory are mentioned.

## I. WHY ALTER QUANTUM THEORY?

What is wrong with quantum theory? Why should anyone want to change it? We will give two reasons. First, quantum theory does not satisfactorily describe an individual system. Second, quantum theory does not explain why an individual event occurs. These reasons will be amplified below.

There are a number of logically consistent and experimentally unrefuted interpretations that can be grafted onto the formalism of quantum theory (see Table 1). An interpretation can be



*S. Diner et al. (eds.), The Wave-Particle Dualism, 457–483.*

succinctly, if oversimply, characterized by its answer to the question, "What is in one-to-one correspondence with the statevector of the theory?" For the four leading interpretations of quantum theory in Table 1, in no case is the answer, "A single system in Nature."

However, many physicists subscribe informally to an interpretation of quantum theory (which we shall call "naive") with the following properties:

a) the statevector is in 1:1 correspondence with a single system in Nature
b) there are things in Nature, called observers, which are *not* described by quantum theory
c) the statevector changes continuously until the instant an observer makes an observation, and then it changes (reduces) discontinuously.

There are three objections to this interpretation that can be raised.
1) Observers are physical objects (whether machine or human or part of human). Therefore, they should be described by physics. There appears to be no good physical reason for leaving some physical objects out of the quantum description (5).[1]
2) The Schrödinger Cat Paradox (6) provides an objection to this interpretation, but to none of the others.[2] Schrodinger envisaged an experiment on a single cat whose end result (before an observation) is described by the statevector $(1/\sqrt{2})|\text{cat alive}\rangle$ + $(1/\sqrt{2})|\text{cat dead}\rangle$. This conflicts with our intuitive feeling that a cat subjected to this experiment is either alive or dead and so should be described by either $|\text{cat alive}\rangle$ or by $|\text{cat dead}\rangle$. Moreover, when an observation of the cat is performed, the statevector must instantaneously jump to one of these two states. Since that is a large change in the statevector, and the statevector is in 1:1 correspondence with the physical state of the cat, the change ought to represent a large change in the physical state of the cat. It is hard to believe that a tiny peek at a cat would have a big effect on the physical state of the cat, nor is it clear what physical change the cat undergoes.[3]
3) If a system is continuously observed, it can be shown that, according to the naive interpretation, the system never evolves. It is worth reviewing this argument because it does not appear to have been previously raised as an objection to the naive interpretation (7). If a system evolves for $\varepsilon$ seconds and then is observed, the probability that it is observed to remain in the original state is

$$|\langle\psi,0|\exp-(iH\varepsilon/\hbar)|\psi,0\rangle|^2=1-\hbar^2\langle\psi,0|(H-\overline{H})^2|\psi,0\rangle\varepsilon^2+\ldots \quad (1.1)$$

This fact, that the initial evolution out of a quantum state is slower than linear with time,[4] is well known from second order

Table 1. Leading Interpretations of Quantum Theory

| Name | The Statevector Describes (Is In 1:1 Correspondence With) | Quantum Theory Doesn't Describe |
|---|---|---|
| Bohr (1) | all possible observations of a single system | reality, independent of observation |
| Einstein (2) | an ensemble of identical systems | a single system |
| Born (3) | knowledge of a single system | nature without mind |
| Everett (4) | many real universes | the observable universe alone |
| Naive | a single system | the observer |

(Note: Reduction is necessary for the first, third, and fifth interpretations, and is not necessary for the second and fourth.)

time dependent perturbation theory. If this sequence of evolution and observation continues N times, the probability that the system remains in the original state after $T=N\varepsilon$ seconds is

$$[1-c^2(T/N)^2]^N \rightarrow \exp-(c^2T^2/N) \tag{1.2}$$

in the limit of large N ($c^2 \equiv \langle(H-\overline{H})^2\rangle$). That such a sequence in this way slows down the evolution from the original state, compared to what it would have been if there were no observations, is also well known (8). As emphasized by Misra and Sudarshan (8), a continuous observation should be considered as the limit of such a sequence as $N\rightarrow\infty$. In that case, Eq. (1.2) predicts that the probability the system remains in its original state is 1, i.e. the system never evolves.[5] Since continually observed systems do in fact evolve, this is an erroneous prediction. It constitutes the most serious objection to the naive interpretation.

We conclude that the naive interpretation cannot be applied to quantum theory as it stands.

Even if it were possible to argue that the statevector describes a single system, one could still not maintain that quantum theory explains the occurrence of individual events. Quantum theory gives no explanation why a certain particle should scatter to the right rather than to the left, or why a particle decays when it does. Quantum theory predicts the probabilities of events, not the events themselves. It is hard to believe that these events occur for no reason at all. It is not unreasonable to suppose that Nature *does* determine the outcome of these events, although human beings and their apparatus, being built on a gross scale, may be incapable of affecting these outcomes.

## II. PROPOSAL FOR A DYNAMICAL REDUCTION THEORY

Thus, if we wish to describe a single system in Nature by a statevector, and to explain the occurrence of individual events, we must alter quantum theory. The objectionable features of the naive interpretation described in the previous section are the omission of the observer from the description of quantum theory, and the instantaneous nondeterministic nature of the reduction of the statevector consequent to an observation. We therefore propose that quantum theory be modified so that it possesses the following properties:

a) the statevector is in 1:1 correspondence with a single system in Nature
b) the observer (and anything else) is part of the single system described by the statevector
c) the statevector changes continuously, following a modified Schrödinger equation

A brief description of the behavior of this modified Schrödinger equation follows. When an experiment is performed on a microscopic system, first the microscopic variables interact with each other, and then they interact with the detecting apparatus. We suppose that there is a *preferred basis*[6] $|\phi_n,t\rangle$ in terms of which the statevector can be expanded

$$|\psi,t\rangle = \sum_n a_n(t)|\phi_n,t\rangle \tag{2.1}$$

One may think of the preferred basis vectors as describing different possible outcomes of the experiment, i.e. each vector will eventually describe a different "pointer position." Their time dependence is given by

$$i\hbar\frac{d}{dt}|\phi_n,t\rangle = H_0|\phi_n,t\rangle \tag{2.2}$$

where $H_0$ describes the evolution of the detecting apparatus, its interaction with the microscopic variables and the free evolution of the microscopic variables, but does not describe the interaction of the microscopic variables among themselves.

The amplitudes $a_n(t)$ are to satisfy the (interaction picture) modified Schrödinger equation

$$i\hbar\frac{d}{dt}a_n(t) = \sum_m H_{Inm}a_m(t) + (\text{nonlinear term}) \tag{2.3}$$

When the microscopic variables self-interact, the matrix elements of $H_I$ are generally nonzero, the nonlinear term is supposed to be negligibly small, and the superposition (2.1) is generated by the Schrödinger equation (2.3) exactly as in ordinary quantum theory.

When the microscopic variables get around to interacting with the apparatus, the matrix elements of $H_I$ have by this time vanished. As the states become more and more *macroscopically distinguishable*[7] (due to their evolution governed by $H_0$) the nonlinear term in (2.3) is supposed to grow in magnitude and *reduce the statevector.*

## III. REDUCING THE STATEVECTOR

How is the nonlinear Schrödinger equation (Eq. (2.3) with $H_{Inm}=0$) to perform the task of reducing the statevector? We will specify three properties its solutions must have. Any equation whose solutions satisfy these three properties we will call a "reducing equation."

For convenience, we write the amplitudes as

$$a_n(t) \equiv \sqrt{x_n(t)}\exp i\theta_n(t) \tag{3.1}$$

The three properties can be written in terms of the magnitudes $x_n$:

bookkeeping property 0: $\sum_n^N x_n(t) = 1,\ 0 \le x_n \le 1.$

reduction property 1: All $x_n$'s eventually vanish, except one which achieves the value 1.

probability property 2: There are many solutions to the equation which have the same initial values $\{x_1(0), x_2(0), \ldots x_N(0)\}$. A fraction $x_n(0)$ of these solutions eventually ends up with $x_n=1$.

There are a number of ways of depicting the dynamics described here. One way, a "configuration space" picture, is illustrated in Figure (1a) where the trajectories of a typical solution of a reducing equation, $x_n(t)$ vs. $\theta_n(t)$, are drawn. A more useful picture is to represent the solution by a single point moving in a "phase space." This is illustrated in Figure (1b), where the phase angle evolution is suppressed.

In terms of this phase space picture, *property 0* ensures that the representative point stays on the "triangular" shaped (N-1 dimensional hyper-) plane illustrated in Figure (1b). The representative point of a typical N-state solution wanders over the (N-1 dimensional hyper-) plane until it reaches an (N-2 dimensional hyperplane) "edge," wanders over that "edge" until it reaches an (N-3 dimensional hyperplane) "edge," etc., until, by property 2, it eventually comes to rest at one of the N "vertices" $(x_n=1,\ x_{\neq n}=0)$, as is required by *property 1*.

When the representative point reaches a vertex, the wave function it describes is said to be "reduced." This by itself is not sufficient to ensure agreement with the predictions of quantum theory. For that, we require *property 2*, a restriction on the *ensemble* of solutions of the Schrödinger equation. (There is an ensemble of solutions with the same initial values $x_n(0)$ because, as we shall see, the nonlinear Schrödinger equation depends on random functions of time. Any particular choice of such functions will determine the specific vertex to which a particular solution reduces.)

Consider an ensemble of "identical" experiments, and suppose that the microscopic interaction takes place during $t \lesssim 0$. Since it is the usual Schrödinger equation which governs the evolution of the statevector in this time interval, what is produced is the usual quantum theory value of the magnitude $x_n(0)$, which is interpreted in quantum theory as the probability of the $n^{th}$ outcome. It is because of property 2 that the continued evolution (for $t>0$) of the ensemble of statevectors (which describes this ensemble of experiments) results in a fraction $x_n(0)$ of these statevectors ending up at the $n^{th}$ vertex ($x_n=1$, $x_{\neq n}=0$). Thus the relative frequencies predicted by this theory are in agreement with the probabilities predicted by quantum theory.

## IV. THE GAMBLER'S RUIN

At this point, we pause to point out a precise analogy between the reduction of the statevector and a gambling game whose description goes by the melodramatic title "The Problem of the Gambler's Ruin. (13)." Consider two gamblers who have between them \$1000. They toss a fair coin: heads means gambler 1 wins \$1 from gambler 2, tails means the reverse. The game proceeds until one gambler loses all his money, and so can play no longer.

If we define

$$x_n \equiv \text{gambler } n\text{'s money}/\$1000, \qquad n=1,2 \tag{4.1}$$

then we can identify three properties of this game.

property 0: $x_1+x_2=1$ (the money in the game remains constant).
property 1: Eventually either $x_1=0$ and $x_2=1$ or vice versa (the game ends).
property 2: A gambler starting with $x_n$ of the money has a probability $x_n$ of winning it all.

To prove property 2, denote by $P(x)$ the probability that gambler 1 wins all the money if he starts with x of it. $P(x)$ obeys the following difference equation

$$P(x) = \tfrac{1}{2}P(x-\varepsilon) + \tfrac{1}{2}P(x+\varepsilon) \tag{4.2}$$

($\varepsilon$ is the fraction of total money bet on each coin toss, 1/1000 in this example). Eq. (4.2) states that on the next coin toss there are two routes to the end of the game. Gambler 1 can lose, but eventually win the game, or he can win and eventually win the game. The boundary conditions are $P(0)=0$ (with no money, gambler 1 can never win) and $P(1)=1$ (with all the money, gambler 1 has won).

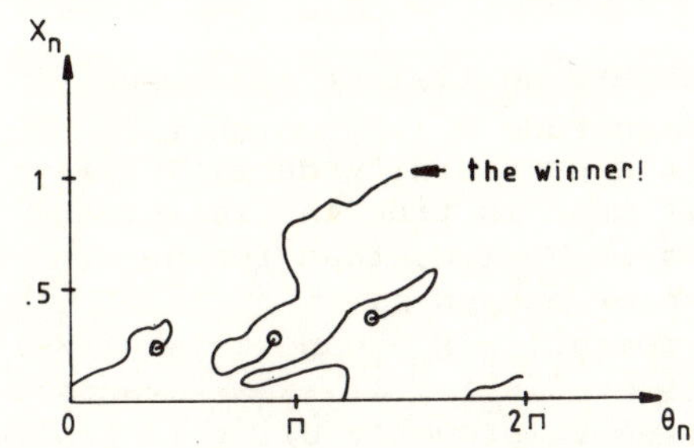


Figure 1a. A "configuration space" picture: trajectories of three competing amplitudes.

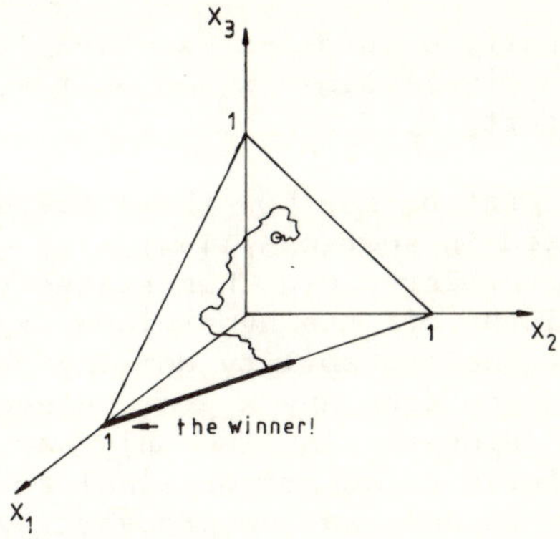


Figure 1b. A "phase space" picture: trajectory of three competing amplitudes.

The general solution of Eq. (4.2) is P(x)=Ax+B, and when the boundary conditions are applied we obtain

$$P(x) = x \tag{4.3}$$

which proves property 2 is satisfied.

Since this is such a simple vivid picture, it is useful to think of the statevectors in a superposition as "playing the game" among themselves until one of them wins. It is not hard to generalize (14) the gambler's ruin game to N players who play continuously (not just at discrete times) and then the description of an ensemble of reducing statevectors and the description of an ensemble of gamblers ruin games are identical descriptions.

## V. ENSEMBLE DESCRIPTION

Consider a description of the ensemble of solutions of a reducing equation. This is conveniently expressed in terms of the probability $\rho(x_1,..x_N;t)dx_1..dx_N$ that a solution lies in the phase space volume $dx_1..dx_N$ (the $\theta_n$ dependence is ignored). Properties 0, 1, 2 can be expressed in terms of the probability density $\rho(\vec{x};t)$. Remarkably, only the lowest few moments $\langle 1\rangle, \langle x_n\rangle, \langle x_n x_m\rangle$ are involved. (The definition of the moments are as usual

$$\langle x_n^r x_m^s \ldots\rangle \equiv \int_V dx_1 \ldots dx_N x_n^r x_m^s \ldots \rho(\vec{x};t) \tag{5.1}$$

where the integration volume V is the "triangular" shaped hyperplane illustrated in Figure (1b), including the "edges" and "vertices".)

In fact, we have

property 0: $\rho \sim \delta(1-x_1-x_2-\ldots-x_n)$, $d\langle 1\rangle/dt=0$
property 1: $\lim\langle x_n x_m\rangle \to 0$ as $t\to\infty$, for $n\neq m$
property 2: $d\langle x_n\rangle/dt=0$

Let us see how these statements are equivalent to those listed in section III.

Property 0: This states that the probability density always vanishes off the hyperplane $\Sigma x_n=1$ if that is true initially, and that the probability density remains in V, i.e. that for no solution does any $x_n$ become negative or exceed 1.

Property 1: The only way the moment $\langle x_n x_m\rangle$ can vanish ($n\neq m$) is for $\rho$ to be proportional to $\delta(x_n)$ or $\delta(x_m)$. If $\langle x_n x_m\rangle$ vanishes for all $n\neq m$, and no probability leaves V (property 0), this implies

$$\rho = \sum_{n=1}^{N} p_n \delta(x_1)\ldots\delta(x_{n-1})\delta(1-x_n)\delta(x_{n+1})\ldots\delta(x_N), \quad \sum_{n=1}^{N} p_n = 1 \quad (5.2)$$

i.e. the probability is at the vertices. Thus the vanishing of $\langle x_n x_m\rangle$ asymptotically is equivalent to the statement that all phase space representative points end up at the vertices. Incidentally, the time at which $\rho$ is reduced (achieves the limit (5.2)) is expressed above as $t\to\infty$ for convenience only. As we shall see, the mean time it takes to reduce the statevector (mean reduction time) is finite, so only a set of measure 0 of solutions take infinite time to reduce.

Property 2: It is most remarkable that agreement with the predictions of quantum theory can be expressed so simply as this statement. Actually, we only need the weaker condition

$$\langle x_n(0)\rangle = \langle x_n(\infty)\rangle \quad (5.3)$$

(which of course follows from the above statement of property 2) to prove this. Since the probability ends up at the vertices, $x_n(\infty)$ either equals 1 or 0. Therefore,

$$\langle x_n(\infty)\rangle = 1\cdot(\text{probability that } x_n(\infty)=1) + \\ + 0\cdot(\text{probability that } x_n(\infty)=0) \quad (5.4)$$

Since $\langle x_n(0)\rangle = x_n(0)$ (because before the reduction commences, the amplitudes are computed according to ordinary quantum theory, i.e. $\rho(\vec{x},0)=\delta(x_1-x_1(0))\ldots\delta(x_N-x_N(0)))$, it follows from (5.3), (5.4) that

$$x_n(0) = (\text{probability that } x_n(\infty)=1) \quad (5.5)$$

which is precisely the probability of the $n^{th}$ outcome predicted by quantum theory.

In a previous paper (14) we remarked that although only the asymptotic behavior (5.3) is necessary for this argument, we need the stronger form of property 2. The reason given was that it would be difficult to construct a reducing equation satisfying (5.3) unless $\langle x_n(t)\rangle$ was constant at all times. A more profound reason for the stronger form of property 2 has emerged, and will be given in section XI. It will be shown that constancy of $\langle x_n\rangle$ causes the predictions of this theory to be identical to those of quantum theory for a rapid sequence of experiments, even though the reduction started by one experiment has not been completed before the next experiment commences.

We believe that the constancy of $\langle x_n(t)\rangle$ during the reduction process, as the mechanism whereby the physical system "remembers" the quantum mechanical predictions, is the most important physical idea in this theory.

## VI. DIFFUSION EQUATION

It is not hard to write down a diffusion equation (15) governing the evolution of the probability density $\rho$, having the required three properties 0,1,2 and so fit to describe the behavior of the ensemble of solutions of some reducing equation. One way of proceeding is to consider the generalization of the gambler's ruin problem to N players, write down the difference equation describing the time evolution of the probability distribution for the gamblers' stakes, and then to pass in a well known way (16) to the continuous time limit. In this way we can obtain the basic diffusion equation

$$\frac{\partial\rho}{\partial t} = \sum_{m,n=1}^{N}\left(\frac{\partial}{\partial x_m} - \frac{\partial}{\partial x_n}\right)^2 \alpha_{mn}(x_m, x_n)\rho \tag{6.1}$$

which describes the "continuous time" gambler's ruin problem (14).

The functions $\alpha_{mn}$ may be thought of as characterizing the rate of play of the $m^{th}$ and $n^{th}$ gamblers (or alternatively, the amount of their wager). For example, consider the simple choice

$$\alpha_{mn} \sim (x_m x_n)^r \tag{6.2}$$

If $r=0$, the $\alpha_{mn}$ are constants and the rate of play is constant regardless of the size of the gamblers' stakes. If $r>0$ in (6.2), a gambler's rate of play slows down as his stake approaches zero.

As will be indicated in section VII, it is possible to construct a reducing equation for any value of $r\geq 0$ in Eq. (6.2). However, we are partial to the case $r=1$, for reasons which will become especially clear in section IX, where a close relationship between the reducing equation for this case and the usual linear Schrödinger equation will be displayed. Thus in the remainder of this paper we will concentrate on the reducing equation whose ensemble of solutions is described by the diffusion equation

$$\frac{\partial\rho}{\partial t} = \sum_{m,n=1}^{N}\sigma_{mn}{}^2\left(\frac{\partial}{\partial x_m} - \frac{\partial}{\partial x_n}\right)^2 x_m x_n \rho \tag{6.3}$$

(the $\sigma_{mn}$ are positive real constants).

In the remainder of this section, we will consider the simplest case N=2. We will illustrate how the solutions of Eq. (6.3) behave, and prove that properties 0,1,2 are obeyed. These results are readily generalized to the N-state equation (10,14,15).

The two-state diffusion equation can be written

$$\frac{\partial}{\partial t}\rho(x_1,x_2;t) = \sigma^2\left(\frac{\partial}{\partial x_1}-\frac{\partial}{\partial x_2}\right)^2 x_1x_2\rho(x_1,x_2;t) \tag{6.4}$$

With a change of variables $x\equiv x_1$, $X\equiv x_1+x_2$, Eq. (6.4) becomes

$$\frac{\partial}{\partial t}\rho(x,X;t) = \sigma^2\frac{\partial^2}{\partial x^2}x(X-x)\rho(x,X;t) \tag{6.5}$$

from which we see that the variable X does not change. Since initially X=1, we may write $\rho(x,X;t)=\rho(x,t)\delta(1-X)$ (thereby verifying part of property 0), and obtain the diffusion equation

$$\frac{\partial}{\partial t}\rho(x,t) = \sigma^2\frac{\partial^2}{\partial x^2}x(1-x)\rho(x,t) \tag{6.6}$$

This equation is solvable as a series (10). The solution at different times is illustrated in Figure 2. The initial condition at t=0 is $\rho=\delta(x-x_0)$. As time increases, the representative points diffuse away from $x_0$ toward the boundaries at x=0 and x=1, where they are trapped. As $t\to\infty$, the probability all ends up at the boundary, and $\rho=x_0\delta(1-x)+(1-x_0)\delta(x)$.

The behavior at the boundary of Eq. (6.6) is quite interesting. Eq. (6.6) is said to possess an *exit boundary* (17), which means that absorbing boundary conditions at x=0, x=1 are the only permissable boundary conditions. This contrasts with the ordinary diffusion equation (Eq. (6.6) with x(1-x) replaced by 1) which possesses a *regular boundary*, for which absorbing or reflecting (or a combination ) boundary conditions are possible, and have to be imposed in addition to the equation. This also contrasts with an equation possessing an *open boundary* (e.g. Eq. (6.6) with x(1-x) replaced by $x^2(1-x)^2$), where the probability always vanishes at x=0, x=1, never actually reaching these points, but jamming up in their open neighborhoods as $t\to\infty$. Indeed, if x(1-x) in Eq. (6.6) is replaced by $x^r(1-x)^r$, there is an open boundary for $r\geq 2$ (for which, strictly speaking, the statevector never completely reduces), an exit boundary for $2>r\geq 1$, and a regular boundary for $r<1$.

The mean reduction time $m(x_0)$ for the representative points which diffuse from $x_0$ to reach the boundary at 0,1 is well known (17) (as "mean exit time") to satisfy the adjoint equation

$$\sigma^2x_0(1-x_0)\partial^2m(x_0)/\partial x_0^2 = -1, \quad m(0) = m(1) = 0 \tag{6.7}$$

This equation is solved by the entropy-like expression

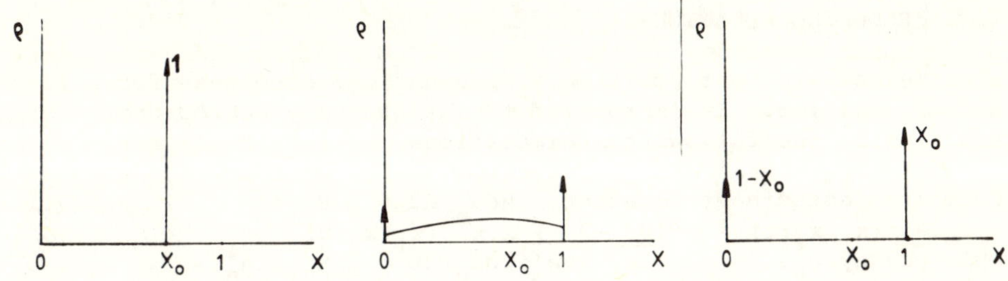


Figure 2. Behavior of the solution of the diffusion equation (6.6) at three different times: $t=0$, $t=\sigma^{-2}$, $t=\infty$.

$$m(x_0) = -\sigma^{-2}[x_0 \ln x_0 + (1-x_0)\ln(1-x_0)] \tag{6.8}$$

which proves the assertion made earlier, that the statevector is reduced in a finite time (except for a set of measure zero).

The solution of Eq. (6.6) may be written as

$$\rho(x,t) = \rho_i(x,t)+p_0(t)\delta(x)+p_1(t)\delta(1-x) \tag{6.9}$$

Substitution of (6.9) into Eq. (6.6) gives separate equations for $\rho_i$ (which is finite but generally nonvanishing at the boundary) $p_0$, and $p_1$ describing the buildup of $p_0,p_1$ as $\rho$ decreases.

However, we do not need the solutions of Eq. (6.6) to prove that it possesses properties 0,1,2. All we need is a differential equation for the expectation value of the moments, which can be obtained from Eq. (6.6):

$$\begin{aligned} d\langle x^m\rangle/dt &\equiv d\int_{0-}^{1+} dx x^m \rho(x,t)/dt = \int_{0-}^{1+} dx x^m \sigma^2 \partial^2[x(1-x)\rho]/\partial x^2 \\ &= m(m-1)\sigma^2\langle x^{m-1}(1-x)\rangle \end{aligned} \tag{6.10}$$

(The last step follows after two integrations by parts.) We then have from Eq. (6.10)

property 0: $d\langle 1\rangle/dt = 0$
property 2: $d\langle x\rangle/dt = 0$
property 1: $d\langle x(1-x)\rangle/dt = -2\sigma^2\langle x(1-x)\rangle$, which implies
$\langle x(1-x)\rangle = x_0(1-x_0)\exp -2\sigma^2 t \to 0$ as $t \to \infty$

Since this argument is readily generalized to show that Eq. (6.3)'s solutions possess properties 0,],2 for arbitrary N, we reach the following conclusion:

*If* ρ describes the solutions of a (nonlinear) Schrödinger equation, *and* ρ can be shown to satisfy Eq. (6.3), *then* that (nonlinear) Schrödinger equation reduces the statevector.

## VII. REDUCING EQUATION

We now are in a position to present our candidate for a reducing equation. It is related to the ordinary Schrödinger equation by the following manipulations:

Ordinary Schrödinger Equation: $i\hbar da_n/dt = \sum_m H_{nm}a_m$ (7.1a)

Multiply by $a_n^*$: $a_n^* i\hbar da_n/dt = \sum_m H_{nm}a_m a_n^*$

Take the complex conjugate of the right hand side, to obtain

$$i\hbar da_n/dt = \sum_m H_{mn}a_m^* a_n/a_n^* \quad (7.1b)$$

We claim that Eq. (7.1b) is a reducing equation, if the Hermitian matrix elements $H_{mn}$ are certain random functions of time. More generally, we can show (15) that

$$i\hbar da_n/dt = \sum_m A_{mn}\dot{B}_{mn}(a_m^* a_n)^r/a_n^* \quad (7.2)$$

(where $\dot{B}_{mn}$ is a Hermitian matrix of white noise with variance $\sigma^2$ (this will be expanded upon shortly) and $A_{mn}$ is a constant (or "slowly" varying) Hermitian matrix) is a reducing equation, because the ensemble of solutions of Eq. (7.2) obeys

$$\partial\rho/\partial t = \sigma^2 \sum_{m,n} |A_{mn}/\hbar|^2(\partial/\partial x_m - \partial/\partial x_n)^2(x_m x_n)^r \rho \quad (7.3)$$

We will content ourselves in this section with sketching the ideas that enable one to prove this assertion.

Let $B(t)$ be a representative of Brownian motion, i.e. a real continuous function of time, where $B(0)=0$, where $\Delta B(t) \equiv B(t+\Delta t)-B(t)$ has a Gaussian (normal) probability distribution and variance $\sigma^2\Delta t$, and where $\Delta B(t_1)$, $\Delta B(t_2)$ are statistically independent if their associated time intervals do not overlap. A crude definition of white noise is the time derivative of Brownian motion (crude because, strictly speaking, Brownian motion is nowhere differentiable) and is denoted by $\dot{B}(t)$. It may be thought of as a wildly randomly oscillating function of time. It is characterized by

$$\langle\dot{B}(t')\dot{B}(t)\rangle = \sigma^2\delta(t'-t) \quad (7.4)$$

where the average is over all representative Brownian trajectories.

Each $\dot{B}_{mn}(t)$ in Eq. (7.2) is a complex number which can be written in terms of real white noise functions $\dot{B}^R_{mn}$, $\dot{B}^I_{mn}$ as

$$\dot{B}_{mn}(t) = \dot{B}^R_{mn}(t) + i\dot{B}^I_{mn}(t), \quad (\dot{B}^R_{mn} = \dot{B}^R_{nm}, \ \dot{B}^I_{mn} = -\dot{B}^I_{nm}) \quad (7.5a)$$

$$\langle\dot{B}^*_{m'n'}(t)\dot{B}_{mn}(t)\rangle = \delta_{m'm}\delta_{n'n}2\sigma^2\delta(t'-t) \quad (7.5b)$$

All the $\dot{B}^R_{mn}$, $\dot{B}^I_{mn}$ are statistically independent of each other, except for the symmetry relations expressed in (7.5a). This is what we mean when we say that $\dot{B}$ is a Hermitian matrix of white noise.

Differential equations like (7.2), where the right hand side is proportional to white noise, are called stochastic differential equations. A simple example of such an equation is

$$dx/dt = F(x)\dot{B} \tag{7.6}$$

It was shown by Itô (17) how to find the diffusion equation that describes the behavior of the ensemble of solutions of Eq. (7.6). The procedure is to first expand (7.6) in a Taylor series in $\Delta t$ about t=0, retaining only terms of first order in $\Delta t$. Itô pointed out that, while the second order term in the Taylor series is $\sim\dot{B}^2\Delta t^2=(\Delta B)^2$, since $<(\Delta B)^2>=\sigma^2\Delta t$ is of first order in $\Delta t$, this term must be retained (and indeed, $(\Delta B)^2$ must be replaced by $\sigma^2\Delta t$):

$$\Delta x = F(x)\Delta B + \tfrac{1}{2}\sigma^2 FF'\Delta t + \ldots \tag{7.7}$$

Itô further showed that x is a Markov process characterized by only its mean and variance. These can be read right from Eq. (7.7)

$$<\Delta x> = \tfrac{1}{2}\sigma^2 FF'\Delta t \qquad \text{(since } <\Delta B> = 0) \tag{7.8a}$$

$$<(\Delta x)^2> = \sigma^2 F^2\Delta t \qquad \text{(since } <(\Delta B)^2> = \sigma^2\Delta t) \tag{7.8b}$$

and inserted into the Fokker-Planck equation for the density $\rho(x,t)$,

$$d\rho/dt = -d[\tfrac{1}{2}\sigma^2 FF'\rho]/dx + \tfrac{1}{2}d^2[\sigma^2F^2\rho]/dx^2, \tag{7.9}$$

*In the same way*, the set of stochastic differential equations

$$i\hbar da_n/dt = \sum_m A_{mn}\dot{B}_{mn}a_m^* a_n/a_n^* \tag{7.10}$$

can be replaced by differential expressions like (7.7), the first and second moments can be read off as in (7.8), and the resulting Fokker-Planck equation becomes (15) (with $a_n, a_n^*$ replaced by $x_n, \theta_n$):

$$\partial\rho(\vec{x},\vec{\theta};t)/\partial t = (\sigma/\hbar)^2 \sum_{m,n} |A_{mn}|^2[(\partial/\partial x_m - \partial/\partial x_n)^2 x_m x_n\rho + \\ + \tfrac{1}{2}\partial^2(x_m x_n^{-1}\rho)/\partial\theta_n^2 + \tfrac{1}{2}\partial^2\rho/\partial\theta_m\partial\theta_n] \tag{7.11}$$

When the $\theta_n$ are integrated over, we obtain Eq. (7.3) (with r=1).

This concludes the demonstration that *the nonlinear Schrödinger equation (7.10) reduces the statevector.*

## VIII. INTERLUDE

We are now able to write down the Schrödinger equation alluded to in (2.3), that describes both the production of a superposition (as in the usual quantum theory) and its reduction:

$$i\hbar da_n/dt = \sum_{m=1}^{N} H_{Inm} a_m + \sum_{m=1}^{N} A_{mn} \dot{B}_{mn} a_m^* a_n / a_n^* \qquad (8.1)$$

Eq. (8.1) may be thought of as the middle of this theory. But a middle requires also a beginning and an end. That is, we must ask, "What is the cause?" and, "What are the predicted consequences?" of the nonlinear Schrödinger equation (8.1). Work toward answering these questions has only begun. In the next three sections, three recent results are presented. We might call the remainder of this paper "Selected Topics in Statevector Reduction."

In section X we give a possible answer to the question: "What causes the random fluctuations $\dot{B}_{mn}$?" We will show how they may be attributed to complicated fluctuations in the preferred basis states which describe a macroscopic apparatus.

It would be very nice if the other features of the nonlinear part of Eq. (8.1) had roots elsewhere in physics. Penrose (18) has recently suggested that a quantum theory of gravity should of necessity be accompanied by a mechanism for statevector reduction, so perhaps it is to gravity we should look for answers to the following unanswered questions.

What is the nature of the coupling $A_{mn}$? This coupling is expected to increase as the states it couples become more macroscopically different: perhaps it is an increasing function of particle separation and particle masses. It is expected to be nonlocal, coupling two states which are spatially separated.[8] It is also what I will call "specific," that is, it only couples states which already appear in the superposition. (It does not introduce transitions into other states not in the superposition.) These are not usual properties of interactions in quantum theory.

What determines the preferred states $|\phi_n(t)\rangle$ that we see around us, and the Hamiltonian $H_0$ that governs their time dependence?[6]

Is there a more profound reason for the peculiar nonlinear dependence of Eq. (8.1) upon the amplitudes than the argument we have just given, namely that it "does the job" (reduces the statevector)? In the next section we point out a suggestive connection between the two parts of Eq. (8.1). We argue that the reducing part is related to an antilinear operator in a natural way, similar to how the usual part is related to a linear operator. But this comment, while provocative, still leaves one groping for a more physical answer to this question.

Finally, in section XI, we remark upon some experimental consequences of this theory. In particular, we consider two measurements which are performed in rapid succession. Experiments of this type are widely expected to give different results for quantum theory than for a dynamical reduction theory, if the second measurement takes place before the reduction due to the first measurement has been completed. Remarkably, it is shown that the predictions of (8.1) are identical to the predictions of quantum theory for the overwhelming majority of such experiments. It turns out that the constancy of $\langle x_n(t)\rangle$ is all that is needed to ensure this result. Possible experiments which can distinguish between quantum theory and this dynamical reduction theory are briefly mentioned.

## IX. NONLINEARITY ARISING FROM ANTILINEARITY

We have previously observed (10,15) that the linear and nonlinear Schrödinger equations (7.1a) and (7.1b) are almost identical when written in terms of "action-angle" variables $x_n, \theta_n$:

$$dx_n/dt = \pm\sum_m 2(x_n x_m)^{\frac{1}{2}}\alpha_{nm}\sin(\theta_n-\theta_m+\beta_{nm}) \tag{9.1a}$$

$$d\theta_n/dt = -\sum_m (x_m/x_n)^{\frac{1}{2}}\alpha_{nm}\cos(\theta_n-\theta_m+\beta_{nm}) \tag{9.1b}$$

(where $H_{nm}=\alpha_{nm}\exp i\beta_{nm}$). The positive sign in (9.1a) corresponds to the nonlinear Schrödinger equation (7.1b), the negative sign to the linear Schrödinger equation (7.1a).

We will present here another, perhaps related, expression of the relationship between the linear and nonlinear Schrödinger equations. Consider the equation

$$i\hbar d|\psi\rangle/dt = H|\psi\rangle \tag{9.2}$$

If H is required to be the most general *linear* operator, consistent with conservation of probability

$$d\langle\psi|\psi\rangle/dt = 0 \tag{9.3}$$

for arbitrary choice of statevector $|\psi\rangle$, it is easy to show that H must be a Hermitian operator. We seek the analog of this statement when H is an *antilinear* operator.

The most general antilinear operator can be decomposed (19) into the product of an antilinear complex conjugation operator K and a linear operator A:

$$H = KA \tag{9.4}$$

A complex conjugation operator is characterized by the "real" basis $|b_n\rangle$ it leaves invariant

$$K|b_n\rangle = |b_n\rangle \tag{9.5}$$

from which its action on any other vector can be determined

$$K\sum_n c_n|b_n\rangle = \sum_n c_n^*|b_n\rangle \tag{9.6}$$

The conservation of probability (9.3) imposes this constraint:

$$\langle\psi|K(A|\psi\rangle) = \langle\psi|K(A|\psi\rangle)^* \tag{9.7}$$

which can be written in terms of $\psi_n \equiv \langle b_n|\psi\rangle$ and $A_{mn} \equiv \langle b_m|A|b_n\rangle$ by inserting $1=\sum_n |b_n\rangle\langle b_n|$ on both sides of A and employing (9.6):

$$\sum_{mn} \psi_n^* \psi_m^* A_{mn}^* = \sum_{mn} \psi_n \psi_m A_{mn} \tag{9.8}$$

Eq. (9.8) cannot be satisfied for arbitrary K, A, any more than conservation of probability (9.3) can be satisfied for an arbitrary linear operator H. If K is not restricted at all, then the real and imaginary parts of the $\psi_n$'s can be chosen separately with the consequence $A_{(mn)}=0$, i.e. A is antisymmetric in the basis $|b_n\rangle$. Since antisymmetry in one basis does not imply antisymmetry in another basis in a complex vector space, we have achieved freedom in choosing K at the expense of a very restrictive constraint on A. We would like, in some sense, to maximize our freedom in choosing both K and A. In what follows, we will make A an arbitrary Hermitian matrix so that the freedom in choosing A is as great as in ordinary quantum theory.[9]

This means we must restrict K. Our constraint on K appears relatively modest: it is that its real basis can be arbitrarily chosen *up to phase factors*. That is, if you give me an arbitrary basis $|\phi_n\rangle$, I will have the freedom to choose the (in general time dependent) phase factors $\theta_n$ to define the real basis

$$|b_n\rangle = \exp i\theta_n |\phi_n\rangle \tag{9.9}$$

so that conservation of probability (9.8) may be satisfied.

If $A_{mn}$ is arbitrary Hermitian, it follows from (9.8) that

$$\psi_n^* \psi_m^* = \psi_n \psi_m \tag{9.10}$$

Then, according to Eq. (9.10), $\psi_n$ must be *real* (or imaginary, but this leads to the same final result (9.13) so we shall ignore it).

Reality of $\psi_n$ sounds like a severe restriction, but in fact it is easily obtained. This is because the expansion of $|\psi\rangle$ in the arbitrary basis $|\phi_n\rangle$ can always be written as

$$|\psi\rangle = \sum_n a_n |\phi_n\rangle = \sum_n |a_n| \exp i\theta_n |\phi_n\rangle \equiv \sum_n |a_n| |b_n\rangle \tag{9.11}$$

where the last step follows because we must identify the phase factors of the amplitudes $a_n$ with the phase factors in (9.9) (up to a phase factor $\exp i\pi$ which doesn't affect the final result (9.13)) in order for $\psi_n$ to be real. Incidentally, this choice of phase factors means $|\psi\rangle$ is real: $K|\psi\rangle=|\psi\rangle$.

To summarize, the Schrödinger equation (9.2) with the antilinear Hamiltonian (9.4) will conserve probability if A is an arbitrary Hermitian operator, and K's real basis is arbitrary up to (time dependent) phase factors which are chosen according to (9.11). It only remains to see what this Schrödinger equation looks like in the basis $|\phi_n\rangle$.[10] From (9.2),(9.4),(9.5) we have:

$$i\hbar d\langle\phi_n|\psi\rangle/dt = \langle\phi_n|K(A|\psi\rangle) = \sum_{m,r} \langle\phi_n|K(|b_r\rangle\langle b_r|A|b_m\rangle\langle b_m|\psi\rangle), \text{ or}$$

$$i\hbar d\langle\phi_n|\psi\rangle/dt = \sum_m \langle\phi_n|b_n\rangle\langle b_n|A|b_m\rangle^*\langle b_m|\psi\rangle^* \tag{9.12}$$

Using (9.9) and (9.11) to express this equation in terms of $a_n$ and $|\phi_n\rangle$ we obtain

$$\begin{aligned} i\hbar da_n/dt &= \sum_m \exp(i\theta_n)\exp-(i\theta_m)\langle\phi_m|A^\dagger|\phi_n\rangle\exp(i\theta_n)|a_m| \\ &= \sum_m \langle\phi_m|A|\phi_n\rangle a_m^* a_n/a_n^* \end{aligned} \tag{9.13}$$

Thus we arrive at the nonlinear reducing part of the Schrödinger equation.

There are some suggestive aspects to this result.

One aspect is the involvement of a complex conjugation operator. In quantum theory, such an operator is associated with time reversal invariance. Here it is associated with a highly time reversal nonivariant behavior, the statevector reduction.

Another aspect is that the complex conjugation operator requires for its definition a preferred basis (the "real basis" (9.5)). However, the theory of statevector reduction also needs a preferred basis, as we have seen. A connection between these preferred bases naturally appears in the argument given here.

## X. WHAT CAUSES RANDOM FLUCTUATIONS?

In this section we are going to make a conjecture concerning the source of the white noise coefficients in the reducing Schrödinger equation. Suppose we start with the Schrödinger equation *without* white noise

$$\begin{aligned} i\hbar da_n(t)/dt = &\sum_m \langle\phi_n,t|H_I|\phi_m,t\rangle a_m + \\ &+ \sum_m \langle\phi_m,t|A|\phi_n,t\rangle a_m^* a_n/a_n^* \end{aligned} \tag{10.1}$$

where the matrix elements $A_{mn}$ are not constant, but are constructed appropriately for the interaction picture, just like the matrix elements of $H_I$. We will argue that when $|\psi\rangle$ describes a complicated system, the time dependence of $A_{mn}$ nicely simulates white noise.

Let us expand the preferred basis in energy eigenstates:

$$|\phi_n,t\rangle = \sum_k c_{nk}\exp-(i\theta_{nk})|\phi_n,E_{nk},k\rangle\exp-(iE_{nk}t/\hbar) \tag{10.2}$$

(the constants $c_{nk}$, $\theta_{nk}$ are real). Consider an experiment whose microscopic interactions occur for $t<0$ ($H_{Inm}\neq 0$, $A_{nm}=0$). We expect the states $|\phi_n,0\rangle$ (which describe the apparatus as well as the microscopic system) to have random phases $\{\theta_{nk}\}$ but reasonably comparable $\{c_{nk}\}$ for the ensemble of repeated performances of this experiment. This is because the apparatus is complicated so its statevector is not under precise experimental control leading to random phases, yet the amplitudes $c_{nk}$ are determined both by the quantum evolution of the microscopic system and the Boltzmann distribution of the apparatus energy levels which are reproduceable.

We now hypothesize that the magnitudes of the matrix elements of A between energy eigenstates $|\phi_n, E_{nk}, k\rangle$ are reasonably independent of $E_{nk}$ and k for the overwhelming majority of states. The supposition is that when the matrix elements of A are large it is because they connect two states which are macroscopically different, and the microscopic differences should be of secondary importance.

Because of this, it follows from (10.2) that

$$\langle\phi_m, t|A|\phi_n, t\rangle \approx \tilde{A}_{mn}\beta_m^*(t)\beta_n(t) \quad (10.3a)$$

$$\tilde{A}_{mn} \equiv \langle\phi_m, E_{mk}, k|A|\phi_n, E_{nj}, j\rangle \exp -i[(\bar{E}_n - \bar{E}_m)t/\hbar] \quad (10.3b)$$

$$\beta_n(t) \equiv \sum_k c_{nk} \exp -(i\theta_{nk}) \exp -i[(E_{nk} - \bar{E}_n)t/\hbar] \quad (10.3c)$$

($\bar{E}_n$ is the average energy in the state $|\phi_n\rangle$, and is inserted for later convenience.)

Our claim is that $\dot{b}_{mn}(t) \equiv \beta_m^*(t)\beta_n(t)$ has all the necessary properties of the white noise $\dot{B}_{mn}(t)$ in Eq. (7.10).

1) It is Hermitian, $\dot{b}_{mn}^* = \dot{b}_{nm}$.

2) Its mean vanishes

$$\langle\dot{b}_{mn}\rangle = \sum_{jk} c_{mj}c_{nk} \exp -i[(E_{nk} - E_{mj} - \bar{E}_n + \bar{E}_m)t/\hbar] * \\ * \langle \exp i(\theta_{mj} - \theta_{nk})\rangle = 0, \quad m \neq n \quad (10.4)$$

since the average is taken over all possible phase angles, and these are randomly distributed.[11]

We emphasize, according to the point of view of this section, that *it is lack of experimental ability to fix the phase angles* $\theta_{nk}$ *which is responsible for the indeterminism in quantum theory.* If these phase angles (and the amplitudes $c_{nk}$) could be precisely known, the results of the associated experiment would be in principle predictable.

3) Its second moments behave correctly, that is, as expressed by Eq. (7.5b). We compute using (10.3c)

$$\langle\dot{b}_{m'n'}^*(t')\dot{b}_{mn}(t)\rangle = \langle\beta_{m'}(t')\beta_m^*(t)\rangle\langle\beta_{n'}^*(t')\beta_n(t)\rangle \\ = \delta_{m'm}\delta_{n'n} f_m^*(t-t') f_n(t-t') \quad (10.5)$$

$$f_n(t-t') \equiv \langle\beta_n^*(t')\beta_n(t)\rangle = \int dE \rho_E c_{n,E}^2 \exp -i[(E-\bar{E})(t-t')/\hbar] \quad (10.6)$$

(where we have replaced the sum by an integral using the density of states $\rho_E$, and $c_{n,E}^2 \equiv \sum_k c_{nk}^2 \delta_{E_k,E}$).

It is expected that $\rho_E c_{n,E}^2$ will have a sharp peak in the neighborhood of $\bar{E}_n$. This will certainly be the case if the apparatus is in approximate thermal equilibrium, where $c_{n,E}^2$ will be given by the Boltzmann factor $\sim \exp -E/kT$. Moreover, because $|\phi_n, t\rangle$

has unit norm, it follows from (10.2) that

$$f_n(0) = \int dE\rho_E c_{n,E}^2 = 1 \tag{10.7}$$

Thus it is a good approximation to write

$$\rho_E c_{n,E}^2 \approx [(2\pi)^{\frac{1}{2}}\Delta E]^{-1}\exp-[(E-\bar{E}_n)^2/2(\Delta E)^2] \tag{10.8}$$

where $\Delta E$ characterizes the energy dispersion of the state $|\phi_n,t\rangle$ and should be essentially independent of n. Thus

$$f_n(t-t') \approx \int_{-\infty}^{\infty}dE[(2\pi)^{\frac{1}{2}}\Delta E]^{-1}\exp-[(E-\bar{E}_n)^2/2(\Delta E)^2] \; * $$
$$* \exp-i[(E-\bar{E}_n)(t-t')/\hbar] = \exp-[(t-t')^2/2(\hbar/\Delta E)^2] \tag{10.9}$$

Finally, we insert the result (10.9) into (10.5), obtaining

$$\langle\dot{b}^*_{m'n'}(t')\dot{b}_{mn}(t)\rangle = \delta_{m'n}\delta_{n'n}(\pi^{\frac{1}{2}}\hbar/\Delta E) \; *$$
$$* \{[(2\pi)^{\frac{1}{2}}(\hbar/2^{\frac{1}{2}}\Delta E)]^{-1}\exp-[(t-t')^2/2(\hbar/2^{\frac{1}{2}}\Delta E)^2]\} \tag{10.10}$$

This result is to be compared with the expression (7.5b) for the second moments of white noise.

The bracketed Gaussian in (10.10) plays the role of $\delta(t'-t)$ in Eq. (7.5b). Now, white noise is always an approximation in physical theories because it is a function that has all frequencies equally well represented, while actual fluctuating physical quantities always have a frequency cutoff. The use of white noise is justified if the time scale for the fluctuations of this physical quantity is much shorter than the time scale for other physical processes in the theory. Therefore, in the present theory, the use of the white noise approximation is justified if

$$\hbar/2^{\frac{1}{2}}\Delta E < \tau_R \tag{10.11}$$

where $\tau_R$ characterizes the reduction time.[12]

Eq. (10.11) gives us a lower bound on the reduction time. We can estimate the size of this lower bound by considering the apparatus to be in thermal equilibrium at temperature T. It is well known (20) that the deviation in energy for such a system is

$$\Delta E = (kT^2C_V)^{\frac{1}{2}} \approx kT(N_P)^{\frac{1}{2}} \tag{10.12}$$

where $C_V$ is the specific heat of the apparatus, and we have employed the estimate $C_V \sim kN_P$ where $N_P$ is the number of particles.[13] Thus, according to Eq. (10.11), at room temperature

$$\hbar/kT(2N_P)^{\frac{1}{2}} \approx 10^{-14}\text{sec.}/N_P^{\frac{1}{2}} < \tau_R \tag{10.13}$$

which is not an unreasonable figure.

In fact, Bohm and Bub (21) suggested the time h/kT as the *relaxation* time for the hidden variables in their theory of the dynamical reduction of the statevector (the *reduction* itself was to take place much more rapidly). Their theory was the first to give dynamical equations for the reduction. (As far as the author is aware, this is the second attempt.) Bohm has repeatedly insisted upon the importance of the apparatus' complexity and temperature in the reduction process (21,22).

There is another side to the coin of this mechanism for obtaining random fluctuations that govern statevector reduction. Consider a system that is described by a superposition of states that are macroscopically different but nonetheless each is close to being an energy eigenstate. In this case one should expect deviations from the predictions of quantum theory. On the one hand, the nonlinear term in Eq. (10.1) should not vanish because the matrix elements of A are indeed taken between macroscopically different states, and so are large. On the other hand, the matrix elements of A will not fluctuate quasirandomly and simulate white noise behavior because the states $|\phi_n,t\rangle$ will be close to energy eigenstates. One therefore has a nonlinear Schrödinger equation where the nonlinear term does not correctly reduce the statevector. In this case, it is possible to devise experiments to test this theory.

Leggett (23) has discussed the possibility of observing a system in a superposition of macroscopically distinguishable states. In particular, he has focussed on "flux tunnelling" in a superconducting ring containing a Josephson junction, the basic element of a so-called SQUID. In coherent flux tunnelling, the statevector is a superposition of two states corresponding to supercurrents generating different amounts of flux in the ring, and the amplitudes of the states oscillate back and forth periodically. We shall merely state here, without giving the details, that if the nonlinear term is regarded as a small perturbation on this process, the tunnelling frequency will be modified. However, testing this theory by this means will not be easy because, while flux tunnelling may have been observed (24), it is probably incoherent tunnelling according to Leggett. Even if coherent flux tunnelling is observed, it is not easy to find an accurate quantum prediction of the tunnelling frequency.

A second experiment involves a search for deviations from the quantum theory prediction in a two-slit neutron interference pattern. The idea is that the two states corresponding to the neutron passing through the two slits may be different enough so that the nonlinear term may be small but measurable. (The spatial separation of the wave packets is "macroscopic," $\approx 10^{-2}$ cm., the neutron mass is as heavy as $\approx$2000 electrons, so one might regard these two states as borderline macroscopically distinguishable.) Then the nonlinear term will slightly alter the interference pattern, and indeed, the existence or nonexistence of random fluctuation behavior in this term are in principle distinguishable. A two-slit neutron interference pattern has in fact been observed by Gaehler, Shull,

Treimer and Zeilinger (25). The relevance of this experiment to this dynamical reduction theory has been discussed by Pearle and Zeilinger (26).

## XI. RAPIDLY REPEATED EXPERIMENTS

An often expressed expectation is that a theory of the dynamical reduction of the statevector is sure to produce predictions different from those of quantum theory when applied to two experiments that follow one another very rapidly (27). The idea is that if the statevector does not have enough time to completely reduce by the time the second experiment is performed, the statevector will be "caught out," that is, more than one amplitude in the superposition will be nonzero. Since a dynamical reduction theory is designed to agree with quantum theory when it describes an experiment that begins and ends with a reduced statevector, it might be expected that the second experiment, which does not begin with a reduced statevector, will produce disagreement between the two theories.

Pearle and Zeilinger (26) have shown that, whereas the above expectation is correct in the case of the Bohm-Bub (21) dynamical reduction theory, remarkably enough it is not the case for the present theory. In particular, the experiment of Papaliolios (28), which is a paradigm of this kind of experiment (the polarization of a photon was measured twice in rapid succession), places a severe restriction on the parameters in the Bohm-Bub theory but is no test of the theory described here. The reason is simply that the $\langle x_n(t)\rangle$ are constant during the reduction process in this theory, and are not constant in the Bohm-Bub theory. Because of the importance of this result we shall repeat the essential argument here, but refer the reader to reference (26) for a detailed discussion of its relevance to the Papaliolios experiment and the Bohm-Bub theory.

Consider two experiments taking place at time 0 and time T (we are assuming that the duration of each experiment is negligible). Immediately after the first experiment, the statevector is given by

$$|\psi,0\rangle = \sum_n a_n(0)\,|\phi_n,0\rangle \qquad (11.1)$$

According to quantum theory, until the second experiment, the statevector is given by

$$|\psi,t\rangle = \sum_n a_n(t)\,|\phi_n,t\rangle, \qquad 0\le t\le T \qquad (11.2)$$

$$|a_n(t)|^2 = |a_n(0)|^2, \qquad 0\le t\le T \qquad (11.3)$$

where (11.3) states that the probabilities of the outcomes do not change during this time interval. The states $|\phi_n,t\rangle$ are not constant (we are dealing with the interaction picture), expressing the continued time evolution of the microscopic system-plus-apparatus states.

When the second experiment is performed at time T, the state $|\phi_n, T>$ undergoes an evolution:

$$|\phi_n, T> \to \sum_m a_{nm} |\phi_{nm}> \tag{11.4}$$

Therefore, immediately after the second experiment, according to quantum theory the statevector is

$$|\psi, T^+> = \sum_{n,m} a_n(T) a_{nm} |\phi_{nm}> \tag{11.5}$$

We shall assume that the statevectors $\{|\phi_{nm}>\}$ are all orthogonal. This is the usual case as when, for example, the results of the first experiment are recorded. Then the probability of the sequential result n,m for the two experiments (which does not change with time) is, according to quantum theory,

$$P_{n,m} = |a_n(T) a_{nm}|^2 = |a_n(0)|^2 |a_{nm}|^2 \tag{11.6}$$

where we have used (11.3).

Now consider the description of this same sequence of experiments according to a dynamical reduction theory. Eq. (11.3) is not obeyed because the amplitudes undergo reduction dynamics during the time interval $0 \le t \le T$. However, for a theory of the type described here, we can at least assert that

$$<|a_n(t)|^2> = |a_n(0)|^2, \quad 0 \le t \le T \tag{11.3r}$$

Immediately after the second experiment, the statevector is likewise given by Eq. (11.5). Thereafter the amplitudes eventually completely reduce so that one amplitude achieves the value 1, and the rest vanish. The fraction of statevectors ending up as

$$|\psi, t> = 1 \cdot |\phi_{nm}, t>, \tag{11.7}$$

which also have the particular numerical amplitude $a_n(T) a_{nm}$ multiplying the state $|\phi_{nm}>$ at time $T^+$, is

$$|a_n(T) a_{nm}|^2, \tag{11.8}$$

according to reduction dynamics.

To find the total fraction of statevectors ending up in the state (11.7), we must multiply (11.8) by the fraction of statevectors possessing this particular amplitude, and sum over all possible amplitudes. But this is

$$P^r_{nm} = <|a_n(T) a_{nm}|^2> = <|a_n(T)|^2> |a_{nm}|^2$$
$$= |a_n(0)|^2 |a_{nm}|^2 \tag{11.6r}$$

where the last step follows from Eq. (11.3r).

Thus the quantum theory prediction (11.6) and the dynamical reduction prediction (11.6r) are identical. We note the importance of (11.3r) in obtaining the result (11.6r). Any dynamical reduction theory in which (11.3r) is not obeyed, such as the Bohm-Bub theory, will not in general produce predictions in agreement with those of quantum theory for such a sequence of experiments.

If the states $|\phi_{nm}>$ are not all orthogonal, i.e. if the second experiment measures interference between the states produced by the first experiment, then the argument following Eq. (11.5) is no longer valid, and deviations from the predictions of quantum theory are expected. For example, this is the case in a two-slit neutron interference experiment, where the passage of the neutron through the slit barrier constitutes the first experiment (although no number is recorded, insofar as it causes the reduction process to begin, it can be regarded as an experiment), and the neutron detection constitutes the second experiment. Present experimental results (25) are in agreement with quantum theory's predictions, and place a lower limit on the reduction time of the neutron states during such an experiment (26).

In conclusion, although the dynamical reduction theory presented here is perhaps more "robust" than the Bohm-Bub theory (harder to experimentally refute), it is not impossible to experimentally distinguish such a theory from quantum theory. Such experiments should be pursued, for it is important to the supporters of quantum theory that it withstand tests from all sides. And, if finally an experiment should appear whose results conflict with the predictions of quantum theory, that would indeed be exciting.

ACKNOWLEDGMENTS

I would like to thank Professor R. Penrose and the Mathematical Physics group at Oxford for their encouragement and hospitality. I also wish to acknowledge a stimulating conversation with Professors D. Bohm and B. Hiley at Birkbeck College.

NOTES

1. In fairness, it should be said that each of the other interpretations of quantum theory is also liable to a similar criticism. In each there is something that quantum theory does not describe that a reasonable person might have expected *would* be described by a physical theory (see the last column in Table 1). If one agrees with that "reasonable person," this could be taken as evidence that quantum theory is incomplete, and should be altered.
2. In no other case does the superposition "alive plus dead" describe a single cat. Thus, for Bohr's interpretation, it

describes all subsequent measurements that could be performed on the cat; for Einstein's interpretation, it describes an ensemble of cats; for Born's interpretation, it describes one's knowledge of the cat; for Everett's interpretation, it describes the cats in different universes.

3. In fairness, it should be said that there is nothing logically inconsistent or experimentally wrong with the naive description of the cat experiment. It just conflicts with a "reasonable person's" intuition.

4. Even if the series does not converge (as in the case of decay: see Misra and Sudarshan (8)), this remark is correct although the right side of Eq. (1.1) must be modified to $1-c^2\varepsilon^k+..$, with $k>1$.

5. This was called "Zeno's paradox in quantum theory" by Misra and Sudarshan (8) and called "watched pot behavior" by Pearle (7) and Peres (8), after the housewife's adage "a watched pot never boils."

6. Although there has been some work done on preferred bases (9,10,11,12), we here take the attitude that the precise prescription for the preferred basis (and its time dependence, i.e. the choice of $H_0$) is still unknown, but is an important subject for further elaboration of the theory.

7. Again, the precise definition of "macroscopically distinguishable," as well as the physical mechanism behind the growth of the nonlinear term, are not given in this paper, but are obviously important topics for future elucidation.

8. Such a nonlocal interaction between macroscopically different states amounts to the statement that Nature obeys a nonlocal hidden variable theory. (The hidden variables are the fluctuating coefficients which determine the outcomes of experiments.) This can be viewed as a satisfactory explanation of why Bell's inequality disagrees with experiment, i.e. Nature doesn't obey a local hidden variable theory because it obeys a nonlocal one.

9. The reader may convince himself that any greater freedom in choosing A results in extremely restrictive constraints upon K. Incidentally, if $H=KA$, then since $H=KAKK\equiv A_kK$, the decomposition of H with K on the right no longer has A arbitrary, independent of the choice of K. That is, AK and KA with A arbitrary Hermitian are two different postulates. It is straightforward to show, with the choice AK, that conservation of probability also implies $K|\psi\rangle=|\psi\rangle$. Thus the Schrödinger equation becomes $i\hbar d|\psi\rangle/dt=AK|\psi\rangle= A|\psi\rangle$, i.e. it has its usual linear form.

10. If $|\phi_n\rangle$ depends upon time according to (2.2), we must consider the full Schrödinger equation $i\hbar d|\psi\rangle/dt=(H_0+H_I)|\psi\rangle+KA|\psi\rangle$ and take the scalar product of this equation with $\langle\phi_n|$. The result is the interaction picture equation (8.1) (with $\hat{B}_{nm}$ replaced by 1) instead of Eq. (9.13).

11. We can assume without loss of generality that the diagonal matrix elements of A vanish. It can be seen from Eq. (10.1) that the diagonal part of the nonlinear interaction is in fact linear (since $a_m^*/a_n^*=1$ if $m=n$), so if it does not vanish, it may be

included in the linear part of the Schrödinger equation. However, we expect the diagonal matrix elements of A to vanish anyway, because A is supposed to have large matrix elements only between states which are macroscopically different, and a state cannot be macroscopically different from itself! Thus we need not concern ourselves with $\dot{b}_{mm}$.

12. According to Eq. (7.3), $\tau_R$ is approximately $\tau_R \approx (\hbar/\sigma|A|)^2$ ($|A|$ is the magnitude of a typical matrix element $\tilde{A}_{mn}$). Also, by comparing (10.10) with (7.5b), we see that $2\sigma^2$ is to be identified as $2\sigma^2 = \pi^{\frac{1}{2}}\hbar/\Delta E$. Now, in section VII, white noise $\dot{B}$ and $\sigma^2$ both have dimension time$^{-1}$, whereas here in section X, $\dot{b}$ is dimensionless and $\sigma^2$ has dimension time$^{+1}$. It is possible to change the dimension of $\dot{b}$ by multiplying $\tilde{A}_{mn}$ and $\beta_n$ respectively by normalization factors $\eta^{-1}$ and $\eta^{\frac{1}{2}}$ without of course affecting $\tilde{A}_{mn}\beta_m^*\beta_n$ which appears in Eqs. (10.3). These factors can be carried through the equations in section X, resulting in a factor $\eta^2$ on the right side of Eq. (10.10), and thus $2\sigma^2 = \eta^2\pi^{\frac{1}{2}}\hbar/\Delta E$. One could then choose $\eta \sim \Delta E/\hbar$, for example, to give $\sigma^2 \sim \Delta E/\hbar$ the dimension time$^{-1}$.

13. Actually, $N_p$ should be regarded, not as the total number of particles in the apparatus, but rather as the number of particles in the apparatus which behave differently in the different superposed states (a number which, of course, grows during a measurement). To see this, suppose we write $|\phi_n,t\rangle = |\Phi,t\rangle|\phi_n,t\rangle'$, where $|\Phi,t\rangle$ represents the apparatus-environment that is unaffected by the measurement, and $|\phi_n,t\rangle'$ represents the affected part. Then, if we reserve the definition $\beta_n(t)$ in (10.3c) and the labels $E_{nk}$, k for the states of the affected part, there should be an extra common factor in the expression for each $\langle\phi_m,t|A|\phi_n,t\rangle$, which should appear on the right side of Eq. (10.3a), namely $\sum_{\alpha\beta}\sum\langle\Phi,t|\varepsilon_\alpha,\alpha\rangle\langle\varepsilon_\beta,\beta|\Phi,t\rangle$ ($\varepsilon_\alpha,\alpha$ are the energy and degeneracy labels for the unaffected part of the universe). However, because of the extraordinary complexity of these states, only that part of the double sum for which the indices $\alpha$ and $\beta$ are equal should make the dominant contribution, the rest of the terms averaging out to near zero at any instant of time. Since $\sum_\alpha|\langle\Phi,t|\varepsilon_\alpha,\alpha\rangle|^2=1$, these states have no effect on the reduction dynamics.

REFERENCES

1. Bohr, N.: 1966, *Essays 1958-1962 on Atomic Physics and Human Knowledge* (Random House, New York); Schilpp, P.A., ed.: 1959, *Albert Einstein, Philosopher Scientist* (Harper and Row, New York). There are a number of versions of this so-called "Copenhagen Interpretation": see e.g. Heisenberg, W.: 1958, *Physics and Philosophy, The Revolution in Modern Science* (World Perspectives, New York); Born, M.: 1964, *The Natural Philosophy of Cause and Chance* (*New York*). A historical and critical discussion appears in Jammer, M.: 1974, *The Philosophy of Quantum Mechanics* (Wiley, New York).

2. Einstein, A.:1959 in Schilpp, P.A., ed. Ibid; Pearle, P.: 1967, Am. J. Phys. 35, p. 742; Ballentine, L.E.: 1970, Revs. Mod. Phys. 42, p. 358; Belinfante, F.: 1975, *Measurements and Time Reversal in Objective Quantum Theory* (Pergamon, Oxford).
3. This was the original idea of Born, though he later disavowed it: see Jammer, M., Ibid, and Einstein's letter to Schrödinger in 1939 in Klein, M.: 1968, *Letters on Wave Mechanics* (Vision, London). A modern version of this position appears in Hartle, J.B.: 1968, Am. J. Phys. 36, p. 704.
4. Everett, H.: 1957, Revs. Mod. Phys. 29, p. 454; Wheeler, J.A.: 1957, Revs. Mod. Phys. 29, p. 463; Dewitt, B.S., and Graham, R.N.: 1973, *The Many-Worlds Interpretation of Quantum Mechanics* (Princeton University Press, Princeton, New Jersey); Deutsch, D.: 1981, "Quantum Theory as a Universal Physical Theory," preprint.
5. Wigner, E.: 1970, *Symmetries and Reflections* (M.I.T. Press, Cambridge, Massachusetts) argues that consciousness may be responsible for the reduction of the statevector. If that is so, then perhaps something which is *not* physical also belongs in the quantum description!
6. Schrödinger, E.: 1935, Naturwissenschaften 23, p. 807; Dewitt, B.S.: 1970, Physics Today 23, #9, p.30; 1971, Ibid 24, #4, p. 36.
7. Pearle, P.: 1981, "Reduction of the Statevector, Continuous Measurements, and the Rules of Quantum Theory," preprint.
8. Bloch, I. and Burba, D.A.: 1974, Phys. Rev. D 10, p. 3206; Misra, B. and Sudarshan, E.C.G.: 1977, Journ. Math. Phys. 18, p. 756; Chiu, C.B., Sudarshan, E.C.G. and Misra, B.: 1977, Phys. Rev. D 16, p. 520; Allcock, G.R.: 1969, Ann. Phys. 53, p. 253; Ibid, p. 286; Ibid, p. 311; Aharanov, Y. and Vardi, M.: 1980, Phys. Rev. D 21, p. 2235; Simonius, M.: 1978, Phys. Rev. Lett. 40, p. 980; Peres, A.: 1980, Am. Journ. Phys. 48, p. 931.
9. Kubler, O. and Zeh, H.: 1973, Ann. Phys. 76, p. 405.
10. Pearle, P.: 1979, Int'l Journ. of Theor. Phys. 18, p. 489.
11. Deutsch, D., (4).
12. Zurek, W.H.: 1981, Phys. Rev. D 24, p. 1516.
13. Feller, W.: 1950, *An Introduction to Probability Theory and its Applications* (Wiley, New York), Chapter 14.
14. Pearle, P.: 1982, Found. of Phys. 12, p. 249.
15. Pearle, P.: 1976, Phys. Rev. D 13, p. 857.
16. Chandrasekhar, S.: 1943, Rev. Mod. Phys. 15, p. 1 [also in Wax, N., ed.: 1954, *Selected Papers on Noise and Stochastic Processes* (Dover, New York).
17. Wong, E.: 1971, *Stochastic Processes in Information and Dynamical Systems* (McGraw Hill, New York).
18. Penrose, R. in Isham, C.J., Penrose, R. and Sciama, D.W., eds.: 1981, *Quantum Gravity 2, A Second Oxford Symposium* (Clarendon, Oxford).
19. Takesaki, M.: 1970, *Tomita's Theory of Modular Hilbert Algebras and its Applications* (Springer-Verlag, Berlin).
20. Huang, K.: 1963, *Statistical Mechanics* (Wiley, New York).
21. Bohm, D. and Bub, J.: 1966, Rev. Mod. Phys. 38, p. 453.

22. Bohm, D., private communication.
23. Leggett, A.J.: 1980, Suppl. Prog. Theor. Phys. 69, p. 80.
24. den Boer, W. and de Bruyn Ouboter, R.: 1980, Physica 98 B+C, p. 185; Prance, R.J. et al.: 1981, Nature 289, p. 543.
25. Zeilinger, A., Gaehler, R., Shull, C.G. and Treimer, W.: 1981, "Experimental Status and Recent Results of Neutron Interference Optics," in the proceedings of the Symposium on Neutron Scattering, to be published.
26. Pearle, P. and Zeilinger, A.: 1982, "Experimental Tests of Dynamical Statevector Reduction," preprint.
27. Pipkin, F.M.: 1979, *Advances in Atomic and Molecular Physics* 14 (Academic Press, New York), p. 281.
28. Papaliolios, C.: 1967), Phys. Rev. Lett. 18, p. 622; Belinfante, F.: 1975, *A Survey of Hidden Variable Theories* (Pergamon, Oxford).

CPT REVISTED : A MANIFESTLY COVARIANT PRESENTATION

O. COSTA de BEAUREGARD

Institut Henri Poincaré, 11 rue Pierre et Marie Curie,
75231 Paris Cedex 05, France

Z , or $\psi \rightleftharpoons \bar{\psi}$, denoting exchange of relativistically adjoint wave equations, particle-antiparticle exchange $\mathcal{C}$ and Wigner motion reversal $\mathcal{T}$ are related to rest mass reversal M and "Racah time reversal" T as $\mathcal{C}$ = MZ and $\mathcal{T}$ = TZ. $\mathcal{P}$ = P denoting "Racah space reversal", $\mathcal{P}\mathcal{T}$ = PTZ is "covariant motion reversal". This synthetizes the requirements of prequantal relativistic electrodynamics and second quantization. MPT = is a "manifestly covariant" expression of the $\mathcal{C}\mathcal{P}\mathcal{T}$= 1 theorem.

A different vista unfolds in the 2-component schemes of screwing fermions of Weyl and Feynman-Gell-Mann, where rest mass reversal is non significant. There $\mathcal{C}$ = $\mathcal{P}$ = P = T = Z whence, of course, $\mathcal{C}\mathcal{P}$ = $\mathcal{T}$ = 1.

A final section discusses the manifest MPT invariance of the Jordan-Pauli and the Feynman propagators, showing that (contrary to the belief of some) no arrowed causality follows from use of the Feynman propagator.

## I. INTRODUCTION

The well-known $\mathcal{C}\mathcal{P}\mathcal{T}$= 1 theorem (1), where $\mathcal{C}$ denotes charge conjugation or, more generally, particle-antiparticle exchange, $\mathcal{P}$ the so called space, or parity reversal, and $\mathcal{T}$ the Wigner (2) motion reversal does have (orthochronous) Lorentz invariance, but does not display it "manifestly". This is mainly because the $\mathcal{P}$ and $\mathcal{T}$ operations have dissimilar definitions. The fundamental aim of this paper is to clarify this situation[1].

By direct inspection of a pair of relativistically adjoint wave equations (e.g. Dirac's, or Petiau-Duffin-Kemmer's) we will show that particle-antiparticle exchange $\mathcal{C}$ is none else than the



*S. Diner et al. (eds.), The Wave-Particle Dualism, 485–497.*

product of rest mass reversal $M^2$ by exchange Z of the two adjoint equations (symbolically, $\psi \rightleftarrows \bar{\psi}$ ) - a manifestly covariant operation.

It is well known that, in the Dirac case, rest mass reversal is obtained via multiplication from the left by $\gamma^5 = \gamma^1\gamma^2\gamma^3\gamma^4$ : $\psi' = \gamma^5\psi$ .

Similarly, we will show that the Wigner (2) motion reversal $\mathscr{T}$ is the product of the so-called Racah (7) time reversal T (the definition of which is similar to that of the "space reversal" P) by Z. Therefore, setting

$$\mathscr{C} = MZ, \qquad \mathscr{P} = P, \qquad \mathscr{T} = TZ, \qquad (1)$$

we rewrite the $\mathscr{CPT}$ theorem as

$$\mathscr{CPT} = MPT = 1, \qquad (2)$$

the latter form of which is "manifestly covariant".

Now, contrary to a widespread opinion, we will show that the operations P, T and PT, respectively defined (in the Dirac case) via multiplication from the left by $\gamma^4$, $i\gamma^1\gamma^2\gamma^3$ , and $\gamma^5$ (respectively) are definitely not geometrical reversal of the axes (1,2,3), 4, and (1,2,3,4). This is because these operations reverse part or all of the scalar product $\gamma^i(i\partial_i - eA_i)$ which is (loosely speaking) that of the 4-velocity and the momentum-energy operators.

These operations have to do with motion reversal, and we show that

$$\mathscr{T} = P\dot{T}Z$$

is none else that "covariant motion reversal".

So "particle antiparticle exchange" $\mathscr{C}$ = MZ and "covariant motion reversal" $\mathscr{PT}$ = PTZ are two physically different aspects of the same mathematical operation, as displayed in Fig.1, presenting the "interaction picture" of the S-matrix scheme.

This same figure shows that the identity operation in formulas (2) is none else that "geometrical reversal of all four spacetime axes".

Similar considerations hold in the Petiau-Duffin-Kemmer case, and for other spinning wave equations.

This presentation displays a synthesis of the requirements of the prequantal relativistic theory and of the second quantization scheme, where, respectively, particle-antiparticle exchange, as equivalent to covariant motion reversal, is represented by rest mass reversal and by exchange of the emission and absorption operators.

New vistas are opened in the light of the two component screwing fermion formalisms of Weyl, for the non-massive or neutrino case, and of Feynman and Gell-Mann, for the massive case, which are operational in weak interactions ("universal V-A coupling").

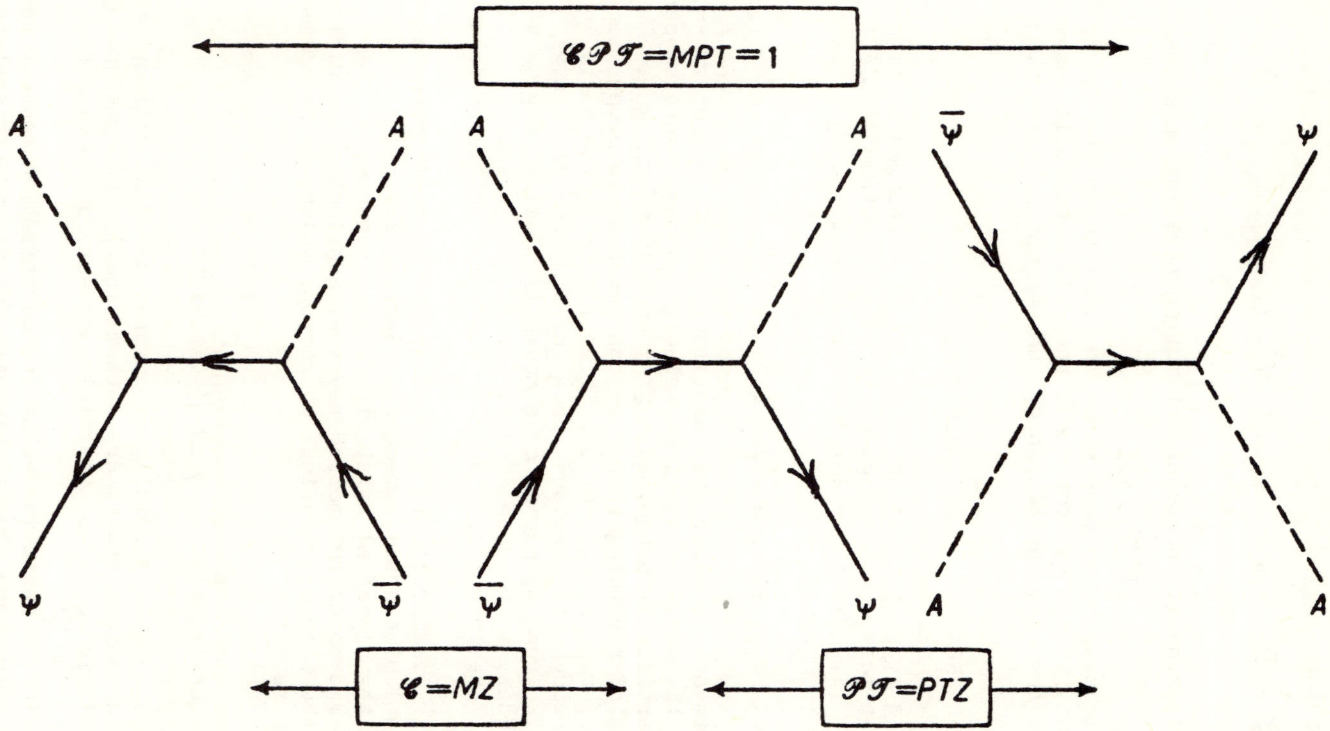


Fig.1. particle-antiparticle exchange $\mathcal{C}$ = MZ and covariant motion reversal $\mathcal{P}\mathcal{T}$=PTZ as mathematically equivalent ; $\mathcal{C}\mathcal{P}\mathcal{T}$ = MPT = 1 is geometrical reversal of all four spacetime axes. The example is annihilation or creation of an $e^+ e^-$ pair into, or from, a 2 $\gamma$ pair.

There, the rest mass reversal concept looses its significance, but the formulas

$$\mathcal{C} = \mathcal{P} = P = T = Z, \qquad \mathcal{C} = TZ = 1,$$

hold, leading naturally to the modified definitions of particle-antiparticle exchange and covariant motion reversal as

$$\mathcal{C}\mathcal{P} = \mathcal{C} = 1.$$

A final Section has been added in order to refute the erroneous feeling, shared by some, that there is a causality asymmetry built in the very definition of the Feynman propagator.

## II. CLASSICAL PREAMBLE : REST MASS REVERSAL

The relativistic equation of motion of a charged point particle of rest mass m and charge e inside an electromagnetic field $H^{ij}$ (i,j,k,l = 1,2,3,4 : c = 1 ; $\hbar$ = 1 ; $x^4$ = it ; $u_i u^i = -1$, $\dot{u}^i = du^i/d\tau$ reads

$$m\dot{u}^i = e\, H^{ij}u_j \,. \tag{3}$$

It is invariant under the substitution e $\rightleftarrows$ -e and m $\rightleftarrows$ -m, and this is the basis of the Stueckelberg-Feynman theory of particle-antiparticle exchange, as m $\rightleftarrows$ -m entails $mu^i \rightleftarrows -mu^i$.

Instead of reversing e or m we can reverse the proper time $\tau$, which reverses $u^i$ but not $\dot{u}^i$. This will bring the particle backwards through its previous space <u>and</u> time coordinates $x^i$, very much like running backwards a movie film will bring back the pictured scenes, in reversed order, through their space and time coordinates, the latter displayed on the clocks pictured in the film.

An other possibility is to reverse the sign of the external field $H^{ij}$, which we shall remember in Section IV.

So, in the classical relativistic electrodynamics, charge conjugation and covariant motion reversal are indeed two different aspects of the same mathematical operation, which can be formalized as rest mass reversal, M.

## III. SECOND QUANTIZATION : $\psi \rightleftarrows \bar{\psi}$ EXCHANGE

AXIOM : Emission of a particle and absorption of an antiparticle (in a given dynamical state) are mathematically the same operation, $\bar{\psi}$ . Emission of an antiparticle and absorption of a particle are denoted $\psi$ .

This axiom is basic in the second quantization scheme, and includes the Dirac and the Stueckelberg-Feynman interpretations of negative energy particles as antiparticles.

THEOREM : Exchange of particle and antiparticle (at emission or absorption) and exchange of emission and absorption of a particle or antiparticle (in a given dynamical state) are mathematically equivalent operations.

COROLLARY : Particle antiparticle exchange and covariant motion reversal are mathematically equivalent.

This is pictured in Fig.1, and is the essence of the $\mathcal{CPT}$ = MPT = 1 theorem, an expression of which is

$$\bar{A} + B + \ldots \rightleftarrows C + \bar{D} + \ldots \tag{4}$$

where (beware !) <u>no upper bar means antiparticle on the l.h.s. and particle on the r.h.s., and an upper bar means particle on the l.h.s, and antiparticle on the r.h.s .</u>

Again, $\mathcal{CPT}$= MPT = 1 is geometrical reversal of all four spacetime axes.

## IV. THE DIRAC EQUATION : MZ = PTZ REVERSAL

THEOREM 1 : "Particle-antiparticle exchange", $\mathcal{C}$ , is the product of rest mass reversal, M, by the $\psi \rightleftarrows \bar{\psi}$ exchange, Z.

Considering the Dirac equation (in units such that c = 1 and $\hbar$ = 1)

$$\{\gamma^i(i\partial_i - eA_i) + m\}\psi = 0 \tag{5}$$

and its relativistic adjoint, written in transposed form

$$\{\tilde{\gamma}^i(-i\partial_i - eA_i) + m\}\tilde{\bar{\psi}} = 0 \tag{6}$$

it is obvious that the $\mathcal{C}$ operation of charge conjugation $e \rightleftarrows -e$ is equivalent to the product of the M operation $m \rightleftarrows -m$ by the Z operation exchanging the two adjoint equations or, symbolically, $\psi \rightleftarrows \bar{\psi}$.

Rigorously speaking the unitary transformation

$$C\tilde{\gamma}^i C^{-1} = \gamma^i \qquad C\tilde{\bar{\psi}} = \psi' \tag{7}$$

would also be required. But, as for all sixteen $\gamma$'s

$$C\tilde{\gamma}C^{-1} = \gamma \tag{8}$$

this is merely a change of representation devoid of physical content[3].

Finally particle-antiparticle exchange is

$$\mathcal{C} = MZ \tag{9}$$

THEOREM 2 : "Covariant motion reversal", $\mathcal{P}\mathcal{C}$, is the product of the so-called "Racah spacetime reversal" PT by the $\psi \rightleftarrows \bar{\psi}$ exchange, Z.

Respectively, the Racah (7) space or parity reversal P, time reversal T, and spacetime reversal PT are defined via multiplication of (5) from the left by $\gamma^4$, $i\gamma^{123}$, and $\gamma^5 = \gamma^{1234}$, and setting correspondingly

$$\psi' = \gamma^4 \psi, \qquad \psi' = i\gamma^{123}\psi, \qquad \psi' = \gamma^5 \psi; \tag{10}$$

we have set by the definition $\gamma^{ij\ldots} = \gamma^i \gamma^j \ldots$ if all indexes are different, 0 otherwise.

So, the P, T, and PT operations respectively reverse the signs of $\vec{\gamma}(i\vec{\partial} - e\vec{A})$, $\gamma^4(i\partial_4 - eA_4)$, and $\gamma^i(i\partial_i - eA_i)$, the former expression being the scalar product of the velocity and momentum operators, and the latter of (loosely speaking) the 4-velocity and 4-momentum operators. Of course, no component of the scalar product of two vectors should change sign under reversal of the corresponding axis. So, strictly speaking, P, T, and PT are not space, time, and space-time reversal. But, remembering the contents of Section II, it is obvious that they have to do with various aspects of motion reversal.

Before going to this we confirm the previous remarks by observing that, under P, T or PT, part, or all, of the components of the Dirac current $i\bar{\psi}\gamma^i\psi$ and the Gordon current $i\bar{\psi}[\partial_i]\psi - eA_i\bar{\psi}\psi$ do not transform the same way (the former only behaving "correctly") ; we have set $[\partial_i] \equiv [\overrightarrow{\partial_i} - \overleftarrow{\partial_i}]/2$. Similarly, part or all of the components of the Tetrode (asymmetric) energy-momentum density i $\bar{\psi}[\partial_i]\gamma_j\psi - eA_i\bar{\psi}\gamma_j\psi$ behave "incorrectly".

Now we show, by paralleling the reasoning of Theorem 1, that PTZ is none else than covariant motion reversal. Obviously, PT = M ; and e $\rightleftarrows$ -e is equivalent to $A^i \rightleftarrows -A^i$. So,

$$\mathcal{P}\mathcal{C} \simeq \text{PTZ}. \tag{11}$$

As for the separate connexions between the factors in formula (11), direct inspection of the accepted definitions shows that

$$\mathcal{P} \equiv \text{P} \qquad \underline{\text{and}} \qquad \mathcal{C} \simeq \text{TZ} \tag{12}$$

Clearly

$$\mathcal{C}\mathcal{P}\mathcal{C} = \text{MPT} = 1, \tag{13}$$

the latter expression (but not the former) displaying "manifest Lorentz invariance".

COROLLARY : "Particle-antiparticle exchange" and "covariant motion reversal" are two physically different aspects of the same mathematical operation , $\mathscr{C} = \mathscr{P}\mathscr{C}$ or MZ = PTZ.

Complete compatibility with the Feynman zigzags scheme is obvious, and exemplified in Fig.1, which (again) shows that the identity operation in formula (13) is geometrical reversal of all four space-time axes.

It thus turns out that there is, so to speak, a preestablished harmony between the Hermitean character of the operator $i\partial$, entailing opposite signs for $i\overrightarrow{\partial}$ and $-i\overleftarrow{\partial}$, and the general scheme of second quantization.

As for the latter, the (Lorentz invariant) commutation relations of free fields

$$\{\psi_\alpha(x), \bar{\psi}_\beta(y)\} = (i\gamma^i\partial_i - m)_{\alpha\beta} D(x-y) \qquad (14)$$

$$[A_i(x), A_j(y)] = \delta_{ij} D(x-y) \qquad (15)$$

display "manifest MPT invariance".

Of course, the source of $A^i$ is the $\mathscr{C}$ invariant current

$$j_i = (e/2)[\bar{\psi}\gamma^i\psi - \bar{\psi}'\gamma^i\bar{\bar{\psi}}'] . \qquad (16)$$

## V. THE PETIAU-DUFFIN-KEMMER EQUATION

Just as an other example we consider this case, the generalization to other spinning wave equations being similar.

Although the algebra of the P.D.K. (8) $\beta$-matrices differs greatly from that of Dirac's $\gamma$'s, statements closely paralleling those of Section IV still hold.

Here, the "space reversal" P of the P.D.K. equation

$$\{\beta_i(i\partial_i - eA_i) + m\}\psi = 0 \qquad (17)$$

is obtained via multiplication from the left by $\eta_4$ belonging to the set

$$\eta_i = 2\beta_i^2 - 1, \qquad (\eta_i)^2 = 1, \qquad (18)$$

which are diagonal, with eigenvalues +1 and -1, and such that

$$\eta_i\beta_j + \beta_j\eta_i = 0 \quad \text{if} \quad i \neq j, \qquad \text{and} \qquad \eta_i\beta_i = \beta_i\eta_i . \qquad (19)$$

The "Racah time T and space time PT reversals" are obtained similarly via multiplication by $\eta_1\eta_2\eta_3$ and

$$\eta_5 = \eta_1\eta_2\eta_3\eta_4 \qquad (20)$$

All other items remain similar to those of Section IV modulo the substitutions $\gamma_i \to \beta_i$ in the wave equation and $\gamma_i \to \beta_i$ for reversing axes (j, k, l).

The Z, or $\psi \rightleftarrows \bar{\psi}$ operation, is defined as before, but, as the $\beta$'s are self transposed, no additional unitary operation is needed.

As the $\eta$'s are diagonal with eigenvalues +1 and -1, each element of $\eta\psi$ equals the corresponding one in $\psi$ up to a sign. So, for being able to handle either the "proper" or "pseudo" scalar, vector and tensor cases, both options $\psi' = \pm\eta_5\psi$ should be retained in the rest mass reversal operation.

Obviously, the two orthogonal projectors

$$S = (1+\eta_5)/\sqrt{2} \qquad\qquad L = (1-\eta_5)/\sqrt{2} \tag{21}$$

respectively select in the $\psi$, as eigenfunctions of $\eta_5$, the scalar and the vector, or the vector and the tensor (or the corresponding "pseudos") contributions. These are not solutions of the P.D.K. equation, but merely of the corresponding Klein-Gordon equation.

These projectors could be used for discussing the puzzling $\mathcal{CP}$ violations in the K meson decays (9).

## VI. TWO COMPONENT SCREWING FERMIONS

Lee and Yang's (10) discovery of $\mathcal{C}$ and $\mathcal{P}$ violation, but $\mathcal{CP}$ conservation, in weak interactions, has prompted a resurrection (11) of the old Weyl screwing neutrino equation, and the Feynman-Gell-Mann (12) proposal of a two component equation for massive spin 1/2 particles, all this leading to the "universal V-A coupling" theory introduced also by other authors (13).

The common recipe in these two component theories is to extract from the Dirac wave function $\psi$ the eigenstates of the $\gamma_5$ matrix

$$\varphi \equiv R\psi = -R\gamma_5\psi, \qquad\qquad \chi \equiv L\psi = L\gamma_5\psi, \tag{22}$$

by means of the orthogonal projectors

$$R = (1-\gamma_5)/\sqrt{2}, \qquad\qquad L = (1+\gamma_5)/\sqrt{2}. \tag{23}$$

In the neutrino case, where the rest mass is zero, $\psi$, $\gamma_5\psi$, $\varphi$ and $\chi$ are all solutions of the Dirac equation. This is no more the case when $m \neq 0$ ; but, of course, $\varphi$ and $\chi$ are then solutions of the Klein-Gordon equation or even, as Feynman and Gell-Mann have shown, of the more general second order Dirac equation if an external electromagnetic field is present.

In both cases, $m = 0$ and $m \neq 0$, this procedure renders meaningless the M operation as previously defined. Only $m^2$ enters the second order Dirac equation, and, of course, the PT operation, multiplication by $\gamma_5$, transforms $\varphi$ and $\chi$ into themselves. So, a radically new vista of particle-antiparticle exchange and of covariant motion reversal emerges when the screwing fermions are introduced.

Clearly, the P and T operations, defined as multiplication by $\gamma_4$ and $\gamma_{123}$, respectively, exchange $\varphi$ and $\chi$ :

$$P\varphi' = T\varphi'' = \chi, \qquad P\chi' = T\chi'' = \varphi. \tag{24}$$

As for the Z operation, $\psi \rightleftarrows \bar{\psi}$, we notice that

$$\bar{\varphi} = \bar{\psi}L = \bar{\psi}\gamma_5 L, \qquad \bar{\chi} = \bar{\psi}R = -\bar{\psi}\gamma_5 R, \tag{25}$$

so that, choosing a representation where $\gamma_5$ is diagonal, and thus reducing $\varphi$ and $\chi$ to two component spinors, and comparing formulas (22) and (25), we get

$$Z\bar{\varphi} = \chi, \qquad Z\bar{\chi} = \varphi. \tag{26}$$

On the whole

$$\mathcal{C} = \mathcal{S} = P = T = Z \tag{27}$$

and

$$\mathcal{C}\mathcal{S} = \mathcal{C} = 1. \tag{28}$$

It is then only natural to redefine particle-antiparticle exchange and covariant motion reversal as $\mathcal{C}\mathcal{S}$ and as $\mathcal{C}$, both being the identity operation.

Such is the paradigm of the weak interactions, where spin 1/2 particles have intrinsic helicity : purehelicity if m = 0 (negative for the neutrino, positive for the antineutrino), helicity tending to pure when the velocity tends to c if $m \neq 0$.

It goes without saying that such a state of affairs lies outside the realm of classical tensorial covariance. This is fairly obvious if the interaction Lagrangians are written in terms of the Dirac $\gamma$'s (13), because, then, they are expressed as sums of a scalar and a pseudo-scalar contribution which, classically speaking, would have been heresy.

## VII. MPT INVARIANCE OF THE JORDAN-PAULI AND FEYNMAN PROPAGATORS

The Jordan-Pauli D and Feynman $D_F$ propagators associated to the Klein-Gordon equation are expressed as follows in terms of the positive and negative frequency contributions, and the retarded and advanced propagators[6] :

$$D \equiv D_+ - D_- = D_R - D_A, \tag{29}$$

$$D_F = D_R + D_- = D_A + D_+. \tag{30}$$

They are of course Lorentz invariant, and have simple transformation properties under P, T and M. Being a scalar, and odd in $x^4$, D is zero outside the light cone, a fact with important consequences[7].

The following table speaks by itself :

| | $D_R$ | $D_A$ | $D+$ | | $D-$ | $D$ | $D_F$ |
|---|---|---|---|---|---|---|---|
| x future timelike | $D_R$ | 0 | $D+$ | | $D-$ | $+D_R$ | $D+$ |
| x spacelike | 0 | 0 | $D+$ | = | $D-$ | 0 | $D+ = D-$ |
| x past timelike | 0 | $D_A$ | $D+$ | | $D-$ | $-D_A$ | $D-$ |

As D+ = D- for x spacelike, one can decide that

$$D_F = D_+ \text{ if } t \geq 0, \qquad D_F = D_- \text{ if } t \leq 0, \tag{31}$$

but, of course, this is a "relative" prescription.

No time nor causality asymmetry whatsoever exists in the Feynman propagator, any more than in the Jordan-Pauli propagator.

What may have induced the erroneous feeling, shared by some, that there is one, is that, as Jauch and Rohrlich (14) have shown, use of the Feynman propagator for describing virtual particles automatically yields an exponential decay law for higher energy levels. But this stems entirely from use of a predictive statistical reasoning. Using instead a (blind) statistical retrodictive reasoning would entail that an excited level "has emerged" according to an exp(+ at) law - a statement quite similar to familiar ones in statistical mechanics.

## VIII. CONCLUSIONS

In classical relativistic electrodynamics charge conjugation is equivalent to rest mass reversal, as e/m is the only significant parameter in this respect.

In the Dirac, or other spinning wave equations, something else turns out : charge conjugation is equivalent to rest mass reversal combined with the Z , or $\psi \rightleftarrows \bar{\psi}$ exchange.

This synthetizes the requirements of classical electrodynamics and of the second quantization scheme, a basic assumption of which is the equivalence between emission of a particle and absorption of an antiparticle (or vice versa). So there is, so to speak, a pre-established harmony between the second quantization scheme and that of the spinning wave equations.

The $\mathcal{CPC}$ = 1 theorem (or MPT = 1, as we have reformulated it), that is, of equivalence between "particle-antiparticle exchange" and "covariant motion reversal", essentially stems from the principle of second quantization we have just recalled.

A different perspective unfolds in the light of the two component screwing fermions either non massive of Weyl, or massive of Feynman and Gell-Mann. There, the rest mass reversal concept is useless, but formulas

$$\mathcal{C} = \mathcal{P} = P = T = Z, \qquad \mathcal{C} = TZ = 1$$

hold for redefining charge conjugation as $\mathcal{C}\mathcal{P} = 1$ and covariant motion reversal as $\mathcal{C} = 1$.

This certainly looked like a very pleasant vista, until the puzzling discovery of the $\mathcal{C}\mathcal{P}$ violations in the K meson decays...

NOTES

1 My motivation for doing this enquiry was that my interpretation (3) of the EPR correlations rests on the S -matrix formalism, together with its Lorentz and its CPT-invariances. A less elaborate version of this thinking has already been published (4).

2 Compare with Tiomno (5) and Sakurai (6).

3 If one proceeds in one step, by-passing rest mass reversal, via the unitary transformation

$$C\hat{\gamma}^{i}C^{-1} = -\gamma^{i}, \qquad \psi' = C\,\hat{\overline{\psi}},$$

then "there is physics" in it, as

$$C\hat{\gamma}C^{-1} = \pm\gamma$$

(depending on which of the sixteen $\gamma$'s is considered).

4 All three definitions are invariant under the continuous Lorentz group, as the $\gamma$'s are kept constant while the $\psi$ is varied.

5 That is, a pure helicity state in the extreme relativistic limit.

6 Authors differ concerning the factor of the various D's. The choice here is very near Jauch and Rohrlich's (14).

7 An interesting one is the liberty of closing the integration contour for $D_R$, $D_A$ and $D_F$ either below or above the real axis when x is spacelike, thus preserving "manifest covariance".

REFERENCES

(1) Lüders G. : 1953, Mat.Fys.Med.Kong.Dan.Vid.Selsk. 28, n°5.
Pauli W. : 1955, Niels Bohr and the Development of Physics, W. Paulied., Pergamon Press, pp. 30-51.

(2) Wigner E.P. : 1932, Gott.Nachr. 31, pp. 546-549.

(3) Costa de Beauregard O. : 1976, Found.Phys. 6, pp. 539-559.
1977, Nuovo Cim. 42B, pp. 41-64.
1979, Nuovo Cim. 51B, pp. 267-279.
1980, Physis 22, pp. 211-242.

(4) Costa de Beauregard O. : 1982, Found.Phys., pp. 861-871.

(5) Tiomno J. : 1955, Nuovo Cim. 1, pp. 226-232.

(6) Sakurai J.J. : 1958, Nuovo Cim. 7, pp. 649-660.

(7) Racah G. : 1937, Nuovo Cim. 14, pp. 322-328.

(8) Petiau G. : 1936, Mém.Acad.Roy.Belg. Cl.Sci. 16, pp.1-15.
Duffin R.J. : 1939, Phys.Rev. 54, p. 1114.
Kemmer N. : Proc.Roy.Soc. A173, pp. 91-116.

(9) Christenson J.H., Cronin J.W., Fitch V.L. and Turlay R. : 1964, Phys.Rev.Lett. 13, pp. 138-141.

(10) Lee T.D. and Yang C.N. : 1956, Phys.Rev. 104, pp. 254-258.
1957, Phys.Rev. 106, pp. 340-345.

(11) Lee T.D. and Yang C.N. : 1957, Phys.Rev. 105, pp. 1671-1675.
Salam A. : 1957, Nuovo Cim. 5, pp. 299-304.
Landau L.D. : 1957, Nucl.Phys. 3, pp. 127-131.

(12) Feynman R.P. and Gell-Mann M. : 1957, Phys.Rev. 109, pp. 193-198.
Brown L.M. : 1958, Phys.Rev. 111, pp. 957-964.

(13) Sakurai J.J. : 1958, Nuovo Cim. 7, pp. 649-660.
Sudarshan E.C.G. and Marshak R.E. : 1958, Phys.Rev. 109, pp. 1860-1862.

(14) Jauch J.M. and Rohrlich F. : 1955, The Theory of Photons and Electrons, Addison Wesley, pp. 408-410 and pp. 419-424.

NOTE ADDED AFTER THE TYPING OF THE PAPER

Recami draws my attention to the fact that he and coworkers (1) have been the first to reach conclusions similar to mine, especially concerning the characterization of particle-antiparticle exchange by rest mass reversal (2) and the equality $\pi\Xi$ = CPT = 1, where $\pi\Xi$ denotes geometrical reversal of all four space-time axes, PT covariant motion reversal, and C particle-antiparticle exchange. This is interesting, because our respective reasonings are very different from each other.
A simple and significant remark is this one : THEOREM : CPT = 1 holds for any solution of a system of covariant wave equations linear and homogeneous in the four gradient operators $\partial_i$ and the rest mass m. Proof : $\partial_i \rightleftharpoons -\partial_i$ is covariant motion reversal and $m \rightleftharpoons -m$ is particle-antiparticle exchange.
REMARK : $\pi\Xi$ = CPT = 1 follows automatically from the Feynman algorithm, via its topological invariance.
Although I agree completely with Recami on these points and a few others, I do not endorse his insistence that causality should be essentially retarded. What the formulas do say is that causality is $\pi\Xi$ = CPT invariant, that is, arrowless. This comes out as favily obvious in the S-matrix formalization of the EPR correlations (3).

(1) Recami E. and Ziino G. : 1976, Nuovo Cim. 33A, pp. 205-215. Recami E. and Rodrigues W.A. : 1982, Found. Phys. 12, pp. 709-718. Pavsic M. and Recami E. : 1982, Lett. Nuovo Cim. 34, pp. 357-362.
(2) Tiomno J. : 1955, Nuovo Cim. 1, pp. 226-232.
(3) Costa de Beauregard O. : 1980, Physics 22, pp. 211-242. 1982, La pensée physique contemporaine, S. Diner, D. Fargue and G. Lochak eds., Editions Augustin Fresnel, Rouffignac de Moulidars, 16290 Hiersac.

# WAVE-PARTICLE DUALITY IN A QUARK MODEL

Stanley P. Gudder

University of Denver

ABSTRACT

A quark model based on finite-dimensional quantum mechanics is presented. Observables associated with color, flavor, charge, and spin are considered. Using these observables, we construct quark and baryon Hamiltonians. Wave-particle dualities in this model are pointed out.

## 1. INTRODUCTION

One of the most exciting recent developments in particle physics is the emergence of various quark models [1, 5, 6]. In this article we consider a quark model based on finite-dimensional quantum mechanics [2, 3, 4]. In this model, pure quark states are represented by unit vectors in the 72 dimensional Hilbert space $\mathbb{C}^{72}$ . Observables associated with color, flavor, charge, and spin are represented by self-adjoint operators in this space.

This model exhibits some interesting examples of wave-particle duality. It is shown that the color of a quark changes in a periodic wave-like fashion given by a color wave function. In this way a particle-like quark is associated with a wave-like motion. Furthermore, these color changes can be described by an emission or absorption of particles called gluons. Next, a quark Hamiltonian is introduced. This Hamiltonian, when viewed as an energy observable gives a Schrödinger wave equation. The Hamiltonian may also be viewed as a mass observable whose eigenvalues correspond to the masses of particles.

Finally, a quark field theory is introduced. On the one hand, the Fock space describes aggregates of particles in which the creation and annihilation of particles are allowed. On the



*S. Diner et al. (eds.), The Wave-Particle Dualism, 499–514.*

other hand, the quantum fields are more characteristic of waves than of particles. Another interesting aspect is that the quark Fock space represents lepton states and thus leptons can be described by aggregates of quarks.

## 2. THE QUARK SPACE

It is now believed that there are six quark flavors (d, u, s, c, b, t) and three quark colors (r, y, b) [1, 6]. Also, quarks are spin 1/2 fermions with two spin states, spin up (↑) or spin down (↓). This gives 36 quark states. We also have 36 corresponding antiquark states, giving a total of 72 quark-antiquark states. Quarks and antiquarks can also be classified according to generation and (electric) charge. This classification is given in Table 1. Each quark in Table 1 also has three color states and two spin states.

Table 1 (Quark Classification)

| Flavor | Generation | Charge | Type |
|---|---|---|---|
| d | 1 | -1/3 | quark |
| u | 1 | 2/3 | quark |
| s | 2 | -1/3 | quark |
| c | 2 | 2/3 | quark |
| b | 3 | -1/3 | quark |
| t | 3 | 2/3 | quark |
| $\bar{d}$ | 1 | 1/3 | antiquark |
| $\bar{u}$ | 1 | -2/3 | antiquark |
| $\bar{s}$ | 2 | 1/3 | antiquark |
| $\bar{c}$ | 2 | -2/3 | antiquark |
| $\bar{b}$ | 3 | 1/3 | antiquark |
| $\bar{t}$ | 3 | -2/3 | antiquark |

We thus see that the basic quark space is the Hilbert space $V = \mathbb{C}^{72}$. We may view $V$ as the tensor product of five spaces

$$V = V_1 \otimes V_2 \otimes V_3 \otimes V_4 \otimes V_5 ,$$

where $V_1 = V_2 = \mathbb{C}^3$ and $V_3 = V_4 = V_5 = \mathbb{C}^2$. We call $V_1$ the *color space*, $V_2$ the *generation space*, $V_3$ the *change space*, $V_4$ the *type space*, $V_5$ the *spin space*, and $V_2 \otimes V_3$ the *flavor space*. If $\{e_1, e_2, e_3\}$ and $\{f_1, f_2\}$ are the standard bases for $\mathbb{C}^3$ and $\mathbb{C}^2$ respectively, a basis for $V$ is given by the set of *quark vectors* $\psi(i, j, k, m, n) = e_i \otimes e_j \otimes f_k \otimes f_m \otimes f_n$, $i, j = 1, 2, 3; k, m, n, = 1, 2$. These basis vectors correspond to the usual quark (antiquark) classification scheme as follows:

$$\psi(1, 1, 1, 1, 1) = d_r\uparrow$$
$$\psi(3, 1, 1, 1, 1) = d_b\uparrow$$
$$\psi(1, 1, 1, 2, 1) = \bar{d}_r\uparrow$$
$$\psi(2, 2, 2, 1, 2) = c_y\downarrow$$
$$\vdots$$

The following observables (operators on V) will be important in the sequel. The *color observable* C is defined by

$$C = \text{diag}\,(c_1, c_2, c_3) \otimes I \otimes I \otimes I \otimes I$$

where $c_1, c_2, c_3 \in \mathbb{R}$ are determined later. The *color-anticolor observable* $\underline{C} = \text{diag}(c_1, c_2, c_3) \otimes I \otimes I \otimes \text{diag}(1, -1) \otimes I$. The *flavor observable* $F = I \otimes \text{diag}(f_1, f_2, f_3, f_4, f_5, f_6) \otimes I \otimes I$ where again $f_j \in \mathbb{R}$, $j = 1, \ldots, 6$, are determined later. The *charge observable*

$$K = I \otimes I \otimes \text{diag}(-1/3, 2/3) \otimes \text{diag}(1, -1) \otimes I\ .$$

The *spin observable* $S = I \otimes I \otimes I \otimes I \otimes \text{diag}(1/2, -1/2)$. It will frequently be convenient to suppress the identities in the above operators. In this case, we place a ^ above the operator. For example, $\hat{C} = \text{diag}(c_1, c_2, c_3)$. The basis vectors $\psi(i, j, k, m, n)$ are eigenvectors of the above operators with eigenvalues given by the diagonal matrix elements. For example,

$$\underline{C}\ \psi(1, 1, 1, 1, 1) = c_1\ \psi(1, 1, 1, 1, 1)$$
$$\underline{C}\ \psi(1, 1, 1, 2, 1) = -c_1\ \psi(1, 1, 1, 2, 1)$$
$$F\ \psi(1, 3, 1, 1, 1) = f_5\ \psi(1, 3, 1, 1, 1)$$
$$K\ \psi(1, 1, 2, 2, 1) = (-2/3)\ \psi(1, 1, 2, 2, 1)$$
$$S\ \psi(3, 2, 1, 2, 1) = (1/2)\ \psi(3, 2, 1, 2, 1)\ .$$

In the sequel, we shall frequently need the spectral form of the above diagonal operators. On $\mathbb{C}^r$ we introduce the operators $Q_j^r$, $j = 1, \ldots, r$ where $Q_1^r = \text{diag}(1, 0, \ldots, 0), \ldots, Q_r^r = \text{diag}(0, 0, \ldots, 0, 1)$. Of course, $Q_1^r, \ldots, Q_r^r$ are mutually orthogonal projections on $\mathbb{C}^r$. If $A = \text{diag}(a_1, \ldots, a_r)$ is a diagonal matrix on $\mathbb{C}^r$, we can write $A = \sum_{i=1}^r a_i Q_i^r$. We shall also require the r-dimensional *Fourier transform*. This is defined to be the unitary operator $F_r : \mathbb{C}^r \to \mathbb{C}^r$ given by

$$(F_r f)(j) = r^{-1/2} \sum_{k=1}^r f(k)\, e^{2\pi ijk/r},\ i = \sqrt{-1},\ j = 1, 2, \ldots, r\ .$$

We also, define the set of mutually orthogonal projections $P_j^r$ on $\mathbb{C}^r$ by $P_j^r = F_r^* Q_j^r F_r$. The *Fourier transform* of an operator $A = \text{diag}(a_1, \ldots, a_r)$ on $\mathbb{C}^r$ is

$$F_r^* A F_r = \sum_{j=1}^r a_j F_r^* Q_j^r F_r = \sum_{j=1}^r a_j P_j^r$$

## 3. THE SPIN OBSERVABLE

We saw in Section 2 that a pure quark state is given by a unit vector in $V = \mathbb{C}^{72}$. In order to describe the dynamics or time evolution of a single free quark we must introduce a Hamiltonian and construct the unitary dynamical group. To illustrate this in a simple case, suppose that we are only interested in the spin of a quark. We then consider the spin observable $\hat{S} = \text{diag}(1/2, -1/2) = \frac{1}{2}Q_1^2 - \frac{1}{2}Q_2^2$ on $\mathbb{C}^2$.

We think of the spin as a kind of position observable in which there are only two possible positions $1/2$ and $-1/2$. We then define the corresponding *spin-momentum* observable, $P_{\hat{s}}$ to be $P_{\hat{s}} = F_2^* \hat{S} F_2$. In the standard basis $\beta = \{f_1, f_2\}$, $F_2$ is represented by the matrix

$$[F_2]_\beta = 2^{-1/2} \begin{bmatrix} -1 & 1 \\ 1 & 1 \end{bmatrix}$$

Hence, the matrix representation of $P_{\hat{s}}$ is

$$P_{\hat{s}} = (-1/2) \begin{bmatrix} 0 & 1 \\ 1 & 0 \end{bmatrix}$$

and its spectral form is

$$P_{\hat{s}} = \frac{1}{2} P_1^2 - \frac{1}{2} P_2^2 = \frac{1}{2} \begin{bmatrix} 1/2 & -1/2 \\ -1/2 & 1/2 \end{bmatrix} -1/2 \begin{bmatrix} 1/2 & 1/2 \\ 1/2 & 1/2 \end{bmatrix}$$

Now $P_{\hat{s}}$ generates a one-parameter unitary group $V(t) = \exp(-i\ t\ P_{\hat{s}})$ which in matrix form is

$$V(t) = e^{-i\ t/2}\, P_1^2 + e^{i\ t/2}\, P_2^2 = \begin{bmatrix} \cos t/2 & i \sin t/2 \\ i \sin t/2 & \cos t/2 \end{bmatrix}$$

At $t_n = (2n + 1)\pi$, $n = 0, 1, 2, \ldots$,

$$V(t_n) = \exp[i\ \pi\ (2n+1)/2] \begin{bmatrix} 0 & 1 \\ 1 & 0 \end{bmatrix}$$

For these values $t_n$, $n = 0, 1, 2, \ldots$, $V(t_n)$ executes a flip of the spin states $f_1, f_2$. Indeed, $V(t_n) f_1 = [\exp i\ \pi(2n+1)/2]\, f_2 = \pm i\, f_2$ and $V(t_n) f_2 = \pm f_1$. In this way, the spin-momentum observable $P_{\hat{s}}$ mearures the "motion" in spin space. We define the *spin-energy* observable to be $H_{\hat{s}} = a P_{\hat{s}}^2$ where $a \in \mathbb{R}$ is a constant. The *spin-dynamical group* is given by $U_{\hat{s}}(t) = e^{-i\ t\ H_{\hat{s}}}$. However, since $H_{\hat{s}} = (a/4)I$, the spin of free quark is constant in time.

Spin is a constant of motion for a free quark because quarks are spin $\frac{1}{2}$ particles. This follows since the spectrum of $\hat{S}$ is

$\{\frac{1}{2}, -\frac{1}{2}\}$ and hence the spectrum of $H_{\hat{S}}$ degenerates to $\{a/4\}$. It is interesting to note that if quarks had a higher order spin, then spin would not be a constant of motion. For example, consider the spin 1 operator

$$S_1 = \text{diag}(1, -1, 0) = 1 \cdot Q_1^3 - 1 \cdot Q_2^3 + 0 \cdot Q_3^3$$

on $\mathbb{C}^3$. We then employ the three-dimensional Fourier transform $F_3 : \mathbb{C}^3 \to \mathbb{C}^3$. In the standard basis $\beta = \{e_1, e_2, e_3\}$ we have

$$[F_3]_\beta = 3^{-1/2} \begin{bmatrix} b & \bar{b} & 1 \\ \bar{b} & b & 1 \\ 1 & 1 & 1 \end{bmatrix}, \quad b = e^{2\pi i/3}$$

The projections $P_j^3 = F_3^* Q_j^3 F_3$, $j = 1, 2, 3$, become

$$P_1^3 = \frac{1}{3} \begin{bmatrix} 1 & b & \bar{b} \\ \bar{b} & 1 & b \\ b & \bar{b} & 1 \end{bmatrix}, \quad P_2^3 = \frac{1}{3} \begin{bmatrix} 1 & \bar{b} & b \\ b & 1 & \bar{b} \\ \bar{b} & b & 1 \end{bmatrix}, \quad P_3^3 = \frac{1}{3} \begin{bmatrix} 1 & 1 & 1 \\ 1 & 1 & 1 \\ 1 & 1 & 1 \end{bmatrix}$$

The spin-momentum observable is

$$P_{s_1} = F_3^* S_1 F_3 = P_1^3 - P_2^3 = 3^{-1/2} \begin{bmatrix} 0 & i & -i \\ -i & 0 & i \\ i & -i & 0 \end{bmatrix}$$

The one-parameter unitary group $V(t) = e^{-i t P_{s_1}}$ has matrix form $V(t) = e^{-i t} P_1^3 + e^{i t} P_2^3 + P_3$

$$= \frac{1}{3} \begin{bmatrix} 1+2 \cos t & 1+ \cos(t-2\pi/3) & 1+ \cos(t+2\pi/3) \\ 1+2 \cos(t+2\pi/3) & 1+2 \cos t & 1+ \cos(t-2\pi/3) \\ 1+2 \cos(t-2\pi/3) & 1+2 \cos(t+2\pi/3) & 1+2 \cos t \end{bmatrix}$$

At $t_n = 2\pi n/3$, $n = 1, 2, 3, \ldots$, $V(t_n)$ will execute various flips between spin states. For example,

$$V(2\pi/3) = \begin{bmatrix} 0 & 1 & 0 \\ 0 & 0 & 1 \\ 1 & 0 & 0 \end{bmatrix}$$

and hence, $V(2\pi/3)e_1 = e_3$, $V(2\pi/3)e_2 = e_1$, $V(2\pi/3)e_3 = e_2$. Similarly,

$$V(4\pi/3) = \begin{bmatrix} 0 & 0 & 1 \\ 1 & 0 & 0 \\ 0 & 1 & 0 \end{bmatrix}$$

and hence, $V(4\pi/3)e_1 = e_2$, $V(4\pi/3)e_2 = e_3$, $V(4\pi/3)e_3 = e_1$. This again illustrates that the spin-momentum observable measures "motion" in spin space. The spin-energy observable is

$$H_{s_1} = a P_{s_1}^2 = \frac{a}{3}\begin{bmatrix} 2 & -1 & -1 \\ -1 & 2 & -1 \\ -1 & -1 & 2 \end{bmatrix}$$

The spin dynamical group becomes

$$U_{s_1}(t) = e^{-i t H_{s_1}} = e^{-i a t} P_1^3 + e^{-i a t} P_2^3 + P_3$$

$$= \frac{1}{3}\begin{bmatrix} 1 + 2e^{-i a t} & 1 - e^{-i a t} & 1 - e^{-i a t} \\ 1 - e^{-i a t} & 1 + 2e^{-i a t} & 1 - e^{-i a t} \\ 1 - e^{-i a t} & 1 - e^{-i a t} & 1 + 2e^{-i a t} \end{bmatrix}$$

The transition probability $P_{kj}(t)$ from the spin state $e_k$ to the spin state $e_j$ after time $t$ becomes

$$P_{kj}(t) = |< U_{s_1}(t)e_k, e_j >|^2 = | U_{s_1}(t)_{jk} |^2$$

$$= \begin{cases} (5 + 4 \cos a t)/9 & \text{if } j = k \\ (2 - 2 \cos a t)/9 & \text{if } j \neq k \end{cases} \qquad j, k = 1, 2, 3$$

We thus see that the particle oscillates from one spin state to another with period $2\pi/a$. This can also be seen by computing the average spin $< S_1 >_1 (t)$ at time $t$ given the initial state $e_1$ :

$$< S_1 >_1 (t) = |U(t)_{11}|^2 - |U(t)_{21}|^2 = (3 + 6 \cos a t)/9$$

## 4. THE COLOR OBSERVABLE

In Section 3, we saw that the spin of a free quark is a constant of motion. We now consider the color observable $\hat{C} = \mathrm{diag}(c_1, c_2, c_3)$ on $\mathbb{C}^3$. To determine $c_1, c_2,$ and $c_3$ we first make the fundamental assumptions that the $c_j$'s are distinct, nonzero, and that $\hat{C}$ is traceless; that is $c_1 + c_2 + c_3 = 0$. The reason for the first assumption is that the three colors are distinct. The second assumption is because 0 will represent the color quantum number for "colorless" particles and quarks have color. The reason for the third assumption is that the observed particle compositions of quarks (in particular, baryons) are colorless. Since a baryon is composed of three quarks, each a different color, this assumption gives a total quantum number zero.

We now consider color motion. The color-momentum operator is

$$P_{\hat{C}} = F_3^* \hat{C} F_3 = c_1 P_1^3 + c_2 P_2^3 + c_3 P_3^3$$

where $P_1^3, P_2^3, P_3^3$ are given in Section 3. The color Hamiltonian becomes

$$H_{\hat{C}} = \alpha P_{\hat{C}}^2 = \alpha c_1^2 P_1^3 + \alpha c_2^2 P_2^3 + \alpha c_3^2 P_3^3$$

where $\alpha$ is a constant to be determined. The color dynamical group $U_{\hat{C}}(t)$ is given by

$$U_{\hat{C}}(t) = e^{-i t H_{\hat{C}}} = e^{-i \alpha c_1^2 t} P_1^3 + e^{-i \alpha c_2^2 t} P_2^3 + e^{-i \alpha c_3^2 t} P_3^3$$

This shows that the color of a quark continually changes in a wave-like fashion and illustrates the wave-particle duality of quarks.

This periodic color motion is best seen by computing the probability $P_{11}(t)$ that the quark is in the color state $e_1$(red) at time $t$ given that the initial color state is $e_1$. This is given by

$$\begin{aligned} P_{11}(t) &= |< U_{\hat{C}}(t)\, e_1, e_1 >|^2 \\ &= \frac{1}{9} |e^{-i \alpha c_1^2 t} + e^{-i \alpha c_2^2 t} + e^{-i \alpha c_3^2 t}|^2 \\ &= \frac{1}{3} + \frac{2}{9} [\cos \alpha t(c_2^2-c_1^2) + \cos \alpha t(c_3^2-c_2^2) + \cos \alpha t(c_3^2-c_1^2)] \end{aligned}$$

It is clear that $P_{11}(t)$ is periodic between 0 and 1 and that $P_{11}(t_0) = 1$ if and only if the three cosines have value 1. We must then have

$$\alpha t_0(c_2^2-c_1^2) = 2 \pi m, \quad \alpha t_0(c_3^2- c_2^2) = 2 \pi n, \quad \alpha t_0(c_3^2- c_1^2) = 2 \pi p$$

where $m$, $n$, and $p$ are integers. We now assume that the period is a minimum. This requires that $m, n, p$ are as small as possible consistent with out previous conditions. The distinctness and nonvanishing of the $c_i$'s implies that $m, n, p$ are nonzero. Indeed, if $m$ say, is zero, then $c_2 = -c_1$ which, by the traceless condition, implies that $c_3 = 0$. For the same reasons, we must have $n \neq p$. The smallest values for $m, n, p$ become $m = n = 1$, $p = 2$. Then solving these equations gives $c_2^2 - 2c_1^2 - 2c_1 c_2 = 0$. Hence, $c_2 = (1 \pm \sqrt{3})\, c_1$. Thus to within a multiplicative constant, $c_1 = 1$, $c_2 = 1 + \sqrt{3}$, $c_3 = -(2 + \sqrt{3})$. The period then becomes $t_0 = 2\pi(2\sqrt{3} - 3)/3\alpha$. We shall determine the value of $\alpha$ later using physical considerations. Using this value gives $t_0 \simeq 1.8 \times 10^{-23}$ seconds.

Substituting the above values for $c_1, c_2, c_3$ into $P_{11}(t)$ we obtain

$$P_{11}(t) = \frac{1}{3} + \frac{2}{9} [2 \cos 2\pi t/t_0 + \cos 4\pi t/t_0]$$

Expressions for $P_{12}(t)$ and $P_{13}(t)$ are the same as for $P_{11}(t)$ except for a difference in phase. We conclude that if a free quark is initially red, then the quark color changes periodically to yellow, then to blue, and then back to red with period $t_0 \simeq 1.8 \times 10^{-23}$ seconds. This gives a wave motion associated with quark particles. This wave motion can also be described in terms of an emission or absorption of particles called gluons. The change of a red quark into a yellow quark can be described by the absorption of an antired-yellow gluon or the emission of a red-antiyellow gluon. We have thus come full cycle. A quark, which is thought of an an elementary particle is associated with a wave motion of color changes; this wave motion, in turn, is described by an emission or absorption of other particles called gluons.

It is believed that a baryon is composed of three bound quarks each having a different color. Presumably, these three quarks change color synchronously so that at any given time the total baryon is colorless. These color changes are mediated by an exchange of gluons which are responsible for the strong force which binds the quarks together within the baryon. Since the period of the color motion is $t_0 \approx 1.8 \times 10^{-23}$ seconds, it takes approximately $0.6 \times 10^{-23}$ seconds to change colors. This indicates that gluons are exchanged each $0.6 \times 10^{-23}$ seconds. If a gluon moves at approximately the speed of light, it will travel about $1.8 \times 10^{-13}$ cm during this time interval. This distance is approximately the diameter of a typical baryon which adds plausibility to the above argument. Similar considerations hold for mesons which are thought to be composed of quark-anti-quark pairs. In this case we have a color and its corresponding anticolor which change synchronously.

We can prove mathematically that the average color value of a free quark is zero over a single period. If a quark is in the state $e_1$(red) at time $t = 0$ , then the average color value at time $t$ is

$$< \hat{C} >_1 (t) = \sum_{j=1}^{3} c_j P_{1j}(t)$$

It is easy to check that

$$\frac{1}{t_0} \int_0^{t_0} P_{1j}(t)dt = \frac{1}{3} , \quad j = 1, 2, 3$$

Hence,

$$\frac{1}{t_0} \int_0^{t_0} < \hat{C} >_1 (t)dt = \sum_{j=1}^{3} c_j = 0 .$$

We now use the values $c_1, c_2, c_3$ found above to compute $P_{\hat{C}}$ and $H_{\hat{C}}$ . These become

$$P_{\hat{C}} = \begin{bmatrix} 0 & -1 - \frac{\sqrt{3}}{2} - \frac{1}{2}i & -1 - \frac{\sqrt{3}}{2} + \frac{1}{2}i \\ -1 - \frac{\sqrt{3}}{2} + \frac{1}{2}i & 0 & -1 - \frac{\sqrt{3}}{2} - \frac{1}{2}i \\ -1 - \frac{\sqrt{3}}{2} - \frac{1}{2}i & -1 - \frac{\sqrt{3}}{2} + \frac{1}{2}i & 0 \end{bmatrix}$$

$$H_{\hat{C}} = \alpha \begin{bmatrix} 4+2\sqrt{3} & \frac{3}{2} + \sqrt{3} - (\frac{\sqrt{3}}{2} + 1)i & \frac{3}{2} + \sqrt{3} + (\frac{\sqrt{3}}{2} + 1)i \\ \frac{3}{2} + \sqrt{3} + (\frac{\sqrt{3}}{2} + 1)i & 4+2\sqrt{3} & \frac{3}{2} + \sqrt{3} - (\frac{\sqrt{3}}{2} + 1)i \\ \frac{3}{2} + \sqrt{3} - (\frac{\sqrt{3}}{2} + 1)i & \frac{3}{2} + \sqrt{3} + (\frac{\sqrt{3}}{2} + 1)i & 4+2\sqrt{3} \end{bmatrix}$$

We then assume that $P_c = P_{\hat{C}} \otimes I \otimes I \otimes I \otimes I$ and $H_c = H_{\hat{C}} \otimes I \otimes I \otimes I \otimes I$ .

## 5. THE QUARK TENSOR SPACE

The quark tensor space is defined as

$$\begin{aligned} TV &= C \oplus V \oplus (V \otimes V) \oplus (V \otimes V \otimes V) \oplus \cdots \\ &= C \oplus V \oplus V^{\otimes 2} \oplus V^{\otimes 3} \oplus \cdots \end{aligned}$$

The normalized vectors in TV represent the vector states of all possible combinations of quarks. For $\phi \in TV$ we write $\phi = \phi_0 \oplus \phi_1 \oplus \cdots$ , where $\phi_n \in V^{\otimes n}$ . Define the projection $P_n : TV \to TV$ by

$$P_n \phi = 0 \oplus 0 \oplus \cdots \oplus \phi_n \oplus 0 \oplus \cdots, \quad n = 0, 1, 2, \ldots$$

If $A : V \to V$ is a linear operator, we define $\Gamma(A) : TV \to TV$ by

$$\begin{aligned} \Gamma(A) &= I \oplus A \oplus (A \otimes I + I \otimes A) \oplus (A \otimes I \otimes I + I \otimes A \otimes I + I \otimes I \otimes A) \oplus \cdots \\ &= \Gamma^0(A) \oplus \Gamma^1(A) \oplus \Gamma^2(A) \oplus \cdots \end{aligned}$$

We call $\phi \in TV$ a *color vector* if $\Gamma(\bar{C}) P_n \phi = \Gamma^n(\bar{C}) \phi_n = 0$ for

$n = 2, 3, 4, \ldots$ . Except for the vacuum and the quarks themselves, normalized color vectors represent vector states of colorless particles. The set of color vectors forms a closed subspace of TV and is denoted by

$$(TV)_c = \mathbb{C} \oplus V \oplus V_c^2 \oplus V_c^3 \oplus \cdots$$

where

$$V_c^n = \{\phi_n \in V^{\otimes n} : \Gamma^n(\bar{C})\, \phi_n = 0\}\ ,\quad n = 2, 3, \ldots$$

Let $\phi_n = \psi_1 \otimes \cdots \otimes \psi_n \in V_{c-}^n$, $n \geq 2$, where the $\psi_j$'s are quark vectors. If $\bar{C}\,\psi_j = -\bar{C}\,\psi_k$ we call $\{\psi_j, \psi_k\}$ a *meson pair* in $\phi_n$. Of course, in a meson pair, one of the vectors represents a quark and the other an antiquark. If $\psi_i, \psi_j$, and $\psi_k$ represent quarks and $\bar{C}(\psi_i + \psi_j + \psi_k) = 0$, then we call $\{\psi_i, \psi_j, \psi_k\}$ a *baryon triple* in $\phi_n$. If $\psi_i, \psi_j$, and $\psi_k$ represent antiquarks and $\bar{C}(\psi_i + \psi_j + \psi_k) = 0$, then we call $\{\psi_i, \psi_j, \psi_k\}$ an *antibaryon triple* in $\phi_n$.

Theorem. Let $\phi \in TV$. Then $\phi \in (TV)_c$ if and only if $\phi_n$ is a linear combination of terms each of which is composed of $m$ mesons pairs, $p$ baryon triples, and $q$ antibaryon triples ($m, p, q \geq 0$ depend on the term, $2m + 3p + 3q = n$), $n = 2, 3, \ldots$ .

Proof. The detailed proof is in [3]. The main part of the proof uses the following idea. Let $p_1, p_2, p_3$ be nonnegative integers with $p_1c_1 + p_2c_2 + p_3c_3 = 0$. Then $p_1 + p_2(1 + \sqrt{3}) - p_3(2 + \sqrt{3}) = 0$. It follows that $p_1 + p_2 - 2p_3 = 0$ and $p_2 - p_3 = 0$. Hence, $p_1 = p_2 = p_3$. □

It appears that the following *color condition* is a law of nature. Of the normalized vectors in TV only the color vectors are observable physical vector states. It follows from the color condition and the above theorem that quarks can only combine as mesons, baryons, antibaryons and collections of these three.

## 6. THE FLAVOR OBSERVABLE.

We now determine the eigenvalues $f_1, \ldots, f_6$ of the flavor observable $F$. We do this using the approximate known masses of various quarks. Some of these masses can be estimated using tables such as Table 2.

Table 2 (Particle Masses)

| Particle | Quark Composition | Mass (Mev) |
| --- | --- | --- |
| p | u u d | 938 |
| $\Sigma^+$ | u u s | 1189 |
| $\psi$ | c $\bar{c}$ | 3100 |

Using Table 2 and estimates for b and t mesons we obtain the rough approximations given in Table 3.

Table 3 (Quark Masses)

| Quark | Mass (Mev) |
| --- | --- |
| d | 320 |
| u | 320 |
| s | 500 |
| c | 1500 |
| b | 4000 |
| t | 16000 |

We assume that the free quark Hamiltonian has the following form:

$$H^{(1)} = \alpha P_c^2 + \gamma F^2 = H_c + \gamma F^2 \tag{1}$$

Instead of viewing $H^{(1)}$ as the energy observable, it is more convenient to regard it as the equivalent mass observable. This is another instance of wave-particle duality. When $H^{(1)}$ is viewed as an energy observable, the corresponding Schrödinger equation $\partial\psi/\partial t = -i H^{(1)} \psi$ corresponds to a wave equation. However, when $H^{(1)}$ is viewed as a mass observable, its eigenvalues correspond to masses of particles. Equation (1) states that the mass of a quark is the sum of its color energy mass and its "rest" mass. The color energy mass results from the fact that the quark's color is continually changing, giving rise to a kind of "kinetic" energy, and the "rest" mass is the mass a quark would have if its color were not changing. The basic assumption in Equation (1) is that the rest mass is proportional to the square of the flavor. If $\psi \varepsilon V$ is a quark vector state, the expected mass in the state $\psi$ becomes

$$< H^{(1)} \psi, \psi > = < H_c \psi \quad \psi > + \gamma < F^2 \psi \quad \psi >$$

Notice that the diagonal elements of $H_c$ are equal and let

$$a = < H_c r, r > = < H_c y, y > = < H_c b, b >$$

We then have the following approximate formulas:

$$320 = \langle H^{(1)} d, d \rangle = a + \gamma f_1^2$$

$$320 = \langle H^{(1)} u, u \rangle = a + \gamma f_2^2$$

$$500 = \langle H^{(1)} s, s \rangle = a + \gamma f_3^2$$

$$1500 = \langle H^{(1)} c, c \rangle = a + \gamma f_4^2$$

$$4000 = \langle H^{(1)} b, b \rangle = a + \gamma f_5^2$$

$$16000 = \langle H^{(1)} t, t \rangle = a + \gamma f_6^2$$

Now the mass of the $\pi^0$ meson, $\pi^0 = u\bar{u}$, is approximately 135 Mev. Since the masses of $u$ and $\bar{u}$ are about 320 Mev, this indicates that about 255 Mev from the quark and also from the antiquark is converted into quark-antiquark interaction energy which we assume consists of color energy. The remaining 65 Mev of the quark and antiquark contribute to the meson mass. For this reason, we assume that $a \approx 255$, $\gamma \simeq 65$ and $f_1 \approx f_2 \approx \pm 1$. Solving the remaining four equations gives:

$$f_3 \simeq [(500 - 255)/65]^{1/2} = \pm 1.94$$
$$f_4 \approx [(1500 - 255)/65]^{1/2} = \pm 4.38$$
$$f_5 \approx [(4000 - 255)/65]^{1/2} = \pm 7.59$$
$$f_6 \approx [(16000 - 255)/65]^{1/2} = \pm 15.56$$

The above equations do not determine the signs of the $f_i$'s, $i = 1, \ldots, 6$. Also, these numbers are only rough approximations. Our later mass formulas give the best agreement if we choose the signs +, +, −, −, +, +. Moreover, assuming that nature gives a regular pattern we postulate that

$$f_1 = 1\frac{1}{32},\ f_2 = 1,\ f_3 = -2,\ f_4 = -4,\ f_5 = 8,\ f_6 = 16\ .$$

We have assumed that $f_1 = 1\frac{1}{32}$ to take account of small mass differences between particles containing $d$ quarks and those containing $u$ quarks.

## 7. BARYON HAMILTONIAN

In Section 6 we proposed a free quark Hamiltonian $H^{(1)} = H_c + \gamma F^2$. In this section we shall propose a baryon Hamiltonian $H^{(3)}$. By computing the mass expectations of known baryons in terms of this Hamiltonian, we shall determine the two constants in Equation (1) as well as four others appearing in $H^{(3)}$. Since $H^{(3)}$ is the Hamiltonian for three interacting quarks, we should expect some interaction terms which do not appear in $H^{(1)}$ .

Our proposed baryon Hamiltonian is

$$H^{(3)} = \alpha\, \Gamma^3(P_c^2) + \beta(P_c \otimes P_c \otimes I + P_c \otimes I \otimes P_c + I \otimes P_c \otimes P_c)$$
$$+ \gamma\, \Gamma^3(F^2) - \delta(F \otimes F \otimes I + F \otimes I \otimes F + I \otimes F \otimes F)$$
$$- \mu\, F \otimes F \otimes F - \nu\, K \otimes K \otimes K$$

The first term is a color energy term, the second is a color motion interaction between pairs of quarks, the third is a rest mass term, the fourth is a flavor interaction between pairs of quarks, and the fifth and sixth are a flavor interaction and a charge interaction, respectively, among all three quarks. The mass expectation for a baryon in the vector state $\psi$ is given by $< H^{(3)} \psi, \psi >$ .

By computing the mass expectations for the six baryons $n$, $p$, $\Sigma^+$, $\Xi^0$, $\Delta^{++}$, $\Omega^-$, we obtain the following values for the constants:

$$\alpha = 35.50 \qquad \delta \doteq 11.56$$
$$\beta = 26.46 \qquad \mu = 2.33$$
$$\gamma = 58.77 \qquad \nu = 7.19$$

Using these values for $\alpha$ and $\gamma$ we obtain the following quark masses:

$$< H^{(1)} d, d > = \alpha < P_c^2 d, d > + \gamma < F^2 d, d >$$
$$= 265.05 + 62.50 = 327.55$$
$$< H^{(1)} u, u > = 265.05 + 58.77 = 323.82$$
$$< H^{(1)} s, s > = 265.05 + 325.08 = 500.13$$
$$< H^{(1)} c, c > = 265.05 + 940.32 = 1205.37$$
$$< H^{(1)} b, b > = 265.05 + 376.28 = 4026.23$$
$$< H^{(1)} t, t > = 265.05 + 15045.12 = 15310.17$$

To get an idea of how the Hamiltonian is used to compute masses, we find the mass of the neutron. The neutron is given by the mixed state

$$n = (2)^{-1/2} (u_r \uparrow \otimes d_y \downarrow \otimes d_b \downarrow - d_y \downarrow \otimes u_r \uparrow \otimes d_b \downarrow )$$

The neutron mass expectation becomes:

$$\begin{aligned}
< H^{(3)} n, n > &= 3\,\alpha < P_c^2\, u_r \uparrow, u_r \uparrow > \\
&+ \gamma[< F^2\, u_r \uparrow, u_r \uparrow > + < F^2\, d_y \downarrow, d_y \downarrow > \\
&+ < F^2\, d_b \downarrow, d_b \downarrow >] \\
&- \delta[< F\, u_r \uparrow, u_r \uparrow >< F\, d_y \downarrow, d_y \downarrow > \\
&+ < F\, u_r \uparrow, u_r \uparrow >< F\, d_b \downarrow, d_b \downarrow > \\
&+ < F\, d_y \downarrow, d_y \downarrow >< F\, d_b \downarrow, d_b \downarrow >] \\
&- \mu < F\, u_r \uparrow, u_r \uparrow >< F\, d_y \downarrow, d_y \downarrow > \\
&\quad < F\, d_b \downarrow, d_b \downarrow > - \nu < K\, u_r \uparrow, u \uparrow > \\
&\quad < K\, d_y \downarrow, d_y \downarrow > < K\, d_b \downarrow, d_b \downarrow > \\
&= 3\,\alpha < P_c^2\, r, r > + \gamma[< F^2\, u, u > + 2 < F^2\, d, d >] \\
&- \delta[2 < F\, u, u >< F\, d, d > + < F\, d, d >^2] \\
&- \mu < F\, u, u >< F\, d, d >^2 - \nu < K\, u, u >< K\, d, d >^2 \\
&= 3(4 + 2\sqrt{3})\alpha + [1 + 2(1\tfrac{1}{32})^2]\,\gamma \\
&- [2(1\tfrac{1}{32}) + (1\tfrac{1}{32})^2]\,\delta \\
&- (1\tfrac{1}{32})^2\,\mu - (2/3)(-1/3)^2\,\nu = 939.55
\end{aligned}$$

For computations of other baryon masses and also meson masses we refer the reader to [3].

## 8. A QUARK FIELD THEORY

Wave-particle duality plays a role in conventional quantum field theory. On the one hand, the Fock space describes aggregates of particles in which the creation and annihilation of particles are allowed. On the other hand, the field operators describe fields that are not sharply localized but are smeared over space-time. These field operators are more characteristic of waves than of particles. The quark field theory we now briefly present has a further interesting aspect. The states of the quark Fock space represent lepton states and thus leptons can be described by aggregates of quarks.

For simplicity, we consider only the three color degrees of freedom for a quark. The other degrees of freedom can be incorporated for a complete theory, but for purposes here, these three degrees of freedom will suffice. Let $V = \mathbb{C}^3$ and let

$$SV = C \oplus V \oplus (V \circledS V) \oplus (V \circledS V \circledS V) \oplus \cdots$$

be the symmetric Hilbert space for $V$ where $\circledS$ denotes the symmetric tensor product. Now $SV$ is unitarily equivalent to $L^2(\mathbb{R}^3)$ in a natural way. Indeed, let $e_1, e_2, e_3$ be the standard basis for $\mathbb{C}^3$ and let $h_n \in L^2(\mathbb{R}^3)$ be the Hermite function

$$h_n(x) = (2^n\, n!)^{-1/2} (-1)^n \pi^{-1/4} e^{x^2/2} (d/dx)^n e^{-x^2},\ n = 0, 1, \ldots$$

Then $h_n$ is an orthonormal basis for $L^2(\mathbb{R})$ and

$$h_{j_1, j_2, j_3}(x_1, x_2, x_3) = h_{j_1}(x_1)\, h_{j_2}(x_2)\, h_{j_3}(x_3),\qquad j_1, j_2, j_3 = 0, 1, \ldots .$$

is an orthonormal basis for $L^2(\mathbb{R}^3)$ . For the basis elements $e_1^{j_1} \circledS e_2^{j_2} \circledS e_3^{j_3}$ define

$$J(e_1^{j_1} \circledS e_2^{j_2} \circledS e_3^{j_3}) = h_{j_1, j_2, j_3}(x_1, x_2, x_3)$$

Extending $J$ by linearity and closure gives a unitary operator $J : SV \to L^2(\mathbb{R}^3)$ . In this way, $L^2(\mathbb{R}^3)$ may be thought of as the symmetric Hilbert space (Fock space) for $V = \mathbb{C}^3$ . In particular, the pure states for the usual nonrelativistic quantum mechanical description on $L^2(\mathbb{R}^3)$ of a lepton (for example, an electron) can be written as a superposition of quark states in $\mathbb{C}^3$ .

Define the *annihilation operators* $A(e_j)$ on $L^2(\mathbb{R}^3)$ by $A(e_j) = 2^{-1/2}(x_j + \partial/\partial x_j)$, $j = 1, 2, 3$ . Their adjoints are the *creation operators* $A^*(e_j) = 2^{-1/2}(x_j - \partial/\partial x_j)$, $j = 1, 2, 3$ . These operators satisfy

$$A(e_1)h_{j_1, j_2, j_3} = j_1^{1/2}\, h_{j_1-1, j_2, j_3}$$

$$A^*(e_1) = h_{j_1, j_2, j_3} = (j_1 + 1)^{1/2}\, h_{j_1+1, j_2, j_3}$$

and similar expressions hold for $A(e_2)$, $A^*(e_2)$, $A(e_3)$, $A^*(e_3)$ . The usual commutation relations

$$[A(e_j), A(e_k)] = [A^*(e_j), A^*(e_k)] = 0$$

$$[A(e_j), A^*(e_k)] = \delta_{jk}\ ,\ j, k = 1, 2, 3$$

hold. The corresponding field operators

$$\pi(e_j) = 2^{1/2}\, [A(e_j) + A^*(e_j)] = x_j$$

$$\psi(e_j) = -i\, 2^{-1/2}\, [A(e_j) - A^*(e_j)] = -i\, \partial/\partial x_j$$

are the usual position and momentum operators. In this way, one can continue to develop a quark field theory. For details we refer to [4].

REFERENCES

1. Close, F.E., An Introduction to Quarks and Partons, Academic Press, (1979).
2. Gudder, S., "A Logical Explanation for Quarks", Found. Phys. (to appear).
3. Gudder, S., "A Finite-Dimensional Quark Model", (to appear).
4. Gudder, S. and Naroditsky, V., "Finite-Dimensional Quantum Mechanics", Intern. J. Theor. Phys., 20, pp. 619-643 (1981).
5. Kokkedee, J., The Quark Model, Benjamin, New York (1969).
6. Lichtenberg, D., Unitary Symmetry and Elementary Particles, Academic Press, New York (1978).

# DESTRUCTION OF COHERENCE IN NONDEMOLITION MONITORING: QUANTUM "WATCHDOG EFFECT" IN GRAVITY WAVE DETECTORS

W. H. Zurek
Theoretical Astrophysics 130-33
California Institute of Technology
Pasadena, CA 91125

ABSTRACT

We show that nondemolition monitoring of a Weber bar may prevent changes of the number of phonons, and thus influence the sensitivity of quantum-counting gravity wave detectors. This effect is similar to the <u>Watchdog Effect</u> which is predicted to delay decays of the monitored, unstable quantum system. Relations between watchdog effect and <u>Environment-Induced Superselection Rules</u> as well as its connections to the fundamental questions of the quantum theory of measurement are briefly considered.

## 1. WATCHDOG EFFECT

The aim of this section is to show that, in the initial moments, the decay rate cannot be time-independent: it must be proportional to the time elapsed since the initiation of the decay process. As a consequence, monitoring of an unstable quantum system may slow down, and even halt the decay process. This so-called watchdog effect interferes with the process of detection of gravity waves by the nondemolition detectors, which are briefly described in the following section.

Several authors have recently discussed slowing down of the decay of an unstable quantum system monitored by an external observer (1-5). Following them, let us consider an unstable state $|u\rangle$ which can decay into one of the decay product states $|d_1\rangle$, $|d_2\rangle$, ..., $|d_k\rangle$, ... . The state of the decaying system t seconds after it was last observed in the state $|u\rangle$ is then represented by a state vector:



*S. Diner et al. (eds.), The Wave-Particle Dualism, 515–528.*

$$|\psi(t)\rangle = a_u(t)|u\rangle + \sum_{k \neq u}^{N} a_k(t)|d_k\rangle \tag{1}$$

where $a_u(t)$, $a_1(t)$, ..., $a_k(t)$, ... give appropriate amplitudes. State vector $|\psi(t)\rangle$ can be calculated from the Schrödinger equation. In terms of the total hamiltonian H, this state vector is given by:

$$|\psi(t)\rangle = \exp(-iHt)|u\rangle, \tag{2}$$

providing that at t = 0

$$p(0) = |a_u(0)|^2 = 1. \tag{3}$$

The probability of finding this system still undecayed at some later time t is then:

$$p(t) = |a_u(t)|^2 = |\langle u| \exp(-iHt)|u\rangle|^2. \tag{4}$$

For small t the right-hand side of Eq. (4) can be expanded in powers of t

$$p(t) = 1 - (\Delta H)^2 t^2 + \dots \tag{5}$$

where:

$$(\Delta H)^2 = \langle u|H^2|u\rangle - |\langle u|H|u\rangle|^2. \tag{6}$$

The key qualitative feature of this survival probability which distinguishes it, during the initial time interval from the exponential is the absence of the linear term: The decay begins always "cosinusoidaly", and merges with the exponential only for $t > \tau = (\Delta H)^{-1}$.

Watchdog effect (known also as "Quantum Zeno Paradox", "Turing Paradox", "Watched Pot Never Boils" effect ...) is a direct consequence of this slow, quadratic start. To exhibit it, let us suppose that an unstable system is monitored by n instantaneous measurements performed at intervals t/n. The probability that it will be found in the initial state $|u\rangle$ in all n measurements is given by:

$$P_n(t) = [1 - (\Delta H)^2 (t/n)^2]^n > 1 - (\Delta H)^2 t^2 = p(t). \tag{7}$$

It follows from the above Eq. (7) that the survival probability of the system which is monitored is different from the survival probability of this very same system when it is left "on its own". In particular,

$$P_n(t) \underset{n \to \infty}{\to} 1. \qquad (8)$$

That is, the initial state under continuous observation does not decay at all!

The physics responsible for this surprising behavior, and its significance for some of the proposed techniques of gravity wave detection are the subject of the rest of this paper. At this point let us already stress one tacit assumption made in the course of the derivation of Eq. (7) and Eq. (8). The idealized measurement performed on the undecayed system is assumed to leave it in the undecayed state $|u\rangle$. Such measurements which do not influence eigenstates of the measured observable are taken for granted by von Neumann's "projection postulate". Pauli has called them "measurements of the first kind", to distinguish them from the "measurements of the second kind", which leave measured system in some arbitrary final state, and which, consequently, have only postdictive, but no predictive, power. We shall see in the following section, that realization of such nondemolition measurements which are essential in many designs of the gravitational wave detectors is by no means simple.

## 2. NONDEMOLITION MONITORING OF A QUANTUM HARMONIC OSCILLATOR

One of the most popular designs for a gravity wave detector is based on the prototype built and operated by Weber and his group. The centerpiece of such detectors is a "Weber bar", a solid cylinder with a weight that may be as large as several tons. Weber bar is built of such materials (special aluminum alloys, niobium, ...) and operated under such conditions (in vacuum and close to absolute zero) to ensure that frictional losses of energy from its vibrational degrees of freedom, measured by quality factor Q are as small as possible. A gravitational wave, passing through the bar, exerts a time-dependent force, which can change amplitude and/or phase of its vibrations. This may result, for example, in deposition of a few extra phonons. It is important to stress that because of the large value of Q ($\sim 10^{10}$ in some cases) and near-perfect isolation, coherent, macroscopic vibrations of this large, certainly macroscopic body can be calculated using quantum model of a harmonic oscillator, and solving appropriate Schrödinger equation. To analyze this type of gravity wave detectors one must therefore consider monitored harmonic oscillator, which is weakly coupled to the classical, time-dependent force F(t), exerted by the passing wave (6-10).

To predict the influence of the gravity pulse on the state of the bar one can then employ the usual formalism of quantum mechanics, with the hamiltonian

$$H = H_0 + V = H_0 - \hat{x}F(t). \tag{9}$$

Here $H_0$ is the self-hamiltonian of the harmonic oscillator. It can be expressed either in terms of the position and momentum operators, $\hat{x}$ and $\hat{p}$, or in terms of the usual annihilation and creation operators $a$ and $a^+$:

$$H = \hat{p}^2/2m + m\omega^2\hat{x}^2/2 = \omega(a^+a + \tfrac{1}{2}). \tag{10}$$

It is straightforward, if somewhat cumbersome, to show that if at $t_0$ the state vector was given by some initial $|\psi\rangle$, at a later time t it is:

$$|\psi(t)\rangle = U(t,t_0)|\psi(t_0)\rangle, \tag{11}$$

where:

$$U(t,t_0) = \exp[-i(t - t_0)H_0]\ \exp(-i\beta + \alpha a^+ - \alpha^* a) \tag{12}$$

and

$$\alpha(t,t_0) = i(2m\omega)^{-\frac{1}{2}}\int_{t_0}^{t} F(s)\ \exp(i\omega(s - t_0))ds, \tag{13}$$

$$\beta(t,t_0) = (i/2)\int_{t_0}^{t} (\alpha^*\dot{\alpha} - \alpha\dot{\alpha}^*)ds. \tag{14}$$

Consequently, if one could monitor the state of the oscillator the resulting record could be used to learn about the properties of the force $F(t)$. This plan hinges on the same tacit assumption that was pointed out at the end of the last section: The measurement should leave the system in the state which is known to the experimenter, so that any change of its state could be attributed either to the (known) $H_0$ or to the $F(t)$.

At first sight the ultimate limit on the sensitivity of such a quantum detector appears to be imposed by the principle of indeterminacy: For, if we attempt to detect, say, both the position and the momentum of the oscillator with the accuracy allowed by the Heisenberg's principle, $\delta p\ \delta x > \frac{1}{2}$, we shall alter the total energy of the quantum harmonic oscillator by an amount:

$$\delta E = (p + \tfrac{1}{2}\,\delta p)\ \delta p/m + m\omega^2\ (x + \tfrac{1}{2}\,\delta x)\ \delta x \tag{15}$$

which is readily seen to be smallest in the ground state:

$$(\Delta E)_{SQL} = \sqrt{2}\,\omega. \tag{16a}$$

In the limit of a highly excited oscillator with approximately n quanta, for which the state is $\sqrt{n}$ units away from the origin on the $(x, p/m\omega)$ phase diagram, a minimum uncertainty measurement will change its energy, on the average, by

$$(\Delta E)_{SQL} = (n + 1/4)^{\frac{1}{2}}\,\omega \tag{16b}$$

and its phase by

$$(\Delta\varphi)_{SQL} = 1/(2\sqrt{n}). \tag{16c}$$

This "Standard Quantum Limit" (SQL) will prevent one from detecting a force that has altered the state of the oscillator by less than the "error box" implied by the principle of indeterminacy. In terms of the previously defined quantity $\alpha(t,t_0)$ in order to be detected, F(t) must satisfy a condition:

$$|\alpha(t,t_0)|^2 = \left[\begin{array}{c}\text{the mean number of}\\ \text{quanta deposited by } F(t)\\ \text{in an oscillator which is}\\ \text{initially in a ground state}\end{array}\right] > 1. \tag{17}$$

It is perhaps worth noting that the standard quantum limit cannot be violated by performing many, m, measurements on the same quantum detector. The usual decrease of the error, $1/\sqrt{m}$, which results from the improved statistics, is exactly canceled out by the $\sqrt{m}$ "random walk" performed by the state of the harmonic oscillator as a consequence of m measurements (10).

Errors resulting from the standard quantum limit could be, of course, decreased by the $1/\sqrt{N}$ factor through the use of N distinct and independent Weber bars. Needless to say, this last method is economically prohibitive and not very practical.

There is a much more elegant, and efficient way to improve accuracy. To arrive at it one must notice that the standard quantum limit on the sensitivity of a detector is not a consequence quantum mechanics, but of the standard nature of an attempt, which involves measuring two noncommuting observables $\hat{x}$ and $\hat{p}$. Rather than fighting against the limit imposed by Heisenberg's indeterminacy relations, one can extract more information about the force F(t) by monitoring only one, suitably chosen observable and letting the other one "do what it pleases".

This idea is, of course, in accord with the principle of indeterminacy. One must, however, make sure that the state of the monitored system will remain unperturbed by this accurate measurement: only then the change of its state can be interpreted as caused by the measured force. The problem which must be solved can be then formulated as follows: How can one measure an observable $\hat{O}$ of a quantum system so that the system which is, before the measurement, in one of the eigenstates of $\hat{O}$, will remain in this very same eigenstate after the measurement is completed. A general answer to this question is quite straightforward: The measurement hamiltonian, $H_M$, which couples measured system with the measuring apparatus, must commute with the observable $\hat{O}$:

$$[H_M, \hat{O}] = 0. \tag{18}$$

This condition, pointed out long ago in a footnote of the paper by Landau and Peierls (11), is justified in more detail in the textbook of Bohm (12). In some practical realizations of the nondemolition measurements it is convenient to demand more than Eq. (18). If the state of the system is to remain unchanged by the measurement, it is essential for the system to be in the eigenstate of the nondemolition observable $\hat{O}$ at all time instants $t_i$ when the measurement is performed. This requirement will be satisfied when:

$$[\hat{O}(t_k), \hat{O}(t_\ell)] = 0, \tag{19}$$

where $\hat{O}(t)$ is the nondemolition observable in the Heisneberg picture, and $t_k$, $t_\ell$ are the instants of the measurement (10).

There are three distinct proposals of an experimental setup which allow a realization of the nondemolition condition Eq. (19) in a quantum harmonic oscillator.

Quantum-counting method is most straightforward theoretically. The measured nondemolition observable is the phonon number operator:

$$\hat{N} = a^+a. \tag{20}$$

$\hat{N}$ obviously commutes with the self-hamiltonian of the harmonic oscillator. Therefore, if one were able to find a coupling hamiltonian which depends only on $\hat{N}$, and does not depend on the phase of the oscillator one could satisfy the nondemolition condition, Eq. (18), (19). $\hat{N}$ would be a constant of motion and it would commute with itself at all times. Possible realizations of this quantum-counting scheme have been proposed. They involve so called "quadratic couplings" (9,10).

The state of the harmonic oscillator which serves as a detector will evolve, in between measurements, under the influence of the full hamiltonian H, Eq. (9). Let us suppose that at $t_0$ this state is given by the eigenstate $|n\rangle$ of the number operator $\hat{N}$, which contains n quanta:

$$\hat{N}|n\rangle = n|n\rangle. \tag{21}$$

In the absence of the force F(t) the oscillator will continue to be in the state $|n\rangle$. However, under the influence of F(t) it will evolve — providing that it is not disturbed by a measurement — into the state:

$$|\psi(t)\rangle = U(t_1,t_0)|n\rangle. \tag{22}$$

This new state is not an eigenstate of the number operator $\hat{N}$. It can be, of course, represented in the eigenbasis of $\hat{N}$:

$$|\psi(t)\rangle = a_n{}^{(n)}(t)|n\rangle + \sum_{m\neq n}^{\infty} a_m{}^{(n)}(t)|m\rangle. \tag{23}$$

Absolute values of the time dependent coefficients, which will be of interest to us in the further calculations, are equal to:

$$|a_n{}^{(n)}(t)|^2 = \exp(-|\alpha|^2 \; [L_n(|\alpha|^2)]^2 \tag{24}$$

$$|a_m{}^{(n)}(t)|^2 = (r!/s!)[L_r^{(s-r)}(|\alpha|^2)]^2|\alpha|^{2(s-r)}\exp(-|\alpha|^2) \tag{25}$$

where L are Laguerre polynomials, s = max (m,n), and r = min (m,n).

The sensitivity of the quantum counting method is in excess of the quantum standard limit, Eqs. (16), (17). This can be seen immediately by expanding the "decay" probability, i.e., probability that a number of quanta will change at all, in the powers of $|\alpha|^2$:

$$p(t) = 1 - |a_n{}^{(n)}(t)|^2 \approx 1 - \exp(-|\alpha|^2)[1 - n|\alpha|^2 + \ldots]^2 \tag{26}$$

and noticing that it becomes significantly different from zero when:

$$|\alpha(t,0)|^2 > (n+1)^{-1}. \tag{27}$$

Clearly, nondemolition counting of phonons allows one to detect forces which deposit only a fraction of a single quantum. It is important to stress that where the nondemolition measurement of $\hat{N}$ is performed on a system in a pure state $|\psi(t)\rangle$, Eq. (23), the state of the system "collapses" into one of the eigenstates of the number operator.

Measurement "collapses" the pure-state density matrix

$$\rho_{PURE} = |\psi(t)\rangle\langle\psi(t)| = \sum_m |a_m^{(n)}|^2 |m\rangle\langle m| + \tag{28a}$$
$$+ \sum_m \sum_{\substack{m' \\ m \neq m'}} a_m^{(n)}(t) a_{m'}^{(n)}(t)^* |m\rangle\langle m'|$$

and transforms it — for those observers which are unaware of the actual outcome, into a mixture of various phonon numbers:

$$\rho_{MIX} = \sum_m |a_m^{(n)}|^2 |m\rangle\langle m|. \tag{28b}$$

Any observer who knows the outcome of the measurement will assign to the oscillator pure eigenstate of $\hat{N}$:

$$\rho'_{PURE} = |n'\rangle\langle n'| \tag{28c}$$

Two other methods allowing a nondemolition measurement to be realized in practice are known as a "Back-Action-Evading" (BAE) and "Stroboscopic" techniques.

In Back-Action-Evading measurements the measured observables of the harmonic oscillator are:

$$\hat{X}_1 = \hat{x} \cos \omega t - (\hat{p}/m\omega) \sin \omega t \tag{29a}$$

$$\hat{X}_2 = \hat{x} \sin \omega t + (\hat{p}/m\omega) \cos \omega t. \tag{29b}$$

They are, therefore, respectively, a real and imaginary part of the oscillators "complex amplitude":

$$\hat{x} + i\hat{p}/m\omega = (\hat{X}_1 + \hat{X}_2)\, e^{-i\omega t} \tag{30}$$

and are conserved in the absence of any interactions between the oscillator and the outside world. Furthermore, they are complementary in the same sense in which $\hat{x}$ and $\hat{p}$ are complementary; they satisfy commutation relation

$$[\hat{X}_1, \hat{X}_2] = i/m\omega. \tag{31}$$

Therefore, they cannot be known with the accuracies $\Delta X_1$, $\Delta X_2$ which would violate the inequality:

$$\Delta X_1 \, \Delta X_2 \geq (2m\omega)^{-1}. \tag{32}$$

One can choose one of them, say $\hat{X}_1$, to be a nondemolition observable, and — by an appropriate measurement — prepare the oscillator in an eigenstate of $\hat{X}_1$. Under the action of an external, classical force F(t), $\hat{X}_1$ would evolve as follows:

$$\hat{X}_1(t) = \hat{X}_1(t_0) - \int_{t_0}^{t} (F(s)/m\omega) \sin \omega s \; ds. \tag{33}$$

Consequently, the system already in the eigenstate of $\hat{X}_1(t)$ will be shifted into a new state, $|X_1(t_0)\rangle \to |\hat{X}_1(t)\rangle$. This new state is, nevertheless, still an eigenstate of $\hat{X}_1$. Therefore, a measurement of $\hat{X}_1$ preserves the "purity of state" of the oscillator! This property distinguishes BAE technique from the phonon-counting, where measurement will "collapse" pure state $|\psi(t)\rangle$ in all "interesting instances", i.e., in all instances when there is a to-be-detected, classical F(t).

Stroboscopic technique relies on a judicious choice of the time instants at which some time-dependent observable of the oscillator — say, position $\hat{x}(t)$ — is measured. These time instants are chosen precisely so as to satisfy the commutation relation, Eq. (19). For a harmonic oscillator such opportune time instants occur every half a period of the oscillator (6).

## 3. WATCHDOG EFFECT AND NONDEMOLITION

The aim of this section is to show that nondemolition monitoring of a gravity wave detector does disrupt its hamiltonian evolution in a manner analogous to the one occurring in the watchdog effect. We shall investigate this phenomenon in some detail on an example of idealized, instantaneous, error-free nondemolition readoffs of the phonon number.

For phonon-counting measurements evolution of the density matrix is caused by the combined application of external classical force and instantaneous, precise readoffs. Each readoff results in the "collapse" of the state vector to some eigenstate $|n\rangle$ of the number operator $\hat{N}$. Therefore, after each readoff the density matrix will be given by the projection operator $\rho = |n'\rangle\langle n'|$, Eq. (28c). To show that such re-collapses can, under some circumstances, halt changes of the phonon number, let us consider a simplified case for which $|\alpha(\tau)|$ increases at an approximately constant rate $\lambda$ in the time interval of interest, and the observations are performed in equal time intervals $\Delta t = \tau/\nu$. The increase of $\alpha$ is exactly proportional to time when the wave is precisely on resonance with the detector, that is when the wave period matches the period of the oscillator $T = 2\pi/\omega$. When the typical frequency of the pulse of gravitational radiation, $\Omega$, differs a little from $\omega$, the rate of increase of $\alpha$ will be linear in time for the time interval of the order of $|\Omega - \omega|^{-1}$. Linear approximation will be also valid over short time intervals, smaller than both the time scale of variations of the gravitational wave and than the period of the oscillator. Under any of the circumstances which ascertain approximate linear dependence of $|\alpha|$ on time the probability of no change in the phonon number in any of $\nu$ consecutive measurements is approximately given by:

$$P_\nu(|n\rangle, 0 \rightarrow |n\rangle, \tau) \approx \exp(-\lambda^2 \Delta t^2)^\nu [1 - n\lambda^2 \Delta t^2 + \ldots]^{2\nu} \qquad (34a)$$

which can be written explicitly in terms of $\tau$ and $\nu$ as:

$$P_\nu(|n\rangle, 0 \rightarrow |n\rangle, \tau) \approx \exp(-\lambda^2 \tau^2/\nu)[1 - n\lambda^2 \tau^2/\nu^2 + \ldots]^{2\nu}. \qquad (34b)$$

The above expansion is accurate when $\lambda\tau/\nu \ll 1/\sqrt{n}$. In the limit of frequent observations $P_\nu$ approaches unity:

$$P_\nu(|N\rangle, 0 \rightarrow |N\rangle, \tau) \underset{\nu \rightarrow \infty}{\rightarrow} 1. \qquad (35)$$

The number of phonons in the Weber bar used as a detector may be fixed by too close surveillance, and the detector will be prevented from working!

The physical rationale behind this counterintuitive and peculiarily quantum prediction hinges on the notion of coherent superpositions: To be detected with a phonon number distinct from the one it used to have at $t = 0$, Weber bar must enter into a coherent superposition of the initial state and of the final states. The creation of this superposition from the initial state can be thought of as a rotation of the state vector in the Hilbert space by an angle which is proportional to the time interval for small t. Therefore, the probability of finding the system, upon a measurement, in the same state, is going to, for small angles, decrease as the square of the angle. It is, after all, given by a square of an absolute value of a scalar product of the old and the new state vectors, which is bound to behave as a cosinus! If the coherence of a superposition is destroyed by nondemolition measurements before the rotation in the Hilbert space can become significant, the system will not be permitted to change. Each measurement will find the system with the initial n phonons. Such continuous monitoring of the system will, in fact, induce <u>effective superselection rules</u> (13-15). It is worth stressing, that the intervention of "classical apparatus" or "conscious observer" is completely unnecessary: Both watchdog effect and environment-induced superselection rules can be understood on the grounds of purely hamiltonian evolutions, which lead to the buildup of correlations and to the resulting transfer of information (14,15).

Destruction of a coherent superposition of the eigenstates of the oscillator does not need to be detrimental to the process of detection. In some cases it may, in fact, be useful. To see how consider Weber bar acted upon by a weak, constant, classical force F. Function $\alpha(\tau)$ is then periodic and:

$$|\alpha(\tau)|^2 = f^2 \sin^2(\omega\tau/2) \tag{36}$$

where f is given by:

$$f = \sqrt{2}F/(m\omega^3)^{1/2}.$$

The survival probability is then also periodic:

$$P(|n\rangle, 0 \rightarrow |n\rangle, \tau) = \exp[-f^2 \sin^2(\omega\tau/2)] \\ [1 - nf^2 \sin^2(\omega\tau/2) + \ldots]^2. \tag{37}$$

In the above situation there is clearly no advantage to wait for $\alpha(\tau)$ to increase beyond $t \approx \omega^{-1}$. However, if the force F is applied for the duration of M periods, and the average probability of transition-per-readoff is:

$$\overline{P} = \langle P(t) \rangle \approx n\, f^2 \ll 1,$$

one may enhance the chance of detection of F by performing observations on a "one per oscillator period" basis. As a result, probability of transition from $|n\rangle$ to any other state will be:

$$P_M(|n\rangle, 0 \rightarrow |n\rangle, \tau) = \exp(-M\overline{P}) \tag{38}$$

which will decay to zero for $M\overline{P} \gg 1$.

Will Zeno effect impair performance of a given nondemolition scheme, and if so, by how much? How should one redesign gravity wave antennas to minimize its adverse effects and make use of potential benefits? These are examples of some of the questions that one is led to consider as a consequence of the above discussion. We shall not attempt to analyze them here. Instead, we shall limit our comments below to a few brief remarks, which are meant to define the questions more precisely rather than to give an answer to them.

Let us first note that the readoff of the detector will have no effect on the probability of finding it in the undecayed state if and only if the decay probability as a function of time is strictly exponential:

$$p(|\psi\rangle, 0 \rightarrow |\psi\rangle, t) = \exp(-\lambda t). \tag{39}$$

At first sight it may appear that detectors satisfying above decay law should be easy to design. This is not the case — exponential decay immediately after the readoff of the initial state would require infinitely high frequencies to be present in the spectrum of the perturbing hamiltonian V. A more practical suggestion is to limit the number of readoffs, so that a significant fraction of the pulse of F(t) may completely fit in between the two consecutive "collapses" of the detector state vector. This suggestion may be difficult to realize in the presence of thermal noise, when one has to ascertain that the change of the detector state was caused by a to-be-detected weak force and not be thermal excitations.

The decrease of sensitivity of the monitored quantum detector can be characterized by the ratio of the response to a given, standard, pulse of the weak force in the presence (density matrix $\tilde{\rho}$) and in the absence ($\rho$) of monitoring:

$$\varepsilon(\tau) = \mathrm{Tr}(\hat{O}\tilde{\rho}(\tau))/\mathrm{Tr}(\hat{O}\rho(\tau)) \ . \tag{40}$$

Above $\hat{O}$ denotes this observable of the quantum detector which is considered to be the best evidence of the measured phenomenon. In case of the gravity wave and of the Weber bar $\hat{O}$ may be, for example, the square of the difference between the initial and the final numbers of phonons; $\hat{O} = (\hat{N} - |n\rangle\langle n|)^2$, where n is the initial phonon number.

The density matrix $\rho(\tau)$ is obtained by evolving $\rho(0)$ with the von Neumann equation

$$-i\ \rho(t) = [\rho(t),\ H(t)] \tag{41}$$

where H is the total Hamiltonian. If the detector state was checked several times between 0 and $\tau$, then its density matrix $\tilde{\rho}(\tau)$ can be obtained by a procedure which alternates between unitary evolution, Eq. (19) and nonunitary "collapses of the state vector" which are brought about by the consecutive readoffs. The effect of each readoff is to erase off-diagonal terms of the density matrix, diagonal of which is defined by the eigenstates of the measured operator $\hat{N}$. It is perhaps worth a word of caution that in general the sensitivity $\varepsilon(\tau)$ will be not only a function of the strength of the pulse and of the distribution of readoff times, but also of the relative phases between the time dependent weak force and the instants of the readoff.

Finally, let us note that observables with continuous spectrum such as $\hat{X}_1$, $\hat{X}_2$ of BAE measurements, or $\hat{x}$ and $\hat{p}$ of the stroboscopic technique, may be immune to the watchdog effect. Physical reason for this immunity arises from the fact that the oscillator — once it is prepared in an eigenstate, of, say $\hat{X}_1$, shall remain in some eigenstate of $\hat{X}_1$ even in the presence of classical force F(t). Therefore, no recollapse of its state vector, which is the source of watchdog effect, shall ever be required.

This conclusion can be, however, guaranteed only in the idealized case of perfect eigenstates of $\hat{X}_1$ (which, as all Dirac deltas, are unphysical). For, if measurement of $\hat{X}_1$ prepares a gaussian approximation to a Dirac delta, with a half-width $\delta$, as is likely to be the case in actual experiments (see, e.g., Ref. 6), then each readoff will force a random shift of an expectation value of $\hat{X}_1$ by an amount of order $\delta$. Furthermore, in M consecutive measurements expectation value of $\hat{X}_1$ will perform random walk of $\delta\sqrt{M}$. This forced random walk is a counterpart of the watchdog effect for the case of an observable with a continuous spectrum, when there are no fixed, discrete, distinct states to "hang on to" — as was the case with $\hat{N}$.

This brief discussion does not allow one to draw conclusions about realistic scenarios, which involve thermal noise, as well as realistic model for a continuous readoff process. It appears, however, plausible, that among the three discussed schemes for the nondemolition detection, quantum counting is most susceptible to the watchdog effect.

The instantaneous readoff is, of course, a convenient but unrealistic idealization of the actual process of monitoring. It is therefore important to enquire how should one modify the treatment presented above in the case when each readoff takes a finite time span and involves a definite and finite coupling constant. Such a modification is straightforward, at least in principle. For, one can simply include important parts of the readout system in the new density matrix $\overline{\rho}$, and the von Neumann equation to calculate $\overline{\rho}(\tau)$. The resulting $\rho(\tau)$ can be obtained by partial trace — over the degrees of freedom of the readout system:

$$\rho(\tau) = \mathrm{Tr}\,\overline{\rho}(\tau). \qquad (42)$$

However, the part of the readoff system which is to be included in the calculation of the evolution of $\overline{\rho}(\tau)$ has not been specified. Therefore, the above procedure may appear to be incompletely defined. This ambiguity is, in the generic case, not as serious as it may appear at a first sight. The readoff system plays in the above discussion a role analogous with the role of the environment in the open quantum systems (16): The size of the "heat bath" does not alter its effects, as long as the heat bath includes all the degrees of freedom that couple to the considered system, and as long as its recurrence time is much in excess of the time intervals which are of interest in the calculation (13-15).

The purpose of this note was to show that even if the nondemolition monitoring does not perturb the state of the quantum detector, it may interfere with the very function such detectors are designed to perform: Consecutive "reductions of the detector wave packet" may result in a possibly significant decrease of sensitivity. Even though above we have focused on the practical questions, it is difficult to escape from a more general conclusion: Quantum theory of measurement, dismissed by some as a part of philosphy, has reemerged in the design of experiments, in the very center of "bread and butter" physics.

ACKNOWLEDGMENTS

This research was supported in part by a Richard Chace Tolman Fellowship at the California Institute of Technology, and by the National Science Foundation [AST79-22012-A01].

REFERENCES

1. Ekstein, H., and Siegert, A. J. F.: 1971, Ann. Phys. 68, p. 509.
2. Misra, B., and Sudarshan, E. C. G.: 1977, J. Math. Phys. 18, p. 756.
3. Chiu, C. B., Sudarshan, E. C. G., and Misra, B.: 1977, Phys. Rev. D 16, p. 520.
4. Peres, A.: 1978, Am. J. Phys. 46, p. 745; 1980, 48, p. 931.
5. Kraus, K.: 1981, Found. Phys. 11, p. 547.
6. Braginsky, V. B., Vorontsov, Yu. I., and Thorne, K. S.: 1980, Science, 209, p. 547.
7. Braginsky, V. B., and Manukin, A. B.: 1977, "Measurement of Weak Forces in Physics Experiments" (University of Chicago Press, Chicago).
8. Thorne, K. S., Drever, R. W. P., Caves, C. M., Zimmermann, M., and Sandberg, V. D.: 1978, Phys. Rev. Lett. 40, p. 667; Thorne, K. S., Caves, C. M., Sandberg, V. D., Zimmermann, M., and Drever, R. W. P.: 1979, in "Sources of Gravitational Radiation," edited by L. Smarr (Cambridge University Press, Cambridge); Caves, C. M.: 1982, in "Quantum Optics, Experimental Gravitation and Measurement Theory," edited by P. Meystre and M. O. Scully (Plenum Press, NY).
9. Unruh, W. G.: 1978, Phys. Rev. D 18, p. 1764; 1979, 19, p. 288.
10. Caves, C. M., Thorne, K. S., Drever, R. W. P., and Zimmermann, M.: 1980, Rev. Mod. Phys. 52, p. 341.
11. Landau, L., and Peierls, R.: 1931, Z. Phys. 69, p. 50.
12. Bohm, D.: 1951, "Quantum Theory" (Prentice-Hall, Englewood Cliffs, NJ).
13. Zurek, W. H.: 1981, Phys. Rev. D 24, p. 1516.
14. Zurek, W. H.: 1982, in "Quantum Optics, Experimental Gravitation and Measurement Theory," edited by P. Meystre and M. O. Scully (Plenum Press, NY).
15. Zurek, W. H.: 1982, Phys. Rev. D 26, p. 1862.
16. Davies, E. B.: 1976, "Quantum Theory of Open Systems" (Academic Press, NY).

# NONLOCAL HIDDEN VARIABLES AND NONLOCAL GAUGE THEORIES

Michel BOITEUX

Laboratoire d'Ultrasons[1], Université P. et M. Curie,
Tour 13, 4 place Jussieu, 75230 Paris Cedex 05, France.

*We propose a possible unification of classical fundamental interactions together with quantum interactions, based on an extension of the concept of local gauge invariance to a nonlocal gauge invariance. As an example this new concept is developped for the particular case of the electromagnetic field.*

"I think it is safe to say that no one understands quantum mechanics. Do not keep saying to yourself, if you can possibly avoid it, 'But how can it be like that ?' because you will go 'down the drain' into a blind alley from which nobody has yet escaped. Nobody knows how it can be like that".
R. P. Feynman

## INTRODUCTION

According to the Copenhagen interpretation, quantum mechanics is supposed only to give recipes to make predictions, not to describe how nature operates. Most physicists do not doubt the validity of quantum theory and see no reason for alternative models based on hidden variables. Many are however genuinely puzzled by the kind of nonlocal influence the Einstein - Podolsky - Rosen (1) correlations seem to imply.
Several ways-out have been suggested in the literature (2) :
(i) maybe current notions about space-time are still too innocent; (ii) maybe we must adhere to a superdeterminism in which the choice of the experimenter is not effectively free ; (iii) maybe the current axions of probability do not capture all the properties of quantum phenomena. The trouble is that these possibilities are not



*S. Diner et al. (eds.), The Wave-Particle Dualism, 529–536.*

easy to come to grip with.

Very early, de Broglie and Einstein advocated deterministic models. Deterministic local hidden variables models have been ruled out by experiments after the discovery of Bell's inequalities (2). There still lacks a first principle which could be used to single out these supposed-to-be (nonlocal ) hidden variables so that they remain truly hidden. The purpose of this paper is to exhibit such a plausible first principle and to analyse some of its implications.

In order to find the constraints this first principle has to fulfill, let us argue as follows. If there exists a deterministic model for describing quantum interactions, it must be a classical field theory, where the up-to-now hidden fields have to be explicited.

On the one hand, since in this way quantum interactions appear to be classical, they must fit into the existing unifying schemes of classical interactions (electromagnetic, weak, strong, gravitational). Still not completed, these schemes of the so-called gauge formalism are strongly believed to bear some truth (4). Anyway, they are the only schemes we have. Indeed, for our purpose, without presuming the final gauge theory, it is enough to consider the simplest case of the well-established gauge theory of classical electrodynamics. In our view, there must exist a kind of gauge field associated with the electromagnetic quantum interactions. Of course, it must have some differences with the classical ones since quantum interactions are of a very peculiar nature.

On the other hand, the corresponding classical field theory must be nonlocal (nonlocal fields) otherwise, it would not violate Bell's inequalities. Therefore, it seems we reach a contradiction since local gauge invariance involves local gauge fields. However it is possible to generalize local gauge theory into nonlocal gauge theory. This means the final theory will display a larger symmetry group in such a way that in the classical limit of well localized particles, we recover ordinary local gauge theory. To each local gauge invariance, we will associate a nonlocal one. Each standard fundamental classical interaction will have its own classical quantum field counterpart. The nonlocal gauge symmetry thus discovered must be considered as the characteristic symmetry of the corresponding quantum interaction in the same way as for example U(1) gauge symmetry is the symmetry of electromagnetic interactions or SU(3) gauge symmetry is the symmetry of strong interactions.

The outline is as follows. In part 1 we recall elementary results in classical abelian and nonabelian gauge theory. In part 2, we develop nonlocal (and nonabelian) gauge theory for electromagnetic interactions. Then in part 3 , we discuss some connections of this classical nonlocal field theory with the quantum theory of photons. Lastly in part 4, we comment on some interesting features of this new approach.

## 1. LOCAL GAUGE THEORIES (5)

Electromagnetic interactions are described by an abelian U(1) gauge group. Weak and strong interactions are adequately described, respectively, by the nonabelian groups SU(2) and SU(3). Even better, electromagnetic and weak interactions have been unified within the framework of flavour dynamics of which the minimal $SU(2)\times U(1)$ gauge theory is the most attractive. There exist even " grand unified schemes" describing a world of color and flavour but without gravity. Finally "grand grand unified schemes" which include gravity in a gauge theory are currently under rapid development.
One of the most attractive features of nonabelian gauge theories without symmetry breaking is that they are the only field theory to lead to a renormalizable quantum field theory displaying asymptotic freedom.

### 1.1. Electromagnetic gauge theory.

Let us consider a complex scalar field $\phi(x)$ describing classically a charged spinless particle with lagrangian density.

$$\mathcal{L} = \frac{1}{2}\left(\partial_\mu \phi^* \partial^\mu \phi - m^2 \phi^* \phi\right) \tag{1}$$

$\mathcal{L}$ is invariant under a phase transformation

$$\phi \to g\phi = \phi' \quad , \quad g = e^{i\alpha} \tag{2}$$

where g is a constant phase factor. g is an element of the abelian unitary group U(1). If now $\alpha$ is a function $\alpha(x)$, x a point in Minkovsky space, then $\mathcal{L}$ will no longer be invariant unless we replace the derivative $\partial_\mu$ with the covariant derivative $D_\mu$ which commutes with the gauge transformation g.

$$D_\mu(g\phi) = g D'_\mu \phi \tag{3}$$

where
$$\begin{aligned} D_\mu &= \partial_\mu + i e A_\mu \\ D'_\mu &= \partial_\mu + i e A'_\mu \end{aligned} \tag{4}$$

From Eq. (3) and (4), $A_\mu$ transforms under a gauge transformation g as

$$i e A'_\mu = i e A_\mu + i \partial_\mu \alpha(x) \tag{5}$$

$A_\mu$ is chosen to be a real valued vector and identified with the electromagnetic potential. The quantity

$$[D_\mu, D_\nu] = i e F_{\mu\nu} \tag{6}$$

is gauge invariant. Morever, $F_{\mu\nu}$, the electromagnetic gauge field vanished if and only if A is a pure gauge field ($A_\mu = \frac{1}{e}\partial_\mu \phi$ for some real valued function $\theta(x)$). The lagrangian density of the electromagnetic gauge field is

$$\mathcal{L}_{em} = -\frac{1}{4} F_{\mu\nu} F^{\mu\nu} \tag{7}$$

1.2. Yang - Mills gauge theory

We may define the covariant derivatives of any field taking values in a representation space of any Lie group. The Yang - Mills gauge field $A_\mu$ takes its values in the Lie algebra of the nonabelian group SU(2).

$$A_\mu = -\frac{i}{2} f A^a \tau_a \tag{8}$$

where $\tau_a$ , a = 1, 2, 3, are the Pauli matrices.
If $\phi$ lies in the fundamental representation of SU(2)

$$D_\mu \phi = \partial_\mu \phi + A_\mu \phi \tag{9}$$

and for $g \in SU(2)$

$$D_\mu (g\phi) = g D'_\mu \phi \tag{10}$$

so that $$A'_\mu = g^{-1} A_\mu g + g^{-1} \partial_\mu g \tag{11}$$

The Yang - Mills field $F_{\mu\nu}$

$$F_{\mu\nu} = [D_\mu, D_\nu] = \partial_{[\mu} A_{\nu]} + [A_\mu, A_\nu] \tag{12}$$

transforms under g as

$$F_{\mu\nu} = g^{-1} F_{\mu\nu} g$$

$F_{\mu\nu} = 0$ if and only if $A_\mu$ is a pure gauge field $A_\mu = g^{-1}\partial_\mu g$ for some $g \in SU(2)$ (as a consequence of the Frobenius' theorem).
The Yang - Mills lagrangian density is

$$\mathcal{L}_{YM} = -\frac{1}{4f^2} \operatorname{Tr}\left(F_{\mu\nu} F^{\mu\nu}\right) \tag{13}$$

Remarks : (i) the coupling constant is usually absorbed in the definition of $A_\mu$ ; (ii) nonabelian gauge theories differ from abelian gauge theories by the nonlinear terms in $F_{\mu\nu}$. This results in different behaviour under quantization. Nonabelian gauge theories only display asymptotic freedom or confinement ; (iii) gauge invariance forbids the existence of the gauge field mass term in $\mathcal{L}$. However, if there are symmetry breaking terms for the particle field in the lagrangian, then the gauge field becomes massive due to Higg's mechanism.

## 2. NONLOCAL GAUGE THEORY

The material of this section is new. Rather than give a general discussion, it is written to be explicit and self-contained. Indeed, every local gauge theory can be extended to a nonlocal one under a procedure to be explained below in the case of the U(1) electromagnetic gauge theory.

### 2.1. Nonlocal gauge transformations.

Instead of the invariance of the lagrangian density (Eq. 1), let us consider the invariance of the action S

$$S = \int d^4x \ \mathcal{L} \tag{14}$$

under linear transformations of the field $\phi$.
Let us introduce the Dirac bra-ket notations for the quantity

$$\int d^4 x \ \phi^*(x) \ \phi(x) = \langle \phi | \phi \rangle \tag{15}$$

$|\phi\rangle \in \mathcal{H}$, $\langle\phi| \in \mathcal{H}^*$ the dual space of $\mathcal{H}$. $\langle\phi|\phi\rangle$ is invariant under the unitary group $U(\mathcal{H})$ with elements

$$\hat{U} = \exp\{i \iint d^4x\, d^4y\, \theta(x,y) \ |x\rangle\langle y| \} \tag{16}$$

where $\theta(x,y) = \theta^*(y,x)$.
The quantities

$$L_{xy} = |x\rangle\langle y| \tag{17}$$

span the r-representation of the Lie algebra of the unitary group on $\mathcal{H}$.
They fulfill the commutation relations

$$[L_{xy}, L_{vw}] = \delta_{yv} L_{xw} - \delta_{xw} L_{vy} \tag{18}$$

and are normalized accordingly to

$$tr(L_{xy} L_{vw}) = \delta_{xw} \delta_{yv} \tag{19}$$

It is not difficult to elevate the U(1) local gauge invariance to a gauge invariance under the group $U(\mathcal{H})$. Under Eq. (16), $\langle\phi'|\phi\rangle$, $|\phi\rangle, |\phi'\rangle \in \mathcal{H}$, is invariant, but not $\langle\phi'|\partial_\mu|\phi\rangle$. This can be cured by introducing a nonlocal gauge $A_\mu$, a linear operator on $\mathcal{H}$. We use the notations

$$\hat{D}_\mu = \hat{\partial}_\mu + i\hat{A}_\mu \tag{20.a}$$

$$\hat{\partial}_\mu = \iint d^4x\, d^4y\ \partial_\mu\big(\delta(x-y)\big) \ |x\rangle\langle y| \tag{20.b}$$

$$\hat{A}_\mu = \iint d^4x\, d^4y\, A_\mu(x,y)\ |x><y| \tag{20.c}$$

We require $\hat{A}$ to transform under $U(\mathcal{H})$ in a special way, so that $\hat{D}_\mu|\phi>$ transforms like $|\phi>$. Hence

$$\hat{D}'_\mu\,\hat{U} = (\hat{\partial}_\mu + i\,\hat{A}'_\mu)\,\hat{U} = \hat{U}\,\hat{D}_\mu = \hat{U}\,(\hat{\partial}_\mu + i\,\hat{A}_\mu) \tag{21}$$

If $\hat{U}$ is close to the unity ($\theta = \delta\theta$ infinitesimal), to first order we get

$$\delta\,\hat{A}_\mu = -\,[\hat{D}_\mu, \delta\hat{\theta}] \tag{22}$$

or

$$\delta\,A_\mu(x,y) = -\frac{\partial\delta\theta}{\partial x^\mu} - \frac{\partial\delta\theta}{\partial y^\mu} - i\int d^4z\,\Big(A_\mu(x,z)\,\delta\theta(z,y) - \delta\theta(x,z)\,A_\mu(z,y)\Big) \tag{23}$$

From Eq. (22), $\hat{A}_\mu$ can be chosen to be hermitian. Thus $A_\mu(x,x)$ is real and does not carry any electrical charge while $A_\mu(x, y\neq x)$ is complex and carries electrical charge.

## 2.2. Nonlocal field strengh

We define the field strengh $\hat{F}_{\mu\nu}$ as

$$\hat{F}_{\mu\nu} = i\,[\hat{D}_\mu, \hat{D}_\nu] = \iint d^4x\, d^4y\, F_{\mu\nu}(x,y)\ |x><y| \tag{24}$$

$$F_{\mu\nu}(x,y) = \frac{\partial A_\mu}{\partial x^\nu} + \frac{\partial A_\mu}{\partial y^\nu} - \frac{\partial A_\nu}{\partial x^\mu} - \frac{\partial A_\nu}{\partial y^\mu} - i\int d^4z\,\Big(A_\mu(x,z)\,A_\nu(z,y) - A_\nu(x,z)\,A_\mu(z,y)\Big) \tag{25}$$

Although $U(\mathcal{H})$ is not simple and may give several coupling constants, at this stage of the theory, especially since the $\Pi\, U(1)_x$ gauge group is not an invariant subgroup and since the $L^x_{xy}$ are considered on an equal footing, we shall consider only one coupling constant $f$.

Then the pure nonlocal gauge field action, a real quantity, reads

$$S_{NG} = -\frac{1}{4f^2}\,\mathrm{tr}\Big(\hat{F}^+_{\mu\nu}\,\hat{F}^{\mu\nu}\Big) \tag{26}$$

Remarks.

(i) If we chose $\theta(x,y) = \theta(x)\,\delta(x-y)$ and $A_\mu(x,y) = A_\mu(x)\,\delta(x-y)$ we recover the electromagnetic U(1) gauge theory. This explains the choice of the sign − in Eq. (26).

(ii) $U(\mathcal{H})$ is nonabelian. Hence the theory is nonlinear. This is true even for a real scalar field (6) in which case there is no local gauge field, but always a nonlocal gauge field. This is a necessary condition since a real scalar field can be quantified and we want the classical nonlocal gauge field to produce quantization.

(iii) $$\hat{F}^+_{\mu\nu} = \hat{F}_{\mu\nu} \tag{27}$$

Therefore

$$S_{NG} = -\frac{1}{4f^2} \iint d^4x\, d^4y\, F^*_{\mu\nu}(x,y)\, F^{\mu\nu}(x,y) \tag{28}$$

### 2.3. Field equation

The classical equations of motion are derived from the stationarity condition of the action $S_{NG}$ given in Eq. (26).
Using the symmetry properties of $\hat{A}_\mu$, $\hat{D}_\mu$ and $\hat{F}_{\mu\nu}$ we get

$$[\hat{D}^\mu, \hat{F}_{\mu\nu}] = 0 \tag{29}$$

which provides a nonabelian generalization of Maxwell equations. The nonlinear and nonlocal character of Eq. (29) makes their general resolution nontrivial.
We introduce the analogs of the electric and magnetic fields

$$\begin{aligned} E^i(x,y) &= F^{io}(x,y) \\ B^i(x,y) &= -\tfrac{1}{2}\,\varepsilon_{ijk}\, F^{jk}(x,y) \end{aligned} \qquad i,j,k = 1,2,3 \tag{30}$$

## 3. SOME CONSEQUENCES

The equations of motions derived from the action $S + S_{NG}$ are nontrivial and different from those derived from a local gauge theory only.

### 3.1. The classical limit

If we consider the galilean limit, in which case

$$\int d^3x\, \phi^*(x)\, \phi(x) = \langle \phi | \phi \rangle$$

can be interpreted as the conservation of the quantum particle described by the wave function $\phi(x)$, we have to restrict our nonlocal gauge transformations to space nonlocal gauge transformations only (local in time). Then if the particle is well localized, the quantity $\langle \phi | D_\mu | \phi \rangle$ involves only the diagonal part $A_\mu(x,x)$ of the gauge field. Hence the nonlocal gauge field has no macroscopical effect. As the field equations are nontrivial, there must be microscopical effects.

### 3.2. Quantization

Let us try to see what a quantized nonlocal gauge field theory looks like using, for example, a path integral technique with vanishing gauge symmetry breaking terms. What do we expect to happen? Since the diagonal part $A_\mu(x,x)$ does not carry any charge we may expect it to show an ordinary quantum electrodynamic behaviour,

that is a nonvanishing renormalized coupling constant e . Since the nondiagonal part $A_\mu(x , y \neq x)$ carries charges and since the theory is nonabelian, we expect it to show quantum chromodynamics behaviour that is a kind of asymptotic freedom. Since the $U(\mathcal{H})$ group is larger (indeed the Casimir operator $C_2 = \infty$) , we expect a vanishing renormalized coupling constant for the nondiagonal part, even at small momenta.
What does this mean ? On the one hand, we have no macroscopical effect. On the other hand, all microscopical effects seem to be taken into account -inside a statistical approach- once we have quantized the local gauge field. The only nontrivial possibility is that the nonlocal gauge field is just the classical field responsible for quantum phenomena.
The above arguments give some hints in this direction. Obviously, a convincing proof would be a derivation of a quantization prescription, at least in some approximation, from a stochastic approach to the classical equations of motion for the nonlocal gauge field.

## 4. CONCLUSION

First, we want to give the physical meaning of the nonlocal gauge invariance. The above analysis shows that our notions about space-time are still too innocent. In a world with only microscopical states, it is impossible to define a reference frame for particles. All the space representations $\hat{U}|r\rangle$ are equivalent. It is only once we have macroscopical bodies (or apparatus) that we can fix the nonlocal gauge and define a representation space.
"One of the main objectives of Science is to explain the many diverse physical phenomena by one underlying theory. But, not only a unification of all forces is needed, also a unification of the principles on which the construction of theories is based" (4).

It is a pleasure to thank J. ILIOPOULOS and G. SOBOTTA for valuable discussions.

[1]Associated with the Centre National de la Recherche Scientifique.

(1) Einstein, A., et al: 1935, Phys. Rev. 47, pp. 777.
(2) Garg, A., and Mermin, N.D.: 1982, Phys. Rev. Lett. 49, p.p. 1220-1223.
(3) Bell, J.S.: 1964, Physics (N.Y.) 1, pp. 185.
(4) Van Nieuwenhuizen, P.: 1964, Supergravity - Phys. Reports 68, p.p. 192-395.
(5) Lee, T.D.: 1981, "Particle Physics and Introduction to Field Theory" in Contempory Concepts in Physics, Vol. 1 (Harwood Academic Publishers, New-York) ;

# A FINITE PARTICLE NUMBER APPROACH TO PHYSICS [1]

H. Pierre Noyes

Stanford Linear Accelerator Center

Starting from a discrete, self-generating and self-organizing, recursive model and self-consistent interpretive rules we construct: the scale constants of physics ($3,10,137,1.7\times10^{38}$); 3+1 Minkowski space with a discrete metric and the algebraic bound $\Delta\varepsilon\Delta\tau \geq 1$; the Einstein-deBroglie relation; algebraic "double slit" interference; a single time momentum space scattering theory connected to laboratory experience; an approximation to "wave functions;" "local" phase severence and hence *both* distant correlations *and* separability; baryon number, lepton number, charge and helicity; $m_p/m_e$; a cosmology not in disagreement with current observations.

## I. INTRODUCTION

The claims made in the abstract will strike most physicists as incredible, as they do the author. They are the consequence of the convergence of a number of lines of research extending over three decades, some of which has been published (1,2,3), This contribution to the Symposium on Wave-Particle Dualism[2] is the first attempt to bring these strands together in a coherent fashion. A more detailed presentation by Amson, Bastin, Kilmister, Parker-Rhodes, Stein and this author is in preparation;[3] errors of omission and commission in this version are the sole responsibility of this author.

That quantum mechanics requires revision at a fundamental level, though often denied by workers in the vineyard of experimental and mathematical physics, is no news to the participants of this symposium celebrating Louis deBroglie's 90th birthday and his fundamental contribution to the new direction taken by physics in this century. The point of view developed here is that the discrete phenomena exemplified by quantum mechanics cannot be consistently



*S. Diner et al. (eds.), The Wave-Particle Dualism, 537–556.*

incorporated within the continuum framework of Euclidean geometry and the Newtonian synthesis. Bohr believed that this haptic language was inescapable and tried to meet the problem by a "correspondence principle" and the profound idea of "complimentarity." Einstein and deBroglie challenged this approach as productive of paradox, or at least incompleteness, but for most working physicists the "Bohr-Einstein debate" left Bohr in possession of the field; they went on to more pragmatic concerns. The existence of the de Broglie Foundation and this Symposium is evidence that the issue is far from dead; indeed it has become a fruitful field for experimental study (4).

One way to see why the foundation of physics on continuum standards of mass, length and time with their implied "scale invariance" is inadequate is to follow the method of operational analysis advocated by Bridgman (5) applied to thought experiments in the tradition of Galileo and Einstein. For example we can consider the double slit experiment using deBroglie waves with particle detectors in the slits (6) as well as in the array where the statistical data showing both single and double slit interference accumulates. Since nonrelativistic quantum mechanics depends only on h and m, it is scale invariant, and it would seem that one could establish these effects at macroscopic dimensions and then extrapolate to arbitrarily short distances. However careful analysis shows (6) that if there is a smallest finite mass (usually assumed to be $m_e$) this extrapolation is frustrated.

This recent analysis is but another version of earlier studies. In discussing the measurability of the electromagnetic field, Bohr and Rosenfeld (7) made use of the fact that quantum electrodynamics rests on only two dimensional constants (h and c) and hence that their thought experiments incorporating the (nonrelativistic) uncertainty principle could be performed with macroscopic apparatus and then extrapolated to arbitrarily small distances. But they conclude their paper by noting that once there is a mass parameter in the theory, the analysis breaks down. This challenge to the operational justification for second quantization has never, in our opinion, been met; thus we distrust current quantum field theories for massive particles.

Of more immediate relevance to our current effort is the profound analysis by Wick (8) of Yukawa's meson theory (9). Since special relativity requires that a region of dimension r can only act coherently during a time interval $\Delta t$ for $r < c\Delta t$ and by Heisenberg's uncertainty principle $\Delta t \approx \hbar/\Delta E$, if two systems within a region bounded by r communicate with a system of mass $\mu$ then if $\Delta E \geq \mu c^2$ (by special relativity again) such a mass can be present. Accepting Newton's third law, total momentum will be conserved but the relative momentum between the two systems becomes arbitrary except for that restriction and hence they will "scatter" if they approach each other closer than

$$r \leq c\Delta t \approx c\hbar/\Delta E \leq c\hbar/\mu c^2 = \hbar/\mu c$$

Extending the argument to shorter distance, if $r < \hbar/n\mu c$ there can be n "quanta" present and concept of "particle" as a point with mass breaks down. This is, of course, one of the sources of the infinities in conventional quantum field theory. Thus special relativity and quantum mechanics when coupled together prevent any consistent continuum description of short distances from being constructed.

Yet another example of the problem is provided by Dyson's analysis (10) of what must happen in *renormalized* quantum electrodynamics when more than 137 charged particle pairs are found within a radius $\hbar/2$ mc. Their electrostatic energy then exceeds the energy needed to create another pair and the problem looses precise definition. By extension, the same analysis applies to $1.7 \times 10^{38}$ gravitating baryons of nucleonic mass (11) (they form a Laplacian "black hole"). Thus we already have within conventional physics a reason to take seriously a theory such as the one presented here which calculates the pure numbers 137 and $1.7 \times 10^{38}$ in such a way as to associate them with these quantum phenomena.

We hope that these examples from within physics will motivate some of our colleagues to consider our claim that fundamental revision might be worth while. Mathematical objections to the continuum have, of course, a much longer history starting at least with the paradoxs of Zeno, as discussed by Aristotle, the adoption of a randomized version of Aristotle's least step by Epicurus, and continuing to the present day in the constructive mathematics of Brouwer, Bishop and Martin-Löf.

Another discipline that contributes to the development of our theory is computer science. Manthey (12) claims that concurrent asynchronous communicating digital systems whose communications are required to be synchronized *necessarily* have an uncertainty principle. He also shows that quantum systems can be viewed as concurrent processes described by finite algorithms. Of course any constructive mathematical theory must in principle be computable; this principle has been very helpful in our research.

For this author one important motivation for this research stems from a desire to provide a conceptual framework for physics that is broad enough to encompass all sciences. Given conservation laws, random background, and heritability of characteristics providing dynamic stability against that random background, neo-Darwinian evolution (for species defined by a statistical ensemble and not by a single genotype (13)) is inevitable. We will find that all these characteristics occur in the theory developed here. We will also see that although we can under appropriate circumstances talk about approximately separated systems (approximately because ultimately all our processes have a common origin) and hence in a sense can talk about "observers" our "measurement theory" does not allow them to obtrude in the way they do in von Neumann's quantum mechanics.

## II. GENERATION AND CONSTRUCTION OF THE ALGEBRAIC MODEL

Although the precise form of the generating operation we adopt is not critical, provided it can start from the empty set $\emptyset$, we take for concreteness that due to Conway (14):

G: If L, R are two disjoint subsets of S, adjoin {L/R} to S.

This gives us (as we will see) the natural numbers 0,1,2,... which become (a) our ordering parameters, (b) the source of our "quantization," and (c) the tool we need for constructing "coherence." But since we do not know at any particular stage just what elements are in play using just this operation, we must define some process by which we can check whether an element generated by G is novel or not. This is done recursively as follows.

Specify a function $f: S \times S \to T$ (here T is some collection which includes S) with the property that there is a fixed subcollection Z of S and $fuv \in Z$ iff u and v are the same element. We also require (i) $fuu \in Z$, (ii) $fuv \in Z \Rightarrow fvu \in Z$, (iii) $fuv \in Z$ and $fvw \in Z \Rightarrow fuw \in Z$. If 0 is the unique element of some one element Z we can then prove[1] that

Theorem 1: Every f is equivalent to some g for which (i) $guu = 0$ (ii) $guv = gvu$, (iii) $gu(gvw) = g(guv)w$.

Proof: (a) Define $\bar{f}uv = 0$ if $u = v$, $= fuv$ otherwise. Evidently $\bar{f} \equiv f$ and $\bar{f}uu = 0$. (b) Suppose the elements of S to have been ordered by any recursive process, and define $\bar{\bar{f}}uv = \min(\bar{f}uv, \bar{f}vu)$ where 0 is counted as the least element and 1,2,3,... are regarded as ordered in the usual way. Easily $\bar{\bar{f}} \equiv \bar{f} \equiv f$ and $\bar{\bar{f}}uu = 0$, $\bar{\bar{f}}uv = \bar{\bar{f}}vu$. (c) [Conway's trick] Define $guv$ = the least element w such that $\bar{\bar{f}}w(g\bar{u}v) \neq 0$, $\bar{\bar{f}}w(gu\bar{v}) \neq 0$ for all $\bar{u}<u$, $\bar{v}<v$. Then, easily, $g \equiv \bar{\bar{f}} \equiv f$ and, since $guv = 0$ iff u and v are the same element, we have $guu = 0$. $guv = gvu$, and also $guv = 0$ and $gvw = 0$ implies that $guw = 0$.

It remains to prove (iii) which, following Conway, is most easily done by explicit construction. From the definition $g00 = 0$, $g01 = 1 = g10$, so $g11$ cannot be 1 but it can be 0: therefore $g11 = 0$. $g02$ cannot be 1 or 0 but can be 2. Then $g12$ cannot be 2, 1, 0, so must be called 3. Next $g22 = 0$, and so on.

It is straightforward, if a little tedious to prove:

Theorem 2: If S is a closed system then $|S| = 2^n$ for integral n and there is an isomorphism $(S,g) \simeq (V_n,+_2)$ where $V_n$ is the vector space of n dimensions over $\mathbb{Z}_2$.

One can then go on to discuss closure under this operation (discrimination), define a minimum labeling rule based on this property (15) and prove

Theorem 3: There is a unique complete hierarchy with more than two levels; it has successively completed levels of 3, 10, 137 and $2^{127} - 1 + 137 \approx 1.7 \times 10^{38}$ elements, beyond which further extension is impossible.

The proof of this theorem by Kilmister (15) is unpublished, but an earlier proof by Amson is given in the appendix to (2).

Since we will have need for explicit representations of the hierarchy later, we will use a more explicit construction. Thanks to Theorem 2, we are allowed to consider bit strings of length n $(x_1,\ldots,x_i,\ldots x_n)$ $x_i \in 0, 1$ and for fixed n replace g by *discrimination* defined by $D_n xy = (\ldots x_i +_2 y_i \ldots)$ where $+_2$ is addition mod 2. We then define a "discriminately closed subset" (DCsS) as any set of nonnull strings such that discrimination applied to any two elements gives another element of the set, and we include also the singletons. For example, if we have two linearly independent strings a, b we have the three DCsS: {a}, {b}, {a,b,a+b}. If we had a third linearly independent element c, then we have in addition {c}, {b,c,b+c}, {c,a,c+a}, {a,b,c,a+b,b+c,c+a,a+b+c}. If we had a third we have j linearly independent strings then we can form $2^j - 1$ DCsS, since this is the number of ways we can take j distinct things 1, 2, ..., j at a time. Thus, starting with the strings (01) and (10) we have {(10)}, {(01)}, {(10),(01),(11)}. If we now ask for $2 \times 2$ matrices for which these sets are the only eigenvectors, which are linearly independent, and nonsingular so that they do not map onto zero, we can represent the information about discriminate closure contained in these $2^2 - 1 = 3$ DCsS at a level of greater complexity. Rearranged as strings of length 4, these matrices can then form a basis for constructing $2^3 - 1 = 7$ DCsS. In this case the maximal set with seven elements is (up to an arbitrary reordering of the bits in all the elements simultaneously) unique: {(1110),(1100),(1101),(0010),(0011),(0001),(1111)}. This can again be mapped (2) by $4 \times 4$ matrices, and a third level with $2^7 - 1 = 127$ DCsS constructed. These can again be mapped by $16 \times 16$ matrices (2) and these in turn used to construct a fourth level with $2^{127} - 1$ DCsS. But the process terminates here since there are only $(256)^2$ linearly independent matrices of "0's" and "1's" and these then cannot be used to map all the elements of level 4.

We see that the hierarchy labeling scheme is exhausted after $2^{127} + 136$ labels of length 256 have been assigned, and after generation and discrimination have filled up the remaining labels of length 256, all we can do is to start increasing the length of the strings. We call this additional part of the string, of length n, the "address;" that is, any string from now on will be of length $256 + n$ with the first 256 bits called the "label" and the n bits which follow called the "address."

## III. CONSTRUCTION OF THE STATISTICAL MODEL

Since, because of discriminate closure, discrimination on the labels in our model will keep on throwing up the same values if we wait long enough, eventually we will come to have a large collection of strings all with the same label and with the same number of bits $n = N_0 + N_1$ of the address where $N_0$ and $N_1$ are the number of zeros and 1's in the address, however distributed. Our task is to find a well defined way to "break into" this, at first sight chaotic, situation using rules that can eventually be related to laboratory experience. We hope, ultimately, to derive these stability conditions from our basic operations; the rules we choose are, we claim, consistent with them but whether they are unique remains an open question.

The laboratory paradigm on which the rules are modeled comes from experience in high energy physics laboratories where "particles" are defined by the sequential firing of counters at fixed geometric positions, velocity defined by the ratio of the distance between counters to the time interval between firings, and their mass ratios measured by momentum-energy conservation. Our first step is to define the rational fraction $\omega b=(N_1-N_0)$. Clearly $-1\le\omega\le+1$, and each labeled address is characterized by $A_L(\omega,b)$. Since a large number of addresses carry the same label, we seek an elementary sampling procedure which will select out an ensemble $\mathbb{A}_L(\nu,b)$ for which $\nu=\langle\omega\rangle$. Consider three such samplings through which the label $L_1$ persists ordered by increasing bit number $b_1<b<b_1'$ and for which at the second sampling $b,\langle\omega\rangle=\nu_1$ while at the third $b_1',\langle\omega\rangle=\nu_1'$. Each sampling will involve additional sequences of labeled ensembles; the persistence of $L_1$ is insured by requiring that in the sequence we are discussing $DL_1L_i=L_j$, $DL_1L_2=L_3$, $DL_1L_i'=L_j'$. Consider now the second sequence characterized by label $L_2$ and ordered by $b_2<b<b_2'$, where the value b is required to be the same by requiring $D_bA_1A_2=A_3=D_bA_1'A_2'$; if the bit number b is not common, then no sampling can occur by our fundamental requirement on discrimination.

The label persistence we have now guaranteed is not sufficient to specify the connection between the values $\nu_1\nu_2$ and $\nu_1'\nu_2'$ before and after the sampling (i.e., ordered by $b_i<b<b_i'$). For this we assume that to each label L we can assign a unique positive value (which may be 0 for some labels), and define two functions $\varepsilon(\nu)$, $\rho(\nu)$ by $\varepsilon^2-\rho^2=\mu^2$, $\rho(\nu)=\nu\varepsilon(\nu)$. Then we require that for a sampling to take place the ensembles are connected by the requirement $\rho_1(\nu_1)+\rho_2(\nu_2)=\rho_1(\nu_1')+\rho_2(\nu_2')$.

In our last step we have clearly defined "momentum conservation" in a way that adumbers Einstein's modification of Newton's third law. We feel comfortable with this because the only obvious structure in our address strings other than $b=N_0+N_1$ is the difference defining a rational fraction bounded in absolute value by unity. We also have available the integer ordering parameters $b_1<b<b_1'$ and $b_2<b<b_2'$ which can serve as a "time ordering". The fact that we must rely on discrimination in constructing our information-preserving structure then provides "synchronization" because b must lie between both limits. What is missing is some construction that will connect these elements to a "metric space" in the geometrical sense. Here we are guided by Stein's insight (16) that an ensemble of bit strings of bit number b can be viewed as a "random walk" which has taken b steps.

With this model in mind, we think of the first sequence as starting when the bit length is $b_1^{(0)}$ and that the generation operation has added b bits on the end of this string. We now construct our ensemble which is sampled at the "intersection" of the two sequences using only these b bits, and define the "distance" from the reference position $\xi_1^{(0)}$ to the second sampling as $\xi_1(\nu_1,b)-\xi_1^{(0)}$ with $b=b_1-b_1^{(0)}$. Clearly in this interpretation the mean position of the distribution is at $\nu_1 b$. But the probability of a step in this direction is $\langle N_1/b\rangle=(1+\nu_1)/2$ while the probability of a step in the opposite direction $\langle N_0/b\rangle=(1-\nu_1)/2$. Thus the standard deviation $\sigma=\langle(N_1-N_0)^2\rangle^{\frac{1}{2}}=(b(1-\nu_1^2)/4)^{\frac{1}{2}}$. We see that even if $\nu_1=0$, there is a 50% chance that the sampling will give a position $\sqrt{b}/2$ or greater from the position of the mean. Hence we are lead to define our distance not by the peak of the distribution but by $\xi_1(\nu_1,b_1-b_1^{(0)})-\xi_1^{(0)}-\nu_1(b_1-b_1^{(0)})=(\sqrt{b}/2)\sqrt{1-\nu_1^2}$. But what we mean by the velocity parameter $\nu_1$ is not yet clear. We specify it by assuming that for the second sequence $\nu_2=0$, and hence that $\xi_2(0,b_2-b_2^{(0)})=\sqrt{b}/2$. But the sign of the velocity is still arbitrary, since it comes from choosing $N_1-N_0$ rather than visa versa in our basic definition. Clearly we can resolve the ambiguity by treating $\nu$ as the <u>relative</u> velocity between the two ensembles. Thus we have derived $\sqrt{b}/2=(\xi_2-\xi_2^{(0)})=[\xi_1-\xi_1^{(0)}-\nu(b_1-b_1^{(0)}]/\sqrt{1-\nu^2}$; $\sqrt{b}/2=\xi_1-\xi_1^{(0)}=[\xi_2-\xi_2^{(0)}+\nu(b_2-b_2^{(0)})]/\sqrt{1-\nu^2}$ and hence also by elementary algebra that $b_1-b_1^{(0)}=[b_2-b_2^{(0)}+\nu(\xi_2-\xi_2^{(0)})]/\sqrt{1-\nu^2}$; $b_2-b_2^{(0)}=[b_1-b_1^{(0)}-\nu(\xi_1-\xi_1^{(0)})]/\sqrt{1-\nu^2}$. Thus our interpretation of the statistical distribution carrying a conserved label as the ordering parameter b increases leads directly, we claim, to the algebraic equivalent of the Poincaré transformation in a 1+1 "dimensional" Minkowski space, with the distinction that the "time" is represented by integers. We will see below that this integer character of our description cannot disappear.

So far we have not had to specify our "step length" in the random walks. From our basic definition if $\mu=0$ we necessarily have $\nu=\pm1$ (i.e., the bit string is all "1's" or all"0's" with <u>no</u> dispersion). But this will also be approximately true if $\varepsilon$ is very large compared to $\mu$. Thus we are forced to assume that our step length $\ell_\varepsilon=1/\varepsilon$. This as we will see below is in fact the Einstein-deBroglie relation. We thus have <u>derived</u> the "quantization of energy" directly from the discreteness of our construction. We are very pleased that this confirmation of deBroglie's basic insight that "particles," like photons, have an "internal" periodicity which lies at the heart of quantum mechanics should have been achieved in the year of his 90th birthday. Further, since our time steps are necessarily integral, $\Delta\tau\geq1$ while since also the number of steps $\ell_\varepsilon$ is integral $\Delta\varepsilon\geq1$. Thus, invoking the Lorentz invariance already established we have $\Delta\varepsilon\Delta\tau\geq1;\Delta\xi\Delta\rho\geq1$. We emphasize that these restrictions are <u>not</u> the "uncertainty principle," which we will discuss in due course, but simply an expression of the restrictions arising from the discreteness of our basic construction.

We now have "coordinate" pairs $(\xi,b)$ where $\xi$ can be any positive or negative number and b can be any positive or negative integer. We also have labeled "lines" $L_1,L_2$ which intersect at a position labeled by $L_3$. Since we have a finite metric coming from the integer structure, we can consider three labeled lines and three labeled intersections, and arrange the velocities such that the length $\xi_1-\xi_2$ and $\xi_2-\xi_3$ is greater than $\xi_1-\xi_3$. Then, since we have a metric, for common $b^{(0)}$ and three intersecting lines we have a triangle and Euclidean geometry in a plane. Further, if our velocity dependent transformation is to hold for arbitrary $\nu$ (bounded of course by $\pm 1$), the coordinates perpendicular to the the lines must not transform. Since we have four levels to the hierarchy, we can extend this argument to three spacial dimensions by constructing a tetrahedron, but the fact that there are only four types of labels we can distinguish at this point in the construction prevents us from going further. Thus we claim that our model requires us to use as our basic space the 3+1 dimensional Minkowski space with coordinates that have (except for the integer character of the "time" parameter) the usual Poincaré transformations. Since $\underline{\nu}$ is now a 3-vector, the definition of $\rho$ as a 3-vector is clearly $\underline{\rho}=\underline{\nu}\varepsilon$, from which the 3+1 momentum space follows.

We have now constructed ensembles of labeled binary distributions (LBD's) characterized by a statistical parameter $\underline{\nu}$ which can be accessed by statistical samplings in a space defined by the coordinates of the samplings containing a discrete internal periodicity $\ell_\varepsilon=1/\varepsilon$. From these we construct an array of LBD's in three dimensions all having the common directed $\underline{\nu}$ in a single direction synchronized to a common $b^{(0)}$ in such a way that at that time all the maxima along this direction are separated by the step length $\ell_\varepsilon$. After some number of steps $\eta$ the maxima will move one step length. Since each step takes a time $1/\varepsilon$, the velocity $\nu=\ell_\varepsilon/\eta(1/\varepsilon)=1/\eta\le 1$. But we can also define a second velocity, that in which each step is taken in the direction $\underline{\nu}$ which is $\nu_\varepsilon=\eta\ell_\varepsilon/(1/\varepsilon)=\nu\ge 1$; clearly $\nu\nu_\varepsilon=1$. Were we dealing with a wave phenomenon we would call $\nu$ the group velocity and $\nu_\varepsilon$ the phase velocity; we emphasize that our construction is discrete and algebraic and should not be allowed to carry this continuum implication. But we now can define a coherence length referring to this synchronization namely $\ell=\eta\ell_\varepsilon=1/\nu\varepsilon=1/\rho$. If we now consider a second synchronized array with a different label and $\underline{\nu}^1$ synchronized to the same $b^{(0)}$, the most probable positions for elementary sampling events to exhibit this periodicity will be separated by this coherence length $\ell$. Clearly $\ell$ plays for us the same role as the "deBroglie wavelength" in a wave theory.

We now consider a situation in which there are an enormous number of LBD's in a plane perpendicular to $\underline{\nu}$ so densely distributed that the probability of the original ensemble of LBD's which will encounter a large number of sampling events in this region, emerging on the other side of this plane is negligible, except for two "slits" a distance $d\ell_\varepsilon$ apart, where they are absent. If we go some large distance S behind this screen in the direction $\underline{\nu}$ and consider a plane perpendicular to $\underline{\nu}$ and a distance x in this plane away from the center line the most probable positions for a sampling taken using some other array to occur will be where the paths from the slits differ by an integral number of coherence lengths $n\ell$. Thus by the usual geometric construction $\ell = xd/nS$, and we claim to have derived "double slit interference" from our discrete model. Note that even when x,d and S are "macroscopic" in the sense that they are all very large compared to our metric length $\ell_\varepsilon$, we can always find an arrangement in which this discrete structure is exhibited, thus relating macroscopic dimensions to the "microscopic" parameter $\ell$.

It might appear that this coherent superposition is a rather special case, but in fact it is an idealization of actual laboratory practice which sets up time scales by using counts in detectors, as we will see in the next section. Thus we can use this "basis" as the starting point for constructing a mathematical description of actual experiments. We will therefore call this coherent ensemble of labeled binary distributions identified by a single label L (and hence value of $\mu$) and a statistical parameter $\underline{\nu}$ an "elementary quantum particle." Note that the old problem of whether the "particle" goes through one or the other hole is resolved by us by saying that the particle is not a single entity but a discrete coherent statistical ensemble which can be divided spacially into two subensembles without losing either its coherence or its uniqueness.

## IV. SCATTERING THEORY AND MEASUREMENT

It is now time for us to show that our constructions are, at least sketchily, directly related to the practice of physics and laboratory experience. Physics is usually taught in terms of basic macroscopic standards of Mass, Length, and Time against which laboratory standards are calibrated, thus allowing pure number ratios to be used in calculations. This was appropriate in a continuum theory which used Euclidean geometry as its mathematical paradigm. But in practice the standard meter is now defined by counting the number of wavelengths of light emitted by a well defined atomic source, the second is defined by counting the number of oscillations of an atomic clock, and the standard kilogram can be defined, if one likes, by counting the number of hydrogen atoms in the volume which would balance it. Thus in practice the standards of physics are already "absolute" pure numbers, and our only task is to show how our integers can be related to these standards. In practice the continuum has already disappeared from physics at the level of the fundamental dimensional units and been replaced by "counting."

The basic device we need for our discussion is a "counter." In practice this can be quite a complicated gadget, but it is assumed that the chain of events which leads to the firing of a counter starts from a single elementary scattering event. All we need do is to identify this initial event with our "elementary sampling event"; we can then leave the discussion of the accuracy to which this event can be localized in the laboratory in the competent hands of the experimental physicist. We also assume him equipped with "rods" and "clocks" calibrated against the primary standards, and "sources" of particles which produce sequential firings of counters separated by macroscopic distances, and by time intervals measured on his clock, thus giving measured velocities. We anticipate from our construction that no matter what source he uses there will be some maximum velocity that is never exceeded, which is borne out in practice, and which we call c in dimensional units. Thus his velocity is related to our statistical parameter by $\underline{v}=\underline{\nu}c$.

The particle physicist cannot measure "mass" directly by a balance, but instead uses the particles from some standard source and measures mass ratios by using relativistic momentum-energy conservation. Particle identification in practice can be quite complicated, but usually starts from the fact that electron and proton carry the same absolute value of electric charge, and the fact that this charge is also the quantized unit carried by other particles. Since we will identify this parameter within our discrete construction in the next section, we can follow his practice, identify his mass ratios $\mu$ with our $\mu$ and the mass in our theory as $m=\mu m_0$ where $m_0$ is some appropriately identified standard. To fix our third dimensional constant we note that we now have $E=\mu m_0 c^2 \varepsilon$, $\underline{p}=\mu m_0 c \underline{\rho}$ and hence can relate our microscopic length $\ell$ to macroscopic standards by introducing a universal constant h of dimensions $ML^2T^{-1}$ and taking $\ell=\lambda=h/p$. Thus we also have the fundamental Einstein-deBroglie relation $E=hc/\lambda$. The final step is then to set up the double slit experiment and relate the macroscopic distances d,S,x to this by $\lambda=xd/nS$ where n is the number of the interference maxima counted away from the center line and the distances are measured using the laboratory standard. Thus, as claimed, we have reduced our basic measurements to the counting of integers. Now that we have made these laboratory identifications, we have no more freedom in making contact between the mathematical structure and laboratory experience (except, as we will see in the next section, some still unresolved questions of how we assign quantum numbers at level 4, similar to the problems now being faced by theorists trying for "grand unification" and "supergravity").

Now that we have identified particle sources and counters, we can conduct scattering experiments and measure the number of particles scattered into a specified solid angle relative to the flux in the incident beam. The cross section so defined is dependent only on particle identification and momentum conservation, so does not directly involve us in specifically quantum effects. For ease of contact with conventional theory, we will treat momentum as a continuous variable $\underline{k}$ and $\varepsilon=(\mu^2+k^2)^{\frac{1}{2}}$, although we will have occasion in the next section to return to the discrete picture for a fundamental calculation. Then we can describe any scattering experiment in terms of a set of basis states $\phi_{\underline{k}}(\underline{k}^1)_\infty=\varepsilon_k\delta^3(\underline{k}-\underline{k}^1)$ which are orthogonal and complete in the sense that $\int d^3k/\varepsilon_k)\phi_k(\underline{k}_1)\phi_k^*(\underline{k}_2)= \int(d^3k/\varepsilon_k)\phi_{k_1}(\underline{k})\phi_{k_2}{}^*(\underline{k})=\varepsilon\delta^3(\underline{k}_1-\underline{k}_2)$. Thus we are able to describe any energy-momentum conserving process as a transition which takes us from some product of these states representing the intial degrees of freedom. What is needed is a dynamical theory describing such processes.

We have seen that, although we can assume momentum precisely conserved in each elementary sampling event (or scattering process), that the uncertainty principle in energy prevents a precise specification of the energy in such processes. Thus what we need is to sum up all momentum conserving processes allowed by the combination of initial and final labels we consider in such a way that the uncertainty principle is respected but only energy conserving processes are described by the boundary states. For simplicity we consider the system with zero total momentum so that the invariant four momentum $P=\varepsilon_1+\varepsilon_2=\varepsilon_1'+\varepsilon_2'$ for the boundary states. Then

$$\psi^{\pm} = \delta^3\left(\underline{k}_1+\underline{k}_2-\underline{k}_1'-\underline{k}_2'\right)\left[\varepsilon_1\varepsilon_2\delta^3(\underline{k}_1-\underline{k}_1') - \frac{g(\underline{k}_1\underline{k}_2)g(k_1'\underline{k}_2')}{\varepsilon_1'+\varepsilon_2'-P \mp i\Delta\varepsilon}\right]$$

This gives a high probability of energy conservation when the boundary states are far enough separated (differ by large values of n) but respects the uncertainty principle. The appearance of the imaginary i is forced on us by the fact that if we used a real linear function we could hit an actual zero in the denominator and produce an infinity that would be incompatible with our philosophy. We could try some nonlinear real structure, but this would introduce aribtrary and unwarranted structure into our statistical space. Thus the introduction of the complex field into our theory comes about basically because we insist on linearity in the basic scattering processes.

This argument does not give us anything more than the singularity and does not define the residue, which is also true in conventional scattering theory. The form we select comes from our zero range or single time scattering theory (17,18) and is chosen to correspond to a particle and quantum forming a "bound state" with the same mass and quantum numbers as the particle, and a particle and anti-particle forming a "bound state" with the mass of the quantum (19,20). This, as we will see more clearly when we discuss quantum numbers below, is precisely the structure of our elementary sampling events. Thus we have arrived at a scattering theory in momentum space which, it is claimed, can be used to obtain the usual results of relativistic quantum mechanics. The further development of this theory, which amounts to summing up all scattering processes consistent with the uncertainty principle and leads to Faddeev and Faddeev-Yakubovsky type integral equations, will be found in the series of papers to which we have just referred.

We now have a full momentum space description of the scattering process and a dynamics which allows us to compute cross sections and make experimental predictions. To connect this with the usual s-matrix formalism for quantum mechanics, all we need do is to make a Fourier transform of our momentum space wave functions to configuration space. This is a smoothing out of our discrete theory, and hence an approximation whose limitations we must keep in mind. In particular, we saw that our initial specification of an elementary sampling event involved a chain of three samplings. In a conventional wave function only the coordinates of the event itself appear, and the coordinates which in classical Hamilton Jacobi theory would represent the constants of the motion have disappeared. This is clearly a mistake, as was first pointed out by Phipps (21). We see that our current development forces us to the same point of view, so that our configuration space wave function becomes (22) $\Psi = \exp\left(-i\sum_k \underline{P}_k \cdot \underline{Q}_k\right)\psi_{sch}(\underline{q},t)$ where $\psi_{sch}$ is the conventional Schroedinger wave funtion and $\underline{P}_k$, $\underline{Q}_k$ are the "constants of the motion"; for us these represent the first in the chain of three samplings (which usually occurs somewhater in the collimatior defining that degree of freedom as it enters the scattering region) needed to define the event at $\underline{q}$. Note that this factor does not affect the predictions of probabilities based on $|\Psi|^2$ nor the distant correlations built into $\psi_{sch}$. However, our approach tells us that the event at $\underline{q}$ (which now is being described only statistically) cannot be specified until a third sampling has taken place for each degree of freedom. Until this has happened, $\Psi$ represents a prediction and not an "element of reality" in the classical sense. In particular it will contain distant correlations which can be checked experimentally by making local measurements which complete the sampling chains, in our paradigm by the firing of a counter. Until all degrees of freedom are accounted for, there will still be correlations; once all relevant counters have fired, the process is completed and joins the fixed past. Of course these firings are themselves the start of new scattering chains. We see that for us there is no overall "collapse of the wave function," only a local severance of phase chains. When all degrees of freedom are accounted for we are describing separated systems. Thus the question of whether or not two systems are separated for us boils down to whether we can account for all relevant degrees of freedom with or without distant correlations. If we make a mistake in this, we may find unexpected correlations in the laboratory; all this tells us is that we have left out one or more relevant degrees of freedom and should start our analysis over using a more articulated system. Thus we claim that our theory leads naturally to a view of measurement that fits both quantum mechanical and classical experience without generating paradoxes.

## V. QUANTUM NUMBERS; $m_p/m_e$; COSMOLOGY

Our next step is to make a specific interpretation of the bits in our labels, calculate the basic mass ratio to which all others are to be referred, and discuss the cosmological structure implied by our theory. In our earlier presentation (2) it was noted that the 16-bit strings at level 3 suggested identification of these bits with the conserved quantum numbers of baryon number, lepton number, charge and spin, and that if we could assign a space-time direction sense to (01) and (10) dichotomous pairs, we could start to interpret discrimination as a Feynman diagram vertex. Now that we have succeeded in constructing addresses for our labels and have showed that $\langle N_1-N_0\rangle$ does indeed define the sense of the velocity in 3+1 Minkowski space-time, we can make this connection precise.

Consider again our elementary sampling event $DS_1S_2=S_3=DS_1'S_2'$ where all strings have the common length 256+b, $DL_1L_2=L_3$ in both cases while $A_1A_2$ have velocities $\underline{v_1v_2}$ referring to two samplings before the event (i.e., $b_1,b_2<b$) and $A_1'A_2'$ have velocities $\underline{v}_1'\underline{v}_2'$ referring to two samplings after ($b_1',b_2'>b$) the event. We see that interchanging "0's" and "1's" in all the address strings reverses all directions, and hence corresponds to the parity operation, while the "direction" of time referes to the $\pm i\Delta\varepsilon$ in the propagator for a scattering process described in the last section. Thus, as in conventional theory, "time reversal" corresponds to complex conjugation. Consider first the labels for level 1. We have three basic processes: $(10)+(01)\rightarrow(11)\rightarrow(10)'+(01)'$, where we have used an obvious shorthand for the addresses, $(11)+(10)\rightarrow(01)\rightarrow(11)'+(10)'$, and $(01)+(11)\rightarrow(10)\rightarrow(01)'+(11)'$. At this level we cannot say whether this quantum number refers to charge, or baryon number or lepton number or helicity for spin 1/2. For concreteness think of it as charge. Then interpret the string (xy) with x=1, y=0 as + charge, x=0, y=1 as - charge and x=1, y=1 as a "bound state" of two particles with positive and negative charge which, under macroscopic examination, behaves as an uncharged particle. Then the three basic processes are seen, externally, to conserve charge.

The "internal" structure requires discussion. Consider the second case in which we have + charge externally and - charge internally. But internally, in an elementary process we cannot assign the sign of time flow meaningfully, since it is connected to only two and not three processes, and these are "vertices" and not sampling events. The obvious convention to adopt is that at a "vertex" all lines are "incoming" or "outgoing," and that a negative particle moving backward is the same as a positive particle moving forward. Then the apparent contradiction with charge conservation in this process is removed. In fact, this suffices to establish the usual Feynman rule, and can be extended to all our dichotomous quantum numbers as we see below. Thus we claim to have replaced the heuristic argument in (2) by a rigorous definition justified by the construction of "space-time" achieved above.

A second point requires care. In the addresses there is a symmetry between the interchange of "0's" and "1's" which we exploited in our derivation of the Lorentz transformation. But this symmetry does not apply to the labels; it would take (11) into (00) which is not allowed because the latter is the null string and has an absolute significance. However, we can make an interchange (xy)→(yx) which interchanges + and - charge (or baryon and antibaryon or lepton and antilepton) in our dichotomous notation. Clearly this corresponds to "charge conjugation" in the conventional theory. Spin or helicity cannot be discussed in quite the same way; we defer it to the discussion of level 3.

Our construction of level 2 above was based on a matrix mapping. If we note, following Vanzani (23), that the minimal number of amplitudes needed to describe N particle scattering is $2^{N-1}-1$, we can reinterpret level 2 in terms of seven processes generated by four particles. If we interpret out 4-bit string as ($a\bar{a}$+−) where a and $\bar{a}$ stand for particle and antiparticle, we have two choices depending on whether the particle has positive or negative charge. Both lead to the unique 7 strings of level 2 as we see:

| a positive | b negative | | | |
|---|---|---|---|---|
| a+(1010) | b−(1001) | (a+,$\bar{a}$0),(b−,$\bar{b}$0) = q−(1101) | (q−,q0) | (0001) |
| $\bar{a}$−(0101) | $\bar{b}$+(0110) | (a0,$\bar{a}$0),(b−,$\bar{b}$+) = q0(1100) | (q−,q+) | (0011) |
| a0(1011) | b0(1000) | ($\bar{a}$−,a0),($\bar{b}$+,b0) = q+(1110) | (q0,q+) | (0010) |
| $\bar{a}$0(0111) | $\bar{b}$0(0100) | | (q+,q0,q−) | |

Clearly if we think of the first case as the charge structure of $p\bar{p}n\bar{n}$, q+q0q− correspond to π+π0π− and the four additional quanta to ρ+ρ0ρ− and ω0. The alternative choice would be $e\bar{e}\nu\bar{\nu}$. In both cases we have the basic charge structure of the mesons that describe phenomena in the few hundred MeV region — nuclear forces for the first and electron-positron scattering in the second, where the ρ is indeed the prominent resonant structure.

This interpretation holds up when we go to level 3 and interpret the 8 basic particles we need to form the 7 string basis by adding the left and right helicity states to baryon number, lepton number and charge according to the schema

| | B | $\bar{B}$ | B+ | B− | BL | BR | $\bar{B}$L | $\bar{B}$R | |
|---|---|---|---|---|---|---|---|---|---|
| pL | (1 | 0 | 1 | 0 | 1 | 0 | 0 | 0 | 00000000) |
| pR | (1 | 0 | 1 | 0 | 0 | 1 | 0 | 0 | 00000000) |
| $\bar{p}$L | (0 | 1 | 0 | 1 | 0 | 0 | 1 | 0 | 00000000) |
| $\bar{p}$R | (0 | 1 | 0 | 1 | 0 | 0 | 0 | 1 | 00000000) |

| | | ℓ | $\bar{\ell}$ | ℓ+ | ℓ− | ℓL | ℓR | $\bar{\ell}$L | $\bar{\ell}$R |
|---|---|---|---|---|---|---|---|---|---|
| eL | (00000000 | 1 | 0 | 0 | 1 | 1 | 0 | 0 | 0) |
| eR | (00000000 | 1 | 0 | 0 | 1 | 0 | 1 | 0 | 0) |
| $\bar{e}$L | (00000000 | 0 | 1 | 1 | 0 | 0 | 0 | 1 | 0) |
| $\bar{e}$R | (00000000 | 0 | 1 | 1 | 0 | 0 | 0 | 0 | 1) |

Here we must use care since a Lorentz transformation can, for finite mass particles, change the sign of $\underline{v}$ without affecting the label, and hence take an L state into an R state when referred to an <u>external</u> coordinate system. Consequently this set contains only four basic particles ($p\bar{p}e\bar{e}$) and we must add to them n,$\bar{n}$,ν,$\bar{\nu}$ as already foreshadowed in our discussion of level 2. We do not have space here to go into details of how this is done and how we arrive at our version of the CPT theorem. We note that by putting together a particle and an anti-particle (aLaR$\bar{a}$L$\bar{a}$R) we get vector quanta VL(1010), $V_0$(1111), VR(0101) and a singlet quantum (0000). Since we have reason to believe from the zero range scattering theory (18) that it may be convenient to take a Wheeler-Feynman view of "photons" and give them a mass below the current threshold of detection, in that case the singlet quantum becomes the "coulomb" photon in the radiation gauge. We do not have to commit ourselves at this point to whether or not the neutrino has a mass as the question is still open experimentally. If it does not then we should use a "two component" theory which in our notation will be $\nu$:(0000000010001000) $\bar{\nu}$:(0000000010000001). The decision will only be made after we have faced the dynamics of level 4, where we must achieve weak-electromagnetic unification, calculation of the Cabbibo angle, quarks, gluons, and heavy leptons. Since we obviously have an 8+8 structure at level 3, this might come about through some version of Harari's $0(8)\times0(8)$ "rishons"; we leave this as a question for future research.

Although we are not ready to discuss dynamical calculations at level 4 where, as we have already seen, the basic structure is tetrahedral, the four corners being labeled by the four levels of the hierarchy and connected to external lines, and hence among other processes to the decay of unstable particles such as the neutron, we can already calculate the basic mass ratio $m_p/m_e$ using only level 3 concepts. Since the calculation has already been published (2,3) starting from slightly different but closely related points of view, and the algebraic details are unchanged, we only summarize the argument here, using the current context. The idea is to calculate the electron mass-energy from its electrostatic energy, an idea that antedates quantum mechanics. If we start from an electron at rest with its level 3 label, it can "open up" via the uncertainty principle fluctuations into two vertices labeled by level 1 and level 2 labels and close again with the electron label forming a closed tetrahedral structure, two of whose corners have the same label and which is therefore not chiral. Since we have seen that charge is conserved, one of these corners will carry xe and the other (1-x)e where x is a variable we estimate statistically and lies between 0 and 1 (inclusive). Assuming that these are separated by a distance r, again a statistical variable, we then have that $m_ec^2=e^2\langle x(1-x)\rangle\langle 1/r\rangle$. Since the only other stable (or at least stable enough for the purposes of this calculation, since decays only come in at level 4) particles are the proton and antiproton, the energy which, in this static calculation, defines our shortest discrete step length is $2m_pc^2$, we have that $\langle 1/r\rangle=2m_pc/h\langle 1/y\rangle$ where y is a discrete random variable ranging from 1 to some large number N which for practical purposes can be taken to infinity. Hence $m_ec^2=(2m_pce^2/h)\langle x(1-x)\rangle\langle 1/y\rangle$.

Since the "coulomb photon" through which the process starts and ends is only one of the 137 "quanta" available to us at level 3, the a priori probability of the process is 1/137, which we identify with $e^2/\hbar c$. After the first step, we keep on until the process eventually closes back on the electron. Since each step will increase the separation of the two (statistically defined) lumps of charge and the steps have three degrees of freedom (either the three dimensions of the space or the three degrees of freedom of the three levels of the hierarchy used in this context) we find that the weighting factor $P(1/y)=1/y^2$ and hence that $\langle 1/y\rangle = \int_1^\infty dy/y^6 \,/\, \int_1^\infty dy/y^5 = 4/5$. Since the probability of the separation occurring is proportional to $x(1-x)$ and the process must close, the weighting factor for the charge is $(x(1-x))^2$. For one degree of freedom this gives $K_1=\langle x(1-x)\rangle_1 = \int_0^1 x^3(1-x)^3 dx \,/\, \int_0^1 x^2(1-x)^2 dx = 3/14$. Once the charge has separated the effective charge is $x^2$ or $(1-x)^2$, so we can write the recursion relation

$$K_n = \frac{\int_0^1 x^3(1-x)^3+K_{n-1}x^2(1-x)^4}{\int_0^1 x^2(1-x)^2 dx} = \frac{\int_0^1 x^3(1-x)^3+K_{n-1}x^4(1-x)^2}{\int_0^1 x^2(1-x)^2 dx}$$

$$= \frac{3}{14}+\frac{2}{7}K_{n-1} = \frac{3}{14}\sum_{j=0}^{n-1}\left(\frac{2}{7}\right)^j$$

Putting all this together we find that

$$m_p/m_e = \frac{137\pi}{\langle x(1-x)\rangle\langle 1/y\rangle} = \frac{137\pi}{3/14(1+2/7+4/49)4/5}$$

$$= 1836.151497 \;....$$

in remarkable agreement with the empirical value of 1836.15152(70). This completes the chain of discrete rules and calculations which allows us to connect our numerical theory with laboratory experience.

It is worth noting that our linear theory forces us (as we saw in the discussion of the Einstein-deBroglie relation) to take as our basic step length $\lambda_E=hc/E$ rather than the $\hbar/p$ that seems more natural in a wave theory. This has the consequence that we expect the discrete aspects of our theory to come into prominence in the nuclear force problem at distances of $h/2m_pc=0.63$ fm, rather than the usual estimate from the Wick-Yukawa mechanism of $\hbar/m_\pi c=1.4$ fm. Actually this is all right since our zero range theory gives us (19,20) the usual one pion exchange "interaction" at these distances. But as is well known, though not always admitted, the conventional theories of the nuclear force based on hadrons break down completely at "distances" of $\hbar/2m_\pi c=0.7$ fm, which is still outside our absolute limit. Current theoretical efforts based on quantum chromodynamics try to meet this with a phenomenological "quark bag" whose radius in some versions is ~1 fm, which may already be in trouble with what we "know" empirically about the nuclear force from nuclear physics. Thus, if our fundamental theory is to be approximated phenomenologically, we are comfortably and unequivocally on the side of the "small bag" theorists in this controversy.

To calculate the central mass values of unstable particles will have to be deferred, as already noted, until we can make a firm assignment of quantum numbers at level 4. At that point we will also have to calculate the correction to 1/137 arising from weak-electromagnetic unification, and a host of other problems. Clearly the theory will at that level face stringent empirical tests, and in spite of this promising beginning, could still fail. But there is one problem we must anticipate already when it comes to gravitation. Using the Dyson argument to relate the gravitational coupling constant to the maximum number of gravitating massive particles we can define within their own reduced compton wave length, we find a rest mass $m_G = [(\hbar c/G)/(2^{127}-1+137)]^{\frac{1}{2}} \approx 936$ MeV/$c^2$ rather than the proton mass $m_p \approx 938$ MeV/$c^2$. This of course does not upset our $m_p/m_e$ calculation since even if $m_G$ should be taken as the fundamental unit of mass, it will cancel out in the ratio. But it does raise the question of why the proton is not the fundamental unit. Perhaps it is in fact unstable, and the difference we encounter here is due to an electromagnetic correction which we should ultimately be unable to compute. The question is open for us at present, but we incline to the view that neither is the fundamental unit. We think it would be best to take the mass unit as the rest mass of the universe, which brings us to cosmology.

Returning to our three fundamental processes, we have seen that G and D and our minimum labeling rule first bring us to $1.7\times10^{38}$ labels which we have now identified with "quanta," then fill up the remaining labels of the $2^{256}-1$ non-null possibilities in the 256-bit string with labels we have been able to identify with "particles" (although so far only a few of them with laboratory definitions), and then go on to add "addresses" of increasing length n. Once this starts we have scattering processes which, because of our conservation laws, will settle down to mainly electrons, protons, neutrinos and photons, and if we get our particle physics right, to nuclei, stars, galaxies, planetary systems, biological evolution, social evolution, and "intelligent" technological societies which can begin to understand these processes. Thus our cosmology, up to a point, is fairly conventional. But it has constraints. Clearly the lepton number and baryon number of the universe cannot exceed, for us, $(1/2)(2^{256}-1)$ and must lie between that upper limit and $2^{127}-1+137$. This is not in conflict with current estimates of this number. We must ultimately be able not only to compute this number with precision, but also the finite energy of the universe. Thus we will eventually face stringent empirical tests coming from cosmological "observations."

A second aspect of our model deserves mention. Because for us the maximum discrete address length at any time, n, refers back to our initial generating operation, the "time since the big bang" is indeed for us an absolute time which (outside of being discrete and statistical) is conceptually analogous to the absolute time of Newton. Further, the address strings of length $n_{max}$ which contain all "1's" or all "0's" provide an "event horizon" which keeps on growing. Unless the rest energy associated with our "big bang" turns out to be unexpectedly large, we anticipate an "open" universe. We find this encouraging, particularly in the light of Dyson's recent analysis entitled "Time Without End" (24) which suggests that the evolution of complex communicating systems can continue forever in such an environment.

## VI. CONCLUSION

We claim in this article to have shown that three well defined recursive processes supplemented by self-consistent rules of interpretation provide a self-generating, self-organizing system into which the current practice of physics fits in a natural, and we hope compelling, way. Thus we claim to provide an "exoskeleton" for our understanding of the universe, based on the conventional dimensional units of physics, — mass, length and time — all of which have "derived." Yet we find it remarkable that only these three dimensional units have sufficed for three centuries to form the basis of physics. Our model is clearly richer than this. Our definitions show how the informational content associated with physics can be accessed, but there could be other stable structures with implications that could be tested in the laboratory. For instance, "naked charm" and other quark flavors have been demonstrated indirectly by means of particle experiments. But the quarks and heavy leptons only scratch the surface of the $2^{127}-1+137$ possibilities we necessarily have in our theory. Perhaps the three "colors" of quantum chromodynamics are a selection from a much richer spectrum which (though ultimately discrete from our point of view) might add a new "dimension" that goes beyond mass, length and time. Although our theory might seem to set rather rigid boundaries to what has been called "physics," it could at the same time open up unexpected possibilities for research. Only the uncertain future can decide.

## VI. ACKNOWLEDGEMENTS

This current paper would not have been possible without the discussions at the Fourth Annual Meeting of the Alternative Natural Philosophy Association (Cambridge 21-23 August 1982), and subsequent discussions with J. C. van den Berg, T. Bastin, C. W. Kilmister, and C. Gefwert. The discussion of measurement theory was stimulated by the profoundly important contribution of D. Aerts to the Symposium on Wave-Particle Dualism. The author is most grateful to the organizers of that Symposium for the chance to participate in the lively discussions and to contribute this paper to the Proceedings. This work was supported, in part, by the Department of Energy under contract DE-AC03-76SF00515.

## VII. FOOTNOTES

1. Work supported by the Department of Energy, contract DE-AC03-76SF00515.
2. A preliminary version of this paper prepared before the symposium is available as SLAC-PUB-2906 (April 1982).
3. The work summarized here will be discussed in detail in a paper with the tentative title "A Reconstruction of Quantum Mechanics using a Discrete, Recursive Model" intended for submission to Physical Review D.
4. This contention is presented in detail in T. Bastin, *A Combinatorial Basis of the Physics of the Quantum* (in preparation).

## VIII. REFERENCES


(1) Bastin, T.: 1966, Studia Philosophica Gandensia 4, pp. 77-101.

(2) Bastin, T., Noyes, H. P., Kilmister, C. W., and Amson, J.: 1979, Int'l. J. Theor. Phys. 18, pp. 445-488.

(3) Parker-Rhodes, A. F.: 1981, *The Theory of Indistinguishables*, Synthese Library 150, Reidel, Dorecht, Holland.

(4) Clauser, J. F., and Shimony, A.: 1978, Rep. Prog. Phys. 41, pp. 1881-1927.

(5) Bridgman, P. W.: 1928, *The Logic of Modern Physics*, Macmillan, New York.

(6) Noyes, H. P.: 1980, in *Studies in the Foundations of Quantum Mechanics*, P. Suppes, ed., Phil. Sci. Assn., East Lansing, pp. 77-108.

(7) Bohr, N., and Rosenfeld, L.: 1933, K. Dan. Vidensk. Selsk. Mat.-Fys. Medd. 12, No. 8.

(8) Wick, G. C.: 1938, Nature 142, pp. 993-994.

(9) Yukawa, H.: 1935, Proc. Phys. Math. Soc. Japan 17, pp. 48-57.

(10) Dyson, F. J.: 1952, Phys. Rev. 85, pp. 631-632.

(11) Noyes, H. P.: (rev. April 1974), "Non-Locality in Particle Physics," SLAC-PUB-1405, unpublished.

(12) Manthey, M. J., "The Computational Metaphor and Quantum Physics," Communications of the Association for Computing Machinery (in press).

(13) Eigen, M., and Schuster, P.: 1978, *The Hypercycle: A Principle of Natural Self-Organization*, Springer, New York.

(14) Conway, J. H.: 1976, *On Numbers and Games*, Academic Press (for the London Mathematical Society).

(15) Kilmister, C. W., "Brouwerian Foundations for the Hierarchy" 1.2.81 (unpublished).

(16) Stein, I.: 1978, seminar at Stanford and papers at the 2nd and 3rd annual meetings of the Alternative Natural Philosophy Association 1980 and 1981.
(17) Noyes, H. P.: "Zero Range Scattering Theory., I. Nonrelativistic Three and Four Particle Equations," Phys. Rev. C (in press).
(18) Lindesay, J. V.: "Zero Range Scattering Theory., II. Minimal Relativistic Three-Particle Equations," Phys. Rev. D (in press).
(19) Noyes, H. P., and Lindesay, J. V.: May 1982, "A Covariant and Unitary Equation for Single Quantum Exchange," SLAC-PUB-2928, submitted to Phys. Rev. Lett.
(20) Noyes, H. P., and Lindesay, J. V.: "Zero Range Scattering Theory., IV." (in preparation).
(21) Phipps, T. E., Jr.: 1969, Dialectica 23, pp. 189-215.
(22) Noyes, H. P.: 1976, Found. of Phys. 6, pp. 83-100.
(23) Vanzani, V.: 1978, *Few Body Nuclear Physics*, G. Pisent, V. Vanzani and L. Fonda, eds., International Atomic Energy Authority, Vienna, pp. 57-112.
(24) Dyson, F. J.: 1979, Rev. Mod. Phys. 51, pp. 447-460.

INDEX